This volume treats the technical research that led to the first atomic bombs. The authors explore how the "critical assembly" of scientists, engineers, and military personnel at Los Alamos collaborated during World War II, blending their traditions to create a new approach to large-scale research. The research was characterized by strong mission orientation, multidisciplinary teamwork, expansion of the scientists' traditional methodology with engineering techniques, and a trial-and-error methodology responding to wartime deadlines.

The book opens with an introduction laying out major themes. After a synopsis of the prehistory of the bomb project, from the discovery of nuclear fission to the start of the Manhattan Engineer District, and an overview of the early materials program, the book examines the establishment of the Los Alamos Laboratory, the implosion and gun assembly programs, nuclear physics research, chemistry and metallurgy, explosives, uranium and plutonium development, confirmation of spontaneous fission in pile-produced plutonium, the thermonuclear bomb, critical assemblies, the Trinity test, and delivery of the combat weapons. Readers interested in the development of the atomic bomb will find many previously unrevealed details in this volume while those interested in the more general history of science will find this volume a crucial resource for understanding the underpinnings of contemporary science and technology.

Critical Assembly

Critical Assembly

A Technical History of Los Alamos during the Oppenheimer Years, 1943–1945

LILLIAN HODDESON
PAUL W. HENRIKSEN
ROGER A. MEADE
CATHERINE WESTFALL

with contributions from
GORDON BAYM, RICHARD HEWLETT, ALISON KERR,
ROBERT PENNEMAN, LESLIE REDMAN, and ROBERT SEIDEL

CAMBRIDGE
UNIVERSITY PRESS

CAMBRIDGE UNIVERSITY PRESS
Cambridge, New York, Melbourne, Madrid, Cape Town,
Singapore, São Paulo, Delhi, Mexico City

Cambridge University Press
The Edinburgh Building, Cambridge CB2 8RU, UK

Published in the United States of America by Cambridge University Press, New York

www.cambridge.org
Information on this title: www.cambridge.org/9780521441322

First published 1993
Reprinted 1995, 1997
First paperback edition 2004

A catalogue record for this publication is available from the British Library

ISBN 978-0-521-44132-2 Hardback
ISBN 978-0-521-54117-6 Paperback

Contents

Illustrations

Preface

The story of the Los Alamos project to build the first atomic bombs has been told often. Why then another history of Project Y, as it was known during World War II? Three features distinguish this account: it is a history of the technical developments; it is based on the full complement of documents, both classified and unclassified, of wartime Los Alamos; and it explores for the first time the methodology by which researchers at Los Alamos succeeded in their wartime mission.

Unlike earlier histories of Los Alamos, this book treats in detail the research and development that led to the implosion and gun weapons; the research in nuclear physics, chemistry, and metallurgy that enabled scientists to design these weapons; and the conception of the thermonuclear bomb, the "Super." Although fascinating in its own right, this story has particular interest because of its impact on subsequent developments. Although many books examine the implications of Los Alamos for the development of a nuclear weapons culture, this is the first to study its role in the rise of the methodology of "big science" as carried out in large national laboratories.

Our primary aim is to recount this technical history, but we have not ignored the social context entirely. Although we largely leave for other historians the problem of analyzing the social community at Los Alamos in wartime – for example, the role of women, of foreign scientists, and of military personnel – we do provide an abbreviated account of the establishment and early years of the unique community that grew around the Los Alamos Laboratory.

The principal reason that the technical history of Los Alamos has not yet been written is that even today, after half a century, much of the original documentation remains classified. With cooperation from the Los Alamos Laboratory, we received authorization to examine all the relevant documentation. The book then underwent a classification review that resulted in the removal from this edition of all textual material judged sensitive by the Department of Energy and all references to classified documents. (For this reason, a number of quotations appear without attribution.) However, the authorities removed little information. Thus, except for a small number of technical facts, this account represents the complete story. In every instance the deleted information was strictly technical; in no way has the Los Alamos Laboratory or the Department of Energy attempted to shape our interpretations. This is not, therefore, a "company history"; throughout the research and writing, we enjoyed intellectual freedom.

Previous histories of wartime Los Alamos fall into three categories: semischolarly or popular histories that deal only in passing with the technical and scientific issues;[1] volumes of participant recollections;[2] and official government-commissioned histories.[3] The books in the first two categories had to side-step most of the technical history because the authors, including those who participated in Project Y, did not have access to the full set of technical documents or a comprehensive overview of the project. They had to rely heavily on memory, which inevitably becomes biased by the distortions of time, feelings, and changing values and images.

Of the official histories, only the volume by David Hawkins is centered on Los Alamos. The laboratory authorized Hawkins, a philosopher and mathematician who held an administrative position on its wartime staff, to prepare in 1946–47 a technical history of Project Y. In his heroic effort, Hawkins provided a unique work that has served as a vital source of information as well as a springboard for the present volume, which, however, goes beyond Hawkins. Unlike his report, this treatment is referenced. Whereas Hawkins was primarily interested in setting down *what* transpired, we have tried also to explain how the developments were shaped by individuals and how they related to earlier efforts. For example, we consider how prior explosives research in England or at the Explosives Research Laboratory in Bruceton, Pennsylvania, created the context for the designing of explosive lenses. Hawkins approached his task analytically, summarizing the work done by each wartime division of the laboratory; we have tried to tell an integrated story. Finally, because we are writing more than four decades after the completion of Project Y, we have had the considerable benefit of historical perspective in our analysis and interpretation.

In its entirety, the book presents a coherent, essentially chronological, account of the wartime technical developments at Los Alamos. The volume begins with an introductory chapter that brings into focus the major themes of the book: the new approach to research that made it possible to build the atomic weapons during World War II, the new style of organization, and the role of the military in the scientific developments at Los Alamos. The methodology used by the "critical assembly" of scientists, government officials, military personnel, and engineers who built the first atomic bombs during World War II is illustrated throughout and recapitulated in the final chapter. The second chapter – which readers familiar with the prehistory of Los Alamos can pass over – provides background details, from the discovery of nuclear fission to the start of the Manhattan Engineer District. Chapter 3 describes the early materials program, and Chapter 4 covers the establishment of the Los Alamos Laboratory and the research carried out there in its first months. The following chapters can be read either sequentially or in separate units: Implosion is covered in Chapters 4, 5, 8, 9, and 14–16. Gun assembly is treated in Chapters 4, 5, 7, and 13. Uranium development at Los Alamos is described in Chapters 11 and 13; plutonium in Chapters 11, 14, and 16; and polonium in Chapters 7, 13, 15, and 16. The confirmation of spontaneous fission in pile-produced plutonium is discussed in Chapter 12. The Super is dealt with in Chapters 4, 10, and 17. Nuclear physics is treated in Chapters 4, 5, 10, and 17. Critical assembly is reviewed in Chapter 10 and Chapter 17. Trinity is discussed in Chapter 18; and Delivery in Chapter 19. The institution and community are treated in Chapters 4, 6, and the epilogue.

This volume grew out of the concern of L. M. Simmons and David Sharp of the Theoretical Division, during the mid-1970s, that valuable historical insights were in danger of being lost and that persons who lived through the wartime events might never be interviewed about their important contributions. In a short proposal written in October 1975, Simmons and Sharp suggested that the laboratory establish both an archives and a history project. In March 1977, asa first step in this direction, Simmons hired Lillian Hoddeson, a physicist and historian of modern science, to work as a consultant to the Theoretical Division on a "pilot" history project. T-Division held a particularly valuable collection of reports and notes by Hans Bethe, Richard Feynman, Edward

Teller, Robert Serber, Emil Konopinski, and other wartime members of the division. Because the documents were disorganized – many papers were not even signed by their authors (and required handwriting identification) – Simmons and Sharp worried that if they were shipped to Washington, they would never be seen again. To extract their historical content would require a dedicated effort possible only at Los Alamos.

In the course of the pilot project during the late 1970s, Hoddeson organized the division's wartime documents, conducted a number of oral history interviews (with Bethe, Nicholas Metropolis, Feynman, Serber, and several others), and in a series of memoranda advised the laboratory about its current archival needs. During this time, archivist Alison Kerr joined the history project. Together, Hoddeson and Kerr coauthored proposals to establish an archives and set up a laboratorywide history project.

These proposals came to the attention of Gilbert Ortiz, leader of the laboratory's Communications and Records Management (CRM) Division. Oritz had independently realized the need for permanent laboratory archives. With Ortiz's enthusiastic sponsorship, the archives were established in 1981 within CRM Division. Nancy Zachariasen became the first Los Alamos archivist; Kerr helped Zachariasen build up the archives. The effort was nontrivial because at this time mounds of documents were scattered in many locations about the laboratory and in need of organization if they were to be accessible for a history effort. Before long, Ortiz began "toying with the idea ... of a project to write the history of the Lab."[4] Hoddeson and Kerr drew up a working plan for such an effort and Ortiz and Kerr subsequently appointed an advisory committee for the project, which met first in August 1983. By fall 1983, the laboratory had agreed to support the project and an official request for proposal was sent out to seven historians working in the area of the twentieth-century history of science. A three-year contract was signed in 1984 with History Associates Incorporated, a Washington-based company, whose proposal was prepared by Hoddeson in collaboration with Richard Hewlett, former Department of Energy historian, and Philip Cantelon, the company's president. When the contract with History Associates lapsed, CRM Division elected to continue the project on its own.

Starting the project in July 1984, Hoddeson selected as junior historians Catherine Westfall and Paul Henriksen, who at that time were both advanced graduate students in the history of science. Duties were divided. Hoddeson took responsibility for implosion; Westfall for prehistory, chemistry, metallurgy, and nuclear physics; and Henriksen for Trinity, Delivery, the institution, and the town of Los Alamos. Three consultants helped with the project: Gordon Baym, professor of physics at the University of Illinois; Robert Penneman, a former Los Alamos chemist; and Robert Seidel, a historian of science, who soon afterward became director of the laboratory's Bradbury Science Museum. In addition to providing scientific expertise, Baym contributed substantially to interviews and the writing and editing, particularly of chapters dealing with implosion and nuclear theory. Penneman contributed to and reviewed sections on chemistry and metallurgy. Seidel worked on the early history of the laboratory, particularly its relation to the University of California. On Kerr's departure from Los Alamos in summer 1985, Roger Meade, who had been hired earlier by Los Alamos to work with Kerr as the laboratory's second archivist, assumed all archival responsibilities of the laboratory as well as Kerr's commitment to the history project. Meade took responsibility for writing the history of the gun program. In September 1986, Leslie Redman, a nuclear scientist who for some years had been responsible for security classification for the laboratory, joined the history project as a technical expert; in addition to his primary contributions concerning high explosives and chemistry, he edited many of the chapters. In the final stages of the preparation of the manuscript, Redman contributed hundreds of hours of his time to checking references and technical details. When the classified manuscript was almost completed, Kerr returned to work as a consultant, editing sections and offering critical suggestions for revision of the manuscript.

Although most of the research was based on documents in the archives, the team benefited from a large number of oral history interviews conducted with individuals who had participated firsthand in Project Y. Unfortunately, as in all histories of modern events, the later the start, the fewer the number of central participants who could be interviewed. These interviews were typically carried out in the Los Alamos Records Center, whose secure environment allowed free discussion of material and enabled interviewers to key their questions to specific documents made available to the interviewees both before and during interview sessions. This mode of interviewing helped to refresh and sharpen the memories of the interviewees, and it sometimes led the historians to documents that might otherwise have been overlooked.

A great number of individuals and institutions contributed to the present manuscript, and it is a pleasure to express our appreciation to them here. We sadly regret that we cannot possibly list in the brief space of this preface all those who contributed in various ways to the project, but we are deeply grateful to them all. We thank the Los Alamos National Laboratory for its generous support of this project. We are particularly indebted to Simmons and Sharp for their pioneering efforts on behalf of the archives and history project; to Kerr and Ortiz for their pivotal roles in setting up both the archives and history project; and to Rosemary Harris for guiding the establishment of these efforts through the administration of the laboratory. We are also grateful to the members (past and present) of the Advisory Committee - Harold Agnew, Robert Bacher, Richard Baker, Hans Bethe, Norris Bradbury, Berlyn Brixner, Charles Critchfield, Darol Froman, Louis Hempelmann, L. D. P. King, Robert Krohn, John Manley, Carson Mark, Nicholas Metropolis, Max Roy, Raemer Schreiber, Richard Taschek, and Robert Wilson – for their help in formulating the history project, and for invaluable aid in developing its contents. Special thanks go to all members of the CRM Division office, particularly Adelia Stewart, for superb administrative support, and Judy Rose Archuleta, who was always able to solve problems imaginatively. At the archives, we are indebted to Mollie Rodriguez for her tireless and cheerful archival and administrative services; she always did her best to track down even the most obscure and difficult-to-find documents. For secretarial support during the hectic last months of preparation of this manuscript, we thank Marsha Perez, who graciously spent more time than we could expect on expediting its completion. For general support in the Records Center we thank Tony Rivera and his staff, who not only provided a very hospitable atmosphere but were always willing to help. We also wish to thank Phyllis Hoffman, as well as Eva Roybal, Betty Cummings, Ann Carlyle, and Ileana Buican for painstakingly transcribing interviews. We are very grateful to Dan Baca and his staff in the Los Alamos National Laboratory Report Library for hours of help in providing technical documents. The library staff also generously assisted in uncovering wartime technical reports and other reference materials. We thank Eugene Sandoval and his staff for their efforts and advice on declassification.

We are especially grateful to the many former Los Alamos scientists who often traveled long distances in poor weather to contribute hours and days to help us with interviews and critical readings of the manuscript. For special help with technical material and for reading early drafts, we thank Robert Bacher, Hans Bethe, Kenneth Bainbridge, Charles Critchfield, George Farwell, Al Florin, Peter Galison, Ronald Rabie, Silvan Schweber, Emilio Segrè, Roger Steuwer, Spencer Weart, and Robert Wilson. The efforts of Hedy Dunn and others at the Los Alamos Historical Museum in providing material on the Los Alamos community are greatly appreciated. Gary Westfall kindly helped during the final months of the project by reviewing sections on experimental nuclear physics. We also thank Richard Hewlett for working with us to create the general working plan of the project and for commenting on several early chapter drafts. The staff of History Associates helped to arrange the contract for the history project and to administer the project between 1984 and 1987; we thank them also for helping to locate numerous documents in Washington, D.C. We are particularly grateful for the heroic efforts of Dan Lewart and Tonya Lillie who enabled us to produce this book in TEX on schedule. We would also like to acknowledge the Command Post of Chanute Air Force Base in Rantoul, Illinois, for hosting Hoddeson's work on the history project during various periods between 1986 and 1990. Finally, we thank Johndale Solem for suggesting the title, *Critical Assembly*, which so well describes the wartime scientific community at Los Alamos, the technical work, the bomb assembly itself, and the story that has played such a crucial role in the modern world.

Lillian Hoddeson
Paul Henriksen
Roger A. Meade
Catherine Westfall

1

Introduction

The Discovery of Spontaneous Fission in Plutonium

It was the spring of 1944. In a secluded canyon in New Mexico, 14 miles from the bustling technical area of the wartime Los Alamos Laboratory, three physics graduate students were working inside a Forest Service log cabin filled with electronics. For the past eight months, they had been driving there each day by jeep to search for evidence of "spontaneous fission," a naturally occurring process in which certain heavy atomic nuclei split of their own accord, emitting neutrons. Anxiously, they puzzled over a startling oscilloscope trace produced by a sample of plutonium. Why were these students studying the phenomenon of spontaneous fission in this canyon? What caused their concern?

The professor in charge of the work, nuclear physicist Emilio Segrè, had fled Italy in 1938 and joined Ernest Lawrence's nuclear physics laboratory in Berkeley, California. In 1943, at the request of theoretical physicist J. Robert Oppenheimer, Segrè had moved several of his Berkeley experiments to Los Alamos to be part of Project Y – the secret project to build the first atomic bombs. Jointly directed by Oppenheimer and military engineer Gen. Leslie R. Groves, Project Y was

* This chapter was written by Lillian Hoddeson, with contributions from Catherine Westfall and Gordon Baym. We are grateful to Les Redman and Paul Henriksen for editorial contributions, and to Robert Seidel and Andy Pickering for major suggestions.

a part of the Manhattan Project (the Manhattan Engineer District). Before World War II, Los Alamos, a small New Mexico town on a high mesa, had been the site of a ranch school for boys. It was suddenly transformed in 1943 into a stark military community that included many of the world's best scientists.

The plutonium isotope of mass 239, ^{239}Pu, was one of two materials that Los Alamos planned to use in its atomic bombs. Both plutonium and the other material, the uranium isotope of mass 235, ^{235}U, fissioned readily when bombarded with neutrons, yielding further neutrons in sufficient numbers to sustain in principle an explosive chain reaction. In both cases, however bomb-size amounts of the fissionable material were difficult to amass. ^{235}U could be separated from natural uranium by a tedious, difficult physical process.[1] ^{239}Pu could be produced by bombarding the more abundant uranium isotope ^{238}U with neutrons, and then chemically separating the created plutonium. Since the density of neutrons needed to manufacture bomb-size quantities of ^{239}Pu occurred only in a nuclear reactor, Groves authorized the construction of several plutonium-generating reactors – a small pilot plant at Oak Ridge, Tennessee, and three production reactors at Hanford, Washington.[2] It took until April 1944 for these reactors to produce and send samples of plutonium to Los Alamos. In the preceding eight months, the physicists had had to make do exploring minute (microgram) plutonium samples prepared in cyclotrons, in the hope that the properties would match those of reactor-made plutonium.

Spontaneous fission was on the Los Alamos research program because the two fission bombs that were to be developed initially – "Little Boy" (uranium) and "Thin Man" (plutonium) – were of the "gun"-type design. In a gun weapon, one subcritical piece of fissionable material is shot into another, to form a supercritical mass that then explodes, yielding huge amounts of energy. The laboratory was optimistic about meeting its military objective – namely, to develop both bombs by summer 1945 (when Hanford was to begin producing sufficient amounts of plutonium for bombs) – because the problems associated with building gun weapons were extensions of familiar problems of nuclear physics and ballistics. However, Thin Man presented a number of difficult challenges. For one thing, the number of excess neutrons in the system had to be kept to an absolute minimum. Because gun assembly is a slow process in comparison with the speed of a nuclear explosion, extra neutrons threatened to set off the explosion too early and cause a "fizzle."

Two processes in plutonium could potentially spray extra neutrons into the system: impurity reactions and spontaneous fission. Hoping to minimize the number of neutrons resulting from interactions with impurities, the project managers set up large plutonium purification programs both at Los Alamos and at the Chicago Metallurgical Laboratory. At the same time, Oppenheimer authorized Segrè's spontaneous fission experiments to determine whether neutrons from spontaneous fission were worth worrying about. This authorization was but a precaution, for theory suggested that even if spontaneous fission occurred, the rate would not be high enough to threaten the plutonium gun. Indeed, when Segrè's group measured the rate of fission in cyclotron-made plutonium, they found it comfortably small.

The Segrè group chose to work in a canyon because the experiments were extremely sensitive to environmental disturbances. The level of data collected was extraordinarily low – less than one count per month! To keep background "noise" to a minimum, the group worked in conditions as free as possible of radiation, loud sounds, electrical surges, and other disturbances. All the equipment was battery operated.

Segrè's students were alarmed in mid-April 1944 because they measured in the first samples of reactor-made plutonium, a spontaneous fission rate *five times* that of the cyclotron-produced samples – a rate far too high for a gun assembly! Every count taken over the next four months confirmed these preliminary findings. By July 1944, Los Alamos had to accept the failure of Thin Man.

A crisis ensued. Groves, wanting to preserve the investment that had been made in plutonium production (hundreds of millions of dollars), ordered a plutonium bomb assembled by other means. The only possible alternative was implosion, an assembly explored thus far at Los Alamos only as a contingency. In such an assembly, a subcritical sphere of fissionable material is collapsed inward by the blast from a symmetrical array of high explosive. This process had the advantage of being so rapid that spontaneous fission neutrons would not have time to interfere with the explosion.[3] But those working on implosion in June 1944 thought it would be virtually impossible to achieve a practical implosion for use in the present war. As a result, Los Alamos was forced to turn its relatively small implosion program into a model "big science" effort involving hundreds of workers.

Resolving the Crisis: A New Approach to Research

This book tells the story of how the Los Alamos scientists responded to the spontaneous fission crisis they faced more than a year after the start of Project Y and how this response to the possible failure of the plutonium weapon motivated them to take a new approach to research that enabled building both of the first two atomic bombs. Los Alamos was able to complete the "Fat Man," as the plutonium implosion bomb came to be called, as well as the uranium gun, in time for combat use because Project Y was reorganized radically, and confronted its problems by a powerful methodology fostered by the wartime context.[4]

Under the new approach, members of the communities that coexisted in wartime Los Alamos – the scientists, engineers, and military personnel – blended their traditions. Such blending of scientific and engineering traditions had already begun (on a smaller scale) at Lawrence's laboratory in the 1930s and in a number of the science-based industries.[5] Under the conditions at Los Alamos, however, the process solidified. Historians are only just beginning to study the consequences for the postwar world – both positive and negative.[6]

Scientific research was an essential component of the new approach: the first atomic bombs could not have been built by engineers alone, for in no sense was developing these bombs an ordinary engineering task. Many gaps existed in the scientific knowledge needed to complete the bombs. Initially, no one knew whether an atomic weapon could be made. Furthermore, the necessary technology extended well beyond the "state of the art." Solving the technical problems required a heavy investment in basic research by top-level scientists trained to explore the unknown – scientists like Hans Bethe, Richard Feynman, Rudolf Peierls, Edward Teller, John von Neumann, Luis Alvarez, and George Kistiakowsky. To penetrate the scientific phenomena required a deep understanding of nuclear physics, chemistry, explosives, and hydrodynamics. Both theoreticians and experimentalists had to push their scientific tools far beyond their usual capabilities. For example, methods had to be developed to carry out numerical hydrodynamics calculations on a scale never bfore attempted, and experimentalists had to expand the sensitivity of their detectors into qualitatively new regimes.

As much as the scientists would have liked to provide technical solutions based on full understanding of fundamental laws, meeting Groves's deadline of building the atomic bombs by summer 1945 precluded traditional, methodical research and analysis. Moreover, the wartime problems were necessarily tied to practical issues, like fitting bombs into the B-29 bomb bay or building components that could withstand the severe conditions of high-altitude drops. However "pure" the scientists wanted their work to be, they were forced by the wartime circumstances to embrace the methodology of Edison.[7] That is to say, their objectives shifted from understanding to use, and from general conceptions to particular materials and apparatuses.[8] This reorientation encouraged them to diversify their methodological toolkits with approaches typically employed by engineers and craftsmen, whose technical problems were anchored in concrete phenomena.

In view of the military application of their work, Los Alamos scientists were also forced to pay strict attention to reliability. Thus they sought alternative assemblies in their early efforts (such as the original implosion program or the thermonuclear bomb, at a time when the laboratory was emphasizing the gun method) and they subsequently overdesigned both the gun and implosion bombs in an attempt to guarantee success. The urgency of the wartime mission, the high cost and possible future accountability of research on the atomic bomb, the frightening consequences of miscalculation, the scientific and technological uncertainties of bomb design, the unusual availability of almost unlimited funding and other resources – all these factors fostered a conservative research strategy aimed at avoiding risk. One result was that multiple approaches were taken in addressing most of the problems.

This strategy paid off. For example, the relatively small-scale implosion studies conducted in the first year of Los Alamos yielded concepts that proved essential in completing the Fat Man (e.g., Alvarez's simultaneous electric detonators, Tuck's three-dimensional explosive lens for focusing shock waves, and Christy's brute-force core design). Most important, when the spontaneous fission crisis hit, the laboratory already had an organized research effort that could be shifted quickly to a crash implosion program. In the second year of the project the strategy of "overkill" continued as major attention focused on refining determinations of critical mass and exploring implosion. Typically, Los Alamos researchers sought the most dependable, rather than the most elegant, solutions.

The tight deadline scientists and engineers faced was a critical constraint. They managed to meet it – the entire Los Alamos project was completed in a scant twenty-seven months – in part because the organization of teams combined scientific practice with management procedures borrowed from industry. A strong mission orientation was im-

Fig. 1.1a. Robert Oppenheimer, theoretical physicist, first director of Los Alamos laboratory. LA Photo, 85 1780.

Fig. 1.1b. Major General Leslie Groves. Commanding General, Manhattan Engineer District. LA Photo, LAR 611.

posed: projects in line with the mission received essentially unlimited funding and material support, whereas others were dropped or starved (as was the thermonuclear bomb) of resources. Because the research at Los Alamos was in the national interest, strings were often pulled to supply high-priority projects with the needed equipment or personnel.

Oppenheimer did his best to ensure research freedom, but to meet deadlines he was forced to manage the program in quasi-military fashion. He was empowered to function like a general positioning his scientific troops. It was not uncommon for a researcher to be switched overnight from one project to another having higher priority. Leaders below Oppenheimer on the organization chart directed their groups with the same militarylike authority. Scientists, engineers, and technicians were expected to work together and communicate effectively, thus pooling their separate experience and knowledge about particular problems. Chemists and metallurgists, who rarely talked with each other before the war, found themselves working together in the same division.

Information flow was aided by many committees at the laboratory. Los Alamos was possibly one of the most introspective research organizations ever to exist. Group and division leaders assessed the work of their units in biweekly or monthly committee meetings, and division leaders participated in various advisory committees, such as the Governing Board and the Coordinating Council, which continually reported on progress and suggested the direction of future research. Numerous outside consultants, advisers, and committees also reviewed the work, often biweekly, making recommendations that the laboratory was obliged to take seriously.

This tightly organized, introspective hierarchical institution brought forth a new breed of scientific leader, one able to negotiate with committees, while managing much larger teams than had ever before existed in science. In the early phases of the implosion program, a practical academic scientist such as Seth Neddermeyer could cope with directing the implosion team of half a dozen scientists. But after August 1944, when some implosion teams included 50 to 100 researchers, the new implosion leaders had to be strong managers. Robert Bacher and George Kistiakowsky fit into this category.[9]

By the time Project Y was under way, the American physics community had matured sufficiently to handle the challenge of building the atomic bomb. It was no longer scientifically and institutionally backward in comparison with Europe. Although the group of approximately 200 practitioners of physics in the United States up to 1895 included such renowned scholars as Henry A. Rowland, Willard Gibbs, and Albert Michelson, the first American to win a Nobel Prize in physics, America's physicists were still too few and too widely dispersed to establish high general standards for research. Most lacked Ph.D.s and few published. American physics was then also without professional societies and journals to disseminate information. Industrial research, already established in Germany, was virtually nonexistent.[10] Nor was there an adequate institutional structure upon which to build a vital physics community; only six universities fully prepared students for graduate work in physics, and not all offered first-rate instruction. America's best physicists traveled to Europe for their education and published in European journals. Furthermore, American universities burdened physicists with large class loads. Finally, in an era that celebrated the useful and profitable, only a few philanthropists sponsored basic research projects. In government, only the Smithsonian Institution had a policy of promoting the pursuit of abstract knowledge, and its budget for such purposes was limited. The National Academy of Sciences (NAS), a private organization founded during the Civil War to advise the government, had little money or influence inside or outside the scientific community.[11]

In the last years of the nineteenth century, progress began to be made in the institutionalization of this area of American science. The American Association for the Advancement of Science established a physics section in 1882. In 1894 the *Physical Review* was published for the first time. Leaders of the physics community founded the American Physical Society in 1899.[12] Aided by growing philanthropic support from such wealthy patrons as Johns Hopkins, Jonas Gilman Clark, and John D. Rockefeller, Sr., graduate-level programs in physics expanded.

Growth continued after 1900 as new agencies, including the National Bureau of Standards and the Carnegie Foundation, funded research.[13] Recognizing that basic research had commercial value, large companies such as General Electric (GE), du Pont, and American Telephone and Telegraph (AT&T) established research laboratories sponsoring pure and applied research on a wide range of technologies, such as vacuum tubes, explosives, chemical dyes, artificial fibers, telephone transmission equipment, and X-ray and radio tubes. By 1930, almost 33,000 scientists and technicians were employed in industrial research laboratories. Industrial laboratories tended to emphasize teamwork.[14]

The privately funded National Research Council (NRC) brought together top scientists and engineers from academia, industry, and government to promote military research. During World War I, astronomer George Ellery Hale promoted, organized, and managed this council under the auspices of the NAS. Although the NRC's influence with the military decreased markedly after the armistice, wartime accomplishments, including the development of submarine-detection equipment and aeronautical instruments, convinced military and government leaders that science could have defense value. Strengthened ties between the physics community and industry further stimulated the employment of physicists in industry.[15]

American physics continued to prosper throughout the 1920s and 1930s, despite the Depression.[16] Advances in quantum theory stimulated interest in the microscopic structure of matter, and in 1923 Robert Millikan of Caltech was awarded the Nobel Prize for his work on electrons. In the 1930s and 1940s, Oppenheimer taught quantum theory to large numbers of students at the Berkeley campus of the University of California as well as at Caltech. Also at Berkeley in the 1930s and 1940s, the entrepreneurial Lawrence gathered chemists, engineers, and physicists together in a laboratory where he built a series of ever-larger cyclotrons and led numerous projects in nuclear chemistry, nuclear physics, and medicine. By bringing together specialists from different fields to work cooperatively on large common projects, Lawrence helped to create a distinctly American collaborative research endeavor – centered on teams, as in the industrial research laboratories, but oriented toward basic studies without immediate application.[17] This approach flourished during World War II.[18]

Ties between the physics community and the government strengthened during the 1930s. The Science Advisory Board was established in 1933 at the instigation of Karl Compton, president of the Massachusetts Institute of Technology (MIT). Although attempts to renew the board's charter failed in 1935 in the midst of antagonisms between scientists and politicians, the board stimulated federal funding of scientific projects and gave leaders like Compton, Millikan, and Vannevar Bush political experience and contact with important members of the Roosevelt administration.[19] Such scientific leaders, whose talents spanned science, engineering, and management, were brought forward by the wartime projects.

As political troubles loomed in Europe, American physics reaped benefit from tragedy. Between 1933 and 1941 more than 100 physicists, mostly Jews from Germany and Austria, fled to the United States to escape Hitler and Mussolini. American laboratories, now capable of providing both economic security and a stimulating intellectual environment, attracted such people as Enrico Fermi, Hans Bethe, John von Neumann, Edward Teller, and Eugene Wigner. These immigrant scientists made important contributions to American science and helped to complete its maturation.[20]

As this book argues, the factors operating in wartime Los Alamos – the pragmatic mission of the laboratory, its ample financial support, strict time pressure, and the imposed risk-averse policy – in combination gave rise to an empirical problem-solving methodology based on systematic trial and error rather than thorough analysis. Traditional analytic methods were simply too slow. Among the particular techniques that the Los Alamos physicists and chemists used frequently, in combination with more traditional scientific ones, were the *Edison approach* of trying, in the absence of good theoretical guidance, one after another system or material; the *shotgun* approach, in which all experimental techniques available and everything known about a particular issue were fired at the problem to be solved, in hopes that one or more techniques would hit on a piece of the problem and reveal some important facet; *overlapping approaches*, in which multiple approaches were taken simultaneously to a specific problem in recognition that any one could be incomplete and uncertain by itself, but that together they might be used to build up a consistent picture; the *small-scale model* study, to save time and precious materials; *iteration*, the systematic generalization of cut-and-try "tinkering," long characteristic of American science,[21] in which empirical models were progressively improved after testing; and *numerical analysis*, now for the first time extensively done by computing machines. Although messy and unaesthetic, numerical methods were more far-reaching than analytical models alone, which were simply too incomplete and idealized to handle concrete problems. However, when combined with analytic methods, numerical ones formed a tool of striking power.

To illustrate, the Edison approach guided the countless implosion shots fired. The shotgun technique was combined with overlapping approaches in the many-stranded implosion diagnostic program, in which seven complementary types of experiments – X-ray, photographic, terminal observations of implosion remains, magnetic, electric "pin," betatron, and the "RaLa" method – were oriented toward gathering a flood of data on implosion. Scale models were used in every implosion diagnostic exercise. Iteration was used extensively in the explosive lens program because explosives exhibited complex phenomena and had been little studied previously. In building lenses, theorists would make educated guesses of the index of refraction, on the basis of which the lenses were cast with approximately correct geometries. Their actual index of refraction and focusing properties were then determined and used by the theorists to improve their guesses for the next iteration toward good experimental lenses. Similarly, the development of detonators required trial and error and redundancy, because there was no other way to meet the deadlines reliably. Numerical methods, carried out with the help of IBM calculators (International Business Machines), were employed extensively in the implosion program. The brute-force Christy design simply circumvented the serious symmetry and stability problems of a more elegant implosion design.

The special conditions that nurtured the new approach could continue only under the wartime pressures to build the atomic bomb. Although quite a few members of the original Los Alamos scientific community remained at the laboratory after the war to complete the unfinished work or to take part in the new science that splitting the nucleus had just begun, most of them, including the director and all the division leaders, chose to return to former academic homes or new ones, since postwar support of the physical sciences opened new opportunities. Those who dispersed transplanted the Los Alamos approach. In turn, the new methodology helped to restructure American science, opening new vistas in both applied and pure science, from the space program to research on subatomic elementary particles, to numerical studies of astrophysics. For example, R. R. Wilson reflects that in creating Fermi National Accelerator Laboratory (Fermilab) in the late 1960s – a laboratory designed to conduct forefront research in particle physics – he initially tried to recreate a kind of science city reminiscent of Los Alamos. The empirically oriented trial-and-error methodology used during World War II proved to be both cost- and time-effective during the building of Fermilab. Wilson admits that he probably had wartime Los Alamos "in the back of [his] mind."[22]

2

Early Research on Fission: 1933–1943

Following the discovery of nuclear fission in 1938, scientists in Germany, France, Great Britain, the Soviet Union, Japan, and the United States began to investigate the possibility of exploiting this energy source for military purposes. The United States alone was able to draw its governmental, industrial, and scientific capabilities into an efficient bomb-building collaboration. It had not only the manpower, materials, and industrial support needed for the expensive project – eventually to cost $2.2 billion – but also a sizable and competent physics community well versed in technology, strengthened by talented emigrés, with ties to government and industry, and close international contacts.[1] Some American scientists, like Ernest Lawrence, were experienced in managing large research efforts. This community also included older scientific statesmen, like Vannevar Bush, with political experience and proven abilities in coordinating government-sponsored applied research projects. On 9 October 1941 Bush persuaded President Franklin Roosevelt to authorize American research on the feasibility of a fission bomb.[2]

* This chapter is based on a draft by Catherine Westfall, with contributions from Paul Henriksen, Roger Meade, and Robert Seidel. We thank Spencer Weart, Roger Steuwer, and Richard Hewlett for helpful comments, and Gordon Baym and Alison Kerr for critical editing.

The Discovery of Nuclear Fission

The events leading to Los Alamos began in 1933, when Frédéric Joliot and Irène Curie produced artificial radioactivity by bombarding aluminum with alpha particles. The next year Enrico Fermi and his co-workers in Rome bombarded a variety of elements with neutrons, the neutral fundamental particles that James Chadwick had discovered in 1932. Upon bombarding uranium, Fermi's group found an unexplained radioactive substance and speculated that they had created a new transuranic element.[3] A number of scientists challenged this notion, among them the German chemist Ida Noddack, who instead suggested the atom might have fragmented into several lighter elements. Little notice was given to Noddack's curious idea, which seemed to contradict current theories of nuclear physics.[4]

As European politics grew more turbulent, scientific interest in the mysterious radioactive phenomenon also increased. Shortly after hearing of Fermi's experiments in Rome, Otto Hahn and Lise Meitner bombarded uranium at the Kaiser Wilhelm Institute in Berlin; Fritz Strassmann soon joined them. In Paris, Irène Curie undertook similar research in 1937. In the next two years a number of others, including Philip Abelson at Lawrence's Radiation Laboratory in Berkeley and Norman Feather and Egon Bretscher at Cambridge, explored the radioactivity of bombarded uranium.[5]

After Meitner fled Germany in late 1938, to escape Nazi persecution, Hahn and Strassmann decided to study the radioactive substances resulting from neutron-bombarded uranium. They took these substances to be radium isotopes but to their surprise, found they could not chemically separate the suspected radium from the barium being used as a carrier. Their subsequent chemical analyses indicated that the decay products indeed resembled barium.[6]

Before Hahn and Strassmann published their paper on this bewildering finding, however, Hahn described it in a letter to Meitner, who was by then in Sweden. Meitner's nephew, Otto Frisch, had come from Niels Bohr's Institute in Copenhagen to spend the Christmas holidays with her, and together they tried to make sense of the puzzle. Upon consideration of Bohr's recent liquid drop model of the nucleus, they wondered whether the barium could have indeed been produced through a breaking apart of the uranium nucleus! They contemplated this possibility the day before Christmas, during an outing in the snow. Sitting down on a tree trunk, they calculated the balance of energies for a liquid drop model of the uranium nucleus bombarded by a neutron. The calculation convinced them that the nucleus could indeed divide.[7]

When Frisch returned to Copenhagen, he explained this finding to Bohr, who was just then preparing to visit the United States. Before Frisch could finish the explanation, Bohr exclaimed, "Oh, what fools we have been!" The idea that had previously seemed curious now seemed obvious.[8] In a series of long-distance telephone conversations in early January 1939, Frisch and Meitner composed a note for *Nature* on their interpretation of the division of a uranium nucleus. Frisch also drafted another note announcing that he had confirmed the new phenomenon by detecting fragments from the split uranium nucleus in an ionization chamber. In the paper with Meitner, he dubbed the process "fission," on the analogy of cell division.[9]

Bohr announced the discovery officially on 26 January 1939 at the annual theoretical physics conference at George Washington University in Washington, D.C. The discovery was soon confirmed by researchers at several laboratories in the United States and Europe.[10] With abundant experimental confirmation and a compelling theoretical explanation, the scientific community accepted the phenomenon immediately.

Throughout 1939 fission research was conducted in many laboratories in Western Europe and the United States. No one knew yet whether fission could be applied to produce power or an atomic bomb. Despite the obvious military implications of such a reaction, most of the earlier fission work was aimed at satisfying scholarly curiosity rather than at producing a divergent (runaway) chain reaction. The possibility of a chain reaction depended on whether fission produced neutrons in sufficient numbers and with the energy needed to trigger further fission reactions. By late February 1939, the Paris team, using methods suggested by Hans von Halban and Lew Kowarski, concluded that neutrons were indeed released in the fission of uranium nuclei. The fundamental condition for a chain reaction was therefore satisfied.[11]

Researchers surrounding Fermi, now at Columbia University in New York, came to the same conclusion after Leo Szilard, a far-sighted, unconventional Hungarian emigré, provided them with a gram of privately procured radium to use as the α-particle emitter in a neutron source based on the (α-particle, neutron) reaction. With this radium and Columbia's radon–beryllium source, Fermi's team conducted experiments analogous to those of Halban and Kowarski, from which they obtained similar results.[12] In February 1939, Fermi's team also reported that fission seemed more likely to occur when the neutrons moved more slowly. This observation implied that a chain reaction might be more readily produced if a moderator, such as heavy water, could slow down the neutrons.

On 18 March 1939, two days after German troops invaded Czechoslovakia, Fermi met with American naval officers and explained in the most conservative terms the possible results of using nuclear chain reactions in bombs: such reactions might be capable of blasting craters more than a mile in diameter or of propelling submarines. The officers were interested enough in Fermi's presentation to keep track of future research, but they saw no need to ask the government to sponsor fission research. However, they did donate $1,500 to Columbia University toward Fermi's fission research.[13]

By this time, Fermi had come to the conclusion that the number of neutrons created per fission (ν) was too small to support a chain reaction. Halban, Joliot, and Kowarski in France disagreed. They calculated ν to be 3.5 and published this result on 22 April 1939, despite fervent efforts by Szilard to keep this and other information concerning chain reactions secret. Even though the French team's value of ν turned out to be mistakenly optimistic (a later, more accurate, measurement yielded 2.6), their work helped keep alive the hope of creating a nuclear chain reaction.[14]

The French publication of the ν for uranium in April spurred Allied research on chain reactions. Struck with the implications, British physicists George P. Thomson and William L. Bragg conferred and concluded that uranium would probably be a good source of power and heat and could perhaps be used to produce a powerful explosion. In response, the British government began to procure uranium and sponsor fission research, efforts that later proved beneficial for both the British and American war efforts. In Germany, ν calculations confirmed by Gottfried von Droste and H. Reddemann caused great excitement. Several German scientists independently informed their government of the military and economic potential of nuclear fission, prompting Hitler's bureaucracy also to procure uranium and undertake fission research.[15]

Through mid-1939, Fermi's group at Columbia and Joliot's in Paris carried out fission reaction studies, although their work with dispersions of uranium oxide had not so far produced such a reaction. Theorist George Placzek, then at Cornell, identified the problem; the concentration and spatial arrangement of the moderator and fissionable material in the mixture had to be just right. With too much uranium, neutrons would be captured as they slowed down, because the predominant isotope in natural uranium, ^{238}U, has a large cross section for capturing neutrons at a particular intermediate energy, less than that of fission neutrons but more than that of slow neutrons. With too little uranium, slow neutrons would be absorbed by hydrogen in the water moderator. Although mixtures of various ratios of water and insoluble uranium oxide were tested both at Columbia and in Paris, no chain reaction was achieved because these nuclear properties were not yet understood. Throughout 1939, the French team searched for the correct mixture. Szilard wanted to try using graphite as the moderator with uranium oxide but had difficulty in obtaining sufficient quantities of these materials. Fermi had several experiments in mind, but he was feeling pessimistic about the possibilities of nuclear energy and was also tired of working with Szilard.[16] He turned to theoretical studies of cosmic rays until enough material could be obtained for a large-scale experiment. The withdrawal of these members of the Columbia team from chain reaction research in effect shut down American work until the spring of 1940.

European efforts also proceeded at a slow pace. The British program in the summer of 1939 included two chain reaction studies, one by Thomson, the other by Mark L. Oliphant, but the results were discouraging. Furthermore, many British physicists were more concerned with developing radar because, like Thomson and Oliphant, they felt that possible applications for fission were remote. The British uranium program made little progress after Britain and France entered the war in September 1939. In late 1939, only the French were reporting advances in chain reaction work, after achieving a convergent chain reaction in a homogeneous mixture of uranium and water. But they published no more results once the war began.[17]

Just as chain reaction research came to an impasse, important new information became available. Two days before France and Britain entered the war, Bohr and John A. Wheeler, a Princeton physicist, published the definitive theoretical analysis of fission. This study implied that the rare isotope ^{235}U – not the more common ^{238}U – was the component of natural uranium most likely to undergo fission. Attention turned to studies of ^{235}U.

Among those who studied ^{235}U were Frisch, now trapped in Birmingham by the outbreak of the war. Although as an alien he was not allowed to participate in highly sensitive radar research, he was able to pursue his studies of ^{235}U. He tried to separate ^{235}U from ^{238}U using a method of thermal diffusion developed by German chemists Klaus Clusius and Gerhard Dickel. Clusius and Dickel had almost completely separated the major chlorine isotopes with this method, but separating ^{235}U would be more difficult, because nuclear mass difference and the amount of ^{235}U in ^{238}U were both small.[18] Early in 1940, Columbia physicist John R. Dunning studied the fissioning of a sample of ^{235}U that had been separated in a mass spectrometer by Alfred O. Nier at the University of Minnesota. By the time Dunning's group published the result, German work on isotope separation had begun in earnest. Werner Heisenberg had worked out a theory for energy production by nuclear fission based on the assumption that ^{235}U was the most likely isotope to fission. Similar work was also under way in the Soviet Union.[19]

By this time, communication on fission research between countries had become guarded, even among allies. The problems noted by Placzek – neutron absorption in water and resonance capture in uranium – had not been resolved; the last published French experiment seemed to show that a homogeneous uranium and water mixture could not be made critical. By late 1939, the solution occurred independently to scientists in the United States, France, and Germany: it would be necessary to devise a heterogeneous reactor made of a moderator and uranium oxide, so that neutrons would be less likely to hit the uranium at speeds leading to capture.[20]

Possibility of a Fission Bomb: The MAUD Committee

Although many physicists in 1940 still doubted that a uranium fission bomb could be built, some of them – including Rudolf Peierls and Frisch – thought about whether an atomic weapon was feasible. Fearing that such a weapon might already be on the drawing boards in Germany, these German emigrés, now in Birmingham, began to calculate features of an atomic bomb.[21]

Peierls performed critical mass calculations on the basis of work done by Francis Perrin in the Paris group. He found that natural uranium had a critical mass of tons, much too large for use in a bomb. In the early spring, however, he and Frisch decided to calculate the critical mass of pure ^{235}U. They assumed – justifiably, as it turned out – that most incident neutrons, of all energies, cause ^{235}U to fission. Although at first their calculations seemed little more than an academic exercise in view of the difficulty of obtaining the pure form of the rare isotope, their interest heightened when their initial estimates indicated that only about a kilogram was needed for a chain reaction.[22]

Frisch and Peierls now calculated that a thermal diffusion scheme with a hundred thousand separation tubes might separate the required amount of ^{235}U from the more abundant isotope ^{238}U in a matter of weeks. The magnitude of the predicted explosion was astonishing: a 5-kg bomb would yield the same energy as several thousand tons of dynamite! Working in Great Britain, where German air attacks were having a devastating effect, the two refugees looked to an atomic bomb as a beacon of hope in a bleak time, when, as Peierls explained recently, the war "didn't look so hopeful."[23]

Peierls and Frisch then presented the first thorough scientific analysis of the feasibility and destructive power of a fission bomb in a memorandum that provided the theoretical explanation for their critical mass calculations. They suggested methods for detonating the weapon and for avoiding premature detonation, proposed a thermal diffusion method for separating isotopes, and evaluated the destructive effects of the bomb and the dangers of radioactivity. This memorandum went to Oliphant, who in March 1940 passed it on to Thomson, whom the British government had put in charge of uranium research. The British immediately took steps to protect their uranium oxide stocks, find out about the level of uranium research in Germany and America, and organize their scientific expertise.[24]

On 10 April 1940, Thomson convened the first official British scientific committee on atomic bomb research. Known by its code name MAUD, this committee included Oliphant, John D. Cockcroft, and Philip B. Moon. It was linked to the Ministry of Aircraft Production (MAP). Peierls, who had just been naturalized, and Frisch, still classified as an enemy alien, were not allowed to take part in the meetings, despite their expertise. When Peierls pointed out the decision was both unreasonable and stupid because the two could make valuable contributions to the research, arrangements were made to include them and other refugee scientists, by way of a technical subcommittee. And when the French government warned that the Germans were interested in seizing a heavy water plant in Norway, the MAUD committee launched an all-out research effort. Although Prime Minister Winston Churchill's scientific adviser, Sir Henry Tizard, had argued that "uranium disintegration" was not "in the least likely to be of military importance in this war," British scientists slowly mobilized for nuclear research aimed at developing an atomic weapon. James Chadwick organized pure research activities at his home institution in Liverpool and invited Bretscher and Feather at Cambridge and a group at Bristol to join in. At Birmingham,

William N. Hayworth directed the chemical work necessary for isotope separation studies, enlisting help from the large industrial firm, Imperial Chemical Industries. Frisch and Peierls contributed to both efforts in a number of capacities.[25]

The American and British Bomb Programs: 1940–1941

American work on a nuclear bomb proceeded more slowly, perhaps because the United States was not yet at war. One of the early decisive steps was the famous Einstein letter, arranged by Szilard with the help of his Hungarian colleagues Eugene Wigner and Edward Teller. Trasmitted to President Roosevelt in late 1939 by economist Alexander Sachs, the letter reported on recent work at Columbia and in Paris and urged the government to support uranium research. In response, Roosevelt established the Advisory Committee on Uranium, chaired by Lyman J. Briggs, director of the National Bureau of Standards (NBS).

American interest in uranium research gradually increased over the next six months. Fermi and others, including Ross Gunn, the technical adviser of the Naval Research Laboratory, discussed the potential of a ^{235}U chain reaction at the April 1940 American Physical Society Meeting. At Gunn's suggestion, the government formed a scientific subcommittee to the uranium committee that included physicist George B. Pegram, an administrator at Columbia; Nobel laureate Harold Urey, a chemist expert in isotope separation; Jesse W. Beams, a centrifuge expert from the University of Virginia; Merle A. Tuve, a Carnegie Institution physicist interested in developing isotope separation with a centrifuge method; and Gregory Breit, a physicist at the University of Wisconsin. In June, Bush, then president of the Carnegie Institution, persuaded Roosevelt to appoint him head of a new organization, the National Defense Research Committee (NDRC). The committee, acting under the authority of the World War I Council of National Defense, broadened U.S. efforts in defense science by enlisting scientists not employed by industry or the government. By this time, the uranium program had been granted funds to buy materials and equipment for Fermi.[26]

By spring 1940, Szilard had obtained four tons of graphite, purer than that used by the French team, with which to continue chain reaction research at Columbia. Thanks to Fermi's careful calculations of the effects of the pile's finite size, the Columbia experiments were more sensitive than similar ones in France and thus required less graphite to produce reliable results. In addition, Fermi and Szilard were considerably ahead of their French counterparts in developing the optimum moderator-to-uranium ratio. Szilard had been making three-dimensional lattice calculations since early 1940. Whereas the French team never got beyond rough ideas about how to arrange the uranium, Fermi and Szilard realized by mid-May the advantages of arranging uranium blocks in a three-dimensional lattice embedded in the moderator. They found compelling evidence that a graphite-moderated pile could be used to produce a chain reaction.[27]

Fermi planned to publish these results, but Pegram, urged on by Szilard, argued successfully against publication. This move proved fortunate for the Allies. Kept ignorant of Fermi's success with graphite, the Germans, who were also conducting sophisticated theoretical analyses of the proper arrangement of uranium in a moderator, subsequently abandoned their work on carbon moderators. Instead they focused on the use of heavy water, a substance that was to become scarce in Germany after the British raid on the Vemork hydroelectric station in occupied Norway.[28]

In June 1940, Abelson and Edwin M. McMillan in Berkeley announced the discovery of element 93, the first transuranic element, which they later named *neptunium.* Although at this time McMillan and Abelson did not see the connection between their work and chain reaction studies, the discovery of neptunium proved to have important short- and long-term consequences for the fission field. After hearing about the new element, Princeton physicist Louis A. Turner correctly guessed that neptunium beta decays to produce an element of higher atomic number. Bretscher and Feather at Cambridge independently came to the same conclusion. Encouraged by this speculation, researchers at Berkeley mounted efforts to create the higher element, later named plutonium.[29] At the same time, the announcement of neptunium's discovery prompted a flurry of protests within the American physics community about the propriety of publishing such work. Breit, whom Briggs would soon appoint to coordinate theoretical work on fast neutron research, organized a successful effort in the United States to persuade both scientists and editors of scientific journals to refrain from publishing papers relating to fission. After the war it became apparent that the German nuclear power project used the article on the discovery of element 93 and others in key U.S. publications to outline the military applications of nuclear fission.[30]

Scientific communication between Great Britain and the United States had almost ceased by this time as the war had begun to take its toll. Through the efforts of Lord Lothian, Great Britain's ambassador to the United States, collaboration between the two countries resumed. The "Tizard mission," a group of British scientists, including Tizard and Cockcroft, arrived in Washington, D.C., in September 1940, bringing a black box containing drawings and other details of British inventions and weapons.[31] This exchange revealed that American researchers were months behind the British in most war-related research, but ahead in a few areas, notably, in their work on the centrifuge method of ^{235}U separation. Cockcroft recognized that the United States, with its immense industrial capability, was better suited than Britain for the research and development of the bomb. As a result, scientific liaison offices were set up in Washington and London, and later in Ottawa. With this machinery in place, a great deal of information was exchanged, although British scientists complained that many more technical reports traveled from Britain to America than in the opposite direction.[32]

Throughout 1940 and 1941, Britain continued to contribute heavily to the basic understanding needed to create a uranium bomb. In the summer of 1940, Halban and Kowarski joined the British effort and, after gathering a small group at Cambridge, began chain reaction studies in November using a sphere filled with a mixture of uranium oxide and heavy water. By December, they had gathered substantial evidence to indicate that it would be possible to create a divergent chain reaction.[33]

Meanwhile, isotope separation studies on both sides of the Atlantic were focusing on gaseous diffusion. In Britain, Frisch and Peierls made a detailed theoretical analysis of a variety of separation schemes and concluded that gaseous diffusion was the most promising method. Peierls recorded their conclusions in a series of papers for the MAUD technical subcommittee. A group of researchers at Oxford were already thinking about the possible design of a ^{235}U separation plant. By the end of 1940 they had produced initial plans for an industrial-sized gaseous diffusion plant.[34] In May 1940, Harvard chemistry professor George Kistiakowsky introduced the idea of gaseous diffusion to Bush at a conference at the Carnegie Institution. Kistiakowsky investigated a diffusion apparatus for separating mixed gases that had been developed by the U.S. Bureau of Mines. At the same time, Dunning and a team at Columbia were exploring alternate gaseous diffusion schemes.[35]

As this work progressed, researchers in Britain grew more optimistic about the prospect of an atomic weapon. Tuve's group at the Carnegie

Institution had just sent their measurements of the fission cross section of natural uranium at various energies to Britain. The behavior of ^{235}U and ^{238}U deduced from these measurements exactly matched the predictions made by Bohr and Wheeler's theory of fission. In March 1941 Peierls wrote, "This first test of theory has given a completely positive answer."[36] He now felt confident that only a few kilograms of ^{235}U would be needed for a critical mass if a scattering or tamper material was used. The British were further encouraged in April by Geoffrey I. Taylor's thorough calculations of the bomb's destructive potential, which indicated that a 10-kg bomb would deliver the power of about 1,000 tons of TNT.[37] When measurements of the fission cross section at Liverpool confirmed those made in Washington, Chadwick became convinced "that a nuclear bomb was not only possible – it was inevitable."[38]

With the discovery of plutonium, the American program gathered more steam. In mid-1940, McMillan had noticed an α-emitting body, while watching the decay of element 93, neptunium. He wondered whether he had produced an isotope of element 94, the new element predicted by Turner, Bretscher, and Feather, and made attempts to separate neptunium from the α activity to make a chemical identification. This work was interrupted when McMillan left Berkeley to join the radar project at MIT. The investigation was assumed by chemists Glenn T. Seaborg, Joseph Kennedy, and Arthur Wahl.[39] Working late into the night, graduate student Wahl made the final oxidation on 25 February 1941. As Seaborg later recorded in his diary, the results made it "clear that our alpha activity is due to the new element with the atomic number 94."[40]

Bohr and Wheeler's theory of fission suggested that the new element plutonium had a fissionable isotope. But could it be created and would it actually fission? At the urging of Fermi and Emilio Segrè, Lawrence was persuaded to order the production of enough plutonium to answer these questions. In a painstaking series of experiments that stretched throughout March, Seaborg, Kennedy, and Segrè created ^{239}Pu at the 60-inch cyclotron, irradiated it with neutrons, and detected fission. Encouraged by plutonium developments and concerned by the worsening outlook of the war, Lawrence strongly advocated a large American bomb effort.[41]

British confidence in the feasibility of a ^{235}U weapon became official in July 1941. Late in the month, the MAUD committee announced it had "reached the conclusion that it [would] be possible to make an effective uranium bomb" and explained why ^{235}U was needed for an explosive chain reaction, what size the bomb should be, why the bomb would be efficient, what the damage would be, how the bomb could be fuzed to prevent premature explosion, and how it would be possible to separate ^{235}U.[42]

Physicist Charles C. Lauritsen of the California Institute of Technology (Caltech) was in London when the MAUD report appeared and relayed the committee's optimistic conclusions to Bush. In the same month, Seaborg and Segrè reported the first rough measurement of the fast-neutron fission cross section of ^{239}Pu, finding the value to be 3.4 times greater than that of ^{238}U. This encouraging evidence, coupled with British optimism, fueled further American bomb research. As information from the MAUD committee was being transmitted through a number of channels to American researchers, Bush worked with chemist James B. Conant, then president of Harvard, to strengthen the existing Committee on Uranium and drafted plans for drastically enlarging the American bomb program. In October 1941, Bush presented these plans to Roosevelt, who subsequently authorized a full-scale effort to explore the possibility of building an atomic weapon.[43]

Bush, Conant, and other scientists who had participated in the World War I mobilization of American science now joined government and military leaders (notably, President Roosevelt and Gen. Leslie R. Groves) as well as industrialists (such as Percival Keith, vice president of the M. W. Kellogg Company, and the contractor for the gaseous diffusion plant development) to create the network of politicians, engineers, industrialists, and scientists that would develop the atomic bomb. The resulting project would draw on prewar models of team research found in scientific laboratories, such as Ernest O. Lawrence's Radiation Laboratory in Berkeley, and in the research laboratories of industrial firms such as GE, du Pont, and AT&T. The scientific community would offer state-of-the-art research and scientific apparatus, as well as a system of assessment by expert review. The military would bring strong focus on a combat mission and a risk-averse approach to solving problems. Industry would add technological experience and further expertise in organization. The project would become the largest, most expensive national research and development effort yet created.

3

The Early Materials Program: 1933–1943

A comprehensive American program of research on plutonium and ^{235}U isotope separation evolved between mid-1941 and mid-1942. Plans were formulated for the construction of plutonium production reactors, uranium separation plants, and centralized bomb research facilities. The program for producing ^{235}U by gaseous diffusion was slowed both by the technical problem of finding a suitable isotope separation filter, or "barrier," and by the difficulty of coordinating Kellogg Company employees and Columbia University researchers. Despite such obstacles, plans for providing fissionable materials were well on the way to being implemented by 1943.

Expansion of the American Atomic Bomb Program

In June 1941, Vannevar Bush persuaded President Roosevelt to form the Office of Scientific Research and Development (OSRD), under the aegis of the Office of Emergency Management. With Bush as director, the OSRD assumed responsibility for mobilizing scientific resources and applying research to national defense. James Conant replaced Bush as

* This chapter is based on a draft by Catherine Westfall, with contributions from Paul Henriksen, Roger Meade, and Robert Seidel. We thank Spencer Weart, Roger Steuwer, and Richard Hewlett for helpful comments, and Gordon Baym and Alison Kerr for critical editing.

chairman of the National Defense Research Committee, which now operated as a unit of the OSRD. Recommendations for research contracts were channeled through Conant and placed by Bush. The Advisory Committee on Uranium, essentially a research organization, became the S-1 Section of OSRD and remained in place throughout the war. Bush and Conant established three subsections of S-1: one on theoretical research under Fermi; one on power production under George Pegram, physicist and dean of the graduate faculties at Columbia University; and a third on heavy water and isotope separation under Columbia chemist Harold Urey.[1]

For technical advice on uranium research, Bush turned to an already formed NAS committee led by Karl Compton's brother, Arthur Holly Compton. This committee, which included Ernest Lawrence and theorists John Slater and John Van Vleck, had the task of calculating the destructiveness of the bomb. Lawrence discussed the problem with Robert Oppenheimer, who calculated the fraction of available fission energy that would be released in a fission explosion. His conclusion, confirmed by Compton and George Kistiakowsky, was contained in the NAS committee report of 6 November. "A fission bomb of superlatively destructive power will result from bringing quickly together a sufficient mass of element ^{235}U."[2] Spurred on by this optimism and aided by the committee, Bush drafted plans for an enlarged bomb effort. The Japanese attack on Pearl Harbor on 7 December 1941 and the U.S. entry into World War II heightened the determination to build an atomic weapon quickly to counter the Japanese attack and preempt a German atomic attack.

Bush recognized that building the bomb would require a great deal of research. Almost nothing was known about the chemical, physical, or nuclear properties of plutonium or about how to design production piles or devise the necessary chemical separation schemes. Although production piles were needed to create plutonium, a divergent chain reaction had not yet been achieved. The great stumbling block to a uranium weapon was isotope separation; no production-scale process existed for preparing enriched ^{235}U. Even after the chain reaction was demonstrated a year later, in December 1942, and after provisions were made for producing ^{235}U and ^{239}Pu, much work remained: designing and building a weapon required navigation of the uncharted physics of fast-neutron chain reactions.

By mid-December 1941, the outlines of the new program had emerged. Bush assigned responsibilities: chain reaction studies and weapon theory to Compton, electromagnetic separation and plutonium research to

Lawrence, and heavy water research and centrifuge and diffusion separation methods to Urey. Plans for building industrial-scale uranium and plutonium production plants proceeded alongside the research program. The task of preparing for the production phase was given to chemical engineer Eger V. Murphree, who formed a planning board for engineering assistance.[3]

Soon after Bush made these assignments, Lawrence and Urey organized their respective research programs. Compton had to solve a serious organizational problem before starting his effort. Research on the chain reaction and weapon theory was being coordinated at an increasing number of universities in the country; this work could not be consolidated in one place because the experimental research could only be done with certain machines by a small number of experimental nuclear physicists, and neither machines nor physicists could be easily uprooted from their universities or government homes. On 3 January 1942, Compton called a meeting in Chicago to discuss the organization of his project. He decided that Fermi would build a pile at Columbia, physicist Samuel K. Allison would build one at Chicago, Wigner would concentrate on the theory of the chain reaction at Princeton, and Oppenheimer would concentrate on theoretical studies of fast-neutron reactions at Berkeley. In late January 1942, however, Compton decided that research for the plutonium weapon should be consolidated at the University of Chicago. He formed a new organization, code-named the Metallurgical Laboratory (Met Lab) and named Richard Doan director. Compton appointed Allison, Wigner, and Fermi to coordinate the research studies at Chicago, and he asked Szilard to procure supplies. Over the next few months, the Columbia and Princeton groups moved to Chicago.[4]

Compton also organized chemical and metallurgical research, naming chemist Frank Spedding of Iowa State University in Ames the director of chemical research. Spedding led three groups; these studied the metallurgy of uranium and plutonium, the separation of plutonium from uranium and fission products, and the health protection of workers. The work at Ames complemented that of four groups at the Met Lab, where Charles Coryell led research on fission product chemistry, George Boyd's group worked on materials research (which included analyses of the purification of materials), Milton Burton's group studied the then poorly known effects of radiation on materials and chemical processes, and Herbert McCoy's group researched the processing of pile materials, in conjunction with an engineering group under Thomas V. Moore. Plutonium separation studies continued at Berkeley under the direction of

Fig. 3.1. Ernest Lawrence, Enrico Fermi, and Isadore I. Rabi. LA Photo, LAT 419.

Kennedy and Wahl. Seaborg, now at the Met Lab, directed the larger plutonium separation effort.[5]

In February 1942, Compton asked Gregory Breit of the University of Wisconsin to coordinate physics research on fast-neutron phenomena.[6] This research was essential to determining the size of critical masses and efficiencies of explosion. Using new fission cross-section data, Breit and Oppenheimer estimated that only 5 kg of ^{235}U would be needed for a spherical critical mass, in contrast to the 2–100 kg estimate given by the NAS committee in November. The new estimate agreed with that indicated in the MAUD committee's report.

The smaller critical mass meant that less fissionable material had to be produced. More precise calculations allowed Compton to recommend sizes of production plants with greater certainty. However, critical mass estimates continued to vary through 1944, even after plans for production plants were well under way. In January, Bush asked Oppenheimer and Kistiakowsky to reexamine the data and make new efficiency calculations. Previous estimates had indicated that the weapon would have the destructive power of 600 tons of TNT. Oppenheimer and Kistiakowsky now concluded that it would have the equivalent power of 2,000 tons.[7]

Bush gave top priority to isotope separation, pile-related chain reaction studies, and plutonium research – areas in which a great deal of work was needed before detailed weapons research could begin. Because of the extensive industrial effort required to set up and operate isotope separation plants, early planning for uranium production was especially crucial. Murphree's Planning Board outlined four immediate goals: build experimental gaseous diffusion and centrifuge units on an industrial scale, design pilot plants for isotope separation, secure sufficient supplies of uranium oxide and other materials needed for the plants, and produce a small supply of heavy water in case graphite moderators proved unsatisfactory. Bush coordinated the efforts; Hugh S. Taylor at Princeton developed a suitable process.[8]

At Berkeley, Lawrence used his mass spectrometer to work on the electromagnetic isotope separation method, in which gaseous uranium ions travel circular paths under a magnet, the ions of the lighter isotope following a tighter circle ending in a collection cup. In early January 1942, Lawrence produced by this method 18 μg of material enriched to 25 percent ^{235}U. The next month, he produced three 75-μg samples containing 30 percent ^{235}U. On the basis of these successes, he planned to use the magnet from the 184-inch cyclotron to provide fields for a number of such mass spectrometers – "Calutrons," named for the University of California (California University Cyclotrons). These devices would provide the material for the initial American and British experiments.[9]

After the early start on gaseous diffusion at Oxford and Columbia, progress lagged. The problem of finding a suitable barrier for separation plagued ^{235}U production efforts. Eventually Taylor at Princeton solved the problem.[10] Also needed was a high-speed pump that incorporated a seal capable of functioning for long periods at sustained high speeds without lubricant in corrosive uranium hexafluoride gas.[11] To make gaseous diffusion practical, Murphree's Planning Board initiated a large research effort in early 1942 to develop first a pilot plant and then an industrial facility. Percival Keith, vice president of the M. W. Kellogg Company, was recruited to direct this effort, which drew on the expertise of many companies and universities. Keith sent Kellogg researchers to

Columbia to help design separation and barrier equipment. He also contracted with several pump manufacturers and enlisted Henry A. Boorse and others at Columbia to work on the pump problem. Because a great deal of preliminary research had to be done on lubricating substances and seals and on the resistance of materials to corrosion, Keith did not begin designing the detailed pilot plant until late spring 1942.[12]

In late 1942, when many of the gaseous diffusion problems were known but not solved, a new development caused further complications. After a review of the project in October 1942, Conant suggested to Bush that it was time to limit the Anglo-American partnership. Noting that the U.S. project had shifted largely to military control and that the United States was doing most of the developmental work, and feeling that U.S. security was best served by closer control of information, Conant saw "no reason for a joint enterprise as far as development and manufacture is concerned." This conclusion was passed on to Roosevelt, who restricted the exchange of information between the United States and Britain. The results of this new policy were unfortunate: the British were developing solutions to the problems plaguing the design of gaseous diffusion plants, but Keith did not discover the details until it was too late for them to be implemented. In part, as a result of this communication gap, gaseous diffusion continued to be a major problem through 1944 (Chapters 11 and 13).[13]

The centrifuge method of separation proved even more troublesome. Scaling up laboratory procedures to production level meant that a plant would need about 50,000 centrifuges with 1-m rotors to produce 1 kg of ^{235}U a day. Because such a high number of centrifuges could probably not be kept in continuous operation, designs using 4-m rotors were investigated. Even with 4-m rotors, 10,000 machines would be needed to barely reach that level. Larger rotors were harder to accelerate through certain vibration frequencies that would destroy the machines. Beams attained inconclusive results in experimental runs at the University of Virginia, and Westinghouse had trouble solving the design and construction problems of an industrial-size centrifuge. When problems with this method continued, Conant advised Bush in October 1942 that the centrifuge process was the weakest of those being investigated, and such research was subsequently abandoned.[14]

The project, now extended to numerous laboratories throughout the country, was in need of better coordination. To solve the problem, Roosevelt turned to the military and placed the Army Corps of Engineers in charge of the huge construction projects needed to supply raw materials.

Army Chief of Staff Gen. George C. Marshall chose Brig. Gen. Wilhelm D. Styer of the Army Services of Supply and liaison to the S-1 Section to follow nuclear developments. On 18 June 1942 Styer ordered Col. James Marshall to form a Corps of Engineers District responsible for atomic bomb research. Colonel Marshall recruited Kenneth D. Nichols, then a lieutenant colonel with a Ph.D. in civil engineering, to assist with the project.

Marshall established the headquarters of the new Corps of Engineers District in the borough of Manhattan in New York City. The district soon assumed the name "Manhattan Engineer District" (MED), or simply, the "Manhattan Project." The district took on the procurement and engineering functions of Murphree's Planning Board, which became inactive. To provide a small group of technical advisers for the military, Bush streamlined the S-1 Section, renamed it the S-1 Executive Committee chaired by Conant; the section included Briggs, A. H. Compton, Lawrence, Murphree, and Urey.

By September, Bush was voicing his concern that the MED needed a more decisive leader. General Marshall and Secretary of War Henry L. Stimson soon responded. Styer would have been the obvious choice to head the new project, because he was already following nuclear developments. But Styer's commanding officer, Lt. Gen. Brehon B. Sommervell, was unwilling to lose his chief of staff. Styer suggested Col. Leslie Groves, deputy chief of construction for the Army Corps of Engineers. Groves had been in charge of several large construction projects, most prominently the Pentagon. A particular advantage was that Groves was already familiar with the Manhattan Project, having worked on several aspects of it earlier that year, such as reviewing plans for the district and examining the site for the ^{239}Pu pilot plant near Knoxville, Tennessee.[15]

On 17 September, Groves assumed responsibility for directing the Manhattan Engineer District, and on 23 September he was promoted to brigadier general. What particularly qualified Groves for his new position was his proven ability to get things done. He followed through with bluster and positive thinking. He was to prove an astute judge of situation and character, with unusual talent for making rapid, sound decisions and choosing effective staff.[16] But his authoritative style would prove hard on the scientists who were accustomed to working and communicating results without restriction, and to idiosyncratic recordkeeping. Perennial concern about future financial reckoning – a concern transferred down to Groves from Under Secretary of War Robert P. Patterson and other superiors – led him to require that meticulous records, particularly of expenditures, be kept in the Manhattan Project.[17]

A few days after Groves's appointment, Stimson formed the Military Policy Committee, which included Bush as chairman, Conant as an alternate, Adm. William R. Purnell from the navy, and Styer from the army. This committee was to serve as an informal board of directors for Groves's district.[18]

The First Chain Reaction

While the administrative structure of the MED was being created, Fermi, now at the Chicago Met Lab, directed the research toward the divergent chain reaction. He expanded the program begun at Columbia using graphite and pure uranium metal. After Groves obtained a high procurement priority, Norman Hilberry, assistant director of the Met Lab, was able to obtain a sufficient quantity of graphite. Szilard worked to free it of neutron-absorbing impurities.[19]

Procuring uranium was more difficult. Because of the minimal prewar interest in uranium, few companies supplied the metal. Uranium compounds extracted from ores had to be treated chemically and mechanically to convert them into the uranium metal required for bomb components. In 1941, samples for experiments at Columbia had come from a Massachusetts firm, Metal Hydrides, which produced a variety of metal powders that were highly pyrophoric. The Westinghouse Lamp Works in Bloomfield, New Jersey, also produced high-purity uranium metal in extremely limited quantities. However, Metal Hydrides had only a few pounds of uranium and Westinghouse only a few grams.[20] To provide the larger supplies needed for the Chicago effort, arrangements were made for Mallinckrodt Chemical to produce uranium oxide. By July 1942, Mallinckrodt was producing one ton of highly pure oxide per day, which Fermi and his team used to supplement the metal obtained from Westinghouse and Metal Hydrides. Because these supplies still did not suffice, arrangements were also made for uranium production at Iowa State University. By mid-1942, Spedding and his co-workers were not only studying the metallurgy of uranium but actually producing uranium metal on a much larger scale than previously attempted.[21]

As the graphite and uranium arrived at Chicago, Fermi's group continued its efforts to arrange them in a series of "exponential piles," creating convergent chain reactions of increasing extent as more material came in to help them determine nuclear constants. Early tasks included measuring the fission cross section of unseparated uranium and the neutron-absorption cross section of graphite. By the time he left Columbia in spring 1942, Fermi had measured neutron intensities using several lattices of uranium oxide and graphite.[22] At the University of Chicago, he supervised the construction of lattices to determine the optimum arrangement of materials and the reactivity (or reproduction factor) k, the number of neutrons produced per neutron absorbed, assuming a large system. The k factor was a measure of whether the neutron population in the system increases ($k > 1$) or decreases ($k < 1$). Fermi made his measurements with a neutron source placed near the bottom of the array and with indium foils placed at various distances from the source on the vertical axis. This experimental setup was called an exponential pile, because neutron intensity decreased exponentially with the distance from the source. Fermi's team built several such piles. From 15 September to 15 November, Herbert L. Anderson and Walter Zinn constructed and measured sixteen exponential piles. By the end of this period, Fermi reported the reproduction factors to be 1.04 for uranium oxide and graphite and 1.07 for uranium metal and graphite, which was barely larger than the minimum value of $k = 1$ for a self-sustaining reaction.[23]

Fermi proposed building the first chain-reacting piles in Chicago's Argonne Forest, 25 miles from the university. However, when construction workers near Argonne went on strike in October, Fermi asked Compton if the pile could instead be constructed under the west stands of Stagg Field on the university's campus. After agonizing over the possible consequences of an accident in the highly populated area and wondering whether he should consult with university officials, Compton took sole responsibility for the decision and authorized construction under Stagg Field on 14 November 1942. Crews under Zinn and Anderson machined 210 tons of graphite and pressed uranium oxide into spheres. A group headed by Volney Wilson built control and measuring devices. Construction began with a general plan but no blueprints. After a frame of wooden blocks was put in place, graphite blocks, which came in varying degrees of purity, were used to make a lattice containing, at regular intervals, uranium oxide or uranium metal. When a substantial delivery of high-purity uranium metal arrived from Spedding, Fermi rearranged the distribution of uranium metal and oxide in the pile to optimize the use of the high-purity metal. Fermi's team also had to decide on the exact placement of control rods. Fermi coordinated the activity, doing much of the necessary calculation himself.[24]

After setting down the eleventh layer of graphite blocks, the group began measuring neutron activity with a boron trifluoride counter and indium foils. On the basis of these results, Fermi plotted out the approach to criticality and determined that the pile would go critical with the fifty-sixth layer. To add a margin of control, the team added a fifty-seventh layer of blocks. On 2 December 1942, following the careful routine Fermi devised, over a period of four hours the team extracted the last cadmium rod bit by bit, in increments of six inches to a foot, with Fermi checking the measurements to make sure they matched his predictions. The pile went critical, as anticipated, on 2 December 1942.[25]

Production of Uranium and Plutonium

Plutonium was produced in a nuclear reactor by causing atoms of ^{238}U to capture neutrons. Several companies explored ways to produce the uranium needed in such reactors – among them, Metal Hydrides, Westinghouse, General Electric, Brush Beryllium, and Imperial Chemical Industries in Britain, as well as the NBS and Brown University. The most successful technique was developed by December 1942 at Spedding's uranium production plant at Iowa State.[26] Spedding found "bomb" reductions (reductions done in closed vessels) of uranium tetrafluoride using calcium to be most promising. He instituted a crash program to find the optimum conditions for purifying metal with high yields and in large quantities. Because of the high price of calcium, the process was modified in March 1943 to use a magnesium reductant. In the modified process, magnesium and UF_4 were mixed and put into a steel bomb lined with high-calcium lime and heated to about 650° C. The resulting exothermic reaction yielded up to 95 percent of high-purity uranium metal. With minor modifications, this process was used by du Pont, Mallinckrodt, and Electro Metallurgical Co., a subsidiary of Union Carbide and Carbon Corporation.[27]

Thousands of tons of uranium ore were needed to produce the metal. Nichols made the first large-scale purchase in September 1942 when he contracted with Edgar Sengier of the African Metal Corporation to buy 1,200 tons of Belgian uranium already in the United States and about 3,000 tons above ground in the Congo. In late 1942, the Army contracted with Eldorado Gold Mines, Ltd., in Canada for some 700 tons of ore, as well as refining services at the company's Port Hope, Ontario, facility. In the following year, it took steps to acquire additional ore and arranged for its processing.[28]

In December 1942, less than a month after du Pont agreed to design and operate a full-scale plutonium production plant, du Pont officials met with representatives of Carbide's National Carbon Company to plan the fabrication of graphite bars for the moderator blocks. The production piles needed much purer graphite than the commercial grade material used in the experimental piles (containing 1–2 ppm boron). Ordinary graphite production – by heating coal tar and petroleum coke in an electric furnace – was not difficult, but making graphite for nuclear reactors required extremely pure raw materials heated to 2,800° C or higher. Strict purity and density requirements limited the size of the graphite bars. The Met Lab investigated machining graphite. Graphite produced at the National Carbon Company to meet a 0.5-ppm boron standard was carefully machined to remove surface impurities and achieve the necessary dimensions (48 inches by 4 inches). By late March 1943, plans were complete for a graphite fabrication plant to be built at the Clinton Engineer Works in Tennessee. By the fall of 1943, the plant had produced 700 tons of graphite for the Clinton pile. The next year, du Pont built a bigger graphite fabrication plant at Hanford, Washington, where graphite for the large pile was finished with precision.[29]

Meanwhile, chemists worked to purify the plutonium emerging from the production piles. Using tracer chemistry, Seaborg and his co-workers at Berkeley discovered that the element had at least two oxidation states and found ways to oxidize to the higher state and reduce to the lower state. They also separated large quantities of uranyl nitrate from the plutonium using ether extraction, and found ways to separate the plutonium from thorium, protactinium, and neptunium. In April 1942, Seaborg left Berkeley for Chicago to spearhead the separation of larger quantities of plutonium.

Despite the progress made in Berkeley, sizable obstacles remained in the production of plutonium. One problem was the scarcity of plutonium. The only feasible way to obtain a few micrograms of plutonium was to irradiate hundreds of pounds of uranium with neutrons produced in a cyclotron. The early plutonium work at Berkeley had used plutonium produced by the Berkeley 60-inch cyclotron. Recognizing the urgent need for another plutonium source, Compton arranged for full-time plutonium production at the Washington University cyclotron in St. Louis. Seaborg still had only micrograms of the precious element.

He quickly devised an industrial-scale process for separating plutonium from uranium and fission products.[30]

Seaborg pursued two objectives: to study micrograms of plutonium, on a scale far below that of ordinary microchemistry, and to obtain a weighable quantity of plutonium. His group devised special capillary tubing capable of measuring volumes of the order of 1 mm^3 to 10 mm^3 and balances that could weigh micrograms of solids. Because nuclear scientists were in short supply, Seaborg recruited other specialists in ultramicroscience, including physiologist Isadore Perlman, microchemist Paul Kirk (who was experienced in criminology), entomologist Robert Patton, biochemist Burris Cunningham, and chemist Michael Cefola.[31] Despite difficult conditions, Seaborg's group made considerable advances. In the summer of 1942, several groups at the Met Lab, and others in Berkeley, investigated a variety of plutonium separation methods. Discussions with du Pont representatives focused on speeding up the industrial application of such methods. By fall, attention had turned to using fluoride and bismuth phosphate as carriers.[32]

Efforts to accumulate weighable quantities of plutonium also proceeded. In August 1942, Cunningham and Louis Werner successfully isolated pure plutonium for the first time. In September they were able to weigh a pure plutonium compound and use it to make the first direct half-life measurement of long-lived ^{239}Pu.[33]

Seaborg soon encountered a complication that threw the whole concept of using ^{239}Pu in a fission weapon into doubt. He reported to Oppenheimer on 3 November 1942 that an excessive number of neutrons might be produced from the (α, n) reactions of even slight amounts of light element impurities in the plutonium. Such neutrons, Seaborg pointed out, would be capable of predetonating a plutonium weapon, causing it to explode too soon and resulting in a low-power fizzle rather than a highly destructive explosion. To avoid this catastrophe, neutron production from light element impurities (such as boron) would have to be reduced, in the worst case, to one part in 100 billion, a task that even the confident Seaborg termed "formidable."[34]

Before Oppenheimer and Seaborg had time to assess the extent of this possible disaster, the complication received further unwelcome attention. During the lunch break of the 14 November S-1 Executive Committee meeting, a member of the British project told Conant that Chadwick had concluded that plutonium could not be used because of the light element impurity problem. Already shaken by the morning session, at which Compton had announced the plans to construct the first chain-reacting pile on the University of Chicago campus, Conant hurried to tell Groves of the dilemma.

The news could not have come at a worse time for the general, who was doing his best to persuade du Pont executives to assume responsibility for the plutonium plant at Clinton, Tennessee. Groves quickly enlisted Compton, Oppenheimer, Lawrence, and McMillan to investigate the seriousness of the impurity problem. Although they concluded four days later that purity requirements could be met, Conant remained pacified only until Chadwick's analysis arrived a few days later. When Conant compared the committee's purity figures with those given by Chadwick, he was appalled to see that British calculations allowed less than one-tenth as much impurity as that permitted by the Americans.[35] As Conant wrote to the S-1 Executive Committee, the situation was both "extremely serious" and "embarrassing." He felt it crucial for each committee member to "reconsider all aspects of the program in which he is in any way an expert so as to discover if there are other hidden and forgotten factors."[36] In view of the new developments and a "none too optimistic report by the du Pont Company," Conant also announced that a new review committee would be convened to assess "the entire Chicago program in the light of the present status of the electromagnetic and diffusion process."[37] The committee, chaired by MIT professor of chemical engineering Warren K. Lewis, would review the plans and make recommendations for constructing uranium and plutonium production plants.[38]

The Lewis Committee

The Lewis Committee visited the various bomb projects in late November and early December 1942. In preparation for the committee's visit to Chicago on Thanksgiving Day 1942, Met Lab scientists hurried to produce a report assessing the feasibility of plutonium separation, the pile weapon, and the plutonium weapon. After some "commotion" over whether the light element problem could cause plutonium to fission prematurely, Met Lab scientists presented an optimistic report on the potential of plutonium. "As produced," the report claimed, "this alloy can be used for making successful super bombs (probability 90 per cent)."[39] On 4 December, the committee issued its recommendation: "While recognizing the many uncertainties of the pile process," work on pile research should proceed at full speed.[40] The decision to continue the plutonium effort was reached before Fermi's impressive display of a chain reaction under Stagg Field.

To produce plutonium of the required purity, Chicago initiated a crash program in plutonium purification. At the May 1943 recommendation of a second committee chaired by Lewis (Chapter 5), this purification work was later shifted to Los Alamos.[41] In its November report, the committee also provided an evaluation of the various competing efforts to separate uranium isotopes. Committee members were impressed by the possibilities of gaseous diffusion but believed that the Kellogg–Columbia team responsible for studying it needed better organization and direction. Conversely, they admired Lawrence's leadership abilities, but doubted that the electromagnetic process would be able to produce significant amounts of fissionable material. The diffusion process was believed "to have the best over-all chance of success, and produce the most certainly usable material." Although the committee members suggested that work on electromagnetic separation continue, they did not see that it presented "a practical solution to the military problem at its present capacity."[42] The S-1 Executive Committee (the former Advisory Committee on Uranium) and the Military Policy Committee approved, and Roosevelt authorized, the Lewis committee recommendations in mid-December 1942.

After learning of the Lewis Committee's vote of confidence for gaseous diffusion, the Military Policy Committee reorganized the effort in December 1942. Keith of Kellogg was put in charge, and Kellex, a new subsidiary of Kellogg, was formed specifically to work on the project. To obtain highly skilled workers, Keith recruited engineers, scientists, and administrators from a wide range of industrial jobs to work for the duration of the war. Keith persuaded Carbide to operate the plant and assist research at Columbia. Through its contract with the corporation, the project called upon other members of the Carbide empire, including the Electro Metallurgical Company, the National Carbon Company, and the Linde Air Product Company. The leader of the research group, Dunning, supervised research at Columbia, Princeton University, the Bell Telephone Laboratories in New York, and the Kellex Plant in Jersey City.[43]

In the meantime, Lawrence promoted the electromagnetic separation method. He insisted, and Conant agreed, that the committee had overemphasized the problems of electromagnetic separation. Although discouraged by the many stages necessary to produce fully enriched ^{235}U with the electromagnetic process, Conant wanted the process to be developed. Because he felt uncertain about the chances of the plutonium bomb, he wanted to do everything possible to ensure the success of the uranium weapon. Also, he anticipated that more material would be needed than originally estimated because the bomb now seemed to require a test explosion. For these reasons he felt it was crucial to build a large electromagnetic plant with about 600 Calutrons so that at least 100 grams of ^{235}U could be produced daily.[44]

The final recommendations to Roosevelt were worked out on the basis of deliberations between the S-1 Executive Committee and the Military Policy Committee. Most of the Lewis Committee's suggestions were incorporated in the recommendations except for two new suggestions: that the pile be developed without an intermediate plant and that a gaseous diffusion plant capable of producing 100 grams of material a day should be built. The report Bush sent Roosevelt in December 1942 explained that the bomb would cost more and take longer to build than previously estimated. It recommended that the following facilities be constructed simultaneously: full-scale plutonium and gaseous diffusion plants costing \$100 million and \$150 million, respectively; an intermediate-sized electromagnetic plant capable of later expansion, costing \$10 million; and plants capable of producing 2.5 tons of heavy water each month, costing \$20 million. Although only \$85 million had been authorized in June, the total proposed effort would cost about \$400 million. Roosevelt approved the report, setting in motion the largest construction program ever conducted by private industry under army supervision.[45]

About the same time, du Pont agreed to build an experimental pile and plutonium separation plant in Clinton, Tennessee, and an industrial scale pile. The prime contractor for the Tennessee Plant and the entire Tennessee Site was Stone & Webster. Hanford, Washington, was chosen as the site for the large pile in January. With these arrangements settled, plans for the separation plants proceeded rapidly in 1943. By March, du Pont engineer Charles Cooper had diagrams of the small separation plant. Crawford Greenewalt, du Pont liaison to the Met Lab, outlined preliminary specifications. In June, du Pont decided on the bismuth phosphate process for both plants. Construction began at Clinton in February 1943 and at Hanford in April. The Clinton pile went critical in November, and by spring 1944 grams of plutonium were being shipped to Los Alamos. The first pile at Hanford went critical in September 1944, and by early 1945 Los Alamos had received kilogram amounts of plutonium.[46]

Arrangements for ^{235}U separation plants, also built at Clinton, proceeded just as rapidly. The electromagnetic plant started going up in February 1943 and the gaseous diffusion plant in June. The electromagnetic plant sent the first samples of enriched uranium to Los Alamos in February 1944. Because difficulties arose in producing sufficient ^{235}U by the electromagnetic and gaseous diffusion methods, a thermal diffusion plant was begun in July 1944 (Chapter 11).[47]

By the spring of 1943, the material program could count the following accomplishments: the first divergent chain reaction had been demonstrated; much had been learned about the physical and chemical properties of plutonium and uranium; designs for the first plutonium production pile were nearly complete; and an enormous research program had been launched to solve the remaining problems of ^{235}U isotope separation. Attention could now turn to the next goal: the building of the actual weapon.

4

Setting Up Project Y: June 1942 to March 1943

By the time a fast-neutron fission laboratory was conceived early in 1942, theorists and experimentalists had made initial calculations for the bomb, including new estimates of critical mass and efficiency. Some progress had been made on designing methods for assembling the weapon and on a program for measuring nuclear constants central to bomb calculations. On the whole, however, research languished because of poor information exchange among the various groups involved, which were scattered throughout the United States. Oppenheimer, who replaced Gregory Breit as coordinator of the fast-fission project, recommended to Groves that the effort be centralized. Groves, who recognized the security benefits of centralization, readily complied, thereby setting in motion plans for establishing the Los Alamos Laboratory.

Groves and Oppenheimer took the first step toward creating the laboratory in early 1943 by recruiting many of the world's best scientists. The temporary nature of the project and the urgency of its mission aided the recruitment effort, but the task was complicated by the delicate issue of whether Los Alamos would be a military or a civilian establishment – an issue never formally resolved. The standard caricature of Oppenheimer as an other-worldly intellectual and Groves as a burly martinet

* This chapter is based on a draft by Paul W. Henriksen, with contributions by Catherine Westfall on fast-fission and nuclear physics research, and by Lillian Hoddeson on the implosion and gun methods of assembly. This chapter also draws on an earlier contribution by Robert Seidel. We thank Alison Kerr, Les Redman, and Gordon Baym for extensive editorial contributions.

highlights the misalignment between the military and scientific communities that joined in Project Y. Surprisingly, Oppenheimer and Groves developed a collaboration that was both congenial and fruitful.

Fast-Fission Research before the Start of Los Alamos

Research on fast-neutron fission was a secondary concern at the time the full-scale atomic bomb effort was launched in late 1941. The primary goal was to develop a self-sustaining slow-neutron chain reaction, both to demonstrate its feasibility and to lay the groundwork for designing the plutonium piles. Only a few theorists, including Oppenheimer and Breit, were immersed in calculations of critical mass and atomic bomb efficiency – problems that built on Compton's work for the November 1941 National Academy of Science report.[1]

By March 1942, Compton was able to communicate encouraging fast-neutron results to Bush, who in turn passed them on to President Roosevelt. Whereas the NAS report had put the critical ^{235}U mass between 2 and 100 kg, Breit and Oppenheimer now narrowed it down to between 2 and 5 kg and at the same time laid to rest the pessimistic predictions of low efficiency. Kistiakowsky's earlier claim that a fission weapon would be only one-tenth as effective as a chemical weapon was shown false by new calculations. The efficiency estimate was raised from the destructive effect of 600 to 2,000 tons of TNT.[2]

This heartening news did little to speed progress by the various groups in the fast-fission programs. The groups included the University of California at Berkeley, Stanford, Rice in Houston, Chicago, Wisconsin, Minnesota, Purdue, Harvard, Princeton, Cornell, and the Department of Terrestrial Magnetism of the Carnegie Institution in Washington, D.C. Compton had assigned the task of coordinating the various fast-fission groups to the eminent theorist Breit. However, Breit had had little experience in managing diverse groups. To add to Breit's problems, the project was given low priority, and Compton had given him little authority. Nervous about security, Breit quarreled with Compton about procedures. He eventually resigned on 18 May.[3]

With the S-1 Committee's consent, Compton replaced Breit with Oppenheimer, who had earlier helped prepare the new estimates of efficiency and critical mass.[4] Thus, Oppenheimer was already familiar with the fission project, which he had joined at Lawrence's insistence in October 1941. In January 1942, he had organized Berkeley's theoretical program on fast neutrons.[5] A distinguished theoretical physicist and exceptionally adept expositor, Oppenheimer added to the project a talent for motivating colleagues and for explaining highly technical theories to nontheorists.

Compton was aware of Oppenheimer's reputation for being unapproachable and out of touch with the details of everyday life, and of his habit of delivering cutting criticism to those whose opinions or standards differed from his own.[6] Compton nevertheless judged Oppenheimer as the best available person to direct the division of the Metallurgical Laboratory charged with estimating the bomb's critical mass and efficiency.[7]

To help coordinate the disparate groups working on fast-neutron studies, Compton engaged John H. Manley, a reserved experimentalist, originally from the University of Illinois, who had worked from 1941 at the cyclotron of the Chicago Met Lab.[8] Despite their almost opposite temperaments, Oppenheimer and Manley would collaborate effectively in establishing the Los Alamos Laboratory.[9]

By this time, information exchange with the British project had begun. Oppenheimer had met with Peierls in Berkeley and learned of the Frisch–Peierls work on critical mass, efficiency, and explosion damage. The meeting confirmed that the Americans and British had considered many of the same points.[10] Data from the two projects reinforced the feasibility of assembling the bomb by shooting one subcritical piece of ^{235}U or ^{239}Pu into another using a gun.[11] The atomic bomb problem appeared in principle almost solved. With great optimism, Oppenheimer wrote Lawrence on 19 May 1942, that, with "a total of three experienced men and perhaps an equal number of younger ones," it shoud be possible to solve the theoretical problems of building a fast-fission bomb. At least, he reflected, "we should be able to get quite far within six months ... unless new things turn up in the course of the work." The particular theorists Oppenheimer mentioned were John Van Vleck, Edward Teller, Robert Serber, and Hans Bethe. "I would think that any two of these working with me could get the job done in good order."[12]

Meetings at Chicago and Berkeley in the Summer of 1942

Oppenheimer realized that solving the problems of the fast-fission bomb would require close collaboration between theorists and experimentalists. In his letter of 19 May to Lawrence, Oppenheimer proposed a joint meeting of theorists and experimentalists, "perhaps in Chicago," adding that having the project in Berkeley "would have the advantage that I could continue some work on your end of the project, and we could maintain close contact with Segrè and Kennedy and their group." They ended up meeting both in Chicago, on 5–6 June, and in Berkeley during July 1942. Oppenheimer put his ideas forward in an outline he drafted in June 1942, discussed at both meetings by the same core group of theoreticians (Van Vleck, Serber, Teller, Emil Konopinski, Stanley Frankel, Bethe, and Eldred Nelson) and experimentalists (Edwin McMillan, Manley, Felix Bloch, and Segrè).[13]

At the Chicago meeting, Oppenheimer and Manley informed Met Lab administrators and some fifteen theorists and experimentalists that they would lead the fast-neutron research. Oppenheimer had already begun to assemble a theory group in Berkeley.[14] The plan there was for theorists Serber, Frankel, and Nelson to work with experimentalists, including Segrè, while at the Met Lab, theorists Teller, Konopinski, and Robert Christy would work with Manley and others.[15]

The Berkeley meeting, which began in the second week of July, centered on theoretical work, although a few Berkeley experimentalists participated in some of the activities. They met on the top floor of Le Conte Hall in a sealed-off room near Oppenheimer's offices.[16] Oppenheimer described the events of that first week to Manley in a letter of 14 July 1942. "There is a terrible lot to do here, and I am sure that my first duty is to help get the theoretical questions, some of which are turning out to be very exciting indeed, in the best possible shape while our galaxy of luminaries is still available." In the first days of the meetings, some important points were raised by several of Oppenheimer's Berkeley students – Serber, Nelson, and Frankel – who had been studying the very issues the meeting had been called to discuss.[17] Much of the data were still sketchy or contradictory. Among those examined were British results, sent over to America months earlier by the MAUD committee but kept locked in a safe by Lyman Briggs in an overzealous security effort.[18]

The first topic discussed was the nuclear physics of a bomb assembled by the gun method, its improved efficiency, and damage estimates.[19] Although it was known that slow neutrons (with energies up to a few keV) could induce fission in ^{235}U and that ^{235}U was likely to sustain a chain reaction, the group recognized that an efficient bomb could not be made with slow neutrons, because the system would tend to disassemble before the chain reaction had progressed far enough.[20] Although there was not yet experimental verification that the ^{235}U fission cross section for fast neutrons (with energies of order MeV) was large enough to give an effective explosion, attention soon focused on this issue.[21]

During the first few days, Serber, Nelson, and Frankel presented their calculations suggesting that ordinary large guns could impart the proper velocity to the uranium pieces. Bethe recalls the participants giving a stamp of approval to Serber's calculations.[22] Using British estimates of the mean free path of neutrons in uranium, they predicted that an 8-inch-diameter uranium sphere would allow a chain reaction, with an efficiency of at best a few percent.[23] They extrapolated damage estimates from the fairly well documented explosion of 500 tons of TNT that had occurred in Halifax, Nova Scotia, in 1917 and took into account the harmful effects of the neutron and gamma-ray bursts.[24]

In their discussion of ^{239}Pu, the participants calculated the number of neutrons that would result from α particles colliding with both light element impurities in the plutonium (e.g., beryllium) and heavier elements (e.g., aluminum and silicon). It was clear that impurities would have to be removed, because collisions with them could yield enough extraneous neutrons to threaten predetonation of the weapon. The conferees estimated the duration of the core fission process to be less than one millionth of a second and worked out how the change of the core from metal to gas would affect fission.[25]

The Idea of a Super Bomb

Teller shifted the attention of the conferees to a subject central to his own thinking in the past several months, a hydrogen bomb, subsequently referred to as the Super. Thermonuclear reactions and their role in energy production in stars had been discussed in the 1930s by George Gamow and others at George Washington University. At Cornell, Bethe developed the ideas further (and later won a Nobel Prize in physics for this work.) But realizing thermonuclear reactions on earth seemed a distant prospect in the 1930s.[26]

The idea of a thermonuclear weapon emerged in the spring of 1942, during a discussion of the fission process in the context of the Manhattan Project. After lunch at the Faculty Club at Columbia University, Teller and Fermi were speculating about the use of thermonuclear reactions in a nuclear bomb. Fermi imagined a weapon in which a fission reaction triggered fusion of two light elements into a heavier element, for example, fusion of two deuterium atoms into helium. This process promised to yield substantially greater energy than fission alone. Although Teller initially doubted that fusion could be achieved with temperatures available on earth, the idea intrigued him.[27]

Shortly afterward, Teller moved to Chicago to work at the Met Lab. During his first days there, before being given a particular assignment, he and Emil Konopinski tried to establish on physics grounds that a fusion bomb could not work, in particular, that deuterium could not be ignited by a fission bomb. But their work revealed that such a bomb might in fact be possible.[28] The idea of surrounding a fission bomb with deuterium, which could be ignited and fused and thereby could produce far more energy than a fission bomb tantalized Teller. He brought up the idea at the Berkeley conference.[29]

Teller pointed out that deuterium would be far cheaper to obtain than ^{235}U or ^{239}Pu and that the explosion could be made indefinitely large by increasing the amount of deuterium placed near the fission bomb. From that point on, although Oppenheimer tried to bring the discussion back to the fission bomb, Bethe and others spent much of their time at the meeting arguing with Teller about his Super ideas.[30] Bethe recalls Teller being so preoccupied with the Super that at one point, in a discussion of the Germans' desire for heavy water (as a moderator in the nuclear reactor), "Teller as usual jumped thirty years ahead of time and said, 'Of course they want heavy water to make a super'."

Sharing a house at Berkeley during the conference, Konopinski, Teller, and Bethe continued their discussions about the Super in the evening, after the formal meetings. In considering how the energy from the fission reaction would be transferred to the deuterium, Konopinski suggested that adding some tritium to the deuterium might lower the reaction temperature and make the deuteron easier to ignite.[31] All three realized that radiation from the fission weapon into or out of the deuterium would create problems and talked about using dense and opaque reflectors (tampers) to keep radiation in the deuterium.

At one point, a frightening idea came up. Teller asked, if the fission bomb could ignite deuterium, could it not ignite the nitrogen in the atmosphere? Serber recalls, "Bethe went off in his usual way, put in the numbers, and showed that it couldn't happen .... Oppy made the big mistake of mentioning it on the telephone in a conversation with Arthur Compton."[32] While Bethe and Konopinski considered this question mathematically, Oppenheimer made a quick trip to northern Michigan to discuss the possibility of such a catastrophe with Compton, then resting at his summer cottage on Otsego Lake. They agreed that the

Fig. 4.1a. Edward Teller, theoretical physicist, initially head of the implosion theory group, was the main proponent of the thermonuclear weapon. LA Photo, LAT 1198.

Fig. 4.1b. Hans Bethe, theoretical physicist, was head of the theoretical physics division. LA Photo, LAT 747.

Berkeley theorists should be authorized to continue their calculations of possible ignition of the atmosphere; if they failed to provide conclusive evidence that an atomic explosion could be contained, the bomb project would have to stop.[33] Few of the participants seriously believed that an uncontrolled nuclear reaction in the atmosphere was possible, and their calculations soon showed that the chances of igniting the atmosphere were very remote given that equilibrium with the radiation limited the temperature inside the bomb.[34] The large electrical repulsion between the nitrogen nuclei made their chance of reacting very small. Bethe later referred to the problem as a "red herring."[35]

By the time the conference ended in late August a thermonuclear reaction seemed possible. However, there were many unknowns, beginning with the rate at which the deuterium would radiate the energy it received from the fission reaction and the rate of energy transfer through the tamper. Teller recalls that Oppenheimer was sufficiently interested in the problem to comment, "Now we really need another laboratory."[36] But everyone involved realized that, although study of the thermonuclear weapon was justified because of its possible explosive power, their first job was to develop the fission weapon.

Enthusiasm for the Super during the wartime period was probably highest in the fall of 1942, when atmospheric conflagration worried only the most pessimistic, and the difficulties with Super operation were not yet obvious.[37] Teller continued to work out possible thermonuclear reactions after returning to Chicago. Oppenheimer, Bethe, and Lawrence helped by requesting a study at the Harvard cyclotron of the cross sections of the reactions of tritium with deuterium and two lithium isotopes. At Bethe's request, and with Compton's and Oppenheimer's approval,[38] important studies of tritium–deuterium (T–D) cross sections were done at Purdue by L. D. P. King and Raemer Schreiber, and later by Marshall Holloway and Charles Baker. Work on the Super became even more secret than that on fission. As Oppenheimer wrote to Manley in July, "If our ordinary conversations are secret this should be secret squared ... knowledge of our interest in this subject [should be] restricted to the absolute minimum and if possible [it should] not ... appear to have any connections with the tubealloy project." Not even scientists on the project would be told about the Super bomb unless it was deemed necessary.[39] Oppenheimer arranged for further Super studies at Harvard, Purdue, and Minnesota, and for an early fall meeting in Chicago to plan further research.[40]

Further Nuclear Physics Research in the Fall of 1942

The main obstacle to a theoretical understanding of the fission bomb was the uncertainty surrounding existing experimental data, in part the result of inadequate instrumentation and a lack of experience in the new field.[41] By the time of the Berkeley theory meeting, a series of experimental studies of fast fission were under way. The measurements required several improved fast-neutron detectors, fission threshold detectors, counterionization chambers, and accompanying electrical circuits.[42] At the University of Wisconsin, Joseph McKibben and others used fission detectors and neutrons from carbon–deuteron and deuteron–deuteron reactions created by Van de Graaff accelerators to study the scattering properties of various tamper materials.[43] At Cornell, Robert Bacher's group developed equipment to study delayed neutron emission. Bacher had been challenged by Luis Alvarez's description of Berkeley efforts to produce a modulated neutron source.[44] Although Alvarez judged that it would be extremely difficult to produce modulated neutrons down to the thermal range because of the short time scale needed, Bacher hoped to produce them in short pulses using the arc neutron source with which Stanley Livingston, Charles P. Baker, and Marshall Holloway had equipped Cornell's cyclotron. This device allowed fast time resolution. Baker and Holloway built a device based on the original Alvarez "time-of-flight" velocity selector to separate low-energy neutrons into energy groups (based on the time of flight between the source and detector over a path several meters long). In his June 1942 outline for the experimental program, Oppenheimer planned to have the Cornell group use this equipment to measure the fission spectrum at low neutron energies.[45]

The fission cross sections needed to estimate critical mass for ^{235}U and ^{239}Pu vary with the energy of the incident neutrons. It was therefore crucial to measure the energy spectrum of the fast-fission neutrons. Various methods were proposed, based on either measuring pulses produced in ionization chambers by recoil protons, or measuring traces made by neutrons in cloud chambers. Under Oppenheimer's plan, several groups would examine the fission spectrum through proton recoil studies in ionization chambers: Felix Bloch and others at Stanford would use neutrons from a cyclotron, John Williams and his associates at the University of Minnesota would use neutrons produced in a Van de Graaff, and Mitchell's group at Chicago would try a graphite pile, cyclotron, and chemical separation. H. A. Wilson and William C. Bennett at Rice were to use a cloud chamber and neutrons produced in deuteron–deuteron reactions at the Rice Van de Graaff.[46]

The energy range around 1 MeV was of particular interest in the study of fast-fission cross sections, because theory predicted a peak in the fission neutron spectrum at nearly that energy. Oppenheimer instructed Norman P. Heydenburg at Carnegie Institution to measure ^{235}U fission cross sections with the Van de Graaff there, which was capable of producing monoenergetic neutrons in the range of 0.025 to 4 MeV from deuteron–deuteron and carbon–deuteron reactions. Segrè could then extend these studies, having access at Berkeley to a variety of natural sources that produced monoenergetic photoneutrons in the range of 0.22 to 0.65 MeV by the photodisintegration of beryllium and deuterium, by yttrium and ^{24}Na γ rays, respectively. Both Carnegie and Berkeley measurements used foils prepared and analyzed at Berkeley with the help of nuclear chemists.[47]

The fast-fission experiment tended to follow Oppenheimer's June outline through the remainder of 1942. By September, Harvard and Princeton were planning fusion-related measurements, and Arthur Snell at Chicago was preparing to measure the fission cross sections of unenriched uranium at various energies. In October, Oppenheimer thought the time ripe for measuring higher neutron-energy ^{239}Pu-fission cross sections, which had only been measured at thermal energies. In addition, by the end of 1942, a group at Cornell was studying the delayed emission of fission neutrons.

Frustration and success mingled in the fast-fission program.[48] At Rice, William Bennett summarized his difficulties in measuring the fast-fission spectrum of uranium. "Our work suffers from bad statistics ... wall effects ... inelastic scattering in 238 [U-238] which gives reduced energies for approximately 35% of the neutrons measured." He suggested switching to photographic plate detectors, a technique then being used by James Chadwick at Liverpool. As a result, Bennett and his co-worker Hugh T. Richards moved to the University of Minnesota, where they carried out preliminary tests using such plates.[49]

The fission spectrum experiments at Stanford ran into problems as well. In extending the ionization chamber scheme employed by Zinn and Szilard, Bloch used the Stanford cyclotron to produce neutrons from the beryllium–deuterium (Be–D) reaction. In October 1942, his group planned to measure ^{235}U's neutron fission spectrum by bombarding ordinary uranium with slow neutrons. The distribution of recoil pulses was to be observed in an ionization chamber filled with hydrogen. Unfortunately, the source was so strong that the energies of a significant number of the primary neutrons overlapped those of the neutrons being investigated. To overcome this problem, Bloch's group built a cubical, double-walled graphite box and used cans of oil to remove energetic primary neutrons. The measurements showed the energy of the neutrons tailing off from about 1 MeV, as predicted. However, the Liverpool group's emulsions showed a sharper maximum at about 2 MeV. Questions then arose concerning the accuracy of all measurements using normal uranium, because tiny amounts of ^{235}U were diluted in a large mass of ^{238}U. The hope for better measurements hinged on using samples having a higher percentage of ^{235}U.[50]

^{235}U fission cross sections were also very difficult to measure. The physicists had trouble developing sufficiently sensitive detectors and producing the uniformly thin foils needed to carry out their experiments. Segrè struggled to calibrate his natural sources. In June, for example, an yttrium–beryllium (Y–Be) source showed a ^{235}U fission cross section of 5.4 barns at 0.22 MeV, whereas in July the same source at the same energy showed 6.9 barns.[51] In October, Segrè reported that his group had determined the number of neutrons "by slowing them in a large bath of manganese sulfate solution and determining the manganese activity produced." With the help of calculations by Frankel and Nelson, the group then tallied "the fraction of the total number of emitted neutrons crossing the uranium sample." In this way they found the fission cross section of ^{235}U to be 2.8 barns at 0.22 MeV and 1.7 barns at 0.43 MeV. Manley judged these measurements to be "in reasonable accord" with other measurements, although one could not be confident of the results, because last-minute calibration changes had increased the new figures by 40 percent. Oppenheimer admitted in September, "we are afraid they will always remain fairly inaccurate."[52]

Heydenburg's measurements with the Carnegie Department of Terrestrial Magnetism Van de Graaff, which covered the crucial energy range just above 0.5 MeV, were even more problematic. Measurements with D–D neutrons, from bombarding deuterium atoms with other deuterium atoms, were difficult because the neutrons were emitted at energies near the fission threshold for ^{238}U. In December 1942, Heydenburg's group finally concluded that "nearly all the fissions in normal uranium" were due to ^{238}U. Determining the fission cross section for ^{235}U required the use of "enriched samples in which the ^{238}U content is accurately known." Thus Heydenburg's measurements using D–D neutrons were suspect.[53] He thus switched to 0.64 MeV neutrons from carbon–deuteron bombardments, but to his dismay, encountered further problems. "C^{13} which is normally present in carbon gives rise to much more energetic neutrons of approximately 1.8 and 5.6 MeV, and these neutrons produce fission in U^{238}." The problem had returned. Similar ^{235}U samples were sent to Heydenburg and to Chadwick, who "found only a quarter as much" ^{235}U. Because of the difficulties arising from the C^{13} neutrons, Heydenburg's measurements merely pointed "the way to later experiments with other neutron sources."[54] Manley and Oppenheimer learned that C–D and D–D neutron sources were to be avoided in fission cross-sectional measurements.

This discovery led to the use of neutrons produced by the bombardment of lithium with protons ("Li (p,n)" neutrons). In spring 1943, Alfred O. Hanson and D. L. Benedict announced the results from a series of experiments using the Wisconsin Van de Graaff to produce monoenergetic Li (p,n) neutrons. They determined the neutron flux of this source by two methods, one drawing on the immersion technique employed by Segrè, which had originated with Fermi, and the other using a new detector, the coincidence proportional counter, developed at Wisconsin. Hanson and Benedict simultaneously detected fissions for ^{235}U and ^{238}U with a double ionization chamber having a central electrode loaded with both enriched and normal uranium.[55] They measured for the first time the fission cross section of ^{235}U as a function of energy and determined the fission threshold of ^{238}U. These results, which showed an approximate inverse relation between energy and fission cross section, are impressively close to present values.[56]

Measuring fission cross sections for ^{239}Pu was even more difficult. Researchers knew little about the chemical or physical properties of the element, which had only been discovered in 1941, and were not even certain of the half-life of ^{239}Pu. Thus they could not be sure of the true weight of their samples (estimated by counting alpha emissions), a factor needed to compute fission cross sections. Another problem was that ^{239}Pu produced copious α particles, which would pile up and imitate fission counts in the detectors. In the face of such difficulties, Heydenburg, in winter 1942–43, measured the ratio of the cross section of ^{239}Pu to that of ^{235}U, from thermal energies to 3.95 MeV; meanwhile, Segrè and others at Berkeley measured this ratio at thermal energies using the Berkeley cyclotron. To no one's surprise, considering the uncertainties of measurement, the ratio for thermal energies varied from 1.99 to 1.26. More precise ^{239}Pu cross-sectional measurements would have to wait.[57]

While work in numerous American laboratories continued on the main line of inquiry – fission spectrum and fission cross-sectional measurements for ^{235}U and ^{239}Pu – the Cornell nuclear physics program changed direction. In early September, several slow-neutron projects were discontinued, including the slow-neutron fission spectrum measurements laid out in the June 1942 outline and a series of slow-neutron cross-sectional measurements that Oppenheimer requested in July. As Bacher later explained, "Robert was not much interested in these slow-neutron things" and wanted to divert personnel to higher-priority projects.[58]

Since summer 1942, Cornell researchers had been investigating delayed neutron emission in fission, although this project had relatively low priority. This work extended investigations begun in 1939 by Norman Feather, who determined in a simple comparative measurement of fast and slow neutrons that the time between neutron capture and fission is no more than 10^{-13} s. His finding left open the possibility, albeit remote, that some neutron emissions were delayed after fission, which would change the efficiency estimate for the bomb. By September 1942, the Cornell group had measured neutron delays at intervals greater than one second with the Cornell cyclotron and an ionization chamber. Although they concluded that the delay could not "be determined in the most interesting region" (the pre-explosion time, 10^{-7} to 10^{-9} s), they suggested a "brief examination" within their experimental capability of 10^{-5} s, because negative results would "strengthen the present assumption that there is no appreciable number of delayed neutrons." On 9 November, Bacher reported to Compton that only "about 1%" of fission neutrons were delayed within the measurable time scale.[59]

By then, however, interest was building to measure the delay of neutron emission more accurately. Bacher showed the Cornell results to Bethe, who spoke with Bruno Rossi about counting techniques, and Baker about possible improvements to the Cornell work. After making a number of calculations, including one assessing the potential accuracy of fission and neutron counting, Bethe told Manley that although it would "probably not be quite easy," he did "not see any reason why the method should not work down to times of about 10^{-7} s." Theorists estimated that neutrons were ejected within 10^{-15} s after fission: even with Bethe's new estimates, such an interval was far outside their capabilities. By late 1942, as plans were accelerating for the Los Alamos Laboratory, Oppenheimer felt more pressure than ever to rule out potential unpleasant surprises. Because the Cornell group was scattered by this time, with Baker and Holloway at Purdue making fusion-related measurements and Boyce McDaniel visiting the MIT Rad Lab to learn instrumentation, Oppenheimer had the Cornell timing equipment moved to Los Alamos. Although they had been assigned low priority throughout 1942, such measurements would be among the first done at the new laboratory.[60]

Technical Progress on the Gun Weapon in the Fall of 1942

In the fall of 1942, the members of Oppenheimer's theory group were again working at their home institutions. Continuing to collaborate with experimentalists, by November they could report considerable progress on efficiency, tamper, shock wave, and neutron diffusion. Oppenheimer wanted to bring the British up-to-date. He wrote Peierls, "Since it will be some time before detailed reports on all phases of this work are available, and since, on some of the questions that have been considered by our British colleagues we are not in complete agreement with their results, we thought that it might be helpful to communicate to you in an informal way some of the findings of our group."

In a detailed memorandum, Oppenheimer summarized the main results, drawing particular attention to the major point of disagreement between the American and British scientists, which concerned the role of radiative energy loss. Oppenheimer also noted that his group had studied the effect of te transparency of air to radiation on the formation of the shock wave in the atmosphere following the explosion, concluding that the fission bomb was "intermediate between a TNT bomb and Taylor's mass-less point source explosive." The group had used "slightly different, and we believe superior, methods" for treating the problem of neutron diffusion to determine critical masses and effective neutron multiplication rates. Employing effective "transport" cross sections (given the "present crude state of our empirical knowledge"), the group employed iterative numerical integration to solve the integral equations for neutron diffusion and found quite marked differences from the differential approximation to neutron diffusion. The memorandum also expressed hope of reducing the risk of predetonation.

In his response, Peierls essentially agreed with the conclusions of Oppenheimer's memorandum. Peierls attributed the minor differences between the American and British work to British reluctance to accept the reported accuracy of the data. The British, he explained, were "not trying at this stage to obtain quantitative data for the main parameters required, but rather to explore the field and develop the methods in such a way as to be able to obtain a reliable answer quickly when definite experimental data will be available." After discussing details of the neutron diffusion problem, he noted, "we had not previously been aware of" the results in Oppenheimer's memorandum on the influence of radiation. Peierls's note also sheds light on P. A. M. Dirac's contributions at the time to the British effort, indicating that Dirac and Klaus Fuchs had extended Maurice Pryce's calculations on the explosion dynamics. He also mentioned the interesting point raised by Dirac that when the number of light quanta per cubic wavelength is comparable to or larger than unity, "the contribution to each process of induced emission should be comparable to that of spontaneous emission." Peierls continued with a discussion of "model experiments" being carried out by Frisch on the question of the critical distance for a gun assembly based on a lateral rather than axial approach of two hemispheres. He closed with a report on the work of his group on the properties of the compressional wave in the air generated by the explosion.

At a meeting held in late September 1942 in Chicago, the same core group that had met in Berkeley the previous summer considered whether it would be possible "to use this new tool (the active bomb) by the time the requisite materials are available." The plan was for the theorists to help guide the experimental program. As Compton explained: "Oppenheimer, Van Vleck, Bethe, Teller and Manley ... will consider the type of experiments that should be performed .... I have also invited Mr. Ed MacMillan [*sic*] to attend this conference, having in mind the probability that MacMillan may take a leading part in the future development of this aspect of our program. It would be most helpful if we are able to include Messrs. Fermi, Allison, and Wigner in this discussion since these men should be concerned with guiding the experiments to be performed."[61]

A late November report by Manley, Oppenheimer, Serber, and Teller, "The Use of Materials in a Fission Bomb" dealt with three main topics: the energy released, the amount of material required, and the construction and detonation of a fission bomb. According to their data, they had "strong reason to believe that any of the known possible materials, 23, 25, and 49 will produce the same energy per gram as any other."[62] In summarizing their knowledge and uncertainties on fission cross sections, however, they concluded that "the minimum amount of material required probably differs" with the material; "2 to 3 times less 49 will be needed than 25; of 23 even somewhat smaller amounts may be sufficient. The masses required are of the order of 10 kg. The exact amount required is uncertain because of the uncertainty in experimental data on fission and total cross sections, the number of neutrons per fission, and the density." The report noted that the gun method based on 49 could conceivably be threatened by predetonation and suggested an alternative method that would avoid this danger.

In a further report of 30 November 1942, Teller made a quantitative analysis of the effects of impurities in predetonation resulting from (α,n) interactions. He addressed two questions: "the chances of predetonations in various stages of approach between fragments leading to explosions with correspondingly varying efficiencies," and, "how great is the minimum efficiency i.e. how much energy is released if a neutron is present at the time when the two fragments first reach critical configuration." To study the efficiency he used the calculations on bombs of low efficiency reported by Oppenheimer to the British and derived the probability of a given efficiency in terms of the number of neutrons present, the distance of approach of the active fragments, and other parameters of the bomb.

The threat of predetonation in gun assembly, where velocity is limited, made the pursuit of other assembly systems desirable. An alternative method was to use a high explosive (HE), rather than a propellant explosive, to push material together, because pressures and velocities in detonating HE are very much higher than in gun barrels. However, HE is not readily confined by structures such as gun barrels. A system would be required that would blow inward on itself, as well as blast outward. Such a system might offer a chance to use fissile material with greater efficiency. In 1942, the group working on bombs felt compelled to explore all feasible designs. Autocatalytic assembly appeared inefficient and unpredictable. In contrast, implosion – which could be achieved by surrounding the fissile material with high explosives – was an attractive concept: the detonation would create shock waves that would collapse the fissile material from a subcritical into a supercritical mass.

The Suggestion of Implosion

Theorists appear to have first entertained the idea of assembling an atomic weapon by implosion during the summer conferences of 1942 but gave it little attention. Serber vividly recalls that Richard C. Tolman, an original member of the National Defense Research Committee and subsequently also scientific adviser to General Groves, visited Berkeley and introduced the idea of implosion as an alternative assembly method. Serber also recalls that he and Tolman wrote a short paper on the idea.[63] However, Bethe, Konopinski, and McMillan do not recall any discussion of it in that period.[64] Oppenheimer, writing in 1945, placed the beginning of implosion at Los Alamos in April 1943, when Seth Neddermeyer suggested the idea to the laboratory.[65]

Tolman continued thinking about implosion during the remainder of 1942 and through the early months of 1943, writing to Oppenheimer on 27 March 1943, "Conant and I have discussed the somewhat modified possibility of starting off by using ordinary explosive to blow the shell of active material into the center. I think that this would be an easy thing to do." Two days later, he mentioned an idea like implosion, although not the word, in a memorandum to Groves describing various methods of "securing explosion": "In the case of the mechanism depending on the deformation of a shell of active material, ... it might be possible to bring this about by explosive charges which would blow fragments of the shell into the interior. This latter possibility appears an interesting one to consider." No calculations accompanied the speculations and no mention was made of the notion of compression, on which efficiency depended. The Tolman–Serber idea was, in essence, to shoot together pieces of the divided-up shell.[66] Although implosion was never realized in the form Tolman and Serber initially envisioned, their concept was a precursor to the implosion assemblies later created at Los Alamos.

Planning Los Alamos

Groves's immediate task after taking charge of the Manhattan Engineer District in September 1942 was to find a director for the atomic bomb laboratory. His first two choices, Lawrence and Compton, were both occupied – Lawrence with directing the electromagnetic separation project and Compton with directing the Chicago Met Lab. As leader of the fast-neutron program, Oppenheimer was the next logical choice. Groves was concerned about Oppenheimer's reputed lack of administrative experience – a new criterion in scientific research that would become increasingly important. In fact, Oppenheimer had had considerable experience as a scientific administrator, having led thriving groups of physicists both at the California Institute of Technology (Caltech) and Berkeley. Groves also feared that since Oppenheimer had not been awarded a Nobel Prize he might not be sufficiently respected by his scientific staff. Oppenheimer's associations with leftist organizations in the 1930s were another drawback. But Groves decided to appoint and clear him anyway.[67]

In visiting the various fast-neutron fission groups earlier in 1942, Oppenheimer had noticed some duplication in their work and surmised that certain ideas were being stifled because researchers had no overview of the project. To achieve collaboration without threatening security, he would have to consolidate the project in one isolated, controlled location.[68] Manley agreed, figuring that such a scheme would give the scientists better facilities. He was also weary by then of coordinating the activities of far-flung groups, a job that required him to be the "information chief, procurement agent, liaison officer, member of the technical committee (of two) and group leader for the Chicago program."[69]

Oppenheimer probably communicated his ideas on setting up the laboratory that would become Los Alamos to the S-1 Executive Commit-

Fig. 4.2. John Manley, a group leader in the experimental physics division, was Oppenheimer's right-hand man in setting up the Los Alamos laboratory. LA Photo, LAT 147.

tee, sometime before a meeting that was held on 13 September 1942 at Lawrence's Bohemian Grove retreat. McMillan, an early member of Lawrence's laboratory, was taking a leading role in the planning. In a cable the next day, at Lawrence's suggestion McMillan asked that Oppenheimer, Manley, Fermi, and Lawrence meet in Chicago one week later to formulate plans for the new fast-neutron laboratory that would house the section of the Manhattan Project code-named Project Y. The conference was held on 19–23 September.[70]

Groves initially wanted the laboratory sited near the plant producing the fissionable material, most likely in Tennessee.[71] By 12 October, he had changed his mind and decided that the laboratory should be located far from Tennessee and separated also from Chicago.[72] That day,

Manley sent Oppenheimer plans for the laboratory that he had submitted several days earlier to the engineering firm of Stone and Webster, the prime contractor for the Tennessee site.[73] Manley had estimated the size of the buildings by adding up the number of rooms already in use for the fast-neutron research at the various university laboratories. The planned buildings were to house six theoretical physicists with six assistants, twelve experimentalists with fourteen assistants, and five secretaries. Laboratories that housed radiation-producing machinery were to be isolated from the administrative building and other laboratories.[74] Oppenheimer added a cryogenics laboratory for research on the Super, along with engineering and shop space, and slightly enlarged the buildings to allow modest expansion.[75]

Site selection was the next major issue in the fall of October 1942. Groves had instructed Maj. John H. Dudley of the Corps of Engineers to search the western United States for a site at least 200 miles inland (to be safe from enemy air attacks), with facilities to accommodate a handful of scientists and a few hundred support personnel. The site was to be isolated, but accessible by a road, and to lie in a natural bowl with nearby hills to contain accidental explosions.[76] Dudley selected Jemez Springs in north central New Mexico. But, on visiting the Jemez Springs site, Oppenheimer objected to the narrow valley in which Jemez Springs was located. In addition, he felt that the access road needed only to be good enough to haul in a few howitzers. He suggested the site of the Los Alamos Ranch School, on the other side of the Jemez Mountains. He had visited Los Alamos during summers spent on his family's nearby vacation ranch. On viewing the Los Alamos site, Groves was unconvinced. However after learning that Oppenheimer had told Dudley he was set on Los Alamos, Groves initiated acquisition procedures.[77]

Recruitment: A Military or Civilian Laboratory?

When Oppenheimer and Manley began recruiting personnel in the fall of 1942, they were told that under NDRC policy no one committed to any of its other projects could be approached. F. Wheeler Loomis, personnel director of the MIT Radiation Laboratory, insisted on a strict interpretation of this policy. But adherence was almost impossible because practically all top scientists in America were already engaged in other wartime projects.[78] Oppenheimer struck a deal with Bush, Conant, and Groves: if he could demonstrate that a person was essential to Los Alamos, one of the others would work to remove the obstacles.[79]

For example, although at first McMillan seemed unavailable, a week later Conant had arranged for his transfer.[80] Often the intervention of Groves or some other high-level member of the military or government was needed, as in the case of Norris Bradbury, then at Dahlgren Naval Proving Ground.[81] Groves also intervened for the release of Norman Ramsey, who was willing to leave Washington and work at Los Alamos, but was serving as an adviser to Edward Bowles in the secretary of war's office. Bowles and Groves both wanted Ramsey on their payroll and worked out a compromise by which Ramsey would continue to work for Bowles but be on permanent loan to the Manhattan Project.[82]

Oppenheimer identified many of the recruits through his extensive network of friends and students.[83] On occasion, the cancellation of other projects worked in his favor. For example, he was able to acquire the entire Princeton group working under Robert Wilson on the Isotron (an NDRC electromagnetic separation project that had been canceled early in 1943).[84] Wilson later recalled: "We became ... a research team without a problem, a group with lots of spirit and technique, but nothing to do. Like a bunch of professional soldiers we signed up, en masse, to go to Los Alamos."[85] Some of the recruits, such as those working on fast-neutron problems at Minnesota, Wisconsin, Chicago, Purdue, and Cornell, were simply asked to continue their work at the new location.[86]

Oppenheimer was allowed to tell the scientists being considered for positions as group or division leaders something about the specific work they would be doing and to ensure them that every effort would be brought to bear on this work. But he could tell other recruits practically nothing more than that the work was important enough to possibly end the war. He emphasized the climate, physical beauty, and recreational possibilities of Los Alamos.

From the very outset, the proposed collaboration with the military created a serious recruiting obstacle. Groves, Conant, and Oppenheimer originally conceived of the laboratory as a military installation in which the scientists would be commissioned officers. Indeed, Oppenheimer had already ordered his uniforms. At the same time, he planned to work around the military in such a way that Los Alamos would be military only in name, but many objected to this concept. Bacher, who had been recruited to lead the experimental physics program, and Rabi, one of Oppenheimer's most trusted advisers, both refused to have any connection with a military project.[87]

Groves compromised. He and Conant agreed to make Los Alamos a civilian laboratory initially, but asked that Groves be free to impose military rule if and when he saw fit – which would be the case only if the work became too dangerous for civilians. In a letter to Manley, to be passed on to those who had already agreed to come, Oppenheimer specified the responsibilities of the scientific director and commanding officer, affirming that although the employees of the laboratory would not have to be military personnel, the military would control the town as well as the entrance to the Technical Area.[88] Bacher tendered his resignation effective the day the laboratory became a military facility. Rabi decided that he could not leave his job as associate director of the MIT Radiation Laboratory. He became an adviser to the project rather than a staff member.[89] Others such as Williams at Minnesota were unhappy about the possibility of military control but agreed to come anyway, at least for the first period.[90]

Completing the Site and Starting Work

Many scientists recruited for Los Alamos in 1943 had difficulty finding the laboratory. Nicholas Metropolis was simply handed a packet of papers containing cryptic instructions on how to get to Santa Fe and then to a small unmarked office on the Plaza. To his dismay, Metropolis found that his train did not stop in Santa Fe, but rather at Lamy, a small village approximately 15 miles from Santa Fe. Following instructions, Metropolis caught a bus to the Plaza and began looking for the small unmarked office.

Metropolis had been instructed first to locate the Bishop Building, but while walking across the Plaza with his suitcases, he noticed a man following him. When Metropolis stopped, the other man stopped, too. Upon entering the Bishop Building, Metropolis looked for the man's shadow, but he was gone. He needed to find room number 9. After searching the building, he found the unmarked room by extrapolating from rooms that were marked. Metropolis knocked. A voice behind the door challenged, "Are you expected?" "I think so," he responded and gave his name, whereupon he was instructed to go to 109 East Palace Street.

At 109 East Palace Metropolis was greeted by a friend – Rose Bethe, who was helping Dorothy McKibben, a lifetime Santa Fe resident, who had been appointed by Oppenheimer to manage the Santa Fe reception

Fig. 4.3. Office of Dorothy McKibbin, the first contact arriving Los Alamos scientists had with the laboratory. LA Photo, LAT 2533.

center. Bethe arranged for Metropolis to stay in Tesuque at Rancho Encantado and later at the Gables Ranch, until a room could be found in Los Alamos. Also staying at the Gables Ranch was the man who had followed Metropolis across the plaza – Rene Prestwood, a chemist recruited from Berkeley. Prestwood had actually been trying to avoid Metropolis while he, too, searched for the Bishop Building.

The problem Metropolis and others hired at the start of Project Y faced was that the Los Alamos Laboratory was not yet operating in March 1943. Research was not yet possible "on the Hill," as the laboratory came to be called, because many buildings were still under construction, equipment was still arriving, and food and housing were not yet available for more than a small group of people. Fuller Lodge, one of the main Ranch School buildings, and the "Big House," a large dormitory, were among the few accommodations; most of the scientists were initially housed at dude ranches in the vicinity. Cold box lunches were brought up each day from Santa Fe. For telephone communication lab staff had to rely on an old Forest Service telephone line.[91] Priscilla Greene Duffield, Oppenheimer's secretary, described Los Alamos as "a pretty appalling place. It was windy, dusty, cold, snowy ... and nothing was finished."[92]

Soon after Groves acquired the site, Albuquerque District Engineer Colonel Lyle Rosenberg selected the M. M. Sundt Company as contractor for the first construction effort. Sundt was available, financially strong, and could work almost without subcontractors. With only a handshake to seal the commitment on 6 December 1942 and without even being given plans for the first building, Sundt agreed to a completion date of 1 February for the technical buildings and overall completion on 15 March 1943. Allison, Oppenheimer, Groves, Santa Fe architect W. C. Kruger, and Albuquerque engineer Elmo Morgan planned the original layout of the Technical Area. Oppenheimer, Manley, and McMillan produced and turned over to the Boston firm of Stone & Webster specifications for the technical buildings from which the firm would create the blueprints.[93] The plans were returned to Kruger to ensure they complied with standard military building requirements wherever possible. He was also asked to plan the remodeling of the Los Alamos Ranch School buildings and to design utilities and streets.[94] Sundt did not wait for plans, but began work immediately.

Numerous difficulties prevented Sundt from completing the work on schedule. Local labor was scarce and hard to keep in residence at Los Alamos. Labor unions questioned the hiring and firing practices of the company. Last-minute modifications were often requested verbally by the scientists. The scientists wanted what seemed to them to be obvious improvements, regardless of contractual problems. But unless the Corps of Engineers accepted the buildings according to the original specifications, the contractor was held accountable. When the Sundt company finished a building, they transferred it to the Albuquerque District Engineers, who then transferred it to the Manhattan District Engineers, who then turned it over to the scientists. Only after the buildings were accepted could they be changed without legal or financial harm to Sundt. Another complication was that construction crews sometimes had to complete their work under the scrutiny of security guards. Although scientists found many buildings unfinished on the promised date, they moved in anyway, often at night.[95]

Fig. 4.4. New Mexico State Highway No. 4, the first major road built leading to Los Alamos, October 1943. LA Photo, 00384.

A remarkable number of buildings were completed by mid-April. The first set were the veterinary hospital, the fire station, the post administration building, three gatehouses for pass checkpoints, and warehouses – all necessary buildings, but not laboratories. By late March 1943, the complete set included several additional military post buildings and forty-two apartments. The main technical building, the chemistry and physics laboratories, the cyclotron building, and the shop facility were all completed before the end of March. By mid-April, the Van de Graaff building, the Cockcroft–Walton facility, forty-four more apartments, dormitories for forty bachelors, a recreation hall, theater, and infirmary were finished. By the end of April, 96 percent of the contract had been completed.[96]

The laboratories soon filled with equipment. Truckloads of electronics parts and tools arrived from Wilson's Princeton Isotron project. Parti-

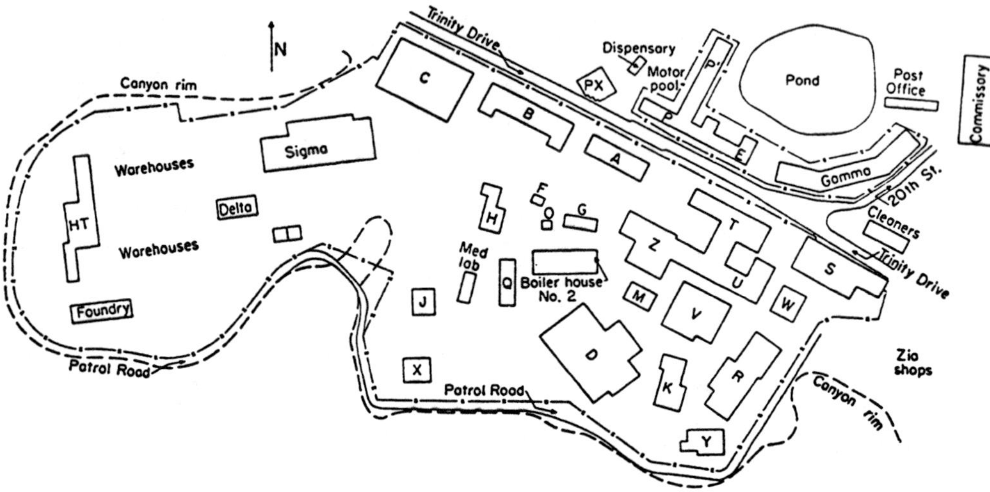

Fig. 4.5. Technical Area Map, ca. 1945.

cle accelerators were among the important larger pieces. Oppenheimer, McMillan, and Manley looked for several accelerators already in use in the fast-fission program. The best Van de Graaffs were pressurized models from Raymond G. Herb's high-voltage laboratory at the University of Wisconsin. After long negotiations, Los Alamos acquired two of them.[97] A Cockcroft–Walton accelerator was "borrowed" from the University of Illinois for two reasons: Manley knew the intricacies of this machine, which he and Leland Haworth had built in 1938. Second, this accelerator was one of the few to incorporate a deflecting system to provide a pulsed neutron source to investigate time-dependent neutron phenomena, such as the capture of neutrons in various substances.[98]

The acquisition of the cyclotron illustrates the absurdities of military requisition in a climate of secrecy. McMillan investigated several cyclotrons and decided that the one at Harvard was most suitable.[99] The military specified that for reasons of security the Manhattan Project had to convince Harvard to turn the machine over to them without offering any explanation for why it was needed. The Harvard physicists suspected the use to which the cyclotron would be put, especially when they saw Wilson as part of the negotiating team. But the army personnel had been instructed to tell Harvard that the accelerator was needed for a medical installation in St. Louis. They were authorized to pay a large sum for it. The army kept up its pretenses while the Harvard group pointed out inconsistencies in their story. As Wilson recalls, the Harvard physicists grew exasperated and in effect told the bargainers they could have the machine for almost nothing if they were going to use it in the fission project. Nonetheless, the army personnel obeyed their orders and stuck to the false story. They finally paid Harvard a large sum for the cyclotron.[100] Procuring both equipment and raw materials at the isolated site without arousing suspicion would be constantly frustrating.[101] Physicist Dana Mitchell was hired to head procurement. One of the earliest consultants on the Los Alamos project, he had handled procurement for the physics laboratory and NDRC fission project at Columbia University.[102]

Priority ratings were a major problem. In 1942, Colonel James C. Marshall, then head of the Manhattan Project, had arranged the highest rating from the Army Services of Supply (SOS) that could be granted to a nonproduction project under the wartime priority system. Projects to produce airplanes, tanks, ships, and guns had highest rating.[103] When Groves assumed command of the project several months later, he pressed the War Production Board (WPB) for the authority to use the emergency AAA priority when necessary. WPB chairman Donald Nelson granted this authority on 26 September 1942, but Groves was still not satisfied.[104] As the massive construction projects at Hanford and Oak Ridge demanded increasing amounts of steel and concrete, Groves sought the highest projectwide rating as well (AA-1). He obtained it on 1 July 1944.[105]

The University of California Contract

Another important administrative task for Oppenheimer and Groves was to select a contractor who could help with procurement and other administrative tasks without knowing why Los Alamos existed. Physical proximity was unimportant, for the contractor only needed to handle business, not scientific work. The University of California at Berkeley, for whom Oppenheimer worked, was a natural choice, because the organization was sufficiently large, was not yet impossibly overburdened with war work, and had ample experience in administering research projects.

The contract negotiations between General Groves, OSRD lawyers, and Robert M. Underhill, secretary-treasurer of the Board of Regents of the University of California, lasted into the early months of 1943. Oppenheimer would not accept the contract until Mitchell (who sat in on the negotiations for Oppenheimer) was satisfied.[106] The two sides came to terms on 20 April 1943 on a contract made retroactive to 1 January 1943 to account for work already done by the university in the past three months.[107]

The contract – still in effect today – established parameters for the participation of the university in the war work. For example, the university agreed to furnish necessary personnel, supplies, materials, and equipment. Oppenheimer depended on UC President Robert Sproul's interpretation of university personnel rules and regulations, as well as more general university operating procedures. The government amended the contract periodically and reserved the right to extend it for the duration of the war plus six months. The university remained almost ignorant of the laboratory's mission. The West Coast procurement office, set up in Los Angeles, used an intermediary, E. J. Workman, a professor of physics at the University of New Mexico, as a blind drop for shipments.[108]

The signing of the contract also signified the formal beginning of the Los Alamos Laboratory.

5

Research in the First Months of Project Y: April to September 1943

As soon as the Los Alamos Laboratory opened its doors, committees were formed to plan the research program and cope with practicalities. Robert Serber offered an indoctrination course early in April 1943 to acquaint scientists with the current state of research on the atomic bomb. Conferences that month laid out specific research objectives. Even though many fission constants were poorly determined and the accuracy of approximations was generally low, Los Alamos physicists were confident that a reasonably efficient gun bomb could be built. Acceptance of the gun as a workable assembly lent optimism to the entire project. As a fallback, Oppenheimer established a small research effort under Seth Neddermeyer to explore implosion assembly.

* This chapter was coauthored by Paul Henriksen, Gordon Baym, Lillian Hoddeson, Roger Meade, and Catherine Westfall. Henriksen wrote on the Planning Board and contributed to the section on the April conferences. Baym wrote on Serber's lectures and contributed to the section on the April conferences. Hoddeson contributed to the section on the April conferences, and wrote the section on implosion. Meade wrote the section on gun assembly, and Westfall the section on nuclear physics. We thank Alison Kerr and Les Redman for extensive editorial contributions.

The Planning Board

Committees helped Oppenheimer make major decisions, with Groves's approval. The first informal committee – Robert Wilson, Edwin McMillan, Oppenheimer, Edward Condon (the associate director), John Manley, and Serber – met on 6 March 1943 and considered practicalities, such as when people and equipment would arrive and who would handle services rendered by the machine and electronics shops.[1]

This initial planning group gave way several weeks later to a larger committee called the Planning Board, which coordinated the technical program over the next month. Oppenheimer, Condon, Dana Mitchell, and Julian Mack provided administrative guidance, while Wilson, Serber, John Williams, McMillan, and Donald Mastick planned the scientific program.

The Planning Board discussed the latest estimates of the critical mass, the availability of materials for the bomb case, and the proper techniques for forming it. They discussed the problem of laboratory space, an issue that would be of special concern for the ordnance groups. According to the plan then in effect, the ordnance effort was to be small, comparable to the metallurgy and cryogenics efforts. Richard Tolman was tentatively penciled in as leader. Most of the meeting was spent on briefing new staff members about arrangements and organization, for example, ordering supplies, procurement, room allocations, ordnance, security, and power allocation. The procurement system was still ill-defined, because the contractor, the University of California, had just been selected. Equipment was not easy to obtain; Oppenheimer's secretary, Priscilla Duffield, bought her own typewriter in Santa Fe and attempted throughout the rest of the war to gain reimbursement.[2]

The Planning Board met again on 2 April, with most of the same personnel and a few important additions: Robert Christy, Richard Feynman, and Emil Konopinski. The emphasis was again on nontechnical matters, including security, electric power, provisions for a town council, relations with the building trades unions, working space, and new buildings. The most important technical decision was to hold a "short course ... *open to scientific personnel only* ... to give a rapid survey" of the technical problems. The date for these lectures, to be delivered by Serber, was confirmed for early April. They planned a more "thorough and systematic set of seminars" to be held after Serber's lectures.

The group at the third and last meeting of the Planning Board, on 8 April, included Edward Teller, Arthur Wahl, Hans Bethe, and

Neddermeyer.[3] As in the previous meetings, the discussion centered on people and working space. The laboratory already had commitments from some 150 staff members, but the available housing was almost filled. The board decided to delay the hiring of personnel not immediately needed. Laboratory leaders agreed that in the future they would need "to be more far-sighted about expansion" and allow time for construction (a plan seldom realized). The remainder of the meeting was used to outline the first three months of the experimental program and the conference that would follow Serber's lectures.

In May, a second committee under Lewis, which had attended the Planning Board meetings, issued its report on the Los Alamos program. The commmittee judged progress to be satisfactory but recommended that the mission of the laboratory be expanded to include plutonium metallurgy and purification, ordnance engineering and development, explosives fabrication, and a variety of community activities. Those recommendations, which Groves accepted in substance, destroyed "the original concept of Los Alamos as a small physical laboratory."[4]

Serber's Introductory Lectures

Serber had remained in Berkeley after the July 1942 conference, continuing his research on the bomb project out of Oppenheimer's office. His five Los Alamos lectures, delivered on 5, 7, 9, 12, and 14 April 1943, were based directly on this research.

Serber's Los Alamos series summarized the state of knowledge on the atomic bomb. As he recalls: "Previously the people working at the separate universities had no idea of the whole story. They only knew what part they were working on. So somebody had to give them the picture of what it was all about and what the bomb was like, what was known about the theory, and some idea why they needed the various experimental numbers."[5] The Planning Board assigned responsibility for the course to Condon and Serber; Condon's principal contribution would be to serve as secretary during the lectures and write a summary of the course.[6]

Serber's lectures reflect the enormous insight that researchers had already achieved into the basic physics of an atomic bomb, as well as the significant amount of work they had done on the problem. The lectures began by defining the objective of the project: "to produce a *practical military weapon* in the form of a bomb in which energy is released by a fast neutron chain reaction in one or more of the materials known to show nuclear fission."[7] He then turned to factors affecting bomb design.

The first factor was the energy release, known to be on the order of 170 MeV per fissioning atom, or 7×10^{17} ergs/gm; thus 1 kg of ^{235}U, or "25," would contain the same explosive energy as 20,000 tons of TNT. The second was the fast-neutron chain reaction. Since a neutron-capture-induced fission of ^{235}U released two neutrons, a fast-neutron chain reaction in a kilogram sample would, as he described, ideally proceed by doubling the number of neutrons in each "generation," requiring some eighty generations "to fish the whole kilogram." Unfortunately, the released energy would heat the material, increase its pressure, and tend to blow it apart before the entire fissioning process was complete.[8]

Whether a bomb would work, therefore, depended on how rapidly neutrons were lost through the surface of the ^{235}U, a loss exacerbated by the expansion of the system, which would eventually stop the chain reaction. "The whole question of whether an effective explosion is made depends on whether the reaction is stopped by this tendency before an appreciable fraction of the active material has fished." Serber optimistically observed that "it is just possible for the reaction to occur to an interesting extent before it is stopped by the spreading of the active material," since the time scale for neutron multiplication was comparable to the time scale for expansion.[9]

Serber turned to the nuclear physics of the "materials in question" – ^{235}U, ^{238}U, and ^{239}Pu. The first factor to consider here was that the fission cross section, σ_f, for the fissionable material on the incident neutron energy. Plotting what was known of the dependence of σ_f for ^{235}U, ^{238}U, and ^{239}Pu, he observed that 25 has a cross section of 1.5×10^{-24} cm^2 for neutron energies above 0.5 MeV, and that this value rises rapidly as neutron energy decreases; the behavior of 49 was presumably similar. By contrast, neutron-induced fission of 28 occurs only for neutron energies above 1 MeV, with a fairly constant cross section above this threshold. As he also pointed out, the spectrum of energies of neutrons released in fission could be interpreted in terms of neutron evaporation from a thermal source at a temperature of about 0.5 MeV, and although the mean energy was about 2 MeV, an appreciable fraction of the neutrons released had energies less than 1 MeV and so were unable to cause ^{238}U to fission. Another parameter was the average number of neutrons, ν, produced in each fission. It was not yet known "whether ν has the same value for fission processes in different materials, induced by fast or slow neutrons or occurring spontaneously," but he gave the best available estimate, $\nu = 2.2 \pm 0.2$, "although a value $\nu = 3$ has been reported for spontaneous fission."[10]

The basis for the production of ^{239}Pu in a slow-neutron fission pile, Serber noted in passing, was the process, $^{238}\mathrm{U} + \mathrm{n} \rightarrow {}^{239}\mathrm{U} + \gamma$, which "acts to consume neutrons." The ^{239}U undergoes two beta decays to ^{239}Pu. The value of ν for ^{239}Pu had not yet been measured although, as he remarked, "there is every reason to expect its ν to be close to that for U." He added that "since it is fissionable with slow neutrons it is expected to be suitable for our problem" and revealed that "another project is going forward with plans to produce it for us in kilogram quantities." He then pointed out that "further study of all its properties has an important place on our program as rapidly as suitable quantities become available."[11]

Why, he asked, was ordinary uranium, containing only 1 part in 140 of ^{235}U, safe against a fast-neutron chain reaction? Because a large fraction of the neutrons produced would not be able to cause component 28 to fission, the effective neutron multiplication number in ordinary uranium is only about 0.4; a value above unity was needed for a chain reaction. Serber judged that an explosive reaction could not occur unless the fraction of 25 in the uranium was increased by at least a factor of 10 from its normal abundance.

Serber estimated the critical mass of a bomb using elementary diffusion theory (which, he recognized, was only valid when the neutron mean free paths are short in comparison with the size of the system, "a condition *not* fulfilled in our case") for a sphere of active material. The rate of buildup of the neutron density in an infinitely large sphere would be $(\nu - 1)/\tau$, where $\tau \sim 10^{-8}$ s is the mean time between fissions in the material; the buildup would be less rapid in spheres of finite size. In this simple theory, the critical radius, R_c, the one for which the neutrons multiply just as fast as they leak out, would be given by

$$R_c^2 = \pi^2 D\tau/(\nu - 1),$$

where D is the diffusion coefficient. The dependence of the latter on the scattering cross section "brings out the reason for measurements of the angular scattering of neutrons in U." With the available numbers, he estimated a critical radius of 13.5 cm, which corresponded to a critical mass of 200 kg. However, he pointed out that the "more exact diffusion theory," recently applied to the problem by theorists at Berkeley, reduced this number to about 60 kg.[12] A professor at heart, Serber posed an exercise for the audience to derive the critical size of a cube of active material.[13]

The advantage, Serber continued, of surrounding the active material by a shell of heavy inactive material – a tamper, such as gold, tungsten, rhenium, or uranium – was that it would reflect inward "some neutrons that would otherwise escape" and thus would allow far less active material to form a critical mass. The tamper would serve the additional purpose of retarding the expansion of the active material. Although the densest materials appeared to be best for reflecting neutrons, he cautioned that "a great deal of work will have to be done on the properties of tamper materials." According to elementary diffusion theory, the critical radius in the limit of a very large tamper would, as he derived for the special case that the neutrons diffuse equally well in the tamper and active material, be just half of that for an "untampered gadget"; in other words, one-eighth the material would be needed. Again he gave the audience an exercise, to study the soluble case in elementary diffusion theory that neutrons diffuse half as fast in the tamper as in the active material, an exercise that showed "it would be very much worth while to find tamper materials of low diffusion coefficient."[14]

Because the correction for a finite neutron mean free path was not as large as for a "bare bomb," more accurate diffusion theory, Serber reported, predicted a decrease in the critical mass by a factor of four, rather than eight, as predicted by elementary diffusion theory. The critical mass of 25 with a normal uranium tamper, weighing about a ton, would be about 15 kg, whereas that for 49 would probably be less, about 5 kg. Serber recognized that the critical masses were still quite uncertain, especially for plutonium. Improving the estimates would, he stressed, require better information on the nuclear properties of both the bomb and tamper materials. Most important, Serber thought the critical masses would have to be determined by actual tests when materials became available.[15]

Serber then turned to the nature of the damage that a bomb would cause. First, the production of radioactive material would severely contaminate the area up to 1,000 yards from the explosion. Second, the blast (or shock) wave produced by the explosion would cause mechanical damage. Serber, here treading on relatively unfamiliar ground for physicists in that period, described the physics of a blast wave qualitatively, rather than introducing the basic theory of shock waves, quoting the result that the maximum pressure in the blast falls off with increasing distance r from the explosion as E/r^3, where E is the total energy released by the bomb (a "scaling" solution). Thus, "if destructive action may be regarded as measured by the maximum pressure amplitude," the radius of destruction would grow with the cube root of E; scaling from the known radius of damage of TNT, he showed that conversion of 5 kg of active material would have a destructive radius of about 2 miles. Because "the one factor that determines the damage is the energy release, our aim is simply to get as much energy from the explosion as we can. And since the materials we use are very precious, we are constrained to do this with as high an efficiency as is possible."[16]

That "efficiency," as Serber defined it, was the "fraction of energy released relative to that which would be released if all active material were transformed." To further explain efficiency, he laid out in a figure the "course of events" in an explosion, showing how the energy released, the neutron density, and the pressure increase exponentially with increasing time. For a mass of active material just above critical, the efficiency would be proportional to the square of $\nu - 1$, divided by the square of τ, the time between fissions, times the cube of the amount that the initial radius of the core exceeded the critical radius. More accurate estimates from the Berkeley studies include the comparably important effect that the buildup of pressure begins to blow off material at the outer edge of the bomb.[17]

Earlier calculations were also made to study the "effect of tamper on efficiency." According to Serber, tamping would always increase efficiency, but not by as much as one might expect simply from the lowering of the critical mass by the tamper. The point is that the slow neutrons that would keep the chain reaction going would spend a long time being reflected by the tamper – in fact, the important time scale would be the neutron lifetime in the tamper, rather than in the bomb, which would be a factor of ten greater. "To get good efficiency" would require a mass well above critical, for which uranium or gold would behave much the same as tampers. An added advantage was that tampers would prevent the edge of the bomb from blowing away.[18]

Next, Serber discussed the detonation of the active material. The problem was how to go from an initial configuration with neutron multiplication less than unity to a final one with a growing chain reaction. To do so would take time, and if the configuration spent too much time being just above critical, "an explosion started by a premature neutron will be all finished before there is time for the pieces to move an appreciable distance." To avoid such a predetonation, "it is therefore necessary to keep the neutron background as low as possible and to effect the rearrangement as rapidly as possible." To calculate the chance of predetonation from the unavoidable neutron background, a problem in probability theory, he appealed to a simple gambling analogy of tossing loaded coins to argue that the effective neutron multiplication number, ν', is the probability that any one neutron would start a chain reaction.[19]

Now, he asked, "what if by bad luck or because the neutron background is very high, the bomb goes off when ν' is very close to zero," that is, it fizzles, leaving the enemy the "opportunity to inspect the remains and recover the material?" This event would not in fact be a concern, since there would also be a small energy release adequate to destroy the bomb.[20]

These worries aside, how would one actually initiate the chain reaction? One possibility would be to make the pieces of the assembly stay in the desired position when shot together. Safer still would be to use a strong neutron source that would become active when the pieces came together.

The actual neutron background could arise from three sources: cosmic rays, spontaneous fission, and nuclear reactions. The number of cosmic-ray neutrons was known to be too small to be of any importance. Similarly, the known limits on spontaneous fission from the elements 28, 25, and 49 were small enough that neutrons from this mechanism in the active material would not be a problem, except possibly with the use of large uranium tampers. The only mechanism that appeared to present a serious problem was that α particles might produce neutrons by interacting with light element impurities in the active material. This problem would be difficult to surmount by producing 49 of extremely high purity; the expected high-neutron background in a 49 bomb would make a high firing velocity very desirable.[21] On the other hand, most of the α's in 25 would come from the rare component 24, and so the neutron background in 25 would be a less critical problem. At this stage, one could not, because of the lack of experimental information, foresee the absolutely overshadowing problem that Los Alamos would face one year later – that of predetonation in the presence of neutrons from the spontaneous fission of ^{240}Pu, a problem that would be unavoidable in a gun-assembled weapon using ^{239}Pu.[22]

Serber turned finally to assembly techniques to bring the active material above criticality – "the actual mechanics of shooting" – admitting that "this is the part of the job about which we know least at present."[23] The basic tool envisioned was the gun, a familiar technology well within the grasp of Project Y.[24] Serber discussed various schemes, among them the Serber and Tolman implosion idea. These schemes raised many questions: How well could guns be synchronized? What were the possibilities of shooting noncylindrical shapes at lower velocities? What were the mechanical effects of the blast wave entering the gun barrel before the projectile? Could projectiles be made to seat properly? And could a piston of inactive material in the gun barrel be used to drive active material together, and thus minimize the effects of the steel gun barrel reflecting neutrons and initiating the chain reaction? The autocatalytic methods, Serber calculated, would require large amounts of active material to be efficient and would be dangerous to handle. In summary, "some bright ideas [were] needed." Those ideas could only come from new studies on "techniques for direct experimental determination of critical size and time scale, working with large but subcritical amounts of active material."[25]

The April Conferences

The next stage in the orientation was the April conferences for new Los Alamos technical staff, held from 15 April through 6 May, for the purpose of analyzing "the scientific problems of the Los Alamos Laboratory and to define its schedules and its detailed experimental program."[26] Oppenheimer began the first meeting with a summary of current views, covering some of the same ground as Serber. He included the estimate that 100 g of ^{235}U would be shipped every day beginning in early 1944, while 300 g of ^{239}Pu could be shipped each day by early 1945.

One means of amplifying the release of energy in the bomb, Oppenheimer stated, was to induce a thermonuclear reaction in liquid deuterium: "This possibility has been considered in detail and it is highly probable that in principle the scheme is feasible. It will need more development than the gadget and is naturally secondary to the development of the gadget. But arrangements should be made that its development follow immediately the completion of the gadget." Thus, while the Super was being discussed by scientists at the laboratory, it was already clear that the project was of secondary importance.

On the second day of the conference, Manley described the upcoming experimental program and the neutron sources it would use: the Harvard cyclotron, the two Van de Graaff machines from Wisconsin, and the Cockcroft-Walton from Illinois. He also mentioned plans for building a pile, dubbed the "Water Boiler." Bethe, on the next day, discussed past measurements of physical constants, time delay, neutrons per fission, fission cross sections, and critical mass and efficiency calculations. On the fourth day, Serber discussed the role and properties of the tamper in the weapon. The fifth day was devoted to experimental methods. Joseph McKibben spoke on the coincidence-proportional counter, Williams and Hugh Richards on other recoil detection schemes, Manley on threshold detectors, Emilio Segrè on the analysis of foils for fission cross-sectional measurements, and Joseph Kennedy on foil preparation and the special problems of detection in spontaneous fission measurements. The following day featured a discussion of the properties of natural uranium, including its potential as tamper material. Oppenheimer led a discussion on detonation by the gun method on the seventh day. Teller spoke on autocatalysis on the eighth day. On the next day, Fermi discussed the development of the pile and its uses. On the tenth and last day of the conference, Bethe discussed ways to approach critical mass gradually in experiments and thereby determine critical mass, time scale, and damage.

From 21 to 24 April the same group discussed the first experiments that needed to be done. More planning was done in the week of 27 April; in line with the earlier discussions, several sessions were devoted to both "differential" experiments, designed to measure the effects of individual nuclear phenomena, and "integral" experiments, designed to duplicate general properties of the bomb. In addition, the group discussed measuring the energy spectrum of fission neutrons, fission cross sections, neutron delay, the water boiler, and thermonuclear reactions. They also laid plans for the chemistry and metallurgy program and problems in ordnance design.[27]

Nuclear Physics Research: April–September 1943

Many problems discussed at the April meetings involved basic measurements in nuclear physics. As Manley noted in his 16 April lecture, only two types of fast-neutron integral experiments had been performed, both at Chicago: a rough measurement of fast-neutron fission using a radium–beryllium source and the Snell experiment, which first measured "the ratio of fission to capture" and later measured "the ratio of fission in 235 [uranium-235] to fission in 238 [uranium-238]." In the second year of the project, neutron multiplication measured in progressively larger assemblies of active material would be done at Los Alamos to determine the critical radius, and, by extrapolation, the critical mass. Since the new laboratory possessed only about 1 g of ^{235}U, obtained from Ernest Lawrence's mass spectrograph in Berkeley, and mere micrograms of cyclotron-produced ^{239}Pu, such experiments could not possibly be done in the first year.

The laboratory could not afford to wait a year for more accurate determination of critical masses. The critical mass, which determines the size of the bomb, was the most vital piece of information in planning the delivery program. The value could affect the plans for plutonium and uranium production at Oak Ridge and Hanford. It fell to the theoretical physics division to estimate this crucial parameter on the basis of imperfect measurements of nuclear constants.

At first, T-Division was not organized into groups, as the other divisions were. Bethe, the division leader, assigned each member to one or more of twelve projects, including the urgent critical mass and efficiency calculations, studies of uranium hydride, the Super, blast waves, and integration methods. He asked some of the theorists to work with individual experiments; for instance, Christy went to the Water Boiler group.[28]

At the 15 April meeting, Oppenheimer estimated the critical mass of a ^{235}U gadget with a tamper to be 25 kg on the basis of the following equation, derived from the best available neutron diffusion theory:

$$M_c = \frac{\pi^4}{2 \cdot 3^{5/2}} \left(\frac{A}{N}\right)^3 \left\{(\sigma_t \sigma_f)^{-\frac{3}{2}}(\nu-1)^{-\frac{3}{2}}\rho^{-2}\right\} \left\{\frac{1+.3(\nu-1)\,\sigma_f}{\sigma_t}\right\}^{-3},$$

"where A is the atomic weight; N is Avogadro's number; σ_t is the transport cross-section of the gadget material; σ_f the fission cross-section; ν the number of neutrons per fission; and ρ the density of the gadget." But the result was only as accurate as the values of the inserted constants; the only available value for σ_t was for ^{238}U, and the only value for ν was from Fermi's "doubtful" calculation of the number of thermal neutrons absorbed and the ratio of fission to capture. Estimates for a ^{239}Pu weapon were even more uncertain, since few cross-section measurements had been made and ν had not been measured. In addition, the diffusion theory used for the estimate was based on a number of simplifying assumptions, one being that all neutrons had the same velocity.

The experimental fast-neutron program had thus far conducted only differential experiments to measure nuclear constants. As discussed at the April meetings, Wisconsin measurements showed that the ^{235}U fission cross section was a constant of about 1.6 barns at energies above 600 keV, and at lower energies it fell approximately inversely with energy. Although few ^{239}Pu cross sections had been measured at fast-neutron energies, measurements at thermal energies indicated that the fission cross section of ^{239}Pu was 1.9 times greater than ^{235}U. Fermi had measured the number of neutrons produced for each thermal neutron absorbed and estimated that the number per fission was 2.2, taking into account the ratio of fission to capture. If the number of neutrons per fission depended on the thermal neutron cross sections, however, Fermi's estimate would be invalid. Fission spectrum measurements with the Rice cloud chamber and the Liverpool photographic plates suffered from inelastic scattering of the neutrons in large samples. Los Alamos had few data on capture and inelastic cross sections. The laboratory clearly had to learn far more about fast-neutron sources and detectors.[29]

The nuclear physics program conducted in P-Division began with two preliminary measurements: the delay in the emission of fission neutrons and a comparison of the neutrons per fission, ν, for ^{239}Pu and ^{235}U. As Wilson recently explained, it was imperative to confirm conclusively that the fission neutrons were not delayed and that ν was large enough "if you're going to spend a billion dollars."[30] Theoretical understanding of fission suggested that most fission neutrons would be ejected shortly after neutron capture. Nonetheless, to provide assurance that the bomb project would not be thrown into jeopardy by unexpected delayed neutrons, which would lower the efficiency of the weapon, Oppenheimer planned more accurate neutron delay measurements. The rationale for measuring ν for ^{239}Pu was similar. Joliot's measurements and Fermi's work on the first chain reaction revealed that at least two neutrons would be emitted in each ^{235}U fission. Furthermore, theory indicated that the ν for ^{235}U should not differ much from that for ^{239}Pu. This system of assumptions had to be checked.

By mid-July, Wilson's group had brought the transplanted Harvard cyclotron into operation and was planning its measurement of ν. Several ν measurements for ^{239}Pu were also under way at the "long tank," the larger of the two Van de Graaffs transported from Wisconsin. What the group actually measured was the ratio of ν for ^{239}Pu to the ν for ^{235}U. As explained in a 1947 report, such comparative measurements were simpler than absolute determinations, "because an ordinary neutron counter may be used in detecting the neutrons from both substances and the efficiency of the counter cancels out in the comparison."[31]

The first physics experiment actually conducted at Los Alamos was a measurement of ν for ^{239}Pu, carried out by the electrostatic generator

Fig. 5.1. Robert Wilson, experimental physicist, replaced Bacher as head of the experimental physics division, after the summer 1945 reorganization of the laboratory. LA Photo, LAT 646.

group. After heated negotiations with the Met Lab, which needed ^{239}Pu for their investigations into plutonium chemistry, 165 μg, an "almost invisible speck," of the precious cyclotron-produced substance was sent from the Met Lab, arriving at Los Alamos on 10 July. Wahl prepared the necessary ^{239}Pu foil, losing a mere 17.6 μg of ^{239}Pu in the process, and passed it on to Williams. Williams and others in the Van de Graaff group then bombarded the ^{239}Pu foil and a ^{235}U foil with Li (p,n) neutrons that had been slowed in paraffin. They detected the number of emitted neutrons by measuring proton recoils in a chamber surrounding the fissionable material. The result, reported to the Governing Board on 15 July by Robert Bacher, was that ν for ^{239}Pu was 2.64 $\pm$ 0.2, about 1.20 times greater than that for ^{235}U.[32] To confirm this heartening result, a second ν measurement, using the same neutron source, detected fast-fission neutrons by counting fissions induced in a thorium foil. By August they knew that ^{239}Pu produced 1.27 ± 0.12 times as many neutrons per fission as did ^{235}U. The proton recoil chamber with the short tank measured a ratio of 1.16 ± 0.13.[33] The assumption that ^{239}Pu was at least as good an emitter of neutrons as uranium had been proven.

In the meantime, Wilson went ahead with his plan to measure ν with the cyclotron. As Oppenheimer succinctly explained in his letter of 3 August to Arthur Compton, Wilson compared for ^{239}Pu and ^{235}U. "the coincidences per fission between fissions and recoils produced in a paraffin lined chamber." Neutrons from a graphite pile irradiated by the cyclotron beam, in turn, irradiated the samples placed on a thin cylinder, the inner wall of the fission counter as well as the outer wall of the recoil counter. At the April meetings, Wilson, who was only twenty-nine when he came to head the cyclotron group at Los Alamos, had vehemently argued the superiority of his method with older, more experienced physicists. They felt the method would be too difficult to implement, because it required tricky improvements to circuits and counters. Although Wilson's experiment was not officially endorsed, he insisted on mounting it anyway, referring to it as the "sub rosa" ^{239}Pu ν measurement in the 15 July progress report. When the small portion of ^{239}Pu arrived, he was able to get a turn at using it.[34] Although equipment problems prevented Wilson from providing reliable results in the summer of 1943, by August the method had won the respect of Oppenheimer, who told Compton about its "considerable promise." He noted that the experiment was "independent of any assay of the quantities of 49 and 25 involved," and gave "an overall count of the fission spectrum which is much less selective for high energy neutrons than the other methods." Although not completely reliable, this method also indicated that "the number of neutrons emitted by the two materials per fission are about the same, with no significant evidence that one is greater than the other."

In the summer of 1943, Los Alamos developed the necessary equipment for the experiments on neutron delay. The results were ready by the fall. As with ν, neutron delay was measured in several different ways. Charles Baker's method, an extension of the work at Cornell using the high speed of fission fragments to measure the neutron emission time, had the highest priority. A ^{235}U foil, wrapped around a cylindrical, paraffin-lined neutron detector capable of detecting only fast neutrons, was placed in a larger chamber and irradiated with thermal neutrons.

Part of the fragments traveled freely when the chamber was evacuated, while almost all of the fragments were stopped near the counter when the chamber was filled with propane. The cyclotron group reasoned that the number of neutrons reaching the central counter would be less, with the chamber evacuated, if a significant number of neutrons were emitted later than 10^{-9}, because of the adverse change in geometry caused by the motion of the fission fragments.[35]

The Cornell velocity separator, when first assembled at Los Alamos, could not make the necessary measurements owing to background problems in the detectors. By the end of the summer, however, promising measurements had been made with natural uranium samples.[36]

The measurements of ν and the neutron emission delay, although not entirely conclusive, served their purpose. As "insurance policies," they were meant to indicate that no problem existed, and they did that job. They would be repeated during later months of the project and not once indicated that the size of ν or the time delay would cause the project to fail.

A side issue at the time was the Super, which had received considerable thought but was officially sanctioned only for part-time work by a few theorists and their helpers, and for experimental study by Earl A. Long, a chemist from the University of Missouri, who set up a cryogenic laboratory and studied the liquefaction of deuterium. Other work was farmed out to Ohio State University, which accepted a subcontract to investigate the properties of liquid deuterium.[37]

The Gun Assembly Program: April–September 1943

The gun gadget offered a sense of security to the laboratory because of its perceived simplicity. Devoid of the technical complexities of implosion, the gun offered scientists an excellent chance to develop an atomic bomb in time to help with the war effort and fulfill the laboratory's mission to build a bomb.[38] Research on the gun method consisted of experiments on the firing range, reduced scale testing, interior ballistics work, and testing of sabots (wrappers for artillery shells that permitted launching a shell smaller than the bore of a gun).

As the laboratory began operations, Oppenheimer and Tolman considered the ordnance problems associated with the gun, and by 15 March had a tentative program in mind. Ordnance research and development could be postponed no longer, because it was "one of the most urgent of our outstanding problems." They had settled certain specifications, but no one knew whether these were practical. Although Oppenheimer believed that the actual gun should not be built at Los Alamos, he did think that firing tests of the projectile should be done on the Hill, in collaboration with the "other workers on the project." The best arrangement would have several experienced ordnance men assigned to the project, to work with physicists "of the proper background." If that could not be arranged, Oppenheimer would settle for having some men from Los Alamos sent to work with ordnance experts elsewhere for a few months and then transfered back to Los Alamos to do the rest of the work. Oppenheimer suggested McMillan, Donald Kerst, Charles Critchfield, and Robert Cornog for such an assignment, because they had either an engineering background or specific training in the desired area.

Meanwhile, Tolman, acting as the unofficial ordnance leader in Washington, D.C., arranged for the services of NDRC ordnance expert, E. L. Rose, head gun designer of the Jones and Lamson Machine Company. Tolman recognized that the engineers and physicists would have to confer on the problems, and he tried to arrange a meeting between Oppenheimer's "ordnance physicists" and Rose in Washington. Realizing that meetings between engineers and ordnance physicists would not be enough, however, Tolman wrote to Oppenheimer on 29 March, asking that Rose be put in touch with one of the theoretical physicists, either Serber or perhaps Oppenheimer himself, because information about "critical masses, case masses, probabilities of predetonation, etc.," was essential for determining the basic properties of the gun. Tolman also wanted physicists from the National Bureau of Standards to work on the ordnance problems, namely Neddermeyer, John Streib, and Critchfield, all of whom would shortly transfer to Los Alamos. Failing that, he again asked Serber to come and visit.[39] Critchfield visited during the period from 21 to 23 May 1943 and discussed ways to reduce the weight of the guns. The reason that standard military guns were so heavy – to keep them from bursting under the pressure created by repeated firings – was of little consequence to Los Alamos.[40]

April and May 1943 were months of uncertainty for the gun program. Tolman, and particularly Oppenheimer, continually took the lead in defining and describing the ordnance problems. Although the scientists had a general knowledge of guns and ballistics, they were uncertain about the specific properties of the gun's nuclear materials and not at all sure what the final product would look like. The neutron-reflecting properties of steel used in gun barrels also had to be determined; if the gun barrel would serve to keep the neutrons in, then the critical mass in the gun barrel would be less.

Because light element impurities in plutonium produced copious neutrons, as noted by Glenn Seaborg in 1942, uranium would be easier to assemble without predetonation than plutonium. The plutonium would either have to be purified far beyond current capabilities or fired at a higher velocity than presently attainable, so that the greater number of stray neutrons would have less time to detonate the weapon prematurely. Because plutonium has a smaller critical mass than ^{235}U, the projectile would not be as large, but higher speed would be needed, which would make it hard to keep the projectile and target together after impact. According to Oppenheimer, the question was whether to "develop a detonation method adjusted for the 25 gadget or to solve at the same time the problems connected with the 49 gadget." The laboratory expected enough ^{235}U for a bomb or two to be ready first, but believed that ^{239}Pu would arrive at a much higher rate once the production problems had been solved.

After the April conference, Tolman and Rose continued their ordnance work, Tolman concentrating on theoretical problems of internal, external, and terminal ballistics, and Rose dealing with gun construction and the basic gun barrel parameters of stress and weight. In his presentation at the conference, Tolman expanded the conventional definition of interior, exterior, and terminal ballistics to include the physical makeup of the gun, because in the weapon design the barrel and projectile were far from standard ordnance. Interior ballistics – the science of the combustion of powder, development of pressure, and movement of a projectile in the bore of a gun – had to include gun barrel design because the Los Alamos guns would only be used once. Exterior ballistics – the science of projectile behavior after leaving the gun muzzle – now included the problem of getting the projectile into the target and venting the gases ahead of it. Terminal ballistics – the science of projectile impact – now included the problem of a uranium projectile hitting a target of whatever shape and material. Rose developed formulas for calculating the maximum stress the gun could withstand and the weight of the gun. With this information, the scientists could design guns to meet the special requirements of an atomic bomb.

During the first months of the project, the ordnance work on the nuclear gun had to move forward in the absence of hard experimental data. Oppenheimer led the activity himself. As he explained to Rose on 17 May, "At the present time our estimates are so ill founded that I think it better for me to take responsibility for putting them forward." He put forth a number of models that he believed would span the range of possible gun types.[41] Only after the arrival of Navy Capt. William Parsons, an ordnance expert, could Oppenheimer delegate much of the responsibility for the development of the gun gadget.

The key people recruited for the gun project were Parsons, McMillan, Critchfield, and Joseph Hirschfelder. These four were responsible for bridging the gap between theory and practice and for laying the groundwork that would enable their successors to bring the gun model to combat readiness. Parsons, at the suggestion of OSRD chief Bush, became the first Ordnance Division leader in June 1943. His first responsibilities were to staff the division and see to the construction of laboratories, machine shops, and firing ranges. Parsons also became the laboratory's liaison with the Navy Department's Bureau of Ordnance, directing the procurement of all special guns.[42]

McMillan had had recruitment and procurement responsibilities at Los Alamos; he recruited Wilson and Feynman, scoured the country for a good cyclotron, and ordered machine tools. Once the laboratory was operating, McMillan performed similar tasks to establish the main ordnance test area, Anchor Ranch. Under Parsons, he became deputy for the gun and eventually assumed the chair of the Steering Committee for the Gun.[43]

Critchfield had worked on sabots before coming to Los Alamos.[44] Because Oppenheimer believed that the projectile critical masses would need sabots, he considered Critchfield vital to the gun effort. Trained as a mathematical physicist, but also adept at ordnance experimentation, Critchfield was an ideal choice to translate gun concepts into experimental models. His contributions to the early gun program exemplify the merging of science and engineering that building the atomic bomb relied on. He became leader of the Target, Projectile, and Source Group, E-4, directing research on the design of gun targets, projectiles, sabots, strippers, and modulated initiators. Under his leadership, E-4 explored the fundamentals of blind targets and developed the impact-absorbing anvil and the capsule-type modulated initiator. Drawing on his practical experience, Critchfield instituted preliminary tests at reduced scale in the early gun program. Full-size gun tubes took approximately six months to design and produce, but 3-inch naval guns and 20-mm antiaircraft guns were readily available in large quantities. Using them saved much time during routine testing. In addition, reduced targets and projectile

Fig. 5.2. Eric R. Jette, head of uranium and plutonium metallurgy and Charles Critchfield, initiator group leader, converse with laboratory director J. Robert Oppenheimer at a part held at the Big House. LA Photo, LAT 446.

shapes were easier and cheaper to produce; thereafter only the most promising designs needed to be tested at full scale.[45]

Because the project's success would hinge on the success of the gun, an interior ballistician was indispensable. Hirschfelder, who began working for the NDRC in 1942, helped make the science of internal ballistics "compatible with the laws of physics – all previous systems were semi-empirical." The new system included the "heat transfer from the powder gas to the bore of the gun," making the theory generally applicable to high powered rifles (not to mention atomic guns), as well as to large naval cannon.[46] Hirschfelder's work came to the attention of Tolman, who brought him into the project.

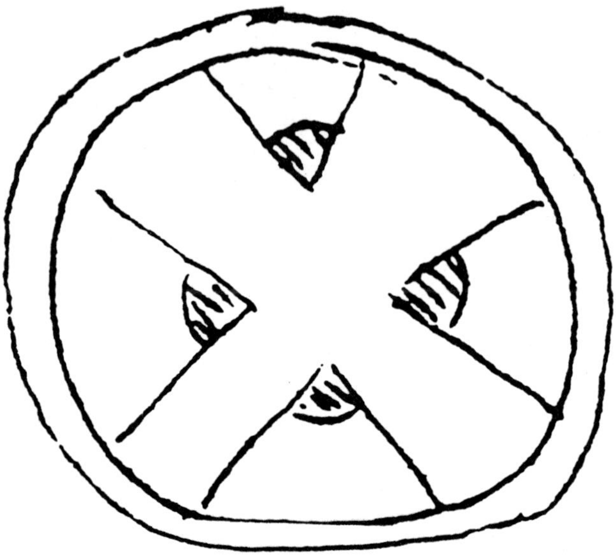

Fig. 5.3. Implosion-like shooting arrangement suggested in Serber's indoctrination lectures. High explosive blows pieces of material mounted on a ring inward. From Serber, *The Los Alamos Primer*, p. 59.

The Implosion Program: April–September 1943

Serber and Tolman's early ideas about implosion were discussed at Los Alamos before the April conference, but it is doubtful that any concrete program would have resulted if Neddermeyer had not taken a confident stance on creating an implosion program at the laboratory. The Planning Board discussed the idea on 30 March and 2 April 1943. At the first meeting, the board supported "the idea of an arrangement in which the material has a hollow case and blows inward to retain its concentration above the critical limit." By the end of the second meeting, they had decided that theoreticians under Serber should concentrate on an analysis of "the Introvert," apparently a reference to an implosion device.[47]

Serber mentioned the implosion idea to the Los Alamos scientists several days later, at the end of his indoctrination lectures in a comment on "shooting" geared to gun methods. Serber's presentation of the Tolman-Serber concept may have stimulated the enthusiastic implosion suggestions then made by Neddermeyer, who was in the audience. In the period between Serber's lectures and the April conference, Neddermeyer carried out preliminary calculations on the velocities that would be reached in an implosion of spherical shells.

He presented these calculations on 28 April, in discussions following either the session "Organization of Ordnance Development," chaired by Oppenheimer, or the session "Design of Gadget," chaired by McMillan.[48] As Neddermeyer explained, the implosion method could achieve higher assembly velocities than could the gun. The novel element distinguishing Neddermeyer's proposal from the earlier one by Serber and Tolman was that of assembling by means of the plastic deformation of a hollow spherical shell, rather than by shooting pieces of fissionable material together. Such plastic deformation seemed easily obtainable with the great forces of the detonation waves emerging from the high explosive. Neddermeyer's suggestion that implosion was "worth investigating further" represents the beginning of this Los Alamos program.

Because of the threat of predetonation owing to light element impurities, Project Y scientists immediately recognized the importance that greater assembly speed would have in assembling plutonium.[49] Thus, "on an exploratory basis," Oppenheimer established a small implosion program under Neddermeyer, on South Mesa, several miles from the main technical area.[50] By June 1943, this effort had become group E-5, which worked on implosion experimentation within E-Division, headed by Parsons. Neddermeyer was assisted informally by Hugh Bradner, John Streib, and Critchfield. However, because the gun method was expected to succeed in assembling both the uranium and plutonium weapons, the new small implosion effort was given low priority within the laboratory; it was viewed as a program to fall back on should unexpected problems arise in developing the gun. Neddermeyer was the only firm proponent of implosion at this time.[51]

The skepticism surrounding implosion in this period is evident in the assessment by L. T. E. Thompson, of the Lukas-Harold Corporation Naval Ordnance Plant in Indianapolis. Thomson was a close acquaintance of Parsons and a frequent visitor and consultant to Los Alamos. Neddermeyer discussed the implosion work with Thompson on 18 June, during the latter's first visit to Los Alamos. After returning home, Thompson wrote Neddermeyer that he doubted that a spherical implosion could ever remain sufficiently symmetric. "There is a fundamental difficulty ... the system is completely unstable and once the collapse is underway I believe it will continue to flatten."[52] Thompson also expressed himself bluntly to Oppenheimer two days later. "A spherical shell under high external pressure ... should begin to collapse ... in about the manner of a dead tennis ball hit with a hammer."[53] Ensuring a symmetric collapse would in fact turn out to be the greatest challenge in the implosion program.

Imploding a metal shell symmetrically required precision in the use of high explosives, an almost unexplored concept in 1943. To become familiar with forefront research on explosives, Neddermeyer and McMillan visited in May the principal American explosives research program. The NDRC Explosives Research Laboratory (ERL) had been established early in 1941 at Bruceton, Pennsylvania, near Pittsburgh, on the grounds of the Bureau of Mines experiment station. George Kistiakowsky of Harvard, the head of this Bruceton program, recalled later that on this visit his staff made the "first implosion charges" for the visitors, who "went away rather pleased with themselves and with us."[54] McMillan recalls running "some experiments with cylindrical implosions ... (using an iron) pipe and making some explosives in a shell around it." Ignition of the explosives wrapped around the pipe "at a few points" set up a convergent wave and one could see clearly that "the pipe had closed in." These experiments demonstrated that one could actually "drive matter in."[55]

The early Los Alamos implosion research was remarkably crude. It was carried out in an arroyo on South Mesa.[56] The first test, using tamped TNT surrounding hollow steel cylinders, was made on the Fourth of July (!) 1943, with Parsons attending. The team centered a piece of steel pipe in a larger piece of stove pipe, and after packing granular TNT into the annular space between the pipes, detonated the implosion using Primacord. Other versions of the experiment used powdered TNT and plastic explosive to squash mild steel pipes into solid bars.[57] Using the "Edison approach" (Chapter 1), Neddermeyer's group repeated this basic experiment many times, varying all the parameters – the explosive arrangement, size of the pipes, and nature of the explosives. The experimental data to be analyzed consisted of a motley collection of bashed-in pipes. These data were subjected to a primitive version of the analysis, which would later in the program be referred to as "terminal observations." The method centered on studying the remains of imploded material after the test shots.

Summarizing the implosion experiments done in July and August, Neddermeyer wrote in one of the earliest technical Los Alamos reports: in tests

> which were of necessity done with meager equipment, the aim has been first to observe the main features of the phenomena when

Fig. 5.4. Data from one of Seth Neddermeyer's earliest implosion tests. The center ring is an untested cross section of the carbon steel tubing used in the experiment. From LA Report 18, August 9, 1943.

> metal shells undergo extreme and rapid plastic flow under external pressure, and to make an empirical determination of the relation between collapse ratio and mass ratio. These experiments are being followed by observations of the velocities and times of collapse, for which several direct methods have been devised.

To cast the needed high explosive for these experiments, E-Division erected a small casting plant at Anchor Ranch. In this early period, Neddermeyer also developed an approximate one-dimensional theoretical model for the implosion process, obtaining for the terminal velocity v of a metal projectile, the expression

$$v = \frac{m_e}{m}\sqrt{\frac{2u_0}{1 + m_e/m}},$$

where m_e/m is the ratio of the mass of explosive to mass of projectile and u_0 is the initial internal energy per unit mass of the exploded gas. Calculations of the collapse ratio of a cylinder as a function of the mass ratio of explosive and cylinder agreed only roughly with the data.[58]

Although Neddermeyer's early work on implosion appears crude in the light of subsequent studies, this first phase in the program exposed many of the difficulties that would plague the Los Alamos Laboratory over the next two years. This small exploratory effort was an attempt to accomplish over several months all the experimental, theoretical, and explosives work that would later be conducted strenuously by the coordinated efforts of several hundred staff members in X-, G-, and T-Divisions. By being "misleadingly hopeful," Neddermeyer's early implosion work set Los Alamos on the road toward its major success, achieved two years later.[59]

6

Creating a Wartime Community: September 1943 to August 1944

The first months on the mesa required drastic adjustments – to Oppenheimer's style of scientific leadership, to Groves's close administration of the town, and to the unusual partnership between scientists and the military. The laboratory grew more rapidly than anticipated because in the early months of the project Oppenheimer and Warren K. Lewis's advisory committee recognized that a scientific community of some 100 scientists was too small to cope with the complexities of producing an atomic bomb. Broadening the laboratory mission, as the Lewis Committee recommended, implied the absorption of new subcommunities, including the Army Special Engineer Detachment (SED) and the British Mission.[1]

Life in wartime Los Alamos was abnormal in almost every respect, but the townspeople strived toward normalcy in their everyday lives, meeting their practical concerns about food, shelter, amusement, and schools with a spirit of adventure. The spartan simplicity and transience of housing and the lack of many community services often turned daily life into a struggle. But as Kathleen Mark reflected, "When one considers that we lived ... closely packed together – aware of every detail of our neighbor's lives – even to what they were having for dinner every night – one can't help but marvel that we enjoyed each other so much."[2] The

* This chapter is based on a draft by Paul Henriksen. We thank Alison Kerr for extensive editorial contributions.

residents worked and also played hard in the isolated military community to which they were restricted.

The Laboratory

Organization and Recruitment

Oppenheimer organized the laboratory, which he administered together with Edward Condon, into five divisions. Except for Administration (A), these divisions corresponded to broad technical areas: Theory (T), Experimental Physics (P), Chemistry (C) (later Chemistry and Metallurgy [CM]), and Ordnance and Engineering (E). Col. J. M. Harman was responsible for the military command of the post. Two committees, the Coordinating Council and the Planning Board, assisted Oppenheimer; the former facilitated communication among division and group leaders, while the latter, which in time gave way to the Governing Board, helped with organization.

Groves and Oppenheimer collaborated in defining the structure of the laboratory. Groves, the post commander and the director of security, controlled town operation and security, and Oppenheimer directed the scientific work. With a talent for assimilating and prioritizing information from a wide range of disciplines, Oppenheimer was able to coordinate decision making so that a wide range of scientists and military personnel were able to exchange information. Oppenheimer and the post commanders shared responsibility for gathering and organizing the personnel and hardware to build bombs.

The scientists and the military both suffered from the juxtaposition of their different traditions. Just as the scientists did not see the purpose of rank and orders, the military could not understand why compartmentalized security was so onerous to the scientists. The fact that Los Alamos had three post commanders in its 2.5 wartime years attests to the frustrations of the joint military and scientific administration of the laboratory and the town of Los Alamos.

Recruitment, although extremely successful, was difficult and time-consuming. Continuing Oppenheimer's practice of inviting students and acquaintances to join the staff, the uncoordinated, and sometimes overzealous, approach of the group and division leaders prompted James Conant to criticize Oppenheimer for the fervor of his recruiting efforts.[3] Oppenheimer acknowledged that it was inevitable his staff would approach some people without going through the proper channels.[4] After June 1943, the recruiting efforts were better coordinated; A. L. Hughes, chairman of the department of physics at Washington University, was hired to oversee personnel matters.[5] Hughes examined housing as well as recruitment, because the scarcity of housing restricted the hiring quotas.[6] Hiring junior-level scientific workers, machinists, and shop workers proved as difficult as finding experienced available scientists. Los Alamos searched for college students having three or more years of physical science courses. A severe housing shortage in 1943 exacerbated the situation. Since dormitory rooms were available but apartments and other family units were not, it became important to hire single people.[7]

Every division of the laboratory expanded between the fall of 1943 and summer 1944. The Engineering Ordnance Division (E) grew to become the largest division. T-Division, although it expanded threefold, remained the smallest division. CM-Division multiplied as it began purification of uranium and plutonium. Administration, which had grown enormously in the first months of Los Alamos, showed only a small increase. Similarly, P-Division increased only slightly as attention shifted from basic research to weapons development.[8]

As the numbers increased, staff composition changed. Until August 1943, employees in the main technical area, dubbed the "Tech Area," were civilians, primarily scientists and their wives. To relieve some of the labor shortage, Groves brought in a contingent of the Women's Army Corps (WAC) in August 1943 and, from late 1943, a group of enlisted personnel in the Special Engineer Detachment. By August 1944, military personnel made up 42 percent of the laboratory.[9]

Administrative realignments occurred frequently during the war period. In general, they were rapid and efficient, unhampered by indigenous populations, tradition, or personal interests – few group and division leaders planned to stay at Los Alamos permanently.

Security

From the start, security was a sensitive issue. Although all workers inside the Tech Area were required to have a security clearance, few had it yet when the laboratory opened, nor had the Tech Area yet been enclosed by a fence. Oppenheimer's early proposal that he, Condon, and Dana Mitchell act as a security committee could not be implemented because he and Mitchell were too busy and Condon left the laboratory after a few months.[10] Consequently, in the initial phase of the project, many scientists were allowed access to the Tech Area while their clearances were pending. Even after the laboratory was in full operation, many scientists (including Oppenheimer) had not been granted clearance. In one of the instances in which Groves put the scientific progress of the laboratory ahead of secrecy, he approved a scheme that allowed several scientists, not necessarily cleared themselves, to vouch for one of their colleagues.[11] This abbreviated clearance procedure was soon streamlined. Oppenheimer vouched for the top-ranking scientists, while statements from three laboratory employees were sufficient to pass junior scientists and technicians. Other employees could be vouched for by their foremen.[12]

By late June 1943, Oppenheimer had clarified information dissemination policies within the laboratory. Access to written materials was given to anyone who could work more effectively or maintain security more easily with the information. Oral communication was not regulated, but everyone was urged to think before speaking. Oppenheimer delegated some responsibility for the dissemination of secret information to the group and division leaders; group leaders could grant access to specific pieces of information, whereas division leaders could grant general access to any properly cleared individual by writing a recommendation. The library maintained a list of personnel who had been given access to all or portions of secret documents.[13]

Oppenheimer tried to define the information policy more clearly in July 1943. Scientifically trained "staff," who had drawn a professional salary before coming to Los Alamos or had been paid on the basis of their academic training, as well as those who would contribute "essentially" to developmental ideas at the laboratory, could see secret documents. "Nonstaff" – people without scientific training, paid at an hourly or monthly rate – could be given access to group seminars with the group leader's approval and could be told the overall purpose of the project if they were completely cleared, had been with the project for two months, and were expected to stay with the project.[14]

To enhance technical communication within the laboratory, Bethe proposed a weekly technical colloquium be held for those working the Tech Area. After the Governing Board accepted this suggestion, Teller was put in charge.[15] The colloquium soon became a weekly tradition and an integrating force in the laboratory, although Tolman and Groves complained that there might be too much discussion about the whole Manhattan Project. Oppenheimer countered that the colloquium boosted morale. They compromised by allowing staff members, but not nonstaff, to attend. Even then Groves forbade discussion of engineering details at

Fig. 6.1. Main laboratory gate. LA Photo, 76 12973.

Chicago or Oak Ridge, and of production schedules, even though such information was vital to Los Alamos.[16]

Groves's concern about security sometimes conflicted with his wish not to waste a day in completing the bomb.[17] But he also exploited secrecy to speed progress. Norman Ramsey recalls that Groves often took advantage of the information restrictions to play one laboratory against another, telling each that the others were ahead.[18]

The army helped enforce security by giving the more prominent members of the Manhattan District code names for use in travel and by enclosing the isolated site with two fences and a small number of guards. Fermi was to be known as Henry Farmer. Niels Bohr became Nicholas Baker, and eventually Uncle Nick at Los Alamos. Similarly, Conant was called Uncle Jim and Tolman Uncle Richard; G. T. Seaborg, G. T. Sutton; Leo Szilard, Leo Samuel; Eugene Wigner, Eugene Winston; Aage Bohr, James Baker; Hans Bethe, Howard Battle; James Chadwick, James Chaffee; Ernest O. Lawrence, Earl Lawson; J. R. Oppenheimer, James Oberhelm; Emilio Segrè, Earl Seaman; Edward Teller, Ed Tilden; and John A. Wheeler, John Woolley.[19] Still other codes were used in the

Delivery and Trinity programs, and scientists' spouses often provided informal nicknames, because they could not use the terms "physicist" or "chemist." (Some called the physicists "fizzlers" and the chemists "stinkers.")[20] Two gates, each staffed by a corporal and two privates, controlled access via the only two roads. Within the site, another fence isolated the Tech Area. The military's jurisdiction ended officially at this second fence; the guards (being uncleared) were forbidden to enter unless they were escorting an uncleared civilian visitor.[21] The almost bewildering initial system of variously shaped and colored security badges was eventually replaced by a simpler scheme of colored badges.[22]

The military and laboratory administration often disagreed about the security procedures. As post commander, Harman was responsible for security in the military population, while Oppenheimer had jurisdiction over security in the Tech Area. Although the scientists monitored their own security, army officers could override this authority if they sensed alarming violations.[23] Security regulations in the community were handled by the military intelligence officer, Capt. Peer de Silva, whose actions – such as refusing to allow residents to visit friends and relatives except in personal emergencies and requiring guns and cameras to be stored in locked vaults – often made him unpopular.[24] To help enforce the limitation of travel to specified areas, a network of army G-2 security men blanketed nearby towns. Groves lifted the restrictions on visits in the fall of 1944, in hope that residents would then be more careful about other security procedures, such as locking up documents. However, nightly patrols continued to confiscate some thirty documents a week.[25]

Security infractions were common. Although typically they were sins of forgetfulness that did not threaten the project, they were of concern because the administration feared that they might lead the military to take control of security. The job of the Security Committee – Hawkins, Manley, and Kennedy – was to remind scientists of their mistakes and discuss infractions at Coordinating Council meetings.[26] In August 1943, the laboratory published a security handbook, which the Governing Board later amended into "guidelines." After the discovery in September 1943 that mail was being opened and resealed, apparently by military intelligence officers, Groves decided that monitoring personal mail was a good idea. A month later he instituted formal censorship.[27]

Fig. 6.2. One of many military inspections at Los Alamos Post. LA Photo, LAT 598.

The SED

To meet the need for skilled technicians, Groves sought army enlisted personnel having technical training or education. He organized them into the SED. More than half became technicians, draftsmen, or scientific assistants. By the war's end, the first group of 39 had grown to approximately 1,600. A few had master's degrees or doctorates, usually in the physical sciences. In the staff hierarchy, some became junior scientists or higher.[28]

Having both technical and military duties, the SED had to negotiate between the often conflicting scientific and military worlds, suffering hardships of military life – inspections and drills, lack of housing for wives and families, restrictions on travel to visit family members, slow promotions, and the like – while working scientists' hours.[29] Civilians

Table 6.1. *British Mission Personnel*

Niels Bohr	Philip B. Moon
Egon Bretscher	Rudolf Peierls
James Chadwick	William J. Penney
Anthony P. French	George Placzek
Otto Frisch	Michael J. Poole
Klaus Fuchs	Joseph Rotblat
James Hughes	Herold Sheard
Derrick Littler	Tony H. R. Skyrme
Carson Mark	Geoffrey I. Taylor
William G. Marley	Ernest W. Titterton
Donald G. Marshall	James L. Tuck

the same age were paid more for the same work without the burdens of army regimentation. Val Fitch, a Los Alamos SED, later a Nobel laureate in physics, recalls,

> We lived in single floor barracks, roughly 60 men to a unit .... We ate in an army mess hall .... We lined up each week to get fresh linen, and once a month to get paid. Reveille came at 6 am, and we had calisthenics from 6:30 to 7:00. We could not leave the barracks for work on Saturday mornings until after inspection of quarters, nominally at 8:00. We worked in ... the Tech Area six days a week. It was the army and still it wasn't the army because in the Tech Area we worked alongside, and were beholden to, civilians.[30]

The SED, like Fitch and Richard Davisson, who were attracted to scientific careers, enjoyed the unusual opportunity to work closely with many of the world's leading scientists.[31] Morale improved among the SED in August 1944, when Maj. T. O. Palmer became commanding officer of the SED; he dropped morning reveille and calisthenics and reduced the rigor of Saturday morning inspections.[32]

The British Mission

James Chadwick (Britain's most prominent nuclear physicist), Rudolf Peierls (who had done some of the earliest work on the feasibility of nuclear weapons), and Marc Oliphant (a member of the MAUD committee; Chapter 2) learned about the Los Alamos project in August 1943, from Conant, Oppenheimer, and Groves. Chadwick responded to the need for senior experimental physicists familiar with engineering, and for experts in hydrodynamics, by bringing to Los Alamos a group of England's best experimental physicists.[33] The first to arrive, on 13 December 1943, were Otto Frisch and Ernest W. Titterton. Nineteen others joined over the next few years, forming the British Mission (see Table 6.1. for a list of British Mission personnel). The last members, George Placzek and Carson Mark, came to Los Alamos in May 1945 from Canada. Four members of the mission became group leaders: Egon Bretscher (Super Experimentation), Frisch (Critical Assemblies and Nuclear Specifications), Peierls (Implosion Hydrodynamics), and Placzek (Composite Weapon).

The most senior members were the Nobel laureates, Chadwick (for discovery of the neutron) – and Niels Bohr (for his theories of the atom). As leader of the Mission, Chadwick was the only Briton informed about the entire Manhattan Project. Bohr joined the British Tube Alloys Project after being flown to Britain following his escape to Sweden from Denmark. As the grand old man of atomic physics, Bohr's occasional presence at Los Alamos was comforting to the other scientists, for whom he sometimes served as a sounding board.[34]

Working Women

Labor was constantly scarce at Los Alamos, because the facilities could support only a limited number of employees, and security was threatened by hiring workers who lived off site. Consequently many scientists' wives were pressed into service; approximately 60 wives were at work in the Technical Area in September 1943. The number increased gradually.[35] By October 1944, approximately 200 (30 percent) of the 670 civilian employees of the laboratory, hospital, and school system were women.[36] About 66 percent of them worked for the administration; the rest were scattered among the other divisions. Many women who intended to work part-time found they could not refuse extra work or leave their jobs before the end of the day.[37]

A few women held scientific positions. For example, Lilli Hornig, working toward her Ph.D. in chemistry, found challenging employment in plutonium chemistry, as did Mary Nachtrieb. When plutonium purification efforts were curtailed in August 1944, Hornig worked on the development of explosive lenses.[38] Charlotte Serber organized and administered the laboratory's scientific and technical library. (The report library she created is still in existence today.) A whole team of women, including Beatrice Langer, helped with numerical calculations in Group

T-5 (Computations), headed by Donald Flanders, who became known as "Moll" Flanders.[39]

Other women assumed responsibility for administrative activities. For example, Rose Bethe and Ruth Marshak (who also taught third grade) handled duties of the housing office. Alice Kimball Smith served on the Town Council. And Shirley Barnett, who came as an army wife, became one of Oppenheimer's secretaries. Still other women taught in the schools, and many raised their children, attempting to keep a semblance of order and a reasonably fulfilling family life in an often chaotic community.

Perhaps the most admired of the project women was Dorothy McKibben, the laboratory's unofficial hostess. McKibben, who lived in Santa Fe, was responsible for meeting newcomers to Los Alamos and arranging transportation to "the Hill." During the war, particularly under the veil of secrecy, travel to Santa Fe was extremely tiring. McKibben's office at 109 East Palace became a haven for tired travelers. In addition, McKibben, who became a close friend of many laboratory scientists and their families, offered her house for weddings. Secrecy made formal ceremonies impossible.

Laboratory Construction and Shops

Between September and November 1943, Sundt crews constructed buildings that expanded the number of square feet of laboratory and office space from 55,000 to 125,000.[40] The laboratory also expanded outside the main Tech Area during the fall of 1943 and throughout 1944. By the end of June 1945, 40 percent of the laboratory buildings were in the outlying sites. One of the largest installations was S-Site, for explosive lens casting and other high explosive work.[41]

The growth of the laboratory intensified the need for shop facilities to produce items such as metal vacuum systems and ionization chambers. The original shop, designated V Shop, also handled such unusual assignments as welding thin stainless steel into envelopes and machining and grinding tungsten carbide.[42] The volume of work increased throughout 1943 and 1944 to approximately 10,000 man-hours per month, nearly double the original load the shop was designed to support.[43] After the Lewis Committee's report, a second shop facility, designated C Shop, was planned and completed in October 1943 to take over some of the demand for shop services.[44] The administration combined the two shops under the management of V Shop leaders Earl Long and Gus Schultz on 14 August 1944.[45] Finding shop workers who were both experienced and willing to postpone their start of work for four to eight weeks while waiting for clearance was difficult.[46] The War Manpower Commission forbade routine recruiting outside New Mexico. Some of the positions were filled by SED. The output of both shops was increased by adding a third shift in January 1945.[47]

The Town

The Los Alamos community – with its security guards, barbed wire fence, and cadre of military personnel – hardly resembled a typical American small town. The population was initially quite homogeneous; most residents were in their twenties or thirties, healthy, and middle-class. People dressed similarly. No one was unemployed. The community structure began to change somewhat with the arrival of the SED, machinists, draftsmen, and technicians.

Remarkably, the members of this special community felt strong bonds of camaraderie even under the stress of the war and the duress of coexistence between the military and the scientists. The common patriotic objective, the condition of physical isolation, and the shared sense of adventure all helped to unite the community. Individual differences in background or status were often unnoticed. Security restrictions forbade anyone to be called a physicist or chemist; the military viewed all technical personnel as engineers.

This unusual aspect of the society extended to the physical facilities. Streets were unnamed, houses unnumbered, and apartment buildings looked alike, as did dormitories, duplexes, hutments, and trailers. Although constant construction changed the detailed features of the town almost daily, the picture remained the same, for the rapidly constructed new buildings tended to resemble those already in place. All were intended to serve only for the duration of the wartime project.

Housing

Because of construction delays and early scheduling problems, new housing was not available when the first employees arrived. Thus they were put temporarily in one of the few Ranch School buildings or in one of the small dude ranches in the Los Alamos vicinity.[48]

Among the original Ranch School buildings were the only distinctive houses on the mesa, a number of stone and log cabins that had been built

Fig. 6.3. Los Alamos Living. LA Photo, 6733.

as residences for married teachers. Known as "Bathtub Row," these were relatively spacious, with large fireplaces, the name-sake bathtubs, and excellent views of the surrounding mountains. The base commander, Oppenheimer, Parsons, and a few other top-level scientists were assigned to them. Several of the larger Ranch School buildings were used as dormitories.

The other family homes in the town were built during three large construction efforts, each known by the name of their contractor (Sundt, McKee, or Morgan), and several smaller ones. The Sundt and McKee companies built the first round of new apartment buildings. Tenementlike in appearance, these were duplexes or quadruplexes, with either wood, tar paper, or asbestos exteriors.[49] They were equipped with fireplaces.[50] Eventually, 334 Sundt and McKee apartments were built in Los Alamos.[51] With more personnel expected after the Lewis committee's recommendation in May 1943 to expand research, Groves approved housing for 320 new employees. While these houses were being built, the neighboring ranches and lodges were used again.[52]

Rents were determined by a military policy based on the salary of the head of the household; they were independent of the size or quality of the house. Rents ranged from 7 percent of monthly income for people at the low end of the salary scale to a maximum of 13 percent for those in the upper middle-income bracket making $6,000 per year.[53] The Town Council later forced a reduction in rents to make some allowance for the type of dwelling, so that, for example, the rent for dormitory rooms would be lower than that for houses.

By August, no more family housing was available, and the Governing Board estimated that the scheduled housing would be roughly half of that needed. The Governing Board directed Hughes, the personnel manager, to urge all new employees not to bring their families and to ask people living in oversize houses to take in people living in the Big House or dormitories. For reasons of security and transportation, they wanted to discontinue using the ranches. Some living in apartments had to be evicted to accommodate "essential" employees, such as power plant operators.[54]

With the completion of the Sundt apartments in mid-October 1943, the housing shortage eased temporarily. But continuing requests for staff increases meant that the present housing would soon be inadequate.[55] Groves urged the post administration crews to complete additional necessary housing to avoid having to hire outsiders.[56] In November 1943, with twenty-eight vacant unassigned houses, the administration felt that 100 additional apartments would be needed. Oppenheimer and the Governing Board discussed moving people around to make optimum use of existing housing.[57] In January 1944, a new housing project was begun by J. E. Morgan and Sons of El Paso.[58] The "Morgan duplexes" marked the nadir of Los Alamos house design, according to one observer.[59] They were ready in two months and improved the housing situation, but only temporarily.

In May 1944, Oppenheimer reopened Frijoles Lodge for overflow housing and the Governing Board held lengthy discussions on how to make better use of the housing. The board even considered how many children should sleep in one room.[60] When only dormitory rooms were available, the board thought of trying to hire only single people.[61] The housing problems of spring 1944 led Groves in June to authorize further new housing construction by the McKee company, which had worked on some buildings for the Tech Area and on an earlier set of apartments of the

Sundt design.[62] Frijoles Lodge continued to operate from mid-July to 5 August 1944, by which time the first McKee houses were ready.[63]

Health Hazards

Life was not safe for those who worked in the Tech Area or at one of the remote sites. High explosives could detonate unexpectedly or cause skin rashes. High voltages, large currents, and toxic chemicals were commonly encountered in routine work. During the war, with its continual atmosphere of crisis, there was little time to worry about contaminated waste. Certainly Oppenheimer and his staff practiced safety and were concerned about the health of all employees. However, there was not a full understanding of the effects of contaminated waste or the ability to know fully how to treat such waste. Radiation from the Water Boiler, the accelerators, or radioactive substances was another danger. Plutonium and polonium emitted ionizing radiation, and uranium was a chemical poison. However, during the first year little fissionable material existed at Los Alamos.

Aware of the danger from radiation, Oppenheimer specified that all staff should have blood tests before there were "any extra neutrons on the Hill."[64] To oversee health concerns, he recruited Louis Hempelmann, M.D., from the Malinckrodt Institute of Radiology at Washington University in St. Louis.[65] He arrived on 24 March 1943. Hempelmann bore responsibility for the safety of all technical operations and for directing the Health Group.[66]

The Health Group consisted mainly of hematology technicians who recorded radiation exposures and monitored potentially dangerous areas. The group was also responsible during 1943 for defining safe exposures to radiation hazards and establishing safe operating procedures.[67] Keeping track of individual exposure proved difficult, for no one knew the size of a dangerous dose, how much radiation had been absorbed, or how to tell if a large dose had been absorbed. They tried to assess bodily damage from radiation by examining the number of white blood cells, but researchers did not know how much variation in the white cell count was normal, and without records taken over a long period of time, the Health Group could not distinguish whether an employee with a low white cell count had a normally low count or a count made low by radiation exposure. To evaluate the situation, the Health Group implored every potentially endangered individual to supply the group with regular urine samples. However the recordkeeping was complicated both by the inadequate number of instruments for measuring radiation dosages and the fact that radiation monitoring devices were a new technology. Film badges and pocket ionization chambers could not reliably record small amounts of radiation.[68]

The arrival of appreciable quantities of plutonium in February 1944 posed a new challenge for the Health Group. Because the element had been discovered so recently, little was known about its physical dangers. Plutonium was believed to be like radium in its effects on the body, but no one knew what risks lay in being exposed to small doses over a number of months. Because plutonium was scarce, little was available for medical experiments. A portion of the first precious gram of plutonium produced at the Met Lab was sent to Berkeley for tests on rats, but the amount was inadequate for all the necessary tests. Hempelmann derived his estimate of a harmful dose of plutonium radiation by comparing its activity with that of radium and applying the data that had been compiled on radium's effects on radium dial painters.[69] In January 1944, the potentially harmful dose was fixed at two micrograms, but a month later, the limit was cut to one microgram.[70]

During much of the time that plutonium was available at Los Alamos in large quantities, only crude tests were performed to detect overexposure. The Health Group made "nose counts" by wiping the inside of workers' nostrils with filter paper and examining the α activity of the paper with a stationary counter. Nose counts continued to be the routine method of testing.[71]

Oppenheimer believed that other parts of the Manhattan Project should address the problem of detecting plutonium contamination. He was opposed to such research at Los Alamos, because of the expense of bringing in additional workers from the outside and providing them with services and housing.[72] By August 1944, it was clear that such research was not going to be performed outside, and Oppenheimer and Groves established a small Los Alamos program for detecting plutonium contamination. By January 1945, Wright Langham, an analytical chemist in CM-Division, had developed a precise method for detecting the presence of 10^{-4} g of plutonium in one day's output of urine. But a contamination-free place to do the testing was not ready.

Testing equipment was hard to acquire and often too insensitive for the monitoring tasks. The Chicago Instrument Group, a branch of the Met Lab, could not supply as many meters as were needed by Los Alamos. Plutonium α emissions could not be measured by the available ionization chambers, because they were insensitive to the soft α particles given off by the plutonium. The more sensitive alpha air counter was unsatisfactory as an all-purpose monitor, since it could not be taken to the sites. To try to fill the void, the Chicago Instrument Group developed portable (but still insufficiently sensitive) α particle counters, the "Pluto" counters. The Health Group continued to wipe surfaces with oiled filter paper and to measure the activity of the paper with stationary counters. In an effort to help, Darol Froman and electronics expert Richard J. Watts began building Pluto counters. Watts's group also developed a methane-filled, thin-windowed proportional counter to monitor the contamination of the hands of workers leaving buildings. Still inadequate, this monitor was all that was available until December 1944, when Chicago and Oak Ridge developed better counters.

The Health Group's greatest problem was a shortage of personnel; its initial staff of seven did not grow in step with the rest of the laboratory. For six weeks during the summer of 1944, the group had no clerical help to keep records on exposures and blood counts. They had difficulty keeping track of all possible hazardous experiments at Los Alamos.[73] The problem was multiplied by the fact that individuals in potentially hazardous environments often failed to keep their exposure and blood records current, or to advise the Health Group when they planned to do dangerous experiments.[74]

The Town Council

An advisory body called the Los Alamos Town Council gave the townspeople a voice in how their community was administered.[75] At the first council meeting, held on 1 April 1943, delegates appointed by the scientists hoped to discuss a variety of practical issues with the post officials, but because of a breakdown in communication, post personnel did not attend.[76] Since many felt that such a group should be elected rather than appointed, the next meeting on 7 May called for the election of a new council.[77] All permanent employees of the laboratory were allowed to nominate candidates, and an election was held. The first elected council consisted of Robert Wilson (chairman), William Dennes (one of Oppenheimer's two administrative aides), Benjamin Diven, Helen Stokes, and William Schafer.[78] Two months later, Oppenheimer suggested that a second election be called to add post personnel; from then on the council became "truly" representative.[79] Among the typical topics it discussed were the milk and water supply, the commissary, housing, the distribution of maid services, restaurant facilities, changes in the hours of movie showings, the lack of sidewalks, heating problems, rents, school, and public conduct.[80]

The by-laws of the second elected Town Council, which met on 27 March 1944, fixed the size of the council at five members having six-month terms of office. Meetings were held weekly.[81] One of the issues discussed by the council was voting in the 1944 federal and state elections. The secret status of Los Alamos made residents ineligible to vote in national elections. The council tried to work out the problem, but the state of New Mexico illegally inactivated the precinct, perhaps out of fear that a group of unknown voters would upset the balance of local elections. An abbreviated federal ballot was finally obtained from the state legislature to allow residents to vote in the national election.[82] Traffic problems were another issue; speeding and reckless driving were especially serious because the town had no sidewalks. The council set up a system of fines, but had no legal jurisdiction to stop either the speeding or dangerous driving.[83] In an effort to protect the safety of hikers and skiers, as well as motorists and pedestrians, the council formed an outing committee to promote outdoor safety and give advice on outing preparations.[84]

One of the hotly debated issues was Post Comdr. Gerald Tyler's restriction on male visitors to certain women's dormitories in the evenings. The restriction was eventually overturned.[85] The council later used this dormitory issue to protest several other actions by the post commander, including his raising of meal prices at Fuller Lodge and restrictions on bus travel to Santa Fe.[86] Such disagreements between the Town Council and the post over minor matters were a continuing feature of Los Alamos life throughout the war.[87]

Quality of Life

Fire protection, hospital, schooling, and communication, as well as food and water, were dealt with as on any military post. The Fire Department was composed of soldiers under a civilian fire chief. Most of their skills were applied to fighting grass and furnace fires. Only one severe fire struck during the war, burning a large portion of C Shop in January 1945 (Chapter 14).[88]

The hospital, started by Hempelmann and surgeon-obstetrician Capt. James Nolan, grew rapidly. Initially equipped with six beds, an operating room, a few small rooms for offices, and a pharmacy, the hospital had 100 staff members by the fall of 1944.[89]

Fig. 6.4. The Los Alamos Post PX. LA Photo, 3232.

In the summer of 1943, Dr. Walter W. Cook and the Governing Board drafted the original plans for the school. The planners envisioned a model school for gifted children staffed by scientists' wives or the scientists themselves, to avoid bringing in teachers from the outside.[90] Cook apparently knew that Los Alamos would be a scientific community, but he must not have been informed about the young age of the scientists, because he designed a curriculum for college-bound students.[91]

The commissary generally received high marks from the townspeople; it often carried specialty items, but could be short of locally grown fruits and vegetables.[92] Meat was usually available.[93] A cafeteria seating 250, somewhat better than the usual messes, opened early in 1945 and satisfied the needs of the community.[94] The expansion of facilities strained natural resources, including the water supply. The water required heavy chlorination after passing through the town's piping system. The first new emergency water line was built during the summer of 1943. When the water situation became serious again a few months later, Oppenheimer approved the stopgap measure of a noninsulated line from Guaje Canyon.[95]

There were few telephones in Los Alamos. Although offices were equipped with telephones, the administration did not want them in the town because they threatened security. A compromise allowed some telephones in the town for emergencies, but the post controlled access. In the early days of the project, the switchboard operated only during business hours because too few WACs were available to keep it open longer. Security overseers had no way to determine whether someone was tapping into the line, so they simply forbade anyone to discuss classified information over the telephone.[96]

Los Alamos had its own radio station operated by volunteers. The broadcasts were transmitted over the power lines and could be picked up at short distances from the site. The military intelligence officer reviewed the programming for classified information. One regular feature was Otto Frisch's classical piano playing.[97]

Before fall 1944, residents could leave the immediate vicinity only in emergencies. After these severe restrictions on travel were lifted, trips away from Santa Fe and Albuquerque, for example, into the Pecos, the Jemez, and Sangre de Cristos mountains were possible by automobile, if gasoline and tires could be found. Bandelier National Monument, which had served as overflow housing several times in the laboratory's existence, was a weekend playground for many residents of the townsite, as well as those living temporarily in the Monument. Picnics in Frijoles Canyon were popular, as was the (now prohibited) hunting for shards of pottery in Indian ruins.[98]

The surrounding Jemez Mountains offered escape from the pressures of the scientific program. Many scientists hiked there on Sunday, their day off. Some enjoyed horseback riding; the golfers built a rough nine-hole golf course; in winter, skiing and ice skating were available.[99] The Post Recreation Committee planned activities with the help of representatives from the town.[100] Parties and dancing were popular on weekends. Alcohol was consumed freely when it could be obtained.[101] The PX was a combination soda fountain, bar and grill, drugstore, and general meeting place that featured jukebox music to supplement the music piped over the power lines. Juvenile delinquency prevention drives usually focused on barring young people from the PX at night.[102]

Indian pueblos were another field for exploration. Some Los Alamos residents developed close relationships with the Indians in nearby pueblos, particularly San Ildefonso. Such friendships sometimes led to invitations to pueblo marriage and birthday ceremonies, or to pueblo feast days. Prices for Indian pottery and jewelry were low; pots by Maria Martinez were especially prized.[103] One high point in relations between the Los Alamos scientists and their Indian neighbors was a party the Indians hosted for Los Alamos residents after the war. Cokes and cookies from Los Alamos as well as Indian bread, fruit tarts, and other native delicacies were served while members of the two communities demonstrated square dancing and ritual dancing.

The lack of amenities at Los Alamos, while inconvenient, tended to reinforce the close community spirit of that society during the war. Neighbors helped each other overcome shortages and other impositions. Both the fence around the town and the common patriotic purpose of the laboratory helped focus attention inward – to the special role Los Alamos would play in World War II.

7

The Gun Weapon: September 1943 to August 1944

Although the problem of producing a gun weapon seemed relatively simple as the laboratory entered the summer of 1943, William Parsons and Charles Critchfield soon recognized that the guns needed to assemble uranium and plutonium would be very different from the ordinary ones that shot high-explosive projectiles at targets. The Ordnance and Engineering Division (E) did not know enough about the critical masses of plutonium and uranium, their metallurgical properties, or the speed required by nuclear constraints to assemble an effective plutonium weapon. Three points were clear: the guns would have to be designed from scratch – standard ordnance cannon could not be adapted to this use; extensive testing of both the plutonium and uranium guns and their mockup target and projectile components would be required; and targets and projectiles of active material would have to be designed. No one had yet tried to create an explosion using two or more pieces of fissionable material. Work on the gun gadget proceeded along three paths – interior ballistics research, experimental testing of designs, and target-projectile-initiator development. Each proceeded more or less independently until the reorganization of the laboratory in August 1944 consolidated all gun work.

Between April 1943 and August 1944, Parsons, Critchfield, Edwin

* This chapter is based on a draft written by Roger Meade. Catherine Westfall wrote the section on polonium procurement. We thank Robert Penneman for his detailed editing of the polonium section.

McMillan, and Joseph Hirschfelder designed, created, and tested all the principal components of the gun gadget, including the gun mechanism, target and projectile geometries, and initiators. Parsons provided a stable, well-run organization, using his navy contacts to expedite the work. He also provided a research environment that promoted individual effort. Critchfield calculated critical masses, spearheaded initiator design, and suggested testing at the 20-mm scale in readily available guns. Reduced-scale testing saved much time and money. Hirschfelder developed the interior ballistics parameters for the plutonium gun, as well as a general theory of gun design that was used to develop the uranium gun. His work was the foundation for the final development of the gun gadget.

Interior Ballistics

During the April 1943 conference, Oppenheimer stated the fundamental interior ballistics problem: to design, build, and test a gun capable of assembling plutonium. Following that, designing a uranium gun would be a straightforward task. Subsequently, Oppenheimer instructed Critchfield, Edwin Rose, and L. T. E. Thompson to formulate a gun program "with emphasis on the physical design of the two types of guns thought necessary." Critchfield and the newly arrived Parsons immediately saw that a good interior ballistician would be needed. While not requesting Hirschfelder specifically, they noted that the gun problems facing the laboratory were "way beyond what has been worked on even by Hirschfelder's group" at the Geophysical Institute.[1] They requested either Hirschfelder or R. B. Kershner, a mathematician at Johns Hopkins University and a member of Hirschfelder's group. Hirschfelder was selected for the Los Alamos position.

Gun design is based on pressure-travel curves and propellant behavior. Pressure-travel curves, which display pressure as a function of the travel of the projectile within the bore of a weapon, are used to attain a desired muzzle velocity. To prevent malfunction or damage, gun designers arbitrarily chose a maximum breech pressure, which is based on the strength of the gun tube steel and the burning characteristics of a particular propellant. They then adjust the propellant to achieve a desired muzzle velocity. Acting in his unofficial capacity as gun coordinator, Richard Tolman asked Hirschfelder's Geophysical Institute group to provide pressure-travel curves for the gun designs that looked promising. The calculations of the curves were completed by early June.[2]

Attention then turned to powder selection. By early August, Critchfield had developed the first full set of parameters for the plutonium gun. Hirschfelder proposed using a hot burning powder, which would give better burning efficiency through the use of a higher density of loading and smaller powder chamber. Even with the smaller powder chamber, weight remained a problem because the bomb would have to be light enough to be delivered by an airplane. Because the gun weapon only had to be strong enough for only one shot, Rose, the veteran gun designer from Jones and Lamson Engineering, realized that a significant amount of weight could be removed by simply reducing the amount of steel in the tube. Tubes to be used in testing would, however, have to be more durable to permit repeated use.[3]

When Hirschfelder arrived in Los Alamos, Parsons placed him at the head of a new group, Interior Ballistics, for which Hirschfelder quickly developed a theoretical and experimental program. The theoretical work included the study of unconventional guns, such as double guns and rocket devices, and the determination of interior ballistics characteristics for guns capable of shooting projectiles at several different velocities. Experimental work included the computation of loading charts for "all conditions of firing"; instrumentation analysis of bore friction, starting pressures, pressure waves, and ignition; and tests of powder, igniter, and primer systems in sturdy closed bombs where their performance as individual components could be evaluated. Hirschfelder also established a team to operate five calculating machines to solve numerical problems in E-Division. Just before coming to Los Alamos, Hirschfelder visited the Ballistics Research Laboratory at Aberdeen Proving Grounds to examine a newly developed ignition powder. He found the Aberdeen powder untested in guns and unusable for the project's purposes.

While still at Berkeley, Serber had thought of using a double gun to assemble active material. Hirschfelder also looked into this concept; in his view, a double gun would have one distinct advantage. "It would be approximately one-half as long and would not involve any exterior ballistics difficulties." The main problem, as Serber had pointed out earlier, was "to obtain simultaneity in the firings."[4] In November, Hirschfelder outlined a simultaneity test using two 20-mm Hispano bores. The bores, connected to one powder chamber, would be mounted parallel and pointed in the same direction. James Serduke, a Russian emigré who had served as a captain in the White Army, was in charge of conducting the test. No records are available concerning this test, and it is uncertain whether it was even conducted. Los Alamos did no further interior ballistics work on the appealing concept of the double-gun system. Rocket devices received even less attention than double-gun systems. After surveying the available literature on rockets during autumn 1943, Serduke concluded that the use of rockets was exceedingly complex, especially in regard to the time of ignition and the time of flight. Since more "standard" ordnance was being modified with success, rockets were quickly forgotten.[5]

Only a small part of the theoretical effort was expended on double guns and rockets. Most work centered on determining the interior ballistic characteristics of guns capable of firing at several velocities using projectiles of assorted mass. The calculations involved guns "of various diameters and operating at low, normal, and high pressures." Critchfield supplied the desired weight of projectiles, which changed periodically as new estimates of critical masses were calculated. Between September 1943 and January 1944, Los Alamos developed the interior ballistics for the plutonium gun, also called the "Thin Man" or the "high-pressure gun."[6]

By January 1944, the theoretical work was completed and Hirschfelder gave Parsons his recommendations on density of loading, the choice of propellant, primer, and powder web (the minimum thickness of the grain between any two adjacent surfaces), the design of powder bags, and the overall design of the gun. These recommendations would be tested on full-scale 5-inch tubes during the spring of 1944.

The Interior Ballistics Group tested several types of powders. They identified those that would work well, but delayed selecting the type to be used in the combat gadget until experimental tests were completed. The group also recommended that density of loading be based on an old ordnance practice of keeping the ratio below the mathematically optimal figure. As Hirschfelder commented, ordnance personnel believed that unusually high densities of loading caused guns to blow up unexpectedly. Hirschfelder's report underscored the many uncertainties that remained, particularly the inability to specify a single powder web. To compensate for these uncertainties, two lots of the same powder were always ordered, one having a smaller web and one having a larger web than the best guess of what was required. By blending the two lots, the group achieved considerable variation in the effective web.[7] A second report to Parsons quickly followed, presenting a general design for any gun that might be needed. This report proved invaluable when it came time to develop the uranium gun and also was used to compute the design variables of a possible second-generation gun gadget using larger amounts of fissile material, the "augmented gadget," in the spring of 1945.

The experimental program of the Interior Ballistics Group began when Hirschfelder recruited John Magee of the B. F. Goodrich Company in October 1943 to help with propellant testing. Hirschfelder sent Magee to the U.S. Bureau of Mines, Bruceton Station, in November to assist with the testing of propellants in "closed bombs," constant-volume testing devices having heavy walls. Bruceton had the only closed bomb available.[8] Magee participated in both propellant and ignition system tests. To determine the maximum pressure that could be obtained before the powder would begin to perform erratically, Magee measured du Pont's pressure parameter curves. Propellant experiments in another Bruceton device, the partial-burning bomb, measured incipient pressure wave formation due to the high densities of loading.

The Bruceton program never lived up to expectations, because problems with the closed bomb prevented extensive tests. The program did, however, provide useful information on the strength of the primers, the tendency of standard navy powders to wormhole, and the effect of low temperatures on powders. The Bruceton work was officially suspended in June 1944 when it became evident that the available gun instrumentation at Los Alamos could answer any questions about propellant performance.[9]

In January and February 1944, the Interior Ballistics Group prepared for the arrival of the first high-pressure tube and several types of powders. Loading charts showed the pressure and muzzle velocity versus charge for each powder, allowing a rapid comparison between observed results and theory. At this time, the group also accepted the final designs for the ignition system and powder bags.[10] The first tube arrived in March and several shots were immediately fired to test instrumentation and propellants. Although the piezoelectric pressure gauge worked satisfactorily, the chronograph proved unreliable. During subsequent analysis in May, strain gauges, a radar bore accelerometer, and piezoelectric pressure gauges, showed no evidence of undue strain or projectile friction even at the maximum test pressure.

Although the work of the Interior Ballistics Group centered on the plutonium gun, some attention was given to the uranium gun. Carlton Green, the navy's chief gun designer, urged the group to settle on the caliber of the uranium gun. Standard naval cannon have whole-integer calibers ranging from 4 to 16 inches. A uranium gun with a nonstandard caliber would have required much additional design work.

The work of the Interior Ballistics Group neared completion in mid-May, and Hirschfelder turned to computational problems for T-Division. By the end of July 1944 and almost in tandem with the reorganization, all interior ballistics work ceased and the group was disbanded. Hirschfelder's last task here was to assess the possibility of building a plutonium gun that could deal with the problems of predetonation, which had recently been highlighted by the discovery in the Experimental Physics Division of significant spontaneous fission by pile-produced plutonium. Hirschfelder found such a gun impractical and vetoed any further work.[11]

Testing

Before the group could actually begin testing any project gun, they had to construct a firing range. A nearly perfect site became available when the War Department purchased an old ranch, near the main laboratory, known locally as Anchor Ranch. Besides having the typical assortment of ranch buildings and equipment, Anchor Ranch had a large flat area for gun emplacements immediately adjacent to a small canyon, which would give natural protection to the control building during test shots.[12]

As constructed, the range consisted of a gun emplacement and an accompanying sand butt to catch projectiles. After a shot, personnel recovered the "yaw card," a piece of cardboard placed in front of the muzzle of the gun. This card indicated the oscillatory movement of the projectile about its vertical axis – which was apparent from the shape and angle of the hole left in the card by the projectile as it passed through. They also dug out the pieces of target and projectile from the sand butt. These recovered pieces showed the effects of the impact. Other more sophisticated equipment and techniques, such as piezoelectric gauges, gave information on tube stress and interior ballistics. A large concrete building, covered with earth and located in the canyon, housed the control room. A large periscope gave the researchers a full view of the firing area during shots.[13]

As a test site, Anchor Ranch had one drawback: it was unsafe for nontest personnel. Since the site bordered the main road, road blocks had to be placed on either side of the range. As more roads were built through the general region, and as the laboratory expanded into this area, safety became more difficult to maintian. A particular hazard was the "large fragments of targets [which] sometimes traveled as far as 75 yards; small fragments went even farther. The fact that there were no accidents was in part due to good fortune, as there were several failures to clear the nearby roads during hazardous shots."

To expedite the testing of other major components, Parsons acquired a 3-inch naval antiaircraft gun, which fired the first shot in the test program. This gun was used until early 1945, when all experimental firing shifted to the full-scale uranium gun. This 3-inch model gun proved invaluable in testing sabot stripper designs, projectile strength, the efficiency of sealing bands, the ejection of absorbing plugs from targets, and target-projectile geometries.

One of the most important additions to test-site personnel was Naval Ordnance Technician Thomas Olmstead, the only one there with experience in loading and firing large-bore cannon. Robert Wilson recalls Olmstead as "a colorful person ... a member of the Black Powder Society ... someone who thrilled me in conversation, a regular guy in an irregular way."[14] Critchfield recalled recently that Olmstead was the only person with enough sense to plug his ears prior to a shot. Olmstead acted as buffer between Parsons and the scientists. Never truly comfortable with the scientists' seemingly undisciplined style of research and experimentation, Parsons relied on Olmstead to keep the big guns in working order and the firing program on schedule.[15]

Meeting for the first time on 25 February 1944, the Steering Committee for Gun Assembly reviewed the progress of the target work. This committee, formed by Oppenheimer at the request of Parsons on 12 February, met every other week through October 1944 under the broad mandate to review the problems of gun assembly. Citing a provisional requirement that the "target assembly stay together after the impact," the committee recommended experiments that would use the strength of steel to hold the target together.[16]

By April 1944, 70 percent of the test shots had investigated blind targets – closed-end systems in which the projectile is stopped. The purpose of these tests was to find the geometry in which the target deformed the least. The factors of interest were the relative importance of strength and inertia in various parts of the target in absorbing energy and restraining unfavorable deformation and the time dependence of deformations during impact, especially the overall time between first collision and final configuration. Solid steel targets and steel projectiles simulated active materials.[17]

The 3-inch test program proved extremely beneficial to the project.[18] During the summer of 1944, however, 3-inch testing became less desirable as the target and projectile designs became more complex and harder to fabricate. An increasing amount of work was done at the 20-mm scale. Since the 3-inch gun was a stopgap instrument and not to scale, little new information could be gained by using it as a primary test device.

The idea to use a 20-mm gun came from Critchfield. While employed by the NDRC, Critchfield had used such guns and found them ideal for scaling experiments. Oppenheimer subsequently approved a 20-mm program to safely test initiators containing polonium, and to test the many expected designs of blind targets. The greater ease of fabrication and the relative ease of heat treating and machining smaller pieces of alloyed steel made the decision an easy one for Oppenheimer.[19] During February and March 1944, Oppenheimer assembled personnel and facilities for the 20-mm project. The most important additions to the new program were a special building to conduct firing tests and the services of David S. Wood, a research engineer, from the California Institute of Technology. Wood assumed responsibility for the mechanical testing of designs and control of hazards.

The construction of the 20-mm laboratory received special attention. It had to be kept small to facilitate instrumentation, the recovery of targets and projectiles, and the containment of all radioactive and toxic materials. Personnel needed special protection from powder gases and the uranium, polonium, radon, and plutonium used during specific tests. To protect against these hazards, the gun, recoil mechanism, and target were enclosed in an evacuated space connected to a special pump and filter system. Gas masks and special clothing provided additional protection for those working in the test chamber or handling radioactive materials. Since the gas masks were designed to remove dust, smoke, and other fine particles from the air, physicist Robert Cornog thought they would effectively remove radioactive dust. The original program for the 20-mm gun classified shots as either inactive or active – on the basis of nonradioactive targets and projectiles requiring no special precautions – or active.

Temporarily head of the 20-mm program, Cornog, a student of L. Alvarez at Berkeley in the late 1930s, tested the similarity of ballistic experiments carried out at different scales. He duplicated many of the more pertinent shots that had already been made at the 3-inch scale. The development of a neutron source for both the uranium and plutonium guns, Cornog's second priority, had to wait for additional equipment and facilities. Cornog planned to develop plutonium projectiles and targets as soon as enough plutonium, or a material resembling plutonium, became available.

The program proved immediately successful. Several targets and projectiles showed enough promise to be scaled up for additional testing. The success of the 20-mm laboratory was mitigated in April by problems with the quality and timeliness of machine shop work. Several assemblies were delivered late and some were defective and had to be returned. As a result, many tests were delayed or abandoned. Parsons assigned physicist Lawrence Langer to work closely with the shop to eliminate as many problems as possible by coordinating personally the shop's 20-mm work, thus bypassing the cumbersome shop administration.[20]

Initiator

The onset of a chain reaction had to be accurately timed to ensure an efficient explosion – a difficult task, since the explosive nuclear fission reaction is very short in comparison with assembly times. During the April 1943 conference, Serber had suggested using an initiator as "extra insurance," a device set inside the weapon that would release a burst of neutrons to help trigger the chain reaction at the optimal time. Developing the initiator required two parallel efforts: Oppenheimer had to procure material for the device, and the Ordnance Division had to create a model rugged enough to withstand being fired from a gun and brought to a sudden stop, yet precise enough to provide neutrons only when the target and projectile had seated.[21]

Procuring Polonium

Serber had proposed using radium–beryllium or polonium–beryllium initiators. Even before the war, ^{210}Po mixed with beryllium was a popular neutron source. The isotope's 140-day half-life was long enough for it to be stockpiled, yet short enough for the α-particle flux to remain relatively high. Moreover, polonium could be obtained by ordinary chemical processes.[22]

Although small amounts of polonium were available, it was not clear that a sufficient quantity of the necessary purity could be obtained in time for use in the program. In June 1943, Oppenheimer informed Groves of the need for procuring polonium, mentioning the two expected sources, neither of them certain: separation from lead residues and pile production. The isotope ^{210}Po, which occurs naturally in the ^{238}U decay series, is ordinarily isolated from lead concentrates. Although entrepreneur Boris Pregel claimed to have access to residues capable of producing 500 curies of polonium, Oppenheimer felt that the necessary separation would be time-consuming and impractical and therefore suggested that this method be used only as a standby.

Oppenheimer believed it would be easier to produce polonium in the plutonium production piles through the capture of neutrons by bismuth. Because of the low neutron-capture cross section of bismuth, large amounts of this element had to be bombarded for a long time to make a small amount of polonium. If 100 pounds of bismuth were placed near the center of the Clinton pile, 9 curies of polonium could be produced every four months, assuming that the pile's power specifications did not change. This would provide enough polonium to meet the immediate needs of the initiator effort and to produce the polonium–beryllium sources for the various experiments. An ample amount for bombs could be provided later when Hanford went into operation; Oppenheimer calculated that 100 pounds of bismuth placed in the center of the Hanford pile could produce 4.5 curies of polonium per day. Of course, this scheme relied on the timely and reliable operation of the piles, which were still in the design phase.

Whatever method would be used for polonium production, considerable work was necessary to improve purification techniques. Oppenheimer promised Groves that the preliminary research would be conducted under the direction of Berkeley chemist Wendell Latimer. Some work had already been done at Berkeley, prompted by the need for polonium for the production of polonium–beryllium sources. For that work, Oppenheimer had obtained what was probably the country's largest concentrated source of polonium from J. G. Hamilton, the Berkeley physician who had pioneered the use of cyclotron-produced radioactive elements and had also used radium to treat tumor patients. Despite the fact that Hamilton had promised several donors that he would personally maintain control of the substance, he agreed in the spring to relinquish his "radium D" (^{210}Pb) solutions. When Rene Prestwood, a young recruit, was commissioned to separate polonium from these solutions, he found separation difficult and his results "disappointing" because of the impurity of the solutions.[23] To determine whether polonium could be separated under better conditions, Oppenheimer asked Latimer in July to begin the separation of bismuth irradiated by the Berkeley cyclotron.[24]

With polonium separation research assigned to Berkeley, plans for irradiating bismuth in the Clinton pile and for exploring natural sources went forward. As summer continued, the researchers encountered several problems. To irradiate bismuth at Clinton, they arranged to place five bismuth slugs, which together weighed 440 pounds, in the pile for 100 days so that 10 curies of polonium could be produced. Obtaining the necessary bismuth proved easy. Samuel Allison found that the American Smelting and Refining Company at Perth Amboy had a large source from which bismuth could be separated. The problem was that any antimony present might remain as an impurity in the bismuth, which would yield radioactive antimony by neutron capture. As Allison explained to Martin D. Whitaker, director of the Clinton pile, this was a pressing problem, "since the presence of a very small amount of antimony will produce a penetrating gamma ray in the irradiated material." Since gamma-ray production would pose a severe health risk to workers, researchers had to be sure that antimony concentrations would be low before irradiating the bismuth.[25]

In the meantime, Hamilton had done some sleuthing about natural polonium sources. In July he reported that nearly 7 tons of lead dioxide were available at Port Hope, Ontario. Assuming current estimates for the amount of radium and lead available per ton, about 2.8 curies of polonium could be separated per ton. It was later determined that the material belonged to the Canadian Radium and Uranium Corporation. Hamilton explained that Pregel had control over the material and warned that he was apparently trying to sell it both to the Canadian refinery and to petroleum companies that could use it to make polonium–beryllium neutron sources for oil-well prospecting.[26] After hearing this news, although he still considered the natural polonium source a "standby," Oppenheimer worked with Groves in July and August to ensure that the lead dioxide would not be sold to other customers.[27]

At this juncture, Oppenheimer made arrangements to assign polonium production to Charles A. Thomas, who had recently been appointed coordinator of the plutonium effort.[28] This appointment was a clever move in more than one way. As central research director of Monsanto Chemical Company, Thomas was in a good position to arrange for Monsanto to assume responsibility for large-scale polonium separation. This was a relief because it was unreasonable to expect the overworked Berkeley group to take on this task after they developed separation procedures.[29] In fact, by late July Los Alamos researchers were worried that Berkeley was not making rapid progress on separation. By appointing Thomas,

Oppenheimer obtained welcome support in his efforts to persuade the overcommitted Berkeley group to work on this task.[30] As he explained to Thomas, polonium separation was not "in the field of Latimer's main interest," and since he was short-handed, some pressure would probably have to be applied "to get this work done on time to be of any use."

By the last day of July, Thomas had taken charge. He offered to encourage and coordinate work with Berkeley. To get the ball rolling, he suggested that a sample of bismuth from the American Smelting and Refining Company be sent to the Berkeley group so that they could work on the same material to be used in large-scale production. It was understood that after separation procedures were developed, Monsanto would translate them into industrial scale. Thomas also reported that the refining company had promised bismuth of 99.99 percent purity with "no detectable antimony."[31] By mid-August this good news was augmented when Latimer reported that the Berkeley group had been able to separate polonium from bismuth with "no trouble."[32]

Even with the encouraging report on antimony, it was not clear that bismuth irradiation would emerge as the preferred method of polonium production. A prime disadvantage was that a considerable number of uranium slugs would be displaced, since large amounts of bismuth had to be bombarded. By fall, Fermi reportedly considered "the nuisance value" of bismuth irradiation to be "quite considerable."[33] At this point, Pregel contacted the War Department to report that he could deliver both polonium and RaD.[34] Although Oppenheimer warned Thomas that Pregel "had a bad reputation" and was "considered fairly unreliable," considerable pains were taken subsequently to investigate the separation of polonium from lead dioxide.[35]

Despite the fact that consultation with Canada was awkward, since the two countries had not yet devised an information-exchange agreement, the Los Alamos scientists decided that a meeting should be called to bring together American scientists, including Segrè, who had organized early polonium separation studies, along with F. Paneth and B. L. Goldschmidt, two scientists from Canada who were experts in separating polonium from lead oxide.[36] Groves agreed to the meeting, although he insisted that it appear that the subject was of interest only to the Met Lab.[37] After the meeting in late October, Segrè reported that the Canadian source "might well compete" with plans to irradiate bismuth in the Clinton pile.[38]

In a November status report to Conant and Groves, Thomas reported new concerns about both polonium sources. He expressed relief that arrangements had been made to secure the Port Hope lead oxide "as an alternate supply," admitting that his group was now concerned about the Hanford irradiation of bismuth. At the full-scale plant, impurities had to be 100 times less than at Clinton, because irradiation intensity was 100 times greater. He also mentioned a new worry that had emerged from the meeting with Paneth and Goldschmidt. The Americans had learned that polonium oxidized much more readily than anticipated, which meant that extra effort would be required to keep oxygen impurities very low to avoid neutron background from (α, n) reactions in oxygen.[39]

Despite these concerns, Thomas announced that success in polonium separation was "reasonably certain." After all, work had been published on polonium. In addition, Paneth and Goldschmidt had reported success in recovering polonium from lead dioxide, and by this time Berkeley had obtained high yields in separating polonium from bismuth.[40]

Thomas also reported that Monsanto was already organizing a laboratory for polonium separation at Dayton, Ohio.[41] In the fall and spring, Los Alamos researchers visited Dayton to advise the Monsanto engineers about the difficulties of working with polonium. Concentrated solutions produced enough radiation to decompose solvents. In addition, the element did not adhere easily and migrated wildly, thereby causing constant contamination problems. Prestwood told the Monsanto engineers that once a plate was covered with polonium, it would contaminate everything near it, unless covered. Even in the process of counting the sample, researchers noticed some of the material was lost because of α-activity recoils that distributed the sample.[42]

In December, a new problem arose. When 3 tons of lead residues arrived in Dayton, the content of radium D was much less than expected and the content of polonium was much higher. This meant that researchers had been misled about the history of the material. The chemists were quite concerned, since a lower radium D content meant that less polonium could subsequently form in the material. Although the lead residues were not expected to supply most of the necessary polonium, they were to be a supplement. At least 10 tons of lead residues would have to be processed to meet delivery schedules at Los Alamos. If polonium production at Clinton did not progress as planned, more lead dioxide processing would be necessary.[43] To help resolve the uncertainty of the Port Hope shipment, a meeting was called with a representative from the refinery. The participants agreed that a mistake had apparently been made and that more care should be taken with the second shipment. Thomas also arranged to have samples taken and measured from each bag in the shipment.[44]

By January, the polonium production outlook was much brighter. Thomas reported that "plans for the large scale extraction of polonium from lead dioxide and from irradiated bismuth were completed." He had arranged for Monsanto and the Firestone Tire and Rubber Company to convert 9 tons of lead dioxide to lead chloride and plate the polonium on nickel foils. Complete recovery would result in approximately 9 curies of polonium from this process. Since the polonium would be deposited on many foils, Dayton researchers had to concentrate the element.[45]

Eventually, Monsanto chemists used the following procedure for isolating polonium from irradiated bismuth:

> After holding long enough for most of the ^{210}Bi [radium E] to decay to polonium, the slugs of metal were melted and dispersed by sieving into water. The pellets were then dissolved in aqua regia and the resulting solution denitrated with formaldehyde. From this solution the polonium was scavengered by stirring in 500 grams 150 mesh bismuth powder and then decanting the solution. The bismuth powder, now carrying the polonium, was dissolved in about one liter of aqua regia and the solution denitrated.[46]

At this point, the polonium was removed by either the tellurium process or the bismuth "Super Scrub" process. In the tellurium process, telluric acid was added to a polonium solution, and tellurium and polonium were jointly precipitated by reducing both with added tin (stannous) chloride. In the bismuth Super Scrub process, bismuth metal powder was used both as a reductant and as a carrier. In both processes, polonium was removed from solutions and concentrated as part of a larger precipitate that included either tellurium or bismuth. The polonium was then separated from these carriers by electrodeposition onto platinum.

Clinton's first batch of bismuth reached Dayton in January. It contained more polonium than expected. To ensure that several curies of polonium would be available to Los Alamos as soon as possible, a large amount of this first batch was separated on a semilaboratory scale. For longer-term production, Dayton researchers constructed a large-scale processing unit capable of processing 110-pound batches of bismuth.[47]

They decided to can (i.e., cover) bismuth slugs in metal containers in February; such canning proved easy. By the end of March, Los Alamos began receiving shipments of polonium, which allowed it to begin crucial physics tests of the initiator. By April, Thomas was able to announce that "we are in the fortunate position of having both horses come through." Dayton began to prepare the final platinum foils containing polonium according to the specifications of Los Alamos.[48]

From this time forth, Monsanto unfailingly provided increasing quantities of polonium, most of which was produced from pile-irradiated bismuth. Starting in April 1944, 2.5 curies per month were produced, and by summer this figure was raised to 6 curies per month. By 1 January, Monsanto was prepared to ship 10 curies per month, the rate Oppenheimer had set in late August as sufficient "for the life of the project."[49]

Initiator Design

Leonard Schiff showed theoretically that the efficiency of the gun gadget could be increased by the use of an initiator. Reviewing Schiff's work, Oppenheimer made two important decisions: in February 1944, he placed Kenneth Bainbridge in charge of the Initiator Committee, which oversaw the initiator program; and, he authorized construction of the 20-mm laboratory as "a wart to Building B" to expedite gun work in general and initiator work in particular. Parsons, ever focused on detail, reiterated Oppenheimer's emphasis on 20-mm testing by formally stating the Ordnance Division's role in initiator development. "It is agreed that the most effective and expeditious method of handling the initiator development is at the 20 mm scale."[50] Oppenheimer requested that serious thought be given to the "physiological dangers and the risks of contamination involved in firing tests with plutonium and polonium."[51] Critchfield assumed responsibility for the experimental work, while Joseph Kennedy and Dodson handled the preparation of polonium.[52]

Through the late winter and early spring, Critchfield made good progress on the initiator design.[53] Tests involving active material, including determination of the necessary source strength, had to wait for the 20-mm laboratory to be completed.[54] The committee settled on the initiator's design in June. The design was accepted in late July.[55] The Initiator Committee then decided on the source strength and other practical aspects. The initiator's structure had to be rugged enough to withstand unreasonably rough handling, accidental drops, airplane vibrations, and the shock of firing. Each unit had to be free from rust and corrosion and remain tightly sealed.[56] Oppenheimer gave final permission to incorporate initiators into the gun gadget on 15 March 1945. He suggested using lower-strength initiators. If the amount of polonium in each gun initiator was reduced, the background activity would remain within a reasonable tolerance, less polonium would be needed for the tests, and some would remain for implosion initiator work.

Later, "forty-odd" initiators were sent to Tinian, the Pacific island airstrip to be used by the U.S. Army Air Forces to launch B-29 bombing raids on Japan, along with equipment to measure neutron background. Sixteen units were selected for testing, and four were ultimately inserted in the combat gadget. All the units tested at Tinian had the same neutron background that they had had in the United States.

Target, Projectile, and Tamper Development

In his April 1943 lectures, Serber had presented general ideas for the target and projectile assembly, but he could not be specific about either design because scientists had no nuclear data for ^{235}U and ^{239}Pu.[57] Even the physical properties of plutonium were by and large unknown. To set limits on the size of the combat unit, researchers had to rely on estimates of the target and projectile sizes and shapes, assembly velocity, and critical masses of active material. The anticipated size of the combat unit was in turn a useful parameter for selecting the type of plane used to deliver the gun weapon. More serious, however, were the conceptual uncertainties they faced in producing a chain reaction at the optimum moment, calculating the number of critical masses that could be placed in each part of the final assembly, and calculating the probability of predetonation on the basis of the manner in which the assembly progressed from a subcritical to a supercritical state. Critchfield's projectile and target group designed the projectiles and targets with the help of McMillan's proving ground group and several metallurgy groups. Cyril Smith's metallurgists proved indispensable in fabricating boron and developing appropriate procedures for heat-treating steel.[58]

By March, the blind target assembly had strong support, especially within the Steering Committee. Parsons, who was becoming increasingly pessimistic about the laboratory's overall progress, took issue with what he saw as the committee's overconfidence in the blind target. Writing to McMillan in early March, he pointed out that it would be practically impossible to conduct full-scale experiments with plutonium to determine its ballistic behavior under high-velocity impact. Given the ballistic uncertainties of both plutonium and the likely tamper, Parsons strongly suggested that the "mechanically simple and reproducible primary assembly" should not be sacrificed to the "violent, difficult, and

Fig. 7.1. Navy Captain William S. Parsons was leader of the engineering and ordnance division and head of the Little Boy program. LA Photo, LAT 717.

non-reproducible method" of the blind target. Critchfield's test program reflected Parsons' concerns.[59]

Unfortunately, the physical properties of plutonium continued to be a problem in the spring of 1944. Smith reported at the 17 March Steering Committee meeting that the melting point of plutonium could be as high as 1,800°C, and its density not less than 17. (These values would be radically revised in the next few months, as explained in Chapter 11.) This high melting point complicated plans for the purification process. Smith proposed making a ternary combination of silver, gold, and copper to approximate the physical properties of plutonium, so that scaled-down target and projectile experiments could begin. Smith did not produce the promised calculations of the effect of steel on the target tamper.

Parsons, still concerned about progress, emphasized the need to speed the project up because gun production at the Naval Ordnance Factory was running ahead of schedule. At the 21 April meeting of the Steering Committee, he noted five key areas in which knowledge was "notably lacking": the nuclear and mechanical problems of plutonium, the critical properties of odd shapes of active material (particularly in connection with the predetonation problem), the strength of the background from active material and initiator, and the ability of a thin gun barrel to withstand 75,000 psi. Only the critical properties of odd shapes seemed hard to solve, but, as Serber pointed out, computational methods were now advanced enough to attack them. P-Division was about to lend a hand on odd-body problems by providing data for pseudointegral and optical analog experiments using a rotating pyramid camera to be set up under Donald Marshall. Parsons was right to be concerned, as became clear when the committee began discussing the very negative issue of McMillan's volumetric theory of assembly. This theory postulated that the criticality of a given mass depended on the volume of tamped space in which it was contained. The rate of change of this volume was proportional to the speed of assembly. If the volumetric theory proved correct, the plutonium gun was doomed.[60]

By early July, well before the August reorganization, Parsons had recognized that the plutonium gun could not work because of the presence of ^{240}Pu and began shifting Ordnance Division activities to the uranium gadget on which all gun work now centered. A suitable caliber was finally chosen. By this time as well, reduced-scale testing had reached its limits. Only full-scale testing would suffice to produce a combat device. This effort could now go forward without reservation. Critchfield noted in his last report as an E-Division member, the uranium gun "is both predictable and satisfactory at full scale."[61] No longer the stepchild in relation to the plutonium gun, the uranium gun could now receive the attention it needed.

8

The Implosion Program Accelerates: September 1943 to July 1944

By late summer 1944, the implosion program was among the laboratory's highest priorities. It had started out as a small, informally run, back-burner effort of a handful of researchers surrounding the reserved Seth Neddermeyer (Chapter 4). Between the fall of 1943 and the summer of 1944, it was transformed into a well-coordinated, multidisciplinary research effort of more than fourteen groups operating within T-Division and the newly created Gadget (G) and Explosives (X) Divisions.

The shift began with a visit in late September 1943 by the great mathematician and physicist John von Neumann. On learning about Neddermeyer's test implosions of small cylindrical metal shells, von Neumann pointed out that their efficiency could be increased using a substantially higher ratio of explosive to metal mass, which would promote more rapid assembly. The suggestion excited leading Los Alamos theorists, including Bethe, Oppenheimer, and Teller, who could now envision an atomic weapon requiring active material having less mass and a lower level of purity than was needed in the gun device – advantages of particular interest to General Groves.

Theorists, particularly Bethe and Teller, spent more and more time on implosion questions, while von Neumann continued to work on theoretical aspects of the implosion in Washington, D.C. The new implo-

* This chapter is based on a draft by Lillian Hoddeson. We are grateful to Gordon Baym and Les Redman for their detailed editing of this chapter and to Robert Penneman for his substantial contribution to the section on the RaLa method.

sion theory group was set up in March 1944 under Teller to develop the mathematical description of implosion. Additional experimentalists joined the program. Neddermeyer's E-Division group expanded from five to roughly fifty. In October 1943, George Kistiakowsky, the leading American explosives expert and director of the Explosives Research Laboratory (ERL) at Bruceton, Pennsylvania, was brought in as a consultant.

The primary experimental concern during the first half of 1944 was to devise diagnostics to measure implosion parameters such as symmetry, the time of collapse, and the degree of compression. In this phase, three techniques were used to examine collapsing cylinders: X-ray photography, optical photography, and "terminal observation" (diagnosing the implosion by studying the remains of test shots). As the months passed, Los Alamos scientists conceived of more exotic diagnostic methods – in particular, the electric "pin," betatron, "RaLa," and magnetic methods. In addition, the group collaborated with the ERL and the Hercules Powder Company in researching the detailed behavior of high explosives, which for the first time had to function as precision tools.

By early 1944, test shots were revealing serious problems: "jets" and "spalling" (explained later in the chapter) were causing asymmetry and turbulence, drastically curtailing efficiency and raising serious doubts about the usefulness of the method. By late spring, the possibility of realizing a practical implosion appeared remote.

Von Neumann's Visit in September 1943

The primitive observations made through September 1943 in Neddermeyer's group in the Engineering Division gave early evidence that simple implosion shots were too asymmetrical to release the nuclear energy required for a usable weapon. But few worried about this implosion problem, because the prevailing opinion at the laboratory was that gun assembly would succeed for both uranium and plutonium.

Oppenheimer thought otherwise. With great foresight, he drew attention to the precarious state of implosion theory. In July 1943, he wrote to von Neumann: "We are in what can only be described as a desperate need of your help .... We have a good many theoretical people working here, but I think that if your usual shrewdness is a guide to you about the probable nature of our problems you will see why even this staff is in some respects critically inadequate." He invited von Neumann to "come, if possible as a permanent, and let me assure you, honored member of our staff," and suggested that "a visit will give you a better idea of this somewhat Buck Rogers project than any amount of correspondence."[1]

Von Neumann was an appropriate choice for such theoretical help. He had previously worked on questions of shock waves and turbulence in fluid dynamics and in the late 1930s had served as one of the original scientific consultants to Aberdeen Proving Ground, where he had studied "shaped charges."[2] A particular strength he could bring to the implosion program was his ability to solve sets of nonlinear partial-differential equations numerically. He had helped place Chapman and Jouguet's classic formulation of detonation waves on a solid foundation, interpreting them as shock waves followed by a chemical reaction zone.[3]

At the time of Oppenheimer's letter, von Neumann was working at the Navy Bureau of Ordnance and was also associated with the Aberdeen Proving Ground, Princeton University, and the NDRC. To draw him to Los Alamos, Oppenheimer, Groves, and Parsons appealed for help to both Richard Tolman and Rear Adm. William R. Purnell (a member of the Military Policy Committee).[4] They agreed that von Neumann would work on theoretical problems in Tolman's office at the National Academy of Sciences in Washington and would make "an occasional visit to Santa Fe."[5]

The implosion program took a decisive turn during von Neumann's first consulting visit to Los Alamos between 20 September and 4 October. Hearing from Teller, one of von Neumann's boyhood friends in Hungary, as well as others about Neddermeyer's implosion studies, the mathematician immediately drew on his recent experience with detonation waves and the Munroe shaped-charge effect to suggest initiating implosion by arranging shaped charges in a spherical configuration around the active material. In this scheme, the jets produced would rapidly assemble the bomb.[6] After thrashing this idea about with Teller, von Neumann, as a hydrodynamicist, took the substantial further step of suggesting a means of achieving a faster kind of implosion assembly based on increasing the amount of high explosive. His conversations with Teller had revealed that the higher pressures generated by such an approach could reduce the amount of fissionable material in the weapon.[7]

Von Neumann's fresh suggestions about implosion, as Charles Critchfield recalls, "woke everybody up .... Johnny ... was a very resourceful man, at least twenty years ahead of his time .... I remember Edward [Teller] calling me and saying 'Why didn't you tell me about this stuff?' I said, ... Seth and Hugh [Bradner] and Streibo [John Streib] and I have

Fig. 8.1. Mathematician John von Neumann, a consultant to the Los Alamos laboratory, suggested the fast implosion concept. LA Photo, LAT 627.

been working on this, and nobody paid any attention to it." Critchfield also recalls Groves scolding Parsons, who had been focusing on the "safe" gun method, for not keeping him adequately informed about the implosion.[8] Bethe, Oppenheimer, and Teller were particularly impressed that an implosion bomb might be a far more efficient nuclear weapon than a gun.[9]

The formal implementation of an expanded implosion program based on von Neumann's ideas grew out of several discussions of the Governing Board in the fall of 1943. The board understood that to develop the new ideas into a useful weapon required a deeper technical understanding of high explosives than was available at Los Alamos. At a meeting on 23 September 1943, Oppenheimer asked whether the board felt that

Kistiakowsky, the eminent Harvard chemist, should be approached for the implosion program. Kistiakowsky was then head of Division 8 of the OSRD, which included the Bruceton Explosives Research Laboratory. Edwin McMillan and Bethe strongly recommended that Kistiakowsky be asked; Rabi suggested hiring him as a consultant and bringing part of Division 8 to Los Alamos. Charles Thomas of Monsanto Laboratories, who was then coordinating studies of plutonium chemistry at Los Alamos, Berkeley, Chicago, and elsewhere, offered to sound Kistiakowsky out informally. The board also discussed expanding the experimental study of implosion test shots, including, as Bethe suggested, X-ray photography.[10]

Further shaping of the expanded implosion program continued in a conference of the scientific staff on 1 October 1943, attended by Oppenheimer and Parsons. Neddermeyer spoke about the existing program centered on cylindrical experiments using small charges – the program whose usefulness von Neumann questioned.[11]

Neddermeyer then proceeded to draft a plan for a more ambitious program, dividing it into what could be accomplished in one month – "work on copper cylinders and spherical shells, including measurement of working stress by calorimeter methods, time measurements and tests of small cylinders of tuballoy, photographic observations"; in the "future and partly paralleling" the work in the first month – "covering X-ray observations with small hemispheres and large charges and full scale tests with steel spheres and hemispheres with large charges"; and, finally, "model experiments on blast effects" requiring "from 2 to 4 extra people, depending on the extent of the program."[12] At this point, the Los Alamos scientists could hardly have anticipated the difficulties Los Alamos would later face in actually building an implosion bomb.

Meanwhile, at the National Academy of Sciences in Washington D.C., von Neumann was exploring the effects of an implosion in different geometries and the means of imparting high velocity to a large mass by means of shaped charges. He pointed out, as Paul Fine reported in a memorandum to Tolman, that one could extrapolate from small-scale blast experiments "at the place where the limit of damage of a given kind is reached. There the effects are the same for large and small charges." He felt that spherical charges "had possibilities. At any rate, the ratio of explosive to mass of projectile was about right in those cases." Fine further reported: "I asked him about the plan that was once considered for having some computers work here. He said that he did not see the need for extensive calculations until a better understanding of the fundamentals is achieved."[13] Von Neumann also suggested a program of experiments on a "two to ten millionth mass scale," to be done at Woods Hole, Massachusetts, over a two-month period under E. Bright Wilson and continued his discussions on implosion with Teller in a series of letters.[14]

Rapid preparations for an expanded implosion program continued. On 19 October, a conference of the high-explosives group – including Oppenheimer, L. T. E. Thompson, Kenneth Bainbridge, Neddermeyer, Felix Bloch, Streib, Bradner, and Parsons – considered "the whole aspect of shaped charge work." The group decided to have small-scale tests carried out at Aberdeen "in the next few weeks," while Los Alamos was setting up for X-ray work.[15] An outline of the program prepared on 25 October, presumably by Neddermeyer, covered photographic, X-ray, and terminal observations, as well as theoretical investigations. Estimates of the number of staff members required in each category – for example, five for the X-ray observations ("1 XR, 1 HE chem, 1 XR T, and 2 T") and for the development of other methods of observation ("1 crackpot") – appear as a handwritten notation.[16]

At a meeting on 28 October, Oppenheimer summarized for the Governing Board "the history of the [implosion] method and the reasons for giving its investigation a high priority at this time." Commenting on the efforts of Neddermeyer's group, he pointed out:

> No experiments were done with very large charge to mass ratio ... Von Neumann was convinced, on the basis of his experience with shaped charges, that the regularity of collapse would improve as the ratio of explosive charge to metal mass went up. He also said that the time necessary for passing from critical to final assembly was close enough to the time necessary for the chain reaction to build up so that there is less danger in pre-detonation.[17]

The board also discussed implosion of both uranium and plutonium. Groves, who was present at this meeting, was particularly interested in the possibility of drastically reducing the quantity of uranium-235 (25) used in implosion, as opposed to gun assembly, "an extremely important consideration from the point of view of the project as a whole." The chemists were impressed with the idea of less stringently purified plutonium-239 (49). "Mr. Kennedy and Mr. Thomas emphasized the great difficulties presented by the purification problem."[18] As a test, they planned to prepare 90 percent enriched 25, and Bethe surmised that "in four to six weeks" the percentage of enrichment needed could be calculated. John Williams "said he thought an intelligent guess could be given on the basis of experimental data in two months, which could be refined in three months."[19] Groves asked Parsons to bring to the policy committee meeting, to be held on 3 November in Washington, "a written statement explaining the advantages and difficulties of the implosion method, and including an outline of the organizational set-up proposed and the personnel needed, if possible with names, and certainly with the qualifications desired."[20] That same day, Neddermeyer prepared a "proposed working schedule for starting the HE program as it was laid out on 25 October," which was to include designing structures for X-ray work, the casting of large charges, procuring shells and miscellaneous equipment, continuing the interim program on X-ray investigation, making photographic observations for engineering of the final models, and testing.[21]

A detailed industrial-like outline of the diagnostic program was in hand by 29 October: The tasks of Phase A, ending 20 November 1943, would be to implode "steel cylinders as organization and construction of future facilities permit," "design and construct X-ray house," and "get 3/4″ steel hemispheres for X-ray use and implode these in X-ray chamber in advance of mounting X-ray machine to determine fragment hazard and develop methods to reduce it"; those in Phase B, ending 30 December 1943, would be to "develop X-ray technique and apply it to small spheres, terminal observations on spheres at several scales and of materials of interest" to "continue imploding cylinders" to "conduct flight test of scale model of complete assembly" to "develop X-ray technique for large charges"; and Phase C, ending 29 February 1944, to "begin full-scale implosions of cylinders and spheres," to "refine implosion initiation on small scale," to "work on detonation control problems," and to "conduct implosion of reduced scale models of final assembly." March and April 1944 were left "for contingencies." The weight of explosive would be assessed experimentally. The weight of active material would be determined "as our knowledge of the physical constants and the theory of the implosion process improves."[22]

One week later, at the Governing Board meeting on 4 November 1943, Oppenheimer reported that "both Groves and Conant seemed very much in favor of pushing the implosion method ... the only one which offers some hope of justification for the electromagnetic method" (the understanding at the time being that the product of electromagnetic separation was not sufficiently enriched to use in the gun). At Groves's request, "a six-month time scale on the development of the implosion method was made up. It was felt, however, that this was a little optimistic." The minutes also report that "the X-ray equipment will be ready fairly soon." Some of the discussion focused on the implosion of uranium hydride, and in this context "Birch suggested calling on Percy Bridgman at Harvard to measure the pressure-density relations up to a hundred thousand atmospheres without involving anyone else in the work. He should be able to get an answer in a few weeks." Indeed, within a few weeks Bridgman's studies of hydride samples showed that the hydride could not be compressed easily.[23]

The intensified interest in implosion during the fall of 1943 also stimulated ideas for new diagnostic methods of studying the phenomenon. For example, the minutes of the 4 November Governing Board meeting report that "[Bernard] Waldman and [Donald] Kerst have turned in a report on the use of the betatron," but it was decided not to develop this diagnostic method at that time. Robert Serber looked into "the use of natural sources," a study that eventually led to the RaLa method (see diagnostics below).[24]

Thus the new "fast-implosion" project was adopted "after much struggle and argument ... with over-riding priority late in 1943."[25] David Hawkins summarized:

> There was an immediate transition from a situation in which it appeared that the problem of implosion could be solved by modest means into one for which there were not even any adequate experimental techniques. As a result the following year was characterized by a succession of new ideas and applications for the experimental procedure, and by a rapid expansion of facilities and personnel.[26]

Kistiakowsky Joins, and the Implosion Program Expands

In late October 1943, the implosion program was still, as the Governing Board minutes of 28 October describe, "being carried on by a group of eight men whose relations with the rest of the engineering division are rather loose," and who "have not yet become accustomed to the idea of large scale operation."[27] Oppenheimer stressed the need to expand the program with men having "practical experience in high explosive techniques," and, furthermore, "to place it under strong leadership." He mentioned in this context Kistiakowsky, who "has intimate knowledge and up-to-date experience in this field, general resourcefulness, and administrative responsibility."[28] By this time, "tentative arrangements had been made to retain Kistiakowsky as a permanent consultant to this work," but Oppenheimer had not yet tried to hire him for a full-time position.[29]

The negotiations on hiring Kistiakowsky at Los Alamos proved to be trying. Kistiakowsky was initially reluctant to come, "partly because I didn't think the bomb would be ready in time and I was interested in helping to win the war," and because he had "what looked like an awfully interesting overseas assignment all fixed up for myself."[30] On 1 November, Oppenheimer approached Conant to exert influence on Kistiakowsky, informing him of the recent decision to pursue the implosion with high priority and of the "reciprocal lack of confidence" between Parsons and Neddermeyer. "We have come to the conclusion that the only step which offers any real immediate promise is the assignment to this work of Kistiakowsky .... In fact, unless possibilities exist of which I have not heard for heading up and staffing the group, I should very seriously doubt whether the implosion method could be developed in time."[31]

By late November, Kistiakowsky was deeply involved in organizing the implosion program. On 26 November, he presented an extensive "preliminary" outline of the program to the Governing Board. This program centered initially on experimental studies of small cylindrical charges, employing flash X rays, flash and rotating-prism photography, and terminal observations. All would give "substantially the same type of information." Extensive experience in both the photographic and terminal observation methods had already been obtained at Bruceton; the X-ray method, as applied to studies of jets from the collapse of shaped charges, was described in documents by Tuck and by John Clark and Leslie Seely (see "The X-ray Method" below). The new fast-implosion program was thus following the principle of exploring multiple lines of inquiry, common to many areas of research during World War II. Kistiakowsky wrote, "It is impossible to predict which of these basic techniques will be the more successful."[32]

Kistiakowsky also set down here the basis of a second principle used extensively in all the Los Alamos programs: to develop extensive experience on small-scale models and then proceed to larger-scale studies. As he wrote in late November, "In the beginning all the three research groups will work on small charges for which the techniques are simpler and are better worked out, so that more shots can be made each day and unsatisfactory designs more rapidly eliminated." For example, some sixty X-ray shots a week were planned. Kistiakowsky predicted that in time

Fig. 8.2. Chemist George Kistiakowsky was head of the explosives division. LA Photo, LAT 726.

> all the three groups will shift to larger charges; the light flash observations with cylindrical charges will be terminated first, while the other two groups will collaborate for a time, until eventually the entire staff will direct its attention to terminal observations on the full scale charges, for which at present no other suitable techniques exist. It is deemed more desirable to start firing large charges only as the result of thorough experience with small ones. The time and effort involved in assembling, firing, and observing a large charge are so great that unless reasonable probability of success exists the shot is not warranted.[33]

Kistiakowsky made specific staffing suggestions for the implosion program: flash photography would be under Streib, "and X (Dr. [Walter] Koski?) is to replace him in January." Rotating prism camera studies would use "equipment designed by [Julian] Mack's group which will do the work"; X-ray study would be under Bradner, with Donald Mueller in charge of the X-ray machines; and terminal observations would be "under [Henry] Linschitz in case of smaller scale program," with a special project related to explosives for the purpose of "development of best possible techniques for preparation of reproducible charges of known and controlled properties and of methods of proper initiation of same. Maybe under [Morris] Patapoff." Kistiakowsky also proposed a data analysis project under Streib "to provide a centralized mechanism for correlation and interpretation of data obtained in the field by the operating crew."[34]

Eventually, Kistiakowsky agreed to become a full-time Los Alamos staff member, joining on 16 February 1944 as deputy to Parsons for implosion.[35] The experimental program of E-Division was then subdivided under two deputy division leaders: McMillan, who headed the gun program, and Kistiakowsky, who headed the implosion effort. In addition, a new group, E-9, was added under Bainbridge to study full-scale implosion assemblies and prepare for the Trinity test.[36]

Although the experimental and theoretical study of implosion was only beginning in this period, certain design features had to be fixed immediately so as not to bottleneck the delivery program. The approximate weight and maximum outer length and diameter were set once it was determined that the bomb would be placed in the B-29 bomb bay, which measured about 5 by 12 feet. The design of the case and tail of the bomb and its release mechanism could now proceed. In order to simplify scaling down charges for the diagnostic program, the pit volume was fixed, as was the size of the bomb and the amount of explosive. The estimated weight was 5 tons. In addition, a decision was made "to work towards an assembly of the charge from small segments cast in individual molds, rather than to attempt to cast the entire charge in the bomb case."[37]

Implosion Diagnostics

By January 1944, the diagnostic program outlined by Kistiakowsky was operating to determine the crucial implosion parameters of symmetry, collapse velocity, and compression. In exhaustive studies of test shots, every possible parameter was varied, including the number of detonation points, the kind and arrangement of high explosive, and the kind of material collapsed. Many of the progress reports describing the tests were pessimistic about the feasibility of an implosion weapon.[38] Little was learned before the following fall about compression, but significant data concerning both the symmetry and velocity were on hand by spring.

All three diagnostic programs immediately observed jets in the implosions of spherical or cylindrical shells – tongues or knives of molten material squirting ahead of the collapsing main shell envelope. These jets drastically upset the symmetry and caused undesirable turbulence in the center of the assembly. Another observed phenomenon that reduced efficiency was spallation, the breaking off of pieces of the imploding metal at its surface, caused by the reflection of the rapid detonation waves at the surface. A related problem was that the measured collapse velocity was smaller than predicted.

Kistiakowsky prepared an ambitious work schedule for March to September 1944 on the basis of X-ray studies, terminal observations, and flash photography to examine parameters such as the time scale and symmetry of the explosion. Rotating prism camera photography of cylinders would primarily "supply accurate measurements of the acceleration of the metal and the velocity of collapse in its relation to the charge/weight ratio." Under Kistiakowsky's plan, molds for charge segments would be "designed and orders for them placed, with emphasis first on reduced scale models." In an eleven-page revision of his initial document, he proposed expanding the explosives fabrication sites (in particular Anchor Ranch and Saw Mill Site), and added a design of the outer case and "large containing sphere" (later called "Jumbo"), and suggested casting charge segments.

In a spring 1944 document projecting work during the last quarter of 1944, Kistiakowsky summarized the conflicting optimism and pessimism of implosion researchers at that time: "During October ... work on all projects enumerated during the previous months will be continued. However, if active material is available, the implosion project staff, it is hoped, will be in a position to recommend a design of the gadget which will have a finite chance of properly functioning." For "November and December," he projected with tongue in cheek, "the test of the gadget failed. Project staff resumes frantic work. Kistiakowsky goes nuts and is locked up."

The X-ray Method

The "flash X-ray method" had been developed during the spring of 1943 to study shaped charges.[39] In this method, single X-ray photographs of a sequence of identical shots, each one taken at a later moment in the shot, produced a "movie" of the implosion, which indicated both the symmetry and compression.

The flash X-ray method was restricted to small-scale shots; to keep scattering of the X rays down, the explosive layer could not be made very large. To allow larger-scale tests, the Los Alamos scientists developed a second X-ray technique early in 1944, in which features of the implosion would be recorded by a bank of electronic Geiger counters oriented perpendicular to the propagating X-ray pulse. Attenuation of the X rays by the explosives was less significant in this method because the Geiger counters were considerably more sensitive than photographic emulsion. The laboratory would later erect a separate X-ray group to explore this "counter" method, but technical problems prevented its realization.

At the time X-ray studies began, the implosion group consisted of Neddermeyer, Bloch, Streib, Bradner, and Mueller, who had recently joined the project.[40] The last three worked on correlating the timing of the implosion with the X-ray discharge and on modifying the electronic control and timing of commercial X-ray machines (developed at Westinghouse). Meanwhile, Parsons was completing the arrangements for adding to the X-ray group the leading American X-ray physicist, Lyman Parratt of Cornell. Parratt visited Los Alamos on 29 November and arrived as a staff member on 21 December 1943. The next day, Trevor Cuykendall, Parratt's colleague at Cornell, joined the X-ray group.

In December and January, the group conducted preliminary studies on small spherical and cylindrical shells aimed at determining the asymmetry, state of compression, and time over which the system stayed compressed. Discrepancies of approximately 20 percent between their results and calculated values of the collapse time underscored the limitations of the available one-dimensional theory. One transparent problem was that the timing control of the X-ray machines, roughly ± 2 μsec, was inadequate, both in setting off the X-ray machine and in synchronizing the flash X rays with detonations. Achieving sufficiently precise timing would be the principal challenge between November 1943 and June 1944.

During February, X-ray shots of small spheres repeatedly revealed the presence of jets, which appeared to arise at the place the Primacord was attached. Neddermeyer immediately recognized the gravity of this problem, noting in a 1 March report: "Whether these jets can be eliminated by improving the symmetry of collapse is not yet known. If not, the jets may be the source of really serious difficulties."[41] Such jets would be seen again and again in similar experiments carried out during the spring and early summer months. The flash X-ray group hypothesized that they arose from the interaction of detonation waves and suggested numerous ways to fix the problem. All aimed at softening the interaction between detonation waves emerging from the individual initiation points by slowing down their transverse propagation.[42] Delaying techniques were also studied in the terminal observation program using lead plates and other materials. But whether any benefits at all were accruing from such tricks was unclear.

The correlated timing of the X-ray flashes with the shots achieved by late February was "good to an average deviation ... of about 1.6 microseconds ... better timing performance than has been recorded from Aberdeen." This was still far from adequate for implosion studies. Bainbridge's instrumentation group, E-2, worked to improve timing control in the X-ray apparatus, but the problem would not be solved until improved X-ray machines arrived in mid-June from Westinghouse.[43] During April and May, Sergeant Gerold Tenney of the SED, an Austrian immigrant Ph.D. with experience in medical radiology, added to the implosion tool kit a method of X-raying explosive charges to study their imperfections.

Like most Los Alamos research groups, the X-ray staff met frequently. Members of T-Division periodically reviewed the X-ray program; for example, Victor Weisskopf examined the difficulties having "to do with the absorption and scattering of the X-rays." An administrative change in late June would further the trend toward well-organized collaborative research in the X-ray program. Kenneth Greisen, a junior researcher in Williams's Van de Graaff group of P-Division, replaced Bradner as head of the Anchor Ranch X-ray research. Greisen would add to the X-ray program an expanded capacity for theoretical analysis.[44]

By this time, two significant technical improvements had been made. Timing control had been refined by installing the new Westinghouse X-ray apparatus, and symmetry had been improved by using more detonation points. Bradner, in his last month as group leader, felt that "pictures thus far obtained do not justify pessimism concerning the possibility of symmetrical collapse of spheres."

During July, Greisen worked on tracing the origin of "nonfundamental asymmetries." He suggested using explosive lenses (see Chapter 9, "Explosives") and electric detonators to improve the implosion parameters. And to improve diagnostic approaches, he suggested using pairs of X-ray machines operating in tandem, increasing the size of the closed chamber to accommodate larger charges, and combining the magnetic method with the X-ray method. Because Greisen was especially worried about the pattern of jets, the X-ray group made numerous attempts to explain the underlying mechanism by drawing on the Tuck–Taylor theory of conical-shaped charges. Zeroing in on the problem of poor detonation synchronization, owing to the use of Primacord, Greisen argued that electric detonation was a "most promising method of eliminating the error in time measurement." By March 1945 he would be one of the principal physicists in the detonator program.

Because many X-ray shots were fired in July and August, Greisen cautioned that "to maintain this firing rate, we must make few or no changes in the type of shots made; this is not in keeping with a research program." Two serious problems were the "difficult shop and procurement situation," and "the haste of the program."[45] Distressed by the "paucity of ideas and suggestions for really critical experiments," he invited more suggestions "from such persons as Neddermeyer and Kistiakowsky, or from the Theoretical Division." He further suggested that "it would be good if the 'audience' at the Monday morning meetings took a more active part in the meetings and accepted more of the responsibility for suggesting and planning experiments, instead of just hearing a report on conditions and plans as they are." According to Greisen, other factors limiting the X-ray studies were "the 1/2 lb charge size limit set by the X-ray machines ... the limit of resolution of structure, set by scattering of X rays, and the size of the focal spot, as well as by the large charge-to-film distance required for protection of the film." For such reasons, the discussions in July and August would turn to methods of superseding the X-ray program, for example, using a betatron to produce much higher energy X rays (see betatron method below.)

Photographic Methods

Two photographic techniques developed at Bruceton – flash photography and the rotating-drum or rotating-mirror method – figured large in the first year of the implosion program. In the "explosive flash method," a series of photographs were taken at particular times, by a simple "brownie" box-camera, armored for protection. The light source was a layer or "cap" of argon gas, which would be excited by a shock wave from an independent HE charge; the light passing through the open axis of the collapsing cylinder (typically about a foot in diameter and 18 inches long) allowed researchers to photograph the cylinder's inner surface.[46] Any jets formed at the inner surface would show up on the pictures. To prevent fogging of the film, the exposure had to be short compared with the duration of the flash; precision timing was achieved using a Primacord shutter, which closed immediately after the beginning of the argon flash. Although a hemisphere could be photographed in the same way (e.g., by reflecting the flash), a collapsing full sphere could not, because there was no hole for the light to pass through. Because only one photograph could be taken of each implosion shot, this diagnostic principally reported on symmetry. As in the X-ray method, however, by photographing a sequence of identical shots at different times, one could deduce the radius-versus-time curve of the imploding material.

The rotating photographic technique allowed scientists to produce a continuous photograph of the implosion. The method consisted of leaving the shutter open while the film advanced on a rotating drum, or the image moved along a length of fixed film by means of a rotating mirror or prism (the centrifuge prism camera). In both variants, the light emanating from an implosion entered the camera through a slit; the data to be analyzed was a streak of exposed film. The unusually high relative speed of the light moving along the film brought the resolving time into the range of a few tenths of a microsecond, making this technique particularly helpful in measuring implosion timing. However, because only slit images were photographed, the technique provided little information about the symmetry.[47]

Neddermeyer's group had been planning to photograph the implosion as early as May 1943, but for almost a year the available technology did not allow them to do so – the flash technique did not begin to operate until January 1944, and the rotating mirror technique was still problematic as late as June 1944.[48] The original rotating mirror camera employed a "pyramid" consisting of three mirrors on a rotor, with the film arranged in a cylindrical track. In the last months of 1943, Berlyn Brixner and Morris Patapoff tried to photograph imploding pipes using this camera, but because they were unable to synchronize the camera, the pictures were extremely poor. Brixner recalls Kistiakowsky being dismayed by the picture quality, insisting that the camera "had optical illusions."[49]

Kistiakowsky consequently directed Streib to try photographing the implosion using A. Wayne Campbell's explosive flash technique and at the same time test whether the mirror camera was indeed beset by optical illusions. Streib and Brixner worked together on this method during December and part of January 1944. Unfortunately, the flash method failed on the first few attempts. Brixner recalls that Streib and he "used to go out on the next mesa every morning, about 2 o'clock when everything was dead in the lab, and we'd just set up on the field. We went back to a shelter, fired it and couldn't get a picture .... I'd develop the film, it was blank. We'd try again and again." He also remembers Kistiakowsky's blunt response: "Well, neither camera is any good. They don't know how to take pictures in this place."[50]

On 19 January, Kistiakowsky turned the flash photography problem over to Walter Koski, who had joined the laboratory on 4 January.[51] Koski was put in Neddermeyer's group, where he was to work with Streib on obtaining a silhouette of an imploding cylinder by means of flash photography. Noticing that in Streib's arrangement the Primacord trailed on the ground, Koski hung the apparatus up using string, and immediately he and Brixner were taking successful pictures, measuring the collapse time, and learning about the symmetry. With the Primacord trailing, the shock wave velocity had increased and caused the shutter to close too early. Because it had a much higher resolution than the X-ray method, the flash technique showed the jetting clearly.[52]

At the same time (early 1944), extensive tests showed that the rotating drum camera was unsuited for implosion studies, because it could not rotate quickly enough and therefore limited the resolution to approximately 5 μs. Patapoff turned to the rotating mirror camera.[53] After numerous technical problems were resolved, Parsons was able to report on 1 February, "Actual velocities have been measured on imploding cylinders, using both the rotating prism camera and by successive, short argon shock-wave flashes." A series of eight shots of steel cylinders yielded collapse times of approximately 15 μs. Despite "uncertainties of timing and definition," he reported that the technique was improving.

The attitude toward the observed jetting grew increasingly pessimistic as winter turned to spring. In late February, although both photographic techniques gave clear evidence of the jets, members of the implosion group still believed that the jetting could be overcome "more or less completely by further refinement of the mode of initiation of the charge and of the design of the charge itself."[54] Attempts to eliminate the jets – for example, by using more detonation points and by placing an air gap between the high explosives and tamper – were still discouraging. Parsons concluded in mid-April that "early jet formation is a phenomenon whose cause and effect must be intensively investigated."[55] Koski carried out this program with determination and precision. In flash photographic studies of small cylinders (3-inch outer diameter and 1/2-inch wall, with a 0.92 mass ratio of Pentolite explosive to steel), he correlated the observed asymmetry "with the blast pattern which is caused by the meeting of two detonation waves starting from the points of initiation." By this means, he discovered that "the points of greatest compression in the metal are directly under the blast patterns."[56]

By late spring, it was obvious that the jetting problems could not be fixed easily. Neddermeyer wrote in May, "The belief stated [in the previous report] that the solution to the difficulties could be found by using a large number of detonation points does not appear to be justifiable." Parsons pessimistically reflected at this time: "Early success of the HE implosion project – which is defined as readiness to use the first available batch of active material as soon as it is delivered here – depends absolutely upon good breaks in the experimental programs. For example, if symmetry is not obtainable in a simply imploded sphere and if trick techiques must be resorted to, the gadget cannot be completed by the time usable quantities of 25 and 49 are available." In a similar vein, Kistiakowsky summarized in June, "The past month has brought a considerable clarification of the nature of the difficulties met in attempts to implode spheres and cylinders symmetrically and at high velocity by Group E-5, but the progress towards their elimination has been slight at best."

Terminal Observations

Terminal observations could be immediately implemented, because, unlike X-ray and optical photography, they did not depend on elaborate instrumentation. Indeed, Neddermeyer's original implosion group had used terminal observations from the beginning of its implosion research (Chapter 5). The shots were done in ruts dug in the ground and lined with steel walls and a cover to confine the fragments; afterward, these fragments were swept up for analysis. So that researchers could recover pieces large enough to afford sensible interpretation (ideally, intact imploded spheres) weak explosive charges had to be used. This meant they had to work under conditions far removed from those encountered in actual weapons. The jets usually did not form under such circumstances, because the imploded metal rarely reached the state of liquid flow. The fact that only the end state, rather than the range of intermediate states, was observed was a further disadvantage, in that no information could be uncovered about the roles played by individual factors.

The more extensive terminal observation program under Harry Linschitz and Walter Kauzmann expanded Neddermeyer's original effort. Explaining their method in February, Kauzmann and Linschitz favored three techniques: (1) simply looking at the recovered fragments; (2) noting any liners and cores attached to spheres; and (3) etching and metallographic study of the products, to explore the effect of implosion on the internal fiber or grain structure of the imploding material. The program focused on macroscopically visible structures, because microscopic structures were far more tedious to observe and were not considered more revealing.

By January, terminal observations had identified a number of factors that might be responsible for the asymmetry, and the group planned to study them using low-power explosives: a mixture of Pentolite, powdered cork, and nitroguanidine. (Robert Wilson recalls that during the war, the Los Alamos scientists in fact referred to almost every explosive as *pientolite* "after the Russian manner of pronouncement used by Kisty.")[57] Experiments would measure jet velocities by inserting plugs of material having different densities into openings made in the wall of the imploding metals; the plugs, which would propagate with the jet, served as markers of particular places in the imploding metal.

One observation reported in late February was that the sizes of implosion fragments "seem definitely to depend on the number of detonation points," and "the percentage of recovery of fragments seems to go down rapidly if shots are made in the open and if mass ratios are much beyond the critical." According to Linschitz and Kauzmann, jets "appeared to be of sufficient importance in affecting the uniformity of the implosion to warrant a detailed investigation." Linschitz presented the overall picture as seen from terminal observations at the end of March in a report that broke down the main forms of asymmetry into three categories: "jets," "folding," and "shear." He hypothesized that at the root of the problem was "the need for a finite number of initiation points, and the finite time required to complete the detonation." As in the other diagnostic programs, various tricks were tried to increase the symmetry, such as placing wooden blocks at places where detonation waves met.[58] By this time, the main emphasis of the terminal observation program had turned to the "use of thick charges of tamped explosive, with TNT or tetryl ... having an explosive density from one-half to two-thirds the normal density of cast Pentolite or Composition B."

The analysis was clearly suggesting that many more detonation points should be used. The principal results of increasing the number of detonation points were that "the degree of collapse increases regularly," and "the regions of interaction of detonation waves on the surface of the sphere widen out very appreciably and show pronounced wide bands over which the shell is uniformly 'etched.'" This pattern suggested "spreading out of the high pressure region at the intersection of the detonation waves." Patterns observed on the outside could be used to check the uniformity and timing of the initiators.[59]

RaLa Method

On 1 November 1943, Serber conceived of a novel procedure for diagnosing implosion based on placing a γ-ray source at the center of a spherical implosion assembly.[60] The emitted γ rays would travel outward radially, through both the collapsing shell and the high explosive. Because increasing compression of the metal caused the γ rays to be increasingly absorbed, the emerging γ rays, monitored by detectors set around the high explosive, would provide information on density changes in the collapsing sphere of metal. The data would indicate the time of collapse, the degree of compression, and the symmetry, by comparing the γ intensity in different directions. Whereas the X-ray and terminal observation methods concentrated primarily on the outside diameter of the collapsing core, this RaLa method (so called, because radiolanthanum would be used as the γ-ray source) would report on the implosion parameters throughout the assembly. Furthermore, unlike the X-ray and flash photographic methods, which took single snapshots of an implosion, Serber's method would yield a continuous record of the progression of the implosion. Taking into account absorption in uranium, Serber estimated the necessary γ source strength to be of the order of 100 curies.[61]

Radiolanthanum-140 (^{140}La), an isotope having a forty hour half-life and strong γ emission at about 2 MeV, was soon found to be a suitable source for the new diagnostic. In principle, large amounts of lanthanum could be obtained from the Clinton reactor at Oak Ridge, because it was made by the beta decay of radiobarium-140, a 12.5-day half-life element which formed plentifully (ca. 6 percent) as a fission product in plutonium production. The radioactive barium–lanthanum pair had been the basis, five years earlier, of the historic discovery of nuclear fission. As it turned out, procuring the needed quantities would be a major enterprise involving many staff members with diverse expertise at both Oak Ridge and Los Alamos.

Serber discussed the γ-ray idea with Bruno Rossi, then cohead with Hans Staub of P-Division's instrumentation group, P-5. Together with James Allen, Rossi had recently increased the sensitivity of large fast linear ionization chambers to ionizing radiation.[62] He immediately recognized that the improved ionization chambers would be ideal detectors for RaLa experiments. The variation of the average ionization current could be recorded by fast linear amplifiers feeding into oscilloscopes. To achieve adequate time resolution in the chambers, the necessary source strength had to be upgraded from Serber's estimate. By 1 December 1943, Rossi was immersed in designing the necessary instrumentation.[63]

Plans for preparing radiolanthanum for this new diagnostic were well advanced at Clinton by late March. Open questions included the size and shape of the external handling container and the method for packing and shipping the source. The original plan was to use the barium parent and lanthanum daughter as a combined source, without separation, even though only the lanthanum provided the needed γ radiation. In a letter to Richard L. Doan, Clinton's director of research, Charles Coryell and Henry A. Levy of the fission products group at Clinton discussed separating the barium from the fission products of the Clinton reactor to produce the lanthanum sources required for the Los Alamos experiments:

> The main job is a preparation on a radioactive scale never before handled, but the chemical and shielding problems are not insurmountable .... We propose to isolate the 12.5-day ^{140}Ba from pile irradiated metal nearly at saturation with respect to this chain. The ^{140}La will grow in the barium sample in 5.5 days to a maximum activity 0.72 times that of the initial barium activity, and eventually the La 140 will decay with the 12.5-day half-life of the barium parent.[64]

In the interim at Los Alamos, Parratt systematically investigated the RaLa method, considering whether radiolanthanum was indeed the correct choice for the source. At a meeting on 8 April 1944, he estimated that the RaLa method using this material would provide fairly accurate information about the density of a tuballoy liner, in contrast to the other diagnostics then in use. Oppenheimer therefore urged those attending the meeting to "request X (Clinton at Oak Ridge) to provide us with 100 cu [curies] of the active material at regular intervals, starting around the middle of July."[65] He had already taken steps to procure this material, confirming in a teletype to Compton on 7 April, "our request for radio barium radio lanthanum since detailed discussion has indicated very great promise in application of this material." Oppenheimer told Compton they would "like to have first delivery of about one hundred cu not before July fifteenth or later than August first." In the future, they "should like subsequent deliveries at three week intervals as suggested by

Coryell's memorandum to Doan although precise timing not important." For use in preliminary studies of RaLa experiments, he also requested smaller "deliveries of order of one curie at early date," adding that "it would be helpful if radiations from these samples could be similar to those from one hundred curie lots." Finally, he informed Compton that the staff at Los Alamos would "fabricate the container ourselves and shall arrange conferences between X and Y representatives for discussion of procedures and handling."[66] Compton promptly responded on 8 April, "We see no reason now why the material cannot be supplied as you request."[67] Within one week, Doan was promising Oppenheimer one curie of the source material by 1 May and 100 "sometime in July," informing him also that "Mr. H. S. Brown is X contact man on job."[68]

To fulfill the Los Alamos request for radiolanthanum, Clinton would have to construct a special extraction (hot) laboratory and a second plant for dissolving radioactive irradiated uranium slugs and recovering barium from the solution. The material would have to be shielded in lead during its transport across 1,200 miles to Los Alamos, in special trucks driven twenty-four hours a day.

In this period, chemists were familiar with working with curie quantities of material, but to work with 100 curies concentrated in a "point source" was unprecedented. Los Alamos chemist Rod Spence remarked recently: "No one ever worked with radiation levels like these before, ever, anywhere in the world. Even radium people normally deal with fractions of grams, fractions of a curie." Making such quantities into a point source had its own difficulties. And the means of handling the sources were both primitive and risky; to protect themselves from radiation, the chemists relied essentially on being far enough away from the source. By today's standards, the exposures they received would be judged unacceptable.

On 10 April 1944, Parratt presented a detailed memorandum on the RaLa experiment, describing the 2-MeV γ-ray source, the fast-ionization chambers, and the recording system. Focusing on the radiation hazard, he cautioned that, owing to scattering of the radioactivity, "the site of each shot will be poisoned for an area of some 3,000 square meters," an effect that "would prevent a man's walking or running on the site, for about six months." He specified that "a site for each shot must be selected which is far removed from desired habitations of plants, animals or humans, so far removed that winds cannot carry the material in dangerous amounts." Nevertheless, he concluded, "the experiment, for all its faults, appears to be the best one yet proposed (which does not use

25 or 49) to determine the final density and the duration of the period of constriction." However, when the radiobarium was removed from the source prior to the actual firings in the fall of 1944 (Chapter 14), the radioactive contamination at the firing site following RaLa shots was in fact insignificant.[69]

The RaLa experimental program would be a prime example of multidisciplinary research at the Los Alamos laboratory. Various groups had to pool their expertise: metallurgists and machinists, who furnished the metal spheres; chemists, who provided the lanthanum; theorists, who helped with planning (e.g., recommending material, shell dimensions, type of explosive) and calculating γ-ray attenuation; electronics engineers, who designed and built the equipment; and explosives specialists, who provided the high explosive. Logistics were crucial. To manage them, Oppenheimer appointed Luis Alvarez head of the RaLa program.[70]

At a meeting on 15 April, Alvarez outlined items that had to be made or dealt with for the experiments, for example, the ionization chambers, the amplifying and recording equipment, the equipment and methods for handling active sources, and the nature of the site, shelters, and tanks.[71] Staub and Rossi were to build the chambers and electronics. Parratt and Weisskopf, who became the theoretical associate for the RaLa program, examined the scale.[72] Kistiakowsky informed Oppenheimer that Rossi and Staub "will need, during the development phase of this work, a steady supply of Ra-Ba-La, amounting to about four to five cu."[73] On 26 April, a RaLa committee under Alvarez, consisting of Bethe, Kistiakowsky, Neddermeyer, Oppenheimer, Parratt, Rossi, Emilio Segrè, Staub, Teller, Richard Dodson, Gerhardt Friedlander, Lindsay Helmholtz, David Nicodemus, and Weisskopf, formed to coordinate the experiments.[74] To protect the experiments from shock, and the workers and the site from radiation contamination, Alvarez proposed, one week later, that the work be done in a "mobile laboratory" made up of several M-4 army tanks, which he had located at Dugway Proving Ground in Utah.[75]

By late spring, Rossi and Staub had completed most of the instrumentation. They planned to surround the HE with several sets of ionization chambers, typically four; just before being destroyed by the explosion, the chambers would register the signals and send them through amplifiers to oscilloscopes set in a bomb-proof shelter. Two calibration circuits, one used before firing and the other after, would be placed in an underground "firing pit," located some 15 feet from the firing point. The chambers were simple enough in design to be produced quickly and in necessary quantities, even by the already overburdened machine shops.

Plans for the RaLa experiments further crystallized during May and June. On 3 May, the Clinton contact Brown asked Alvarez for a design of the RaLa shipping container, "so that we can proceed with the design of the shielding and at the same time assure ourselves as to its feasibility from the point of view of the chemical process anticipated." He suggested that they meet together soon in Chicago.[76] Alvarez waited a week to respond, for he was just then devising an important modification in the RaLa method.

Alvarez's new idea was to separate the lanthanum daughter from its barium parent and use only lanthanum in the test shots. One advantage was that more shots could be performed using the lanthanum obtained by successive "milking" of the longer-lived barium "cow" parent. Extracting the lanthanum from a combination of several aged sources would prove important early in 1945, when RaLa material would be in short supply. A second advantage was that the site contamination resulting from the dispersal of the 40-hour lanthanum would be far less objectionable than contamination resulting from the 12.5-day parent. A third advantage, not recognized at that time, was that the long-lived strontium-90, a fission product accompanying the barium, would be removed, but not the lanthanum. This strontium would have led to dangerous, long-term contamination.

Members of the Governing Board gave Alvarez's lanthanum milking idea a positive reception at their meeting on 4 May. To avoid the construction of a new shielded building, they proposed performing the experiments at a sufficient distance from all personnel using automatic equipment. Later that day, CM-Division head Joseph Kennedy recorded his concern about shielding the RaLa radioactivity in his diary: "The beautiful hot labs at X [Clinton] were inadequately shielded against 100 curies. We could not easily build here an adequate hot lab. However, I felt that semiautomatic equipment (plus distance for protection) could be used. I discussed this with Dodson, and we evolved a probably workable scheme." To provide the necessary distance, Alvarez and Maj. W. A. Stevens selected, on 22 May, a remote steep-walled site for the experiments in Bayo Canyon, about two miles from the townsite.[77]

Alvarez's suggestion to separate out the lanthanum also implied that the size and shape requirements placed on the Oak Ridge product could be relaxed, because a much smaller amount of total material would be used. On 11 May, he responded to Brown: "We are now investigating the separation of La from Ba. If this can be done at site Y, our problem will be much easier .... The first curie of La will be welcomed as soon as it arrives, which we hope will be soon. Size and shape here are not important but we hope it can be in a small container with a 3-foot string attached for handling."[78]

On 12 June, Alvarez assessed the preparations for the RaLa experiments and concluded that there were "no obvious bottlenecks which will prevent the first delivery of Radio Lanthanum from being used in an implosion gadget." He also mentioned that "the equipment, except for that which is expendable, will be installed in 2 M-4 tanks at a distance of 150 ft." He described the tanks, which were inspected at Salt Lake City, as "quite roomy if all equipment is removed." The tanks were expected to arrive on 1 July. He also mentioned who would be handling the various jobs: Helmholtz and Friedlander in Dodson's group would take care of the chemical work; Rossi and Staub would design the electronics and build the prototype; Rossi's group would build chambers and preamplifiers, "which are expendable"; and Mack would supply cameras to record the traces. He also tried to respond in this memorandum to "the asymmetry difficulties predicted by Mr. Taylor" by outlining a possible "new type of Ra-La gadget" that would avoid the asymmetry-producing situation. The British physicist, G. I. Taylor, on his visit to the Laboratory in May, had pointed out that a light layer impinging on a heavy one would lead to instabilities (see "Theoretical Studies of Implosion" below). However, Alvarez noted, "the consensus was that nothing further should be done [about the instability] at the moment on this suggestion."

During June 1944, chemists Helmholtz, Friedlander, and Watkins in Dodson's group, C-4, worked out a chemical procedure for separating 100 curies of ^{140}La from $BaCl_2$. They proudly reported, "The completeness of precipitation of the La is 97 to 98%, the net recovery on the filter about 90%, when the operations are carried out by remote control." They planned to place the lanthanum precipitate in a small cone-shaped source tip and send it to the firing site in a heavy box designed by Alvarez. There, a technician would extract it by remote handling and, using a 10-foot-long rod, "like a fishing pole," place it into a hole in the center of the imploding system, and afterward insert a closure plug with the rod. The geometry was chosen so that the inserted source tip and plug would complete the sphere of explosive and metal.[79]

Mock-up tests of the anticipated lanthanum–barium separation went smoothly. However, the first small source shipment from Clinton, which arrived at Los Alamos early in June, was found to contain iron and other impurities.[80] These impurities – which likely originated from the irradiation of the stainless steel lining of the shipping container – made the separation of the lanthanum from the barium, by means of phosphate precipitation, slow and inefficient, because they caused the precipitated lanthanum phosphate to form a gel.[81] The chemists had to stand ready to perform the separation immediately after the material arrived at Los Alamos. Spence remarked ruefully that the shots carefully scheduled for the afternoon were nearly always performed in the wee hours of the morning.

On 25 July, Alvarez's group E-11 fired a preliminary RaLa test shot in Bayo Canyon, to test the "technique of assembling the charge and detonating it in a sort of dress rehearsal," and of measuring collapse times and velocities. Alvarez recalled recently that "the separation equipment was on the rear right-hand side of an army truck, so that after the separation was done it could be driven away."[82]

The main RaLa shipment, promised originally for 15 August, did not arrive until mid-September, delaying the RaLa program by approximately a month.[83] The first RaLa shot was fired on 22 September.

The Betatron, Magnetic, and Electric Pin Methods

The Los Alamos scientists considered various other implosion diagnostic methods between November 1943 and July 1944. Three were eventually realized: the betatron, magnetic, and electric pin methods.[84] The betatron method, a diagnostic similar to the X-ray method, but employing 20-MeV γ rays rather than several hundred keV X rays, was brought up in November 1943. Donald Kerst, the inventor of the betatron, pointed out that using γ rays made in a betatron, rather than X rays, would make it possible to study implosion on a larger scale, perhaps even full-scale, because the scattering and absorption loss in the high explosive would be much lower for γ rays than X rays. However, Parratt and others who analyzed the suggestion with the idea of using counters or ion chambers for detecting the γ rays pointed out that, because of low intensity, "the observation would at best be exceedingly difficult and that it was better to use the 200 kV X ray on a smaller scale gadget." The idea of using a cloud chamber as the detector was raised but dropped, because shielding the expensive chamber from the explosion appeared too difficult. In deference to this skepticism, work on the betatron method was tabled.[85]

Seven months later, however, the betatron method was seriously reexamined in discussions following the crisis over spontaneous fission. Kerst recalls that at a seminar on the implosion program, probably held early in June, Neddermeyer strongly urged the researchers to improve their diagnostics. "We kicked around these various things and walked away from it in groups to our offices ... and we tried to figure out what there was to the various suggestions that we all had in mind." Kerst then revived his suggested betatron method.[86]

The feasibility and possible advantages of the method were debated on 28 June 1944, at a meeting in Oppenheimer's office. Rossi wondered whether the shadow cast by the sphere would be sufficiently sharp. Oppenheimer concluded that this shadow would be so fuzzy as to require the use of "many counters, obtaining a statistical interpretation from the fraction of them which are discharged." Rossi proposed using ionization chambers, but Parratt noted that a suitable resolution could not be achieved without severely limiting the size of the chamber. Oppenheimer listed other possible difficulties with the design. Using cloud chambers instead of counters was considered feasible "if the detectors can be protected from destruction." The discussion was not conclusive. Tuck thought that it might be more productive to spend the time it would take to develop the betatron method on "extending the X-ray method."[87] Thus, by early July the betatron method was again under active discussion, but no commitment to developing it had been made.

In the same period that the RaLa and betatron methods were conceived, Joseph Fowler, a "ruddy physicist with a Southern accent" working in group E-2, proposed an electromagnetic implosion diagnostic.[88] The basic idea was to place the implosion assembly in a static magnetic field, and, by means of a magnetic pickup coil placed roughly a meter away from the imploding sphere, detect the mechanical motion of the sphere.[89] The induced voltage in the pickup coil (on the order of 10 mV), amplified and displayed on an oscilloscope as it changed in time, would indicate the collapse velocity of the outer surface of the shell. The same method could also indicate the changing magnetic dipole moment induced in a magnetized collapsing shell. This diagnostic showed promise of working on full-scale assemblies and indeed would be the only one to be tried at full scale.

Fowler and his co-workers, especially Roger White, began to plan for the magnetic method during December 1943. The first shot took place on 4 January 1944. The main difficulty was in adjusting the oscilloscope to display the interesting signal; subsequent work focused on constructing a proper amplifier and purchasing a synchroscope from MIT that had a long persistent image on the screen. Fowler was optimistic that the method would soon "get useful records of the implosions."

Through the remainder of the winter and subsequent spring, Fowler's team continued to improve the instrumentation, with help from William Elmore in P-Division on fast amplifiers. New equipment included a 35-mm camera for photographing the oscilloscope traces.[90] By June, they were reporting that "implosions of cylinders and spheres, both magnetized steel and copper, have given reproducible scope traces indicating promise in this method of investigating implosions .... The times obtained for the collapse of the metal agree well with measurements by other methods." Kistiakowsky wrote in mid-July:

> The developmental phase of this method is substantially completed and the results are most encouraging .... Using copper spheres in relatively weak fields of some 10 gauss it is possible to obtain the complete time-radius curve of the collapsing sphere, to measure fairly accurately the velocity of collapse of its outer surface and to determine the final radius with some degree of accuracy.

He also pointed out that the magnetic method was "admirably suited to be used in conjunction with other tests, which (as the flash X rays) provide good information on the symmetry at a particular instant of the process, but are very laborious when used to determine all stages of collapse." They planned to install magnetic method equipment in large-scale X-ray and RaLa tests. At the 13 July 1944 meeting of the Technical Board, "it was agreed that this magnetic technique was very promising and should be vigorously pushed."[91] An electric diagnostic known as the "pin method" grew out of the suggestions of several researchers, including Otto Frisch, Darol Froman, Peierls, Ernest Titterton, Philip Moon, and Alvin Graves. In this technique metal pins were erected in the space near the imploding object and connected to circuits. As the implosion proceeded, these pins were struck and an oscilloscope displayed accurately timed current bursts. Erecting a series of pins at the same distance from the imploding shell, one could determine the symmetry.[92] The pin method ultimately gave the most accurate implosion timing information and was particularly useful in studying jets.

Although the available documents do not precisely indicate when the pin method was first suggested at Los Alamos, the concept was definitely under discussion before the end of June.[93]

Theoretical Studies of Implosion

Extensive theoretical studies supported the experimental implosion program. As mentioned in Chapter 5, Neddermeyer's original E-5 group developed primitive one-dimensional theoretical implosion models during the summer of 1943. After von Neumann's visit in late September 1943, both he and Teller worked on developing the implosion theory - von Neumann, working in his National Academy of Sciences office on the East Coast, discussed the research at length with Teller during visits to Los Alamos.

Von Neumann's correspondence with both Teller and Oppenheimer in the last months of 1943 provides insight into the early theoretical work on implosion that preceded the more detailed studies undertaken by Los Alamos T-Division after March 1944. In a letter to Oppenheimer on 30 December 1943, von Neumann elaborated on his "struggling with 'Phase B' - i.e. the pure high-density, incompressible phase." He hoped "to get the story at least as far as the disappearance of the central cavity in a few days," along with "a better estimate of the continuation in the collapsed and the rebound - or pre-rebound - stage." In the same envelope, von Neumann included two other letters that would be directly relevant to subsequent pre-Trinity work on Jumbo. One, to Teller, contained von Neumann's suggestion of a test implosion within a closed container:

> It may be desirable to produce at some future time "fizzles" with slightly "too small" quantities of active material ... this may be a method to test a great part of the hydrodynamics, compression, and nuclear efficiency under much more realistic conditions than anything else - except the full-size gadget. Such an arrangement would be of such a low nuclear efficiency as to be a nuclear explosion which is negligible compared to the imploding HE .... If this could be confined within a box of reasonable dimensions and thickness, *which would not break*, then the active material would be recovered by washing the inner surface of the box. I believe now, that a box of the diameter of 10 ft, made of 11 in. armour plate would do the trick - possibly even a thinner one. The whole operation would seem very worth while.

By early January, Los Alamos saw the theory of implosion as a problem sufficiently important to be undertaken systematically by theoreticians on the laboratory's permanent staff. Thus, on 11 January, the Theoretical Division set up a small implosion group under Teller's direction.[94] The division was then organized as follows:

> 1. Implosion - Teller and his group (Konopinski, Metropolis and Roberg), 2. Gun - Serber and his group (Frankel and Nelson,

> also Richman if needed), 3. Hydride – Feynman with Ashkin and Ehrlich, 4. Scattering Experiments and Detectors – Weisskopf with Olum and Richman if needed, 5. Effects – Christy, 6. Water Boiler – Christy and Serber, 7. Radiation – Weisskopf, and 8 Super – Teller and his group.[95]

Teller took action immediately. Three days after the implosion group was set up, he sent Kistiakowsky ten detailed suggestions for cylindrical test shots: (1) use long enough cylinders so as to make end effects negligible in the interior; (2) work with materials that on collapse retain "well defined and smooth" surfaces; (3) measure accurately the position versus time of the inner as well as outer cylinder radius; (4) select a small ratio (such as three) between the initial wall thickness and inner radius; (5) in the interest of drawing conclusions "about the acceleration due to the forces to which the outer wall of the cylinder is subject at various times," conduct "experiments on the time dependence of the pressure of the exploded gases on a metal plate"; (6) make observations on the expected deviation from cylindrical symmetry in the "short time before the inner radius becomes 0"; (7) explore various ratios between the explosive mass and cylinder mass, stressing that "the main interest lies in the region where the mass of the high explosive is greater, preferably more than twice as great as the mass of the cylinder"; (8) investigate various cylinder materials, varying both the density of the cylinder and its compression; in addition to iron, uranium, and aluminum, consider lead, as well as "one solid which is particularly highly compressible"; (9) look at "cylinders composed of two concentric layers of different materials"; and (10) consult frequently with Emil Konopinski and Nicholas Metropolis, who have begun "calculations on the behavior of cylinders during implosion," to inform them "about the dimensions and materials of the cylinders for which calculations should be made, and also for the purpose of fast evaluation of the experimental results."

From a theoretical point of view, fast implosion seemed more efficient than gun assembly, but the details were yet to be developed. Teller's group focused first on calculating the time of assembly for large amounts of high explosive. Initial calculations showed that one could assemble a device in times not much longer than nuclear chain reaction times, ruling out the threat of predetonation and making real the possibility of initiation by using the neutrons arising from impurities. The pressures in fast implosions were calculated to be in the range of millions of atmospheres; these would compress the material, decreasing the mean free path of the neutrons and therefore the critical mass. The initial experimental work on the compressibility of various fissionable materials, subcontracted to Bridgman at Harvard, supplied the data needed to calculate theoretical equations of state. In order to extrapolate equations of state to very high densities, Metropolis and Teller introduced the simplification of describing the electron distribution in the matter by the Thomas–Fermi approximation, well known from atomic theory. Later, with Richard Feynman, they improved the calculations to include relativistic effects; the Feynman–Metropolis–Teller equation of state for uranium and plutonium was subsequently used in all hydrodynamic calculations.[96]

However, a major problem was that the partial-differential hydrodynamic equations employing realistic equations of state applicable to high temperatures and pressures were insoluble by hand computation. This difficulty stimulated attempts to devise simplified equation-of-state models. Von Neumann suggested a "multiphase model" in which the system is divided into various phases in which the state changes discontinuously from phase to phase.[97] However, as Bethe commented in February, the theorists encountered "unexpectedly great difficulties" using this model. He also mentioned that some ten machines were on order from International Business Machines (IBM) to help in calculating critical masses of odd-shaped bodies. Following a visit by Peierls that month, T-Division began using these punch-card machines on a large scale to solve the implosion hydrodynamic equations.[98]

Peierls, who was then working in New York City on the diffusion method of separating uranium isotopes, visited Los Alamos from 8 to 10 February 1944. The expressed purpose of the visit was to consult with Oppenheimer, Bethe, and Chadwick on "what parts of the work now being carried out in Britain should be continued." Its main result was an agreement that "British analytical work on critical masses, multiplication rates and efficiencies could best be discontinued," and "that their work on the optical analogy [explosive lenses] ... might profitably be continued and should preferably be transferred to this laboratory [Los Alamos]." They also agreed that the detailed numerical calculations on the blast wave now in progress in Britain should be continued for some time as a "reliable check."

Peierls's visit had two fortunate implications for the implosion program. First, in discussing with Bethe how to solve the implosion differential equations and upon learning about the expected use of the IBM machines, Peierls pointed out that a step-by-step method of solving differential equations that he had developed earlier for blast wave calculations in air could be applied using the new machines. The Los

Alamos implosion equations were of the same form as the blast wave equations.[99] Bethe and Oppenheimer recognized the potential value of this suggestion. Oppenheimer informed Groves on 14 February that "the methods used by the British for integrating the blast wave equations in all their complexity are applicable to the physically different but formally similar problem of the hydrodynamics of implosion . . . [We] are planning to attack the implosion problem along these lines with the highest possible urgency."[100] By late February, T-Division was calculating the initial conditions for numerical integration of the implosion differential equations using Peierls's precedent on the IBM machines.

The second important implication of Peierls's visit in February 1944 would be realized five months later, when Bethe replaced Teller with Peierls as head of the theoretical implosion group. During March 1944, Bethe reorganized T-Division to meet "the great and increased urgency of the implosion program." He himself assumed "special responsibility for the work on implosion," although nominally Teller remained head of the implosion group. As Bethe reported in March, "a large part of the efforts of the entire division" was now aimed at "preparation of the numerical integration of the implosion hydrodynamics on the IBM machines."

Bethe also acknowledged receipt of the IBM machines on 4 April, mentioning that, with the help of Peierls, who was not yet a Los Alamos staff member, the machines had been put to use in implosion calculations. Stanley Frankel and Eldred Nelson were assigned responsibility for the IBM calculations. To check the program, Metropolis and Feynman made parallel calculations using hand-operated Marchant machines, staffed by a group of women who were part of the work force of the laboratory. Like the components of a computer, each carried out a particular step. Feynman later explained: "We worked out all the numerical steps that the machines were supposed to do – multiply this, and then do this, and subtract that." He recalled, "*She* was the multiplier, and *she* was the adder, and this one cubed, and we had index cards, and all she did was cube this number and send it to the next one. We went through our cycle this way until we got all the bugs out." This human computer actually developed speed – the same as that predicted for the IBM machines. But as Feynman noted, "the IBM machines didn't get tired and could work three shifts. But the girls got tired after a while."[101]

By the end of April, Bethe could report optimistically:

> The calculations had progressed almost to complete collapse of the sphere .... The plans for treating the shock wave contemplate the

> use of two different methods; namely, 1) von Neumann's mehod of thermal agitation and 2) Peierls' method of treating the shock wave classically. The latter procedure is now being worked out by Metropolis and is expected to be no more complicated than von Neumann's except insofar as it will require some hand calculations going parallel to the machine calculations.[102]

The calculations were beginning to reveal new quantitative and qualitative details of how implosion would occur. One month later, work on "IBM problem 1," "the first quantitative results on implosion," were described in Bethe's monthly progress report as "very satisfactory as regards density and duration of high densities." These "very favorable" results, discussed in greater detail in June, represented "the first numerical integration of the hydrodynamics of implosion." Bethe also discussed in this later report the efficiency of metal and hydride gadgets and the possibility of increasing the efficiency using a "modulated source which will become active only at the time of collapse ... (or) a weak source which will give a certain probability to detonate only after complete collapse." He cautioned, "It should of course be borne in mind that these results are only valid if the high initial velocity can be reached and if at the same time the implosion is symmetric."

A grim British judgment fell on the implosion program in late May. During a visit to Los Alamos on 24 May, the eminent physicist Geoffrey I. Taylor questioned the stability of the interface between core and tamper in the implosion assembly. Whereas the interface between a light material and a heavy one is generally stable if the heavy material is accelerated into the lighter, in the opposite case in which the lighter is accelerated into the heavier, he observed, the interface becomes unstable, much as fresh paint drips from a ceiling. Thus, although an ordinary untamped implosion might be stable, one in which a heavy core is pushed by a tamper having smaller density would not be. Morever, as Bethe noted, "the expansion of a gadget in the nuclear explosion should in general be unstable ... (and) have considerable influence on the efficiency obtainable from a gadget which will almost certainly be lowered by this effect." Peierls recalls that Taylor's assessment of stability "worried us right to the end."

As a result, Taylor forced the Los Alamos theorists to become more cautious in designing the implosion system. The tamper sphere was redesigned in the summer of 1944.[103] In September 1944, Christy carried this approach to the limit in proposing a solid configuration for the metal components of the implosion system.

In the spring of 1944, administrative difficulties compounded the problems facing T-Division. To the great annoyance of both Bethe and Oppenheimer, Teller, head of the implosion theory group, was working less and less on implosion and increasingly on the thermonuclear bomb, the Super. Both Bethe and Oppenheimer had given the Super low priority, because there was no chance of developing this weapon for use in World War II. Meanwhile, implosion was rapidly becoming a first priority, particularly with early news of the high rate of spontaneous fission of plutonium observed by Segrè's group (Chapter 12). At the exasperated Bethe's request, Oppenheimer in June moved Teller out of T-Division and made him head of an independent group principally aimed at the Super (Chapter 10). In Teller's place, Bethe put Peierls, whom Bethe had known since the late 1920s, when they studied theoretical physics together under Arnold Sommerfeld in Munich. Peierls joined Los Alamos on 3 June as a member of the British Mission.[104]

At this dramatic moment in the implosion program, the laboratory had to face the crisis stemming from spontaneous fission in reactor-produced plutonium. The outlook on implosion was extremely gloomy. Although Bethe could report that "the case of a perfectly symmetrical implosion is now fully understood," the evolving diagnostics, combined with Taylor's pessimistic considerations, indicated that this idealized situation would be difficult, perhaps even impossible, to achieve.

9

New Hopes for the Implosion Weapon: September 1943 to July 1944

Two developments in the spring of 1944 offered hope for overcoming the asymmetry problems of the implosion weapon. The asymmetry arose in part because the detonation waves diverging from the various initiation points met and interacted to produce small regions of markedly increased pressure. Furthermore, the multiple detonations of the surrounding explosive were not adequately simultaneous. The first problem would be dealt with by the three-dimensional explosive lens, suggested by James Tuck in May and given its basic design by John von Neumann. The second problem could be solved, as Luis Alvarez had suggested in May, by replacing the original inherently variable Primacord detonation distribution systems with electric detonators of highly superior reproducibility, thus providing a means of achieving adequately simultaneous detonation at several points. Developing and producing practical explosive lens and electric detonator systems would require a concerted research and development effort right up to the Trinity test in July 1945.

In view of these difficulties, it seemed wise to try testing the device. The decision to do so was made early in 1944. By the fall of that year, the site selection committee had fixed on the Jornada del Muerto region of south central New Mexico.

* This chapter is based on a draft by Lillian Hoddeson, to which Les Redman contributed a piece on explosives and Paul Henriksen the section on the decision to test the implosion device. We are grateful to Gordon Baym and Redman for their detailed editing of this chapter.

Explosives

The research and development of high explosives – materials that detonate at supersonic speeds by a process involving chemical reaction and a shock wave – was (arguably) the most pivotal and problematic component of the implosion program. At the start of World War II, the study of high explosives was still in its early stages. The wartime program would contribute to it substantially.

The history of "low" explosives that deflagrate or burn extremely rapidly, such as black powder, dates back to at least A.D. 1200. Documents from that period reveal that fireworks and black powder were known to the Chinese and Arabs. Roger Bacon recorded a formula for black powder in A.D. 1249, which was probably based on Arabic sources. The discovery around A.D. 1300 in both China and Europe that black powder could do mechanical work in guns was a milestone. Various individuals in England, Germany, France, and other countries experimented on the proportions of ingredients most desirable for use in guns, and by 1900 most countries had adopted the ratio used today (15 parts saltpeter, 3 parts charcoal, and 2 parts sulfur). Smokeless powder, based on nitrated cellulose or guncotton, came into use in the latter half of the nineteenth century. Nitrocellulose was also used in blasting. Double-base propellants, guncotton plus nitroglycerine, were developed in 1888 and found extensive use in Europe. However, the United States continued to use guncotton alone as the base for smokeless powder.

"High" explosives, which are characterized by much higher "brisance" (shattering power), were discovered more recently. Nitroglycerine (glyceryl trinitrate) was prepared in Italy in 1846; a powerful high explosive, it was too sensitive to shock for extensive applications. Twenty years later, in Sweden, Alfred Nobel stabilized nitroglycerine by absorbing it on diatomaceous earth (*kieselguhr*). His work gave rise to a series of dynamites that are still in use. Trinotrotoluene (TNT), used for half a century in the dye industry, became generally available after 1902 when an economical process for its manufacture was developed in Germany. Thereafter, TNT was adopted worldwide by the military.[1]

During World War II, interest turned to two more powerful explosives with a high melting point: PETN (pentaerythritol tetranitrate) and RDX (*sym*- cyclotrimethylenetrinitramine). Both would play special roles in the Los Alamos explosives program. PETN, a relatively sensitive secondary explosive developed for military use between the World Wars, was used commercially as a booster explosive in blasting caps.[2] At Los

Alamos, PETN's most important application would be in bridgewire detonators.[3]

Research Department Explosive, or RDX (referring to the Research Department at Woolwich in the eastern part of London), a compound first prepared in Germany before the turn of the century and known as Cyclonite in the United States, was much more readily manufactured than PETN. It had come into general use by the time World War II began. This explosive, although more potent than TNT, was initially considered too difficult to synthesize and also too sensitive for military use. By the beginning of World War II, however, a number of British researchers working under the auspices of the Advisory Council, took up study of RDX. Shortly afterward, such study took hold in the United States, where the high explosive of primary interest at the time was PETN.[4]

Over the next few years, the United States built an RDX manufacturing plant, the Wabash Ordnance Plant in Indiana, which employed a direct nitration (Woolwich) process. Several university laboratories also worked on methods of synthesizing RDX. Notable success was achieved at the University of Michigan by Werner Bachmann, who put together the elements of the British Woolwich process with acetic anhydride. The Bachmann method produced twice as much RDX from a charge of hexamethylenetetramine as did the Woolwich process. Subsequently, the Holston Ordnance Works of Tennessee Eastman (a subsidiary of Eastman Kodak, then the leading manufacturer of acetic anhydride) in Kingsport, Tennessee, developed Bachmann's method as a continuous process. John Russell, a Los Alamos explosives specialist, who had worked in the early war years at Tennessee Eastman, remembers Bachmann as "a test-tube-and-beaker guy ... a laboratory chemist working with minute quantities." At Tennessee Eastman, they scaled the process up by using "huge pumps and tanks"; as the largest RDX plant in the world, the Holston Ordnance Plant ultimately produced some 15 pounds a minute in each of twenty reactors. This gave 300 pounds per minute of the most powerful explosive available for practical use in World War II.[5]

Research on high explosives was the explicit mission of the Explosives Research Laboratory under Division B of the National Defense Research Committee (Chapter 5). The development of RDX was one of the general missions of Division B.[6] Neither the military nor industry had yet needed, or conceived of, high precision in their use of explosives, and the subject of explosives had drawn little scientific attention. ERL's head,

Kistiakowsky, was one of the few American scientists in the early part of World War II to recognize that high explosives could be made into precision tools. Pioneering explosives research by ERL scientists was to be pivotal to the implosion program.

The basic studies of explosives at ERL between 1942 and 1944 included experiments on the fragmentation of shells, shaped charges, and cavity jets.[7] Other research was concerned with the physical properties of such particular explosives as Comp B,[8] the effect of various wrappings on the behavior of explosives,[9] studies of the effects of different preparations of high explosives,[10] explosion initiation,[11] tracers,[12] and the determination of detonation temperatures for various explosives.[13] Although it was the leading American group conducting explosives research in the early part of World War II, ERL was a relatively small operation, with groups of four to ten people working in buildings only about 16 feet square.[14]

Hiring Kistiakowsky gave Los Alamos direct access to the ERL studies. Several Bruceton researchers transferred to Los Alamos, among them Henry Linschitz, who arrived in mid-November 1943, and Walter Kauzmann, who arrived a month later.[15] In July 1944, the Bruceton program would be joined formally to Los Alamos with the formation of "Project Q," based at ERL under Duncan P. MacDougall, George Messerly, and Eugene Eyster. MacDougall, who had joined Bruceton in January 1941, was then its deputy research director; as head of Project Q, he would periodically visit Los Alamos to coordinate research at the two laboratories.[16]

ERL procedures entered the program along with ERL staff. In his postwar history of the Los Alamos implosion program, Kistiakowsky lists three constraints he transferred to the explosives program: to limit HE fabrication to pieces of about 100 pounds, for safety, ease of handling, and to aid in achieving needed homogeneity; to use only explosive chemicals that were already in industrial production, so as to avoid procurement difficulties or delays; and to make components by casting HE slurries, rather than by pressing.

Like every component of the implosion program, the Los Alamos explosives effort began as a small enterprise. Housed at Anchor Ranch, the program encompassed the production of uniform Pentolite castings at a small plant, as well as quality control, employing X rays and density measurement.[17] Although workers at other laboratories had not experienced difficulties in detonating small (1-1/2-inch-diameter) cylindrical charges of Pentolite, researchers at Anchor Ranch soon ran into worrisome and unexpected failures in detonating higher-density Pentolite reliably, even when great care was taken in preparing the explosives.[18] Subsequent studies underlined the fact that existing knowledge of high explosives was inadequate for the implosion problem.

As the explosives program expanded, Anchor Ranch, which housed the gun program as well, became seriously overcrowded and the Governing Board, in early December 1943, authorized the development of Sawmill (or S-) Site, to provide more physical space for the explosives program.[19] Originally scheduled for completion in February, with full operation in April 1944, S-Site came into partial use during May 1944.[20]

David Busbee, a Naval Ordnance civilian whom Parsons brought to Los Alamos to head explosives manufacture, designed the first S-Site plant. Unfortunately, Busbee and Kistiakowsky did not see eye-to-eye on S-Site. Kistiakowsky later recalled: "The issue very soon became who, Busby [*sic*] or Kisty, knew more about explosives. Busby was a little difficult because when you disagreed with him about what was safe and what was unsafe, he would say 'and have you ever picked up a man on a shovel?'."[21] Reflecting the beliefs of Parsons and following his own naval ordnance concepts, Busbee built a small plant directed toward pouring large castings for the full-scale implosion assembly. However, as Kistiakowsky knew from his NDRC experience, such castings did not reach adequate homogeneity levels and were unsafe to handle. It was far better to make many small segments of 100 pounds or less. A hemisphere of Comp B for the nonlensed design of the period would have weighed about a metric ton; such a massive piece could only be handled by heavy equipment, and even then it had to be cast into a sturdy structural shell. Kistiakowsky recalls that the first S-Site plant "was a monstrosity from our point of view," and "never used afterwards."[22] A few cylinders of Comp B, ranging from 30 to 500 pounds, were cast using the original S-Site equipment during the wait for equipment appropriate to implosion research.[23]

S-Site was almost ready to begin production by mid-June, but it did not yet have the necessary equipment and personnel to bear the work load dictated by the implosion diagnostic program. Staffing consisted largely of SEDs, because men experienced in explosives casting or handling were almost impossible to find. Nevertheless, in August, with the Anchor Ranch casting room still responsible for casting Pentolite,[24] S-Site turned to casting small components of Comp B and various special explosives, including Torpex and Baronal.[25]

The most challenging and decisive problem for the Los Alamos explosives program was to develop the explosive lens, a device composed of explosives that was shaped so as to focus the explosion. Previous work on such a device had been done in England by Tuck, who joined Los Alamos in May 1944 as a member of the British Mission. He had already been thinking about how to focus detonation waves using different explosives, and he brought these ideas to Los Alamos. By the end of the first week of June, Tuck was head of an experimental X-ray program devoted to studying such explosive lenses.

Various attempts had been made to develop two-dimensional explosives lenses before Tuck introduced the concept of the three-dimensional explosive lens to Los Alamos. In England, eight months before Tuck's proposal, M. J. Poole prepared a complete description of a crude two-dimensional lens to generate a plane detonation wave.[26] Tuck, as a scientific assistant to Lord Cherwell, Churchill's science adviser, probably saw Poole's report on this lens before coming to the United States.[27]

About six weeks after Tuck arrived in Los Alamos, Bethe and Peierls began to search for a suitable design for the slow lens component, but without success. The breakthrough occurred shortly afterward when von Neumann proposed a workable design. Elizabeth M. Boggs of ERL had demonstrated a similar lens scheme somewhat earlier, in a memo that MacDougall sent to Los Alamos.[28]

Explosive lens testing was under way in the implosion program by mid-July 1944, when Linschitz fired the first two-dimensional lens model. In August, Joseph Fowler fired the first implosion shot using two lenses; he took a magnetic record with "encouraging features." By this time, Koski, in the photographic program, was reporting on implosion tests made using explosive lenses, but with unsatisfactory results. Lens construction would be among the most frustrating aspects of the implosion program (Chapters 15 and 16).

The overwhelming difficulties of preparing accurate lens molds for casting three-dimensional lenses were obvious by the summer of 1944. Thus, on 26 June 1944, the laboratory recruited Lt. Comdr. Norris Bradbury from his Navy Ordnance research post at Dahlgren Proving Ground, Virginia, to help with lens research.[29] Until then, little work had been done on explosives having low detonation velocities, because they generally had lower detonation pressures and destructive ability. Although the ultimate goal was to develop three-dimensional lenses, for practical reasons most of the experimental work on lenses had focused on two-dimensional lenses, using terminal observations. Reporting on studies that had used hand-tamped TNT as the slow component and

Pentolite as the fast component, Kistiakowsky boasted, "The results to date have been exceedingly encouraging, in that it has been possible to produce convergent waves of predicted spherical curvature and in that two adjacent lenses have been shown to form a single convergent wave without disturbances at the point where a wave from one lens joins that from the other." He complained, however, of "extremely serious problems in the manufacture of the casting molds with our inadequate shop facilities and in the casting of intricate shapes of explosives in them." Although he thought that "rapid progress [was] not very probable," he felt that "the basic questions of lens design and functioning have already been answered and what remain are the developmental problems only."

In contrast, Parsons was pessimistic about the explosive lens problem. "The major problem is this program begins with design and procurement of the molds .... Since these shapes are unusual and awkward to cast, and since nothing but top quality castings can be accepted, it is extremely optimistic to expect that even the second mold designs will be fully satisfactory." He predicted: "As seen in mid-August, 1944, with extremely good breaks ... this development might be ready for field test at full scale by February 1945. With reasonably bad luck ... this development could well occupy most of 1945."

Nonetheless, Kistiakowsky remained optimistic. In a report in August he stated, "The problem must be considered as basically solved, even though the difficulties in the path of their application to the gadget are many and their solution will require long and intensive work." This counterpoint between Parsons and Kistiakowsky on the lens program would characterize the implosion program throughout the fall and winter of 1944.

Electric Detonation

Another important component of the effort to solve the asymmetry problems was the development of consistent electric detonators, on which extensive work began in May 1944. At the start of World War II, both industry and the military were well acquainted with the electric detonator, or "blasting cap," a device for setting off explosions invented by Nobel in the late nineteenth century. In the "bridgewire" version of such a detonator, a high-voltage, low-current electric discharge would heat the bridgewire, which in turn set off a small amount of sensitive primary explosive (e.g., mercury fulminate or lead azide, explosives that one could ignite by a glowing wire or gunpowder fuse) penetrated by a fine bridgewire. In typical designs, the resulting detonation wave would be amplified by an adjacent small "booster" charge made of a relatively sensitive high explosive such as PETN or tetryl, and this amplified detonation wave would set off the main charge, usually a less sensitive explosive such as TNT, RDX, or dynamite. A second kind of electric detonator, the "spark-gap" type, used a spark discharge rather than heat from the bridgewire to ignite the primary explosive.

Electric detonators were initially ignored in the implosion program, because the variations in their timing were too large. An additional problem was that previously they had been used only individually, whereas most implosion configurations required many simultaneous detonations. Instead, implosion shots fired in the first year-and-a-half of the Los Alamos program were detonated by branched Primacord (flexible tubing filled with PETN), which carried the detonation from a single electric detonator to various points on the high explosive.[30] The Primacord (some 100 m in all for a full-scale gadget) was arranged in a harness ending on a system of tetryl booster pellets arranged symmetrically on the HE. Even after the crucial advance of May 1944 that would make electrical detonator systems feasible for implosion gadgets, Primacord would remain essential to the diagnostic programs, because electric detonators with adequate simultaneity would not become generally available for experimental shots until early 1945.

Primacord systems used in implosions were problematic. The Primacord often had nonuniform sections, which caused small random variations in the detonation velocity, on the order of 6,400 m/sec. The variations prevented the detonations of the booster pellets from occuring simultaneously. Furthermore, variations in the PETN batches used in producing the Primacord could cause large sudden variations in the detonation velocity. To eliminate the second problem, Los Alamos arranged, through ERL, to have the Ensign-Bickford Co. produce a few spools of Primacord made from a single PETN batch. But small statistically distributed variations due to nonuniform sections remained a serious problem.

The news in late April 1944 that British researchers had fired lead azide spark-discharge electric detonators gave the implosion program the impetus to replace Primacord with electric detonator systems. Kistiakowsky asked Max Roy, then in Division 8, to obtain whatever information on the electric spark detonators was available in England. He also asked MacDougall at ERL to procure from du Pont a number of

"SSS Detonators," then considered to be the best type of bridgewire electric detonator. He pointed out the need "to make some sort of a formal arrangement" with du Pont, Hercules, Picatinny Arsenal, or some "other organization familiar with the making of detonators" and asked Parsons to "decide on the type of contract ... and take the steps to have such a contract actually made." Kistiakowsky explained to Parsons that he considered it "highly desirable ... to use either an entirely electric or, more probably, a mixed electric and Primacord initiation system," and suggested that testing the delays in initiation should be done partly at Los Alamos and partly at Bruceton. He explained, "While we will be forced to study actual time delays in detonators, we are not well situated to make the detonators ourselves, and this part of the job will have to be farmed out under sub-contract, or in some other arrangement, to an industrial firm, or to Picatinny Arsenal."

Parsons immediately ordered several hundred electric detonators of varying design for experimental use. Kistiakowsky felt that "a total of 1,000 experimental detonators will be desired in five to ten different designs, some of which will involve bridgewire, others using specially designed spark gaps." At this time, he accepted the generally held belief that only primary explosives could be detonated sufficiently rapidly and reliably for the implosion program. A contract was drawn up specifying that the Hercules Powder Company would produce a variety of experimental detonators for Los Alamos, most loaded with lead azide. Although this kind of detonator was trouble-free, it was soon superseded and the contract eventually dropped.

In late May 1944, a series of decisive experiments, conceived of by Luis Alvarez and conducted by Lawrence Johnston, transformed the Los Alamos detonator program. Alvarez, another of Lawrence's "boys" at the Berkeley Radiation Laboratory, had recently arrived at Los Alamos from the MIT Radiation Laboratory. Initially he served in E-Division as Kistiakowsky's "right-hand man." Kistiakowsky brought to his attention the problem of the inadequate simultaneity of multipoint detonations in implosion shots. Although interested in electric detonators, Alvarez recalls that at this time Kistiakowsky was highly doubtful that they could improve the simultaneity.[31]

Alvarez was intrigued by the question of how to make electric detonators work in implosions. Examining what was then known about high explosives and multipoint detonation, he considered arrangement:

> I learned a lot about the five regular polyhedra during this period. The largest number of points that can be spaced equally

> on the surface of a sphere is twenty, a number corresponding to the centers of the twenty triangular faces of an icosahedron. The next largest number is twelve, corresponding to the centers of the twelve pentagonal faces of a dodecahedron. It's possible to interleave a dodecahedron with an icosahedron, as Plato showed, to get the nearly regular faces, alternately pentagons and hexagons that shape an object so familiar as a soccer ball.

Alvarez's problem was to detonate all points electronically with sufficient precision.[32] Alvarez raised the crucial question of whether it was in fact necessary first to ignite a sensitive primary explosive before detonating a less sensitive secondary one. Could one not, he asked on or about 25 May, vaporize the bridgewire using the discharge of a powerful capacitor and detonate the PETN directly by the shock wave produced by the sudden expansion of the vaporized metal? Alvarez recalls Kistiakowsky being skeptical of this scheme, but saying nevertheless, "We've got to try anything, so you go ahead and give it a shot."[33]

Johnston, who had been Alvarez's student and had worked with him at MIT on radar ground-controlled approach, arrived in Los Alamos on 25 May and agreed at once to try out Alvarez's idea.[34] The first test employed a simple method of measuring the simultaneity of multiple detonations. On 28 May on South Mesa on the site of Neddermeyer's first implosion shots performed almost a year earlier, Johnston mounted a 1-foot length of Primacord on a wooden two-by-four, with an electric bridgewire detonator taped at each end, and a lead block set underneath the middle portion of the Primacord. (In later experiments, more accuracy and better time resolution would be achieved using an aluminum rather than lead block.)[35] Discharging a 1-μf capacitor at 15 kV through the two bridgewires of the standard electric detonators connected in series, he set up detonation waves in the two ends of the Primacord. When the waves collided, the increased pressure from the interaction of the two detonation fronts caused a crease in the lead just below this point. This crease would lie at the precise center of the Primacord if the detonators fired simultaneously, and a 1-μsec difference would move the mark off center by about 4 mm.[36]

Johnston recalled recently that the original "shot made a satisfying report and when I surveyed the damage, I found the plank shattered and the lead plate 10 feet away. Every little string in the Primacord had left its impression in the lead and sure enough, there was a deep dark notch cut about an eighth-inch from the center of the plate," indicating a timing difference of 1 μsec. By improving the technique, Johnston "soon got the high explosives going with less than a microsecond delay and then all the explosive experts helpfully suggested how to uniformly load explosives around the bridge wires to improve their performance." Soon they were able to set off the appropriate number of detonators simultaneously to the high degree of precision that was needed.[37] Later, Alvarez's group would bring the timing spread for a system of several hundred detonators down to several hundredths of a μsec.[38]

The importance of Johnston's results was immediately appreciated at Los Alamos. Beyond offering a great improvement in timing, the new PETN detonators were safe from accidental detonation for they contained no primary explosives. This characteristic particularly pleased Parsons because it allowed one to avoid military requirements for using mechanical safety gates between the detonators and the explosive train.[39] From an engineering point of view, safety gates would have been extremely troublesome in the final weapon.

Kistiakowsky then asked Parsons to "obtain for further work a large supply of the best electric detonators available," suggesting "that 2,000 SSS DuPont electric detonators, or equivalent short delay Hercules detonators be procured through the Bureau of Ordnance with a high priority." A research and pilot production contract was drawn up with the Hercules Powder Company. From this point on, extensive research would be focused on the electric detonator.[40]

To achieve the simultaneous firing of many detonators, Los Alamos would have to develop an appropriate switching device and high-voltage power supply. In mid-August, Kistiakowsky separated the responsibility for the research and manufacture of detonators for experiments, which he gave to group E-11, from the development of detonators for the gadget and the development of associated firing circuits, which he gave to group E-9. He put group E-2 in charge of the design and construction of electrical timing sets.[41]

To move from an experimental detonator system to a production device, the laboratory would have to arrive at designs for both the detonator and firing circuit and make both components work effectively and reliably. This effort would take more than six months, and the slow production of X units almost delayed the testing and use of the implosion gadget.

Decision to Test the Implosion Bomb

The idea of testing the implosion bomb, perhaps prompted by von Neumann's suggestion to Teller at the end of 1943, had immediate support when it was raised at Los Alamos in January 1944, even though such a test could waste precious plutonium. Groves agreed to the test on the condition that the active material be recovered.

That meant the test would have to be conducted as a scaled-down explosion in a containment vessel.[42] The problem was, how to build the container and recover the active material once a decision was made on the size of the explosion.[43] In January 1944, the Governing Board asked Norman Ramsey, head of the delivery effort, to evaluate methods of testing the nuclear chain reaction and implosion mechanism, at either full or reduced scale.[44]

In a report to the Governing Board early in February, Ramsey argued for building the containment vessel and moderating the fissionable material with hydrogen to scale down the blast. He pictured a test bomb liberating only twenty to forty-six generations of neutrons, with a chain reaction that stopped before it could unleash much energy.[45] Oppenheimer, however, said they could not accurately predict the amount of fissionable material needed to yield the specified number of generations, and in any case a controlled small-energy release would not simulate the thermal effects of the bomb. Instead, he suggested carrying out the full test in a remote location inside a containment vessel. This vessel soon received the code name "Jumbo." Oppenheimer stressed that it would be easier to predict performance for a full-scale explosion.[46]

After Oppenheimer communicated these ideas to Groves, they decided to proceed quickly with plans for the full-scale blast and with the design of the container, which was to be ready by September 1944. Although still not yet agreeing unequivocally to the test, Groves decided in March 1944 to allow laboratory staff to study the possibilities.[47] Soon afterward, Parsons formed Group E-9 (High Explosive Development) to look into the specifics of the test, with Harvard physicist Kenneth Bainbridge as group leader.[48]

Initially neither Groves nor Oppenheimer were very enthusiastic about the test. Groves envisioned a bomb containing only enough fissionable material to set off a chain reaction, but Oppenheimer thought it would be too difficult to determine the exact amount of material needed for such a test. Furthermore, a small yield test would still entail the risks of a large explosion. Kistiakowsky remembers that in the fall of 1944 he and Oppenheimer were still trying to gain approval for the test from Groves, who was grumbling that he would face the wrath of a Senate committee if he lost a billion dollars worth of plutonium.[49] Even after he had approved Trinity construction plans, Groves wrote Oppenheimer on 1 November 1944, "I feel that you must limit attention to this to the absolute minimum personnel and effort by that personnel."[50]

The Implosion Program in Early Summer 1944

Witnessing the tremendous expansion and diversification of the implosion program during the spring and summer of 1944, Oppenheimer and Kistiakowsky recognized that the program needed a major administrative change. Although Neddermeyer had been just the right person to launch the small-scale implosion program, his academic style of research management was unsuited for a program involving hundreds of researchers.[51]

On 15 June, Oppenheimer relieved Neddermeyer of his leadership of E-5, the Implosion Experimentation group, a post he had held since the start of the Los Alamos Laboratory. Almost apologetically, Oppenheimer explained in a letter, "In view of our conviction that the work of E-5 must be rendered far more effective than it has been in the past and that this must be done with the highest possible urgency, the only step which has appeared possible to me is to ask Kistiakowsky to undertake the direction of E-5." Oppenheimer informed Neddermeyer that this change "meets with my full endorsement." He further explained:

> According to this scheme you will be a technical advisor attached to the division and a member of the steering committee which determines the H.E. program. I believe that the essential point of the organization is that you should be in a position where you are not specifically answerable to anyone, and do not specifically have authority over any of the three H.E. groups.[52]

One month later, Kistiakowsky's new position as head of E-5 would fall to Bradbury.[53] On the same day that Oppenheimer relieved Neddermeyer of his leadership of E-5, Kistiakowsky, now associate leader of E-Division, unveiled the new organization of the implosion program. The technical advisers, with group leader status, were Neddermeyer and Alvarez. The HE Steering Committee consisted of Alvarez, Bainbridge, Kistiakowsky (chairman), Neddermeyer, and Parsons. E-5, the implosion research group, was to be directly under Kistiakowsky, with "Mr. X, Associate Group Leader (Lt. Comdr. Bradbury?)." E-9, the high-explosives development group was to be headed by Bainbridge. E-10, a new group devoted to maintenance and construction for the implosion project as well as operation of S-Site, was placed under Major Stevens. Parsons added in a handwritten note on a copy of this memorandum, also dated 15 June 1944, "I believe this proposal is the best and most workable under the circumstances. I recommend its approval."[54]

Technical overviews of the implosion program written by Parsons and by Kistiakowsky during the summer of 1944 illustrate the ambivalence with which the effort was viewed at the time. Parsons, ever skeptical, wrote that the experimental implosion program "must overcome present handicaps and must make good use of favorable breaks, if the time schedule is to be met." He lamented, "At present, and more seriously in the future, progress is impeded by lack of experienced personnel of group and assistant group leader caliber." As a military officer, he emphasized that the engineering development required by the project was

> extremely difficult to focus from a point so far removed from centers of military and industrial control. Two thousand miles of distance is a factor which is hard to overcome even with adequate mail, telephone, and teletype communication facilities .... It is felt that only by repetition of statements like the above can we be placed sufficiently on guard to offset the handicaps imposed by distance, isolation and security. Another major handicap now imposed by security is the fact that only two top officers of the Air Forces are now fully informed of this development .... However, success or failure in this project will depend upon adaptation of airplanes and tactics to deliver a bomb which is radically different in characteristics and effect from any hitherto carried.

In discussing the jetting problem, he concluded,

> use of Composition B and any practicable number of detonation points will not lead to a satisfactory solution of the implosion problem. Avenues now in sight include: a) development of explosive lenses designed to produce a converging spherical detonation wave, b) preventing interaction between converging waves by use of lead spacers, c) use of aluminized explosives, such as Torpex, to prolong the positive pressure pulse, and thus increase the total impulse given to the collapsing tamper. It is obvious that developments like (a) and (c) above require exhaustive field tests at several scales, and also require versatility on the part of our explosive casting and shaping facilities.

Kistiakowsky, although more optimistic than Parsons, was also deeply worried about the implosion problems. He noted in mid-August, on the positive side, "The urgency for reaching much higher velocities than those previously recorded has been greatly reduced by the calculations carried out by the Theoretical Physics Division." Unfortunately, symmetry was clearly not being achieved by the fixes of the moment; the recent study of collapsing spheres showed asymmetries to persist "even when timing is perfect." The overall state of implosion research in the summer of 1944 was discouraging. Theoretical studies had shown that implosion would be an effective means of assembling the bomb, but only if the implosion were symmetrical. Extensive diagnostic work in the X-ray, photographic, and terminal observation programs and recent theoretical studies of instability gave little hope that a symmetrical implosion could be achieved in practice. Experimentalists had demonstrated compression, but to date on too small a scale. The effects of interactions between divergent detonation waves were considered destructive even in small-scale systems. Primacord would clearly be unusable because of timing inconsistencies inherent in the manufactured product. Furthermore, there was not yet any prospect of making the high-explosive part of such a weapon with adequate homogeneity, and no way yet to see that proceeding with lens systems and electric detonators would eventually succeed. Finally, plutonium could not yet be fabricated according to the design. Indeed, basic plutonium metallurgy procedures were still being devised, since large-scale batches of plutonium had only recently become available. With this complexion of problems, the laboratory had to face the spontaneous fission crisis, which demanded rapid completion of the implosion bomb.

10

The Nuclear Properties of a Fission Weapon: September 1943 to July 1944

Between the summers of 1943 and 1944, the Theoretical Division, under Hans Bethe, and the Physics Division, under Robert Bacher, collaborated in studying the nuclear physics of the atomic bomb. T-Division's responsibilities included calculating critical mass and efficiency. The lack of hard nuclear-constant data was particularly troubling. While P-Division worked to improve the experimental data using available detectors, accelerators, and other devices, T-Division developed flexible models based on the changing set of available data. To cross-check their results, researchers often used different methods to solve the same problem. For example, the Water Boiler, a nuclear pile using enriched uranium in a water solution, provided a means of checking critical mass calculations. As a backup, Richard Feynman made calculations on uranium hydride, then being considered as a potential active material. Teller's investigation of the hydrogen bomb (the Super) was an alternative approach to a nuclear weapon. The opportunity to conduct physics research on a larger scale than had ever before been attempted gave the Los Alamos physicists the experience of working in well-funded multi-

* This chapter is based on a manuscript by Catherine Westfall and Paul W. Henriksen, in which Westfall wrote on experimental and theoretical nuclear physics and Henriksen on the Water Boiler and Super. We are grateful to Gordon Baym for editing the technical content of the section on nuclear theory, Gary Westfall for editing the technical content of the section on experimental nuclear physics, and Les Redman for his general technical review.

disciplinary groups, which included both experimentalists and theorists, as well as electronics experts, chemists, and metallurgists.

Nuclear Theory: Critical Mass and Efficiency

In September 1943, T-Division was refining its critical mass and efficiency predictions and calculating the damage the bomb could cause. Up to this point, the division had remained somewhat informal in its organization to accommodate changing priorities, but by October it had begun to subdivide into groups: Bethe took on the problem of implosion, Victor Weisskopf led the calculations of efficiency, Robert Serber spearheaded diffusion theory, Edward Teller assumed responsibility for both the Super and implosion, Feynman led the uranium hydride calculations, and Donald Flanders headed the computational effort. In Bethe's reorganization of the division in March 1944, Teller remained in charge of implosion theory, while Feynman's responsibility changed from theory of the hydride to neutron diffusion. In June, however, Bethe put Rudolf Peierls in Teller's place as implosion leader, allowing Teller to work exclusively on the Super, a project now separate from T-Division.

Most of the theorists lent a hand in the difficult and crucial task of modeling neutron diffusion, which was important to both critical mass and efficiency calculations. This effort had begun at Berkeley before the start of Project Y, with the extrapolated end-point method, devised by Stanley Frankel and Eldred Nelson for modeling the movement of neutrons in the bomb. But the assumptions were too simplified for anything other than a rudimentary model: all neutrons had the same velocity, the tamper and core were stationary, every neutron collision was elastic, neutrons were scattered isotropically, and neutrons in the core and tamper had the same mean free path. In October 1943, Bethe assigned top priority to finding a more realistic description of neutron diffusion through the core of the bomb.[1] That meant taking into account the fact that the neutrons had a distribution of velocities, that the neutrons did not scatter isotropically, and that mean free paths were different in the tamper and the core.

In an attempt to deal with neutrons with many velocities, the Los Alamos theorists tried a "multigroup method," which exemplifies the trend toward numerical approximation in the theoretical work of this rushed period. The neutrons were divided into several groups, each containing neutrons of the same velocity. The physics problem was therefore

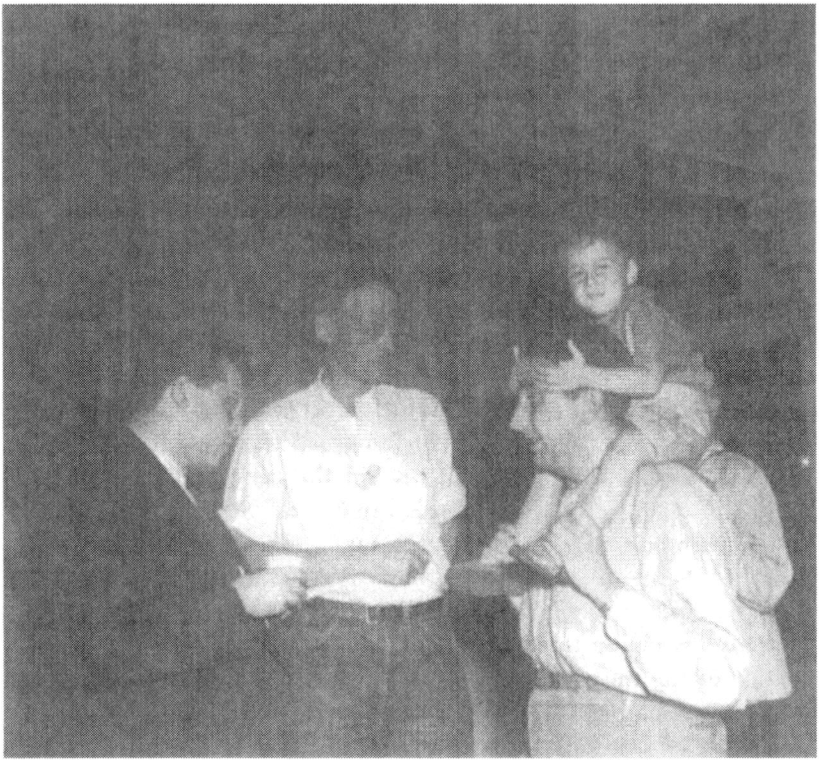

Fig. 10.1. Theoretical physicist Edward Teller, carrying young Paul Teller on his shoulders, speaks with theoretical physicists Julian Schwinger and David Inglis. LA Photo, LAT 416.

reduced to a series of one-velocity problems. By treating many groups having small velocity differences, this model could better approximate reality.

A crucial problem in fall 1943 lay in calculating the critical mass for the gun assembly. Several approaches were taken, all based on approximations. Bethe developed a perturbation theory based on differential transport theory. In another attack on the problem, David Inglis, Frankel, and Nelson applied variational methods to study critical masses of long cylinders. They described the spatial distribution of neutrons in terms of a few parameters and varied the "trial function."[2] To check on their calculations, they compared the critical masses of cylinders with those of tamped spheres, which could be calculated more accurately, and compared variational results with those obtained using the extrapolated end-point method. That method, to the theorists' satisfction, generally yielded excellent agreement, to within a fraction of a percent.

The critical masses of finite tamped cylinders were calculated first by assuming that the neutron mean free paths in the tamper and core were the same. But extending the calculations to a weapon having a tamper and core with different mean free paths proved difficult. Frankel and Nelson looked, unsuccessfully, for "a simple recipe" to extrapolate integral theory results for special cases to general mean free paths. By spring 1944, Chaim Richman, Kenneth Case, and Roy Glauber developed approximate methods for determining the neutron distribution near the interface of the core and tamper. These solutions were then used to calculate the effects of differing mean free paths of the core and tamper, with the influence of shape taken into account by perturbation theory. When the cylinder in their calculations was replaced by an equivalent spherical distribution, taken as the unperturbed distribution, they noted in April that, "the unperturbed distribution already gives very nearly the right answer." An important step forward was Robert Marshak's idea of solving the neutron Boltzmann equation by expanding the neutron distribution in a small number of spherical harmonics.[3] Besides allowing for more realistic conditions in critical systems, the theorists also calculated subcritical systems, to determine, for example, the optimum distribution of material in the target and projectile of a gun assembly and to determine how criticality varied as the projectile entered the target during assembly. They also analyzed various autocatalytic systems.[4]

At Teller's suggestion, T-Division investigated uranium hydride. On one hand, this compound and plutonium hydride seemed to have considerable potential as active materials because the presence of hydrogen would slow the neutrons, cause more fissions in the core, and hence reduce the amount of active material needed in a bomb. On the other hand, they would reduce efficiency because of the slower reaction rates with slower neutrons. The hope was that the savings in critical mass would more than offset the loss in efficiency. Unfortunately, the uranium hydride efficiency was, according to Bethe, found "negligible or less, as Feynman would say."[5] The prospects of the hydride gun grew steadily worse during the project. By August 1944, interest in the hydride gun had disappeared.

The breakthrough in calculating the velocity distribution of the neutrons came in July, when Feynman's group developed an exact many-velocity method applicable to the case of purely elastic scattering. In this method, all the cross sections in the core vary inversely with velocity, whereas cross sections in the tamper remain constant. By comparing the results of this method with approximate several-velocity-group calculations, the researchers sensed how to make such several-velocity calculations accurate enough for practical purposes, or at least for the case of pure elastic scattering in the tamper. With calculations involving only three or four velocity groups, they were able to include inelasticity in the scattering. The basic strategy in calculating with more realistic velocity distributions was, as Bethe summarized, "to reduce problems involving neutrons of several velocities to one-velocity problems so that we shall soon be able to give reliable values for the critical mass of metal with any tamper."[6]

As the experimental results improved, the critical mass calculations were revised. In late September 1943, for example, Fermi in Chicago provided a more precise measurement of ν, prompting a revision of Serber's April estimate of the critical mass of a tamped ^{235}U sphere from 15 to 23 kg. Estimates for the critical mass of a tamped ^{239}Pu sphere also increased, from 4 to 10 kg.[7] In the spring of 1944, the theorists would be presented with an opportunity to test their calculations empirically in the Water Boiler reactor, the first critical assembly at Los Alamos. It used a solution of uranyl sulfate in water, with 14.7 percent ^{235}U (see "The Water Boiler" below). As Robert Christy recently recalled, his original calculations for the critical mass for the water boiler were "quite far off." Just before the reactor went critical, however, he performed "a very quick and dirty correction" to allow for new cross-section data, and the calculation then "turned out to be within a few grams" of the value, 565 grams, measured in June 1944. Although Christy admits that the correction "was not very reliable," the Theory Division nevertheless "got a great deal of glory" for producing the "right number before the experiment."[8]

By July 1944, Bethe announced further progress in determining the critical mass of a tamped metal core. Owing to calculations by Serber's group for cores and tampers with different mean free paths using Marshak's spherical harmonic method, both multiplication rates and critical mass were "completely under control for the case of one neutron group." The following critical mass formula emerged:[9]

$$M \sim \lambda_t^{0.6} \lambda_c^{0.4} \left[\lambda_f/(\nu - 1)\right]^2 ,$$

"where λ_t is the mean free path in tamper, λ_c is the scattering mean free path in the core and λ_f the fission mean free path in the core."

In view of the work of Feynman's group on taking inelastic scattering from the tamper into account, and treating several neutron velocity groups, Bethe was able to write in August: "The problem of neutron diffusion has now been solved in almost all possible cases .... The main uncertainty remaining [is] the experimental constants." Bethe was optimistic. "A reasonable assurance about these constants is given by the agreement obtained," by comparing tamper scattering results from the Cockcroft–Walton ("D–D") group and theoretical estimates. Although in the last year of the program the theoretical groups would refine their estimates by extrapolating from sphere multiplication experiments, the essential problems of calculating neutron diffusion and critical masses had been solved by the summer of 1944.

A parallel development took place in the calculations of the device efficiencies. Efficiency was much more difficult to determine than critical mass, however, because it depended on the evolution of the assembly in time. Here the breakthrough came at the beginning of the project. As Bethe recalls, after Serber discussed efficiency in his introductory lectures, he and Feynman returned one evening after dinner to the Technical Area. Finding few other people around, they fell to thinking about efficiency and "the physical parameters which matter." At this point, they did not know how to solve the complicated diffusion and hydrodynamical equations for supercritical systems. "Clearly the initial multiplication of neutrons [was] very important, and also important must be the expansion of the material before the multiplication stops" – the "endpoint" of the expansion. "And I think we guessed" that the rate of decrease of multiplication during the expansion, assuming known beginning and endpoints of the expansion, would be proportional to the relative expansion. To fix the overall constant, Bethe and Feynman used Serber's result (as given in his lectures) for small excesses over the critical mass, a case that was relatively simple to analyze. Using this approach, Bethe and Feynman developed a formula for efficiency.[10]

T-Division was also concerned with estimating the damage from a nuclear bomb. Two particularly important issues during the summer and fall of 1943 were the effect of shock waves in damage calculations and the optimum height for detonating the weapon. For small explosions, the damage clearly increases with duration of the pulse of the explosion. What was the situation for large explosions? Now, as at other times, the laboratory benefited from the advice of von Neumann. On his visit to the laboratory in late September 1943, he explained that current nonnuclear explosions were "just about at the limit at which further increase of the duration of the pulse has no further advantageous effect on the damage." In an explosion as powerful as a nuclear bomb, the duration of the blast is much longer than the typical periods of vibration that would destroy buildings. Thus, in such explosions "only the peak pressure matters." Damage would result whenever the overpressure produced by the explosion exceeded some fraction of an atmosphere. Bethe was able to estimate from his calculations of the propagation of shock waves in air that a 10-kiloton (equivalent explosive power of TNT) bomb would produce an overpressure of 0.1 atm at 3.5 km away from the explosion and thus would cause severe damage within that radius.

In his September visit, von Neumann made another important contribution, this time to the determination of the height at which the bomb would be detonated. Explosives researchers knew that when shock waves were reflected by solid objects, pressure increased. Von Neumann showed that this effect was much greater "than previously believed" if the angle of incidence of the shock wave was between 90° and some limiting angle. Thus, the theorists deduced that "considerable improvement of the damage radius can ... be expected by detonation at an appreciable height (1 to 2 kilometers)."

Experimental Nuclear Physics

Organizing the Experimental Program

While preliminary experiments were being conducted in the summer of 1943 (Chapter 5), Bacher was busy making longer-term plans for P-Division. In view of the imperfect theoretical understanding of the fast-fission chain reaction, the uncertainties surrounding the available measuring techniques and their limited accuracy, the shortage of time available for development, and the serious consequences of a miscalculation, P-Division faced a challenging task indeed. Work on critical assemblies could not begin until more fissionable material was available, yet it was important to obtain accurate nuclear constant measurements quickly to help move bomb design along.

When Bacher arrived in July 1943, five P-Division groups had already been formed: a group under Robert Wilson to perform experiments at the cyclotron, a group under John Williams to perform Van de Graaff experiments, a group under John Manley to perform Cockcroft–Walton experiments, a group under Darol Froman to develop electronics equipment, and a group under Emilio Segre to perform experiments with nat-

Fig. 10.2. Experimental physicist Robert Bacher was the first head of the experimental physics division and subsequently of the weapon physics division. LA Photo, LAT 612.

ural sources, including some on spontaneous fission. In August, Donald Kerst was appointed to direct work on the Water Boiler, and in September Bruno Rossi was assigned to direct work on detectors. Rossi took control of a group previously led by Hans Staub.[11]

In early August, Bacher made a preliminary list of desirable nuclear constant measurements. By early September, in the midst of numerous discussions with both theorists and experimentalists, a six-page, single-spaced list of proposed projects had been compiled in order of priority. In many ways, the new plan was an extension of the line of inquiry set out by Oppenheimer and Manley in 1942. As in the past, considerable emphasis was put on developing equipment, primarily improving detectors. In addition, fission spectrum and fission cross-section measurements were again given high priority. Oppenheimer and Manley had already set the precedent of using different techniques to measure each nuclear constant. With more funding, workers, and equipment, Bacher planned even more redundancy in the program. The ambitious experimental program he developed was aimed at quickly producing the maximum amount of reliable nuclear constant information in the shortest possible time.[12]

The program called for two high-priority fission neutron spectrum measurements, along the lines of those begun earlier. The photographic plates made by William Bennett and Hugh Richards in Minnesota to investigate the fission spectrum were being measured, and new plates were being constructed. Williams's group planned to use the Van de Graaffs as a source of Li(p,n) neutrons for this work.

Emphasis was also placed on obtaining more complete measurements of ^{235}U fission cross sections from 100 keV to 2.5 MeV, where previous experiments had shown most of the fission spectrum to lie. Manley's group would make ^{235}U cross-section measurements at the upper end of the neutron energy spectrum, using the deuteron beam at the Cockcroft–Walton. Williams's group would cover the lower end of the neutron energy spectrum, building on the promising techniques devised at Minnesota by D. L. Benedict and Alfred Hanson, who was now at Los Alamos. For this work, they would use the proton beams of the two Van de Graaffs, which had complementary capabilities – the "short tank" could carry higher currents while the "long tank" could produce higher-energy beams. Measurements using natural neutron sources augmented measurements with the accelerators.[13]

Other important tasks were further exploration of the spontaneous fission limits of various isotopes and the completion and extension of the study of the neutrons per fission (ν) and delayed neutron emission begun in the summer of 1943. Delayed emission measurements devised by Charles Baker were expected soon, and work had begun on another coincidence method conceived by Wilson. They planned a new series of comparative $\nu(^{239}\text{Pu})/\nu(^{235}\text{U})$ measurements that also used a coincidence method. Although Groves resisted close collaboration between Los Alamos researchers and those outside the project, Bacher insisted that a series of absolute $\nu(^{235}\text{U})$ measurements also be performed by Fermi at Chicago, where a high flux of thermal neutrons was available from the pile. Bacher, as he recently explained, felt it imperative for the work to be coordinated so that Chicago slow-neutron and Los Alamos fast-neutron ν measurements "could be tied properly together."[14]

High priority was also given to work that had previously warranted only scant attention. Tamper materials would be investigated by integral experiments that would simulate the scattering of fission neutrons on various potential dense tamper materials, including gold, rhenium, iridium, lead, platinum, and uranium. These experiments would use neutrons from natural sources, from Li (p,n) reactions at the Van de Graaffs and from D–D reactions at the Cockcroft–Walton. Plans were also laid for another integral experiment, the Water Boiler.

It was important to investigate the relation of the energy of incident neutrons and the ^{235}U fission cross section down to thermal energies. Although Wisconsin measurements from 0.5 to 1 MeV showed an approximate inverse relation between energy and the fission cross section of ^{235}U, physicists did not know whether this relation held at lower energies. Los Alamos had both the equipment and the personnel to investigate this question further. The modulated cyclotron, which was already being used for ν and delayed neutron emission measurements, was a good source of monoenergetic neutrons below 1 keV. Boyce McDaniel, the young Ph.D. assigned to designing equipment for experiments checking the relation of fission cross section and energy, was unusually qualified for the task. In 1942 he rebuilt the time-of-flight equipment at Cornell, completed his thesis, and afterward, at Bacher's instigation, spent time at the MIT Rad Lab to learn more about the electronics innovations developed there.[15]

Although full-scale experimentation with ^{239}Pu could not begin until more material was available (pile-produced plutonium would not arrive until April 1944), measurements were planned to check the relationship between the energy and fission cross section for ^{239}Pu, in addition to the comparative $^{239}Pu/^{235}U$ cross-section measurements mentioned above. Lower priority was given to fission cross-section measurements around 1 MeV of protactinium and thorium (elements in use, or being considered for use, in detectors). Lower priority was also given to measurements from 100 keV to 2.5 MeV of the cross section for (α, n) reactions on boron, an element useful as a neutron source. Later the series of fission cross-section measurements was extended to include ^{237}Np and ^{238}U. In addition, the cross section for (α, n) reactions on lithium was measured.[16]

Equipment for Nuclear Physics Experiments

To perform the experiments that Bacher outlined in September 1943, P-Division had to develop new electronics equipment, obtain optimum performance from laboratory accelerators, and improve the techniques used for detection. Much of P-Division's responsibility for electronics fell to Froman's group, which also supplied equipment to the entire laboratory. This assignment was challenging because prewar electronics equipment was quite slow for the necessary neutron measurements. William Higinbotham, a graduate student with electronics expertise at Cornell when World War II began, notes that amplifiers were "rather crude and non-linear," and that "the counting circuits were terrible." In addition, the prewar version of oscilloscopes used to monitor the signals of circuits ran only to about 20 kc and had triggered sweeps that ran repetitively at a fixed rate. This frustrated physicists at both the MIT Rad Lab and Los Alamos, who, to develop faster circuits, needed an oscilloscope with a fast triggered sweep capable of producing randomly placed pulses.[17]

To provide the necessary electronics equipment at Los Alamos, Froman was aided by a particularly talented group of recruits, including physicist William C. Elmore, Matthew Sands, and SED's Val Fitch. Fitch and Sands were among those who came to Los Alamos with little physics training and later obtained a physics Ph.D. Another gifted worker was Higinbotham, who brought to the project much-needed electronics knowledge. Like McDaniel, Higinbotham served on the MIT radar project before coming to Los Alamos. After working for two years as an electronics technician, Higinbotham was recruited by Bacher and came to Los Alamos in the fall of 1943. Thanks to his MIT experience, Higinbotham knew how to build a variety of new circuits and improved oscilloscopes that would aid in the further development of fast circuits. In addition, he had a sound knowledge of ready-made electronics equipment and suppliers, including the fledgling company now called Radio Shack. On the basis of this knowledge, he quickly compiled a stock catalog to aid in electronics procurement.[18]

The group expanded around a small nucleus of those skilled in electronics – including Froman, Elmore, Sands, and Higinbotham, who replaced Froman as group leader in fall 1944. As the demand for equipment increased, the group mushroomed from about ten members constructing a dozen pieces of equipment in a two-week period in January 1944, to sixty-nine members constructing 100 pieces of equipment during a two-week period in the following January. At the time that the workload reached its peak in early 1945, the group had eighty people, including six women who spent all their time making cables. Although some electronics equipment was designed and built by experimenters, the "well stocked and well staffed electronics laboratory," Wilson noted in 1947, "played an implicitly important role in all experiments by supplying the fast, stable amplifiers, reliable scaling circuits, multi-discriminators ... without which many of the experiments could not have been made."[19]

Meanwhile, the other groups in P-Division struggled with balky accelerating equipment and the challenge of developing adequate detectors. In October, just as the experimental groups were launching the ambitious experimental plan outlined by Bacher, the cyclotron and both of the electrostatic generators suffered operational problems. Difficulties with the cyclotron ion source "considerably cut down the efficiency of that instrument," while vacuum difficulties temporarily stalled the long tank. Although vexing, these problems could be fixed with minor repairs, and both accelerators were operating by 1 November. Problems with the short tank, which was not designed to operate in the keV energy range required by Los Alamos experiments, were not so easily resolved. After experiencing difficulty in exceeding the threshold of the Li(p,n) reaction with the short tank at low energies, Williams concluded that "it would be necessary to remodel the equipment if it was going to be used for this reaction." This remodeling took longer than expected, and the short tank did not begin operating until February 1944, a frustrating delay in light of the pressure to produce results quickly. Richard Taschek, a member of the Van de Graaff group, remembers that even after the accelerators were operating reliably, experimenters were kept busy with the target and ion source development projects necessary for individual experiments, projects that moved rapidly "from invention, to development, to use." Radical modifications were made in the summer and fall of 1943 at the cyclotron, where McDaniel improved the Cornell time-of-flight equipment, and other group members constructed and then restacked a large graphite block to produce an optimum neutron flux. The cyclotron group also installed target collectors, which were sent from the Berkeley model mass spectrograph.[20]

Each group also spent a great deal of time developing suitable detecting techniques, because fast-neutron measurements strained the limits of both detector and electronic capability. As Taschek recalls, at the beginning of the project all techniques "were just very, very slow ... the electronics was slow, the detection equipment was slow, and the way the detectors worked was slow." Higinbotham, who worked closely with detector specialists such as Rossi, explains that although detector development at Los Alamos was "a continuation of [prewar] work on radiation detectors," managing to "squeeze information" from experiments "required ... a thorough understanding of the basic physics of the operation of all these devices, and furthermore, some really extremely clever ideas."[21]

One useful type of detector was the "long counter" developed by Hanson in early 1944. In such a detector, a BF_3-filled proportional counter lined with a ^{235}U foil lies in the axis of a paraffin cylinder. Neutrons incident on the open end of the cylinder were degraded to about 10 keV before diffusing into the proportional counter to bombard the ^{235}U foil. Hanson found that if the paraffin cylinder was sufficiently long, this detector had a response to fast neutrons that was nearly uniform from about 10 keV to 2 MeV. As a result, the long counter was useful for calibrating neutron sources (even if their energy spectra were somewhat different), an impossible task for other detectors. The long counter was also useful in measuring neutron multiplication because of its wide energy range, and in measuring cross sections for neutrons emitted from fission over a particularly large energy range. The standard method for measuring fission cross sections, in which neutrons were counted by extrapolation from the number of proton tracks, was unfeasible below 350 keV because at such energies too small a fraction of recoils gave a signal above the noise background of available detector-amplifier systems. Because the long counter slowed all neutrons down to about 10 keV, it could make cross-section measurements in this previously undetectable range.[22]

Another particularly useful detecting technique at Los Alamos was "electron collection." As Clyde Wiegand, then a graduate student working with Segrè, recalls, standard "proton recoil measurements in ionization chambers were done in air" by detecting the trail of positive ions left by a charged particle. As explained in "Nuclear Physics," a 1947 report summarizing wartime nuclear constant measurements, the "inherent difficulty with positive ion collection is the extreme slowness with which the ions move. This necessitates extremely slow counting rates, long time constants of the amplifier, and consequent troubles of piling up of pulses and a prohibitive acoustical noise level." Before the war, physicists at Berkeley, Cornell, and other universities used "electron collection," an improved technique that collected electrons, which move much more quickly than positive ions because of their low mass. However, electron collection cannot be accomplished with an air-filled chamber, since electrons become attached to the oxygen in air. Thus Segrè devised a nitrogen-filled chamber to detect neutron flux. Further development of nitrogen-filled chambers was important to the work done on plutonium spontaneous fission chambers in 1943 and 1944 (Chapter 12). Electron collection was modified at Los Alamos. Theodore Hall, Philip Koontz, and Rossi constructed an improved, argon-filled electron collection device used in other nuclear constant measurements. A striking example of how this device could be used in experiments was a series of ^{235}U measurements in the 1-MeV range made by Hall, Koontz, and Rossi in 1944. By using a chamber containing both an argon-filled electron collection device and a fission counter, they could calculate the absolute number of fissions per fission neutron and from known values, such as the neutron–proton scattering cross sections, they were able to calculate the ^{235}U fission cross section. As Wilson noted, when used with the fast amplifiers developed by the Electronics Group, electron collection "enabled us to decrease resolving times in counting by several orders of magnitude."[23]

Experimental Results

Just weeks after the September outline of experiments was finished, it appeared that Bacher had been wise to insist upon collaboration with Fermi. As Bacher reported to the Governing Board on 7 October, Fermi's measurement indicated that ν of ^{235}U was 2.0, not 2.2, as previously measured. As Oppenheimer noted in a letter to Groves a few days later: "This is well within the limits of error of the earlier value .... Nevertheless, even this small change means an increase by 40% in the amount of material required."[24]

The incident underlined a distinguishing feature of the Los Alamos experience. As Wilson noted in "Nuclear Physics," "We were constantly plagued by worry about some unpredicted or overlooked mechanism of nuclear physics which might render our program unsound."[25] Uncertainty about the number of neutrons to be expected in the fast-neutron chain reaction was particularly worrisome. The bomb project had already been hit with the unpleasant discovery in late 1942 (Chapter 3) that higher-than-expected neutron production from (α, n) reactions in impurities meant that plutonium could not be used in a gun weapon unless stringent plutonium purification techniques were developed. That further unwanted neutrons would be produced by spontaneous fission was a still untested threat. Oppenheimer responded to the disconcerting uncertainty of changing ν results by emphasizing the need for overlapping approaches to the problem. He pointed out to Groves that "even the most careful experiments in this field may have unsuspected sources of error" and insisted that ν measurements be checked "by all promising means."

Although it was against standard policy to allow Los Alamos researchers to perform experiments elsewhere, Oppenheimer convinced Groves that Segrè should be sent to Chicago to exploit the greater neutron flux available from the Metallurgical Laboratory pile, arguing that no one at Chicago had the expertise to make the measurement he wanted. Segrè planned to calculate ν for ^{235}U from α, the ratio of the radiative neutron capture cross section of ^{235}U to the neutron fission cross section.[26]

Oppenheimer's push for Segrè's Chicago measurement, which would be augmented by a series of capture cross-section measurements at the cyclotron in Los Alamos, signaled the beginning of increased interest in α measurements. Naturally, competition from capture, if severe, could greatly decrease the number of expected neutrons in the fast-fission chain reaction – a danger that had not gone unnoticed. Oppenheimer, Fermi, and other leaders had previously assumed, however, that no other process would compete with fission, a reasonable assumption given the relative rapidity of the fission process in comparison with absorption processes. Although one α measurement for ^{235}U was listed on Bacher's September outline and a series of capture cross sections of gold and other nonfissionable materials had been started, no thorough series of capture cross section or α measurements had been planned for ^{235}U or ^{239}Pu. P-Division did not have enough time to check thoroughly every apparently reasonable assumption or to explore every possible disastrous result. As Oppenheimer's response to autumn ν measurements shows, however, he was always ready to divert effort to solve a problem that suddenly appeared pressing.[27]

Groves honored Oppenheimer's request, and preparations began for Segrè's α measurement, along with other ν measurements at Chicago. By the end of October Segrè had built an ionization chamber and amplifier and left for the Metallurgical Laboratory, where he planned to irradiate a very thin foil of fissionable material in order to calculate α.[28]

As he was setting up his equipment in October, results came from the first round of neutron scattering experiments from various tamper materials. Manley's group had begun these measurements in August, well aware that reliable results would be difficult to obtain given the available detectors and the limited understanding of scattering as a function of neutron energy. Data from preliminary scattering measurements made in 1942 at Wisconsin proved difficult to interpret because they showed "very large differences in the scattering cross sections at relatively similar angles." Because the primary purpose of scattering experiments was to provide comparative scattering measurements of potential tamper materials rather than produce independently verifiable scattering cross-section measurements, Manley's group first made straightforward backscattering measurements at the Cockcroft–Walton of neutrons scattered from 130° to 180° on various materials. As Bacher reported in October, the group used disks "of scattering material 10 inches in diameter and approximately 1 inch thick" and a detector that consisted of "a thick paraffin layer mounted on one side of a spherical chamber which has been coated with gold." This experimental arrangement allowed observation of backscattering "for neutrons scattered without energy loss and for some neutrons with reasonably small energy loss." By unit weight, lead was the most efficient element at backscattering, although its relatively low density made it an unlikely tamper material.[29]

In November, Bacher decided that future scattering measurements, which would include both transmission and backscattering measurements, would be performed using the Van de Graaff, because the energy of incident neutrons was higher at the Cockcroft–Walton than that expected in the fission spectrum. In addition, the number of materials under investigation was reduced to those considered most promising: uranium, tungsten carbide, lead, beryllium oxide, and iron.[30]

In May, as transmission and backscattering measurements continued, the Van de Graaff group began using neutrons from an yttrium–beryllium source to bombard spheres of uranium, tungsten, and other materials. The resulting measurements, although much harder to interpret than experiments using disks, showed how elastic and inelastic scattering and capture (conditions closer to those in the actual bomb) affected scattering properties.[31]

As Manley and Henry Barschall described in "Nuclear Physics," the "most extensive scattering data" were obtained with measurements of disks, a method that "has the advantage that the intensity is good, but the need to take differences in order to get angular distributions decreases the accuracy of such results." To obtain the best data, measurements were taken at several angles, "appropriate corrections" were made on the basis of known factors such as inelastic and elastic scattering, and the total scattering cross section was then calculated. Such measurements showed a particularly high scattering cross section for tungsten carbide.[32]

As tamper scattering experiments were being conducted in the fall and winter of 1943–4, considerable excitement and tension were building in response to higher-priority nuclear constant measurements. Good news came from delayed neutron emission measurements. In November 1943 the cyclotron group completed measurements using the time-of-flight method suggested by Baker (Chapter 5), despite the troubles experienced in October with the cyclotron. As Bacher announced to the Governing Board on 18 November, the time-of-flight method, which was capable of detecting emission delays greater than about 10^{-9}sec, showed that "there is no significant delay." In January, Wilson's coincidence measurement confirmed this negative result for ^{235}U and provided the first evidence that there was no delay for ^{239}Pu, which included measurements of $\nu(^{239}\mathrm{Pu})$. In July 1944 Wilson confirmed the results for ^{239}Pu using his coincidence method. Near the end of the project, Wilson "showed that all fissions take place within a time of about 10^{-10} or 10^{-11}[sec]," using "a more refined experiment" capable of measuring "the recoil of the compound nucleus" with a double ionization chamber.[33]

Although the accumulating evidence laid to rest the distressing possibility of a delay in the emission of neutrons, P-Division now had other worries. On the heels of changing ν measurements in November 1943, McDaniel made the first Los Alamos ^{235}U cross-section measurements at low energies using a "very ingenious" experimental setup, as Bacher later noted. Bacher explained in a late 1943 progress report that measurements from "a multi-plate fission ionization chamber ... coated with approximately 1 gram of 25 ... were compared with those from a boron chamber over the same energy range" at the cyclotron. The fission cross section of ^{235}U could then be calculated from the known properties of boron. Such measurements were possible because the Cornell timing equipment had been "reconstructed with a new arc modulation set."[34]

The results were unexpected. Preliminary measurements in mid-November indicated "a number of rather narrow fission levels at low energy which appear to give a large increase in the cross section at about six volts." By mid-December, Bacher reported "four resolved resonances," from 3.2 eV down to 0.3 eV, and throughout 1944 more low-energy resonances were found. By the end of the project, nine resonances had been identified, at 0.3, 1.1, 2.0, 3.3, 4.8, 5.7, 9, 12, and 40 eV. Bacher recalls that the news of the first resonances came as a shock:

the day that he planned to announce the first solid evidence from McDaniel's measurements, the theory group concluded that "there would be no fission resonances in the thermal region."[35]

Discomfort prompted by the lack of agreement between low-energy-fission ^{235}U cross-section results and theoretical predictions intensified when Segrè began capture cross-section measurements after returning from Chicago. He irradiated a number of elements using neutrons from the large graphite block at the cyclotron and found that his results differed from what the Chicago experiments had predicted. On 15 January 1944, Bacher reported that the observed discrepancies did "not show either a regular or an understandable variation from the total cross section determined by the neutron beam method in Chicago." Worried that Los Alamos results might be skewed by a larger-than-expected α, Segrè irradiated ^{235}U at the cyclotron and looked, unsuccessfully, for ^{236}U, which he thought might be produced from slow-neutron capture. The notion that ^{236}U production might signal trouble did not die, however. In March, in the midst of emerging evidence for narrow low-energy ^{235}U resonances, Oppenheimer noted that since the resonances are "quite narrow ... the rate of fission is slow." It therefore "occurred to many people that the process of radiative capture of the slow neutron forming 26 might compete with fission." An additional problem was that available α measurements disagreed: Chicago measurements showed 0.30, whereas Los Alamos measurements showed 0.07. Given the accuracy of the measurements, the discrepancy was "significant" and "a great deal of effort" would be expended in tracking it down.[36]

By this time, Oppenheimer faced more inconsistent measurements. On 3 February, Bacher gave the status of ν measurements. Two absolute $\nu(^{235}\mathrm{U})$ measurements at Chicago, the one made by Segrè and another by Fermi, in which ν was calculated from the behavior of enriched material when inserted in a pile, yielded 2.15 and 2.18, respectively. The $\nu(^{239}\mathrm{Pu})/\nu(^{235}\mathrm{U})$ measurements were not so consistent, however. After the summer measurements, two further series were made at the cyclotron, one with a hydrogen-filled proportional counter and one with an argon-filled proportional counter. The measurements yielded a ratio of 1.03 and 1.12, respectively, but when repeated, both gave 1.15. Bacher admitted that P-Division "did not understand this discrepancy, which is somewhat large for the statistical errors," and warned against "giving too much importance" to results until more measurements could be made with larger amounts of plutonium, which were expected from Clinton in the next few months.[37]

In the meantime, two more problems emerged with plutonium measurements. In determining the energy spectra of ^{239}Pu fission fragments with the photographic plate method, Williams's group found that "the spectra of 25 and 49 are slightly different in the region from 1 to 2.5 MeV .... Some difference is also observed between the spectra obtained by the photographic plate method and that obtained with an ionization chamber method, these new results accentuating the differences." Bacher said he was not sure "what significance should be attached to these observed differences due to the difficulty of interpretation." When reviewing wartime fission spectrum measurements in a recent interview, Bacher admitted, "We didn't know the spectrum of neutrons ... very well ... there were a lot of unknowns." The inability to discount the possibility that the fission spectrum for ^{239}Pu was signficantly different from that for ^{235}U was doubly disturbing. The uncertainty limited understanding of the ^{239}Pu spectrum, which was crucial to the evolving understanding of the ^{239}Pu fast-fission chain reaction. The uncertainty also cast $\nu(^{239}\text{Pu})$ measurements into doubt, since they hinged on $\nu(^{239}\text{Pu})/\nu(^{235}\text{U})$ measurements that were based on the assumption that the fission spectra of the two isotopes were similar.[38]

In the same month, problems also emerged with plutonium fission cross-section measurements. On 20 April, Bacher told the Governing Board it appeared "certain that the plutonium cross sections must be revised downwards by at least 15% because of incorrect lifetimes assumed." A week later, Williams wrote to Bacher, "In view of our increased knowledge and improved technique it would be wise to repeat the whole series of measurements to give more accuracy in absolute values and more detail in the dependence of the [fission cross section of ^{239}Pu] on the neutron energy."

By the time Bacher amended the plutonium cross-section values at the April Governing Board meeting, much greater trouble was brewing with plutonium measurements. As we discuss in Chapter 12, evidence was beginning to mount for a high neutron background in pile-produced plutonium resulting from spontaneous fission, an effect that eventually led to the greatest crisis wartime Los Alamos was to face. Although the enormity of this crisis was not yet clear, Los Alamos leaders had become deeply worried by spring 1944, not only about these measurements, but also about the growing list of conflicting results and persistent uncertainties of nuclear constant measurements. As Wilson remembers, P-Division realized "there were mistakes in the measurements," some of which were "very large." The spontaneous fission measurements were the "principal" worry, but Wilson was also upset by the status of fission cross-section and ν measurements, which were, in his opinion, "a disgrace." Wilson reacted with a "tremendous furor." So many important measurements seemed to be "based on shifting sand .... I mean, how could we call ourselves physicists!"

Despite Wilson's understandable frustration in the spring of 1944, by the end of the summer, impressive progress had been made in understanding both ν and fission cross-section measurements. To obtain more reliable ν measurements, Manley assigned Robert Walker, a graduate student from the University of Chicago, to implement an idea suggested by Otto Frisch. As Bacher explained in mid-January, Walker was charged with making "an absolute source calibration," using a method "previously explored in England .... The method essentially uses a large tank of borated water and determines the integrated boron activity through the intermediary of indium foils." From the known properties of boron, Walker was able to calculate accurately the number of neutrons per second produced by a radium–beryllium source, which he called source 43. At the same time, Thoma M. Snyder and R. W. Williams made a series of ^{235}U and ^{239}Pu ν measurements using "cadmium-covered indium foils placed at various distances from the fissionable sample ... and in various directions," first using the carbon block of the cyclotron as a neutron source and then using source 43. From such measurements, they could "determine the number of fission neutrons in terms of the standard source strength." After determining the number of fissions from the known fraction of fissionable material in the sample, Snyder and Williams were able to calculate ^{239}Pu and ^{235}U ν measurements with a probable error of 5 percent. In June 1944, they reported that $\nu(^{235}\mathrm{U})$ was 2.44 and $\nu(^{239}\mathrm{Pu})$ was 2.86, values impressively close to the 1979 values of 2.43 and 2.89, respectively.[39]

Measurements by Benedict and Hanson of the fission cross section of ^{235}U around 1 MeV made at the Wisconsin Van de Graaff are close to modern values. Although these measurements seemed reliable in 1943, the Van de Graaff group, in cooperation with the detector and Cockcroft–Walton groups, took careful steps to ensure their reliability, both because these measurements were themselves important, and because the ^{235}U cross section would be used in evaluating other fission cross section measurements, including those for ^{239}Pu. In addition, Los Alamos experimenters determined fission cross sections over a wider range of fast-neutron energies than had previously been measured.

By early fall 1944, J. Williams reported that an extensive series of ^{235}U cross-section measurements had been made with the long counter. To test its accuracy, he measured fission cross sections at 1 MeV with the long counter using a radon–beryllium source that had been calibrated by Alvin Graves. The long counter measured the cross section as 1.37 barns, which was "in excellent agreement" with measurements taken by Hall, Koontz, and Rossi using the argon-filled electron collection device and a fission counter. These results greatly increased "confidence" in measurements in the 1-MeV range. The long counter was then used to measure fission cross sections from 5 keV to 1.9 MeV. To gain further confidence in ^{235}U fission cross-sectional measurements below 400 keV, the Van de Graaff group used two identical high-pressure ionization chambers to make relative measurements at 50 and 500 keV. After assessing all the data obtained by these methods, Williams concluded: "The general form of σ_f 25 as a function of neutron energy is fairly well established .... We believe σ_f is trustworthy to approximately 15% throughout the entire energy range above 15 keV and is known to be between 5 and 10% from 200 keV to 2 MeV." At 1 MeV, the ^{235}U cross section was measured as 1.33 barns; the modern value is 1.22 barns.[40]

To find the fission cross section of ^{239}Pu, the Van de Graaff group made comparison measurements with ^{235}U. These measurements benefited not only from the proven reliability of ^{235}U fission cross-section measurements, but also from advances that aided counting, which had always been a problem with ^{239}Pu because of its copious α production. As Williams explained, by fall 1944 they could "use larger quantities ... in a simple comparison chamber. This improvement results from the design of faster amplifiers by the electronics group and the development of an electroplating technique for 49 by Dodson's group." Thin foils of electroplated ^{239}Pu were put in a comparison chamber with ^{235}U foils and the samples were irradiated simultaneously with neutrons from the Van de Graaff. The variation of the ^{239}Pu fission cross section with energy bore "only a rough similarity to that of σ_f 25." In particular, Williams found that the fission cross section was less than that for ^{235}U "between the lowest measured point for fast neutrons and 100 keV." He judged that "statistical errors of counting" could contribute "less than 3%" errors, and "uncertainties in the masses of the foil – possibly another 3%." Fission cross-section measurements of ^{239}Pu would be extended from fall 1944 to summer 1945 (Chapter 17).[41] Because P-Division had provided reliable ν and fission cross-section measurements by summer 1944, the remaining crucial gaps in nuclear constant information were the fission spectrum and α measurements. Unexplained discrepancies between ionization chamber and photographic-plate fission spectrum measurements remained until the end of the project.[42]

The Water Boiler

The Water Boiler, a reactor using ^{235}U in a water solution, had been proposed in April 1943 as part of a program to measure critical masses from the neutron multiplication of chain-reacting systems. Bacher had argued that the laboratory should "start with something that would have a smaller critical mass, like a water solution" and could be built sooner than a ^{235}U metal assembly. Although the Water Boiler could not indicate the critical mass of a weapon with a ^{235}U metal core (because critical mass depends on the chemical, physical, and geometric conditions), it could check the theory for calculating critical masses, determine the effect of various tamper materials on critical mass, and provide experience in assembling a supercritical system. Bacher recalls his "battle with some of the theorists on this," since they considered critical mass studies with chain reacting systems unlike the one used in the weapon to be a "waste of time." But Bacher prevailed, and plans for the Water Boiler proceeded in summer 1943. This early step in critical mass studies would be followed, as material became available in 1944 and 1945, by neutron-multiplication measurements of uranium hydride cubes, and ^{235}U and ^{239}Pu metal assemblies (Chapter 17).[43]

Kerst, a University of Illinois physicist known for his invention of the betatron, was appointed to head the Water Boiler project. He had had no previous direct experience with building a chain-reacting pile, and could draw on little expert help, for the only scientists experienced in critical assemblies were at Oak Ridge and Chicago; they were not available to Los Alamos on a full-time basis in 1943. Fermi could offer advice only intermittently. Thus, Los Alamos had to develop its own critical-assembly expertise.[44]

The Water Boiler crew came primarily from Purdue University, where they had been working on measurements for the Super. As Raemer Schreiber, one early crew member, recalls, "There weren't any experts in nuclear reactors, so they were picked up from cyclotron people and nuclear physicists." The key crew members involved in assembling and operating the low-power water boiler were Kerst (group leader), Baker, Gerhart Friedlander, Lindsay Helmholtz, Marshall G. Holloway, L. D.

P. King, and Schreiber.[45] Help with general theory came from Robert Christy, whose calculation of the Water Boiler's critical mass came to be known as the "Bible" among the Water Boiler staff.

In early summer 1943, Kerst sketched out a design for the world's first chain-reacting system using enriched material: a simple hollow stainless steel sphere about 1 foot in diameter filled with a solution of a sulphate of ^{235}U and water, surrounded by a BeO reflector, and neutron shielding. No one doubted that it would "go critical" when enough enriched uranium from Oak Ridge was put into it. The main question was the amount of enriched ^{235}U that would be needed.[46] The location of the boiler was also an important consideration, for it had to be accessible from the Technical Area, but also far enough from the town to avoid contamination from radiation leaks or other disasters. Christy calculated the probable area that radioactive materials would cover if an accidental explosion occurred; Oppenheimer's favored site, in upper Los Alamos Canyon just down the cliff from the town and downstream from the water supply, provided that margin of safety. The Governing Board approved this site, designated Omega, at its 19 August 1943 meeting.

The most important technical decision concerning the Water Boiler in the summer of 1943 concerned its operating power. As originally conceived, the Water Boiler would have operated at 10 kW and provided a strong source of neutrons.[47] Although a strong neutron source would have been useful for some experiments, members of the Governing Board objected to high-power operation, in part because it would be difficult to protect operators from a radiation level as high as several thousand curies. Moreover, the ^{235}U would become contaminated with radioactive fragments, which would need to be chemically removed; a complicated cooling system would be required (the enriched uranium was so scarce that it could not be kept in large reservoirs outside the reactor); all of the uranium had to be used at once; and the bubbling of the uranium "soup" would make the neutron output fluctuate even more unpredictably than usual.[48] The proponents for a high-power level won temporarily, as tentative plans were made to operate at 10 kW. But the push for high-power operation collapsed a few weeks later at the 30 September meeting. Samuel K. Allison and Fermi pointed out difficulties in operating at high power and mentioned that the Met Lab no longer needed a high neutron flux to study poisoning effects.[49]

Using Christy's calculation that 600 g of pure ^{235}U would be critical in an infinite water tamper, the Water Boiler group moved ahead with a more detailed, though still mechanically simple, design. Schreiber and

King developed a stainless steel fluid handling system.[50] The "soup" rested in a conical stainless steel reservoir, the shape of which provided a safe geometry, and allowed access to the solution, so that its concentration could be increased as ^{235}U arrived at Los Alamos. The soup was pumped by air pressure into the sphere surrounded by a tamper where the reaction took place. The entire system was closed to the atmosphere, so that any radioactive gas generated during the reaction would not be expelled accidentally. A control rod – similar to, but smaller than, the control rods on the Chicago pile – extended into the tamper shell and could be dropped in quickly to quell a runaway reaction.[51]

An important chemical decision concerned the type of radioactive uranium salt to use; the enriched salt had to be completely soluble within the stainless steel sphere. Chemist Helmholtz's experiments with various compounds narrowed the choice to nitrate and sulfate.[52] In mid-November, the sulfate was chosen because it was more soluble and absorbed fewer neutrons.[53]

Although simple in theory, the Water Boiler was not simple to build. The stainless steel sphere was difficult to construct because the hemispheres could not be joined using solder, which the acid soup would have corroded. Unfortunately, the company that made the sphere for Los Alamos did not understand the requirement and soldered it anyway. Los Alamos ultimately arc-welded the sphere.[54] The size of the sphere was also crucial. If too small, a critical mass could not fit in; if too large, the geometry might not allow a critical mass with minimum ^{235}U. When a 12-inch-diameter sphere was finally selected, a new contour had to be ground into the BeO tamper blocks, which had been fabricated to a different spherical diameter. The tamper that surrounded the sphere to reflect neutrons back into the soup was made from pure beryllium oxide powder pressed into molds to form dense bricks. However, pure BeO proved to be difficult to procure and fabricate into bricks, in part because of backlogs in the shop. The neutron detectors, the most complex part of the Water Boiler system, required collaboration from everyone in the group. The team had time to devote to this problem because the shipment of ^{235}U, expected by late 1943, was delayed.[55] Indeed, sufficient ^{235}U did not arrive until April 1944.[56]

The new building to house the Water Boiler, Omega, became usable (although still incomplete) on 1 February 1944, and the reactor materials came together in March: the BeO bricks arrived acceptably pure, the stainless steel spheres were properly welded, and the fluid-handling equipment was installed.[57] By 1 April, the sphere had been

Fig. 10.3. The Water Boiler was the first reactor to use enriched uranium and the third reactor ever to go critical. LA Photo, 12784.

fitted into the tamper and five different neutron-counting devices were placed in different positions.[58] The Water Boiler crew began to fine-tune their equipment by running experiments with unenriched uranium and a radium–beryllium source. By 1 May, enough enriched material was on hand to begin subcritical experiments.[59] With its reactivity controlled by cadmium rods, the Water Boiler went critical on 9 May 1944, first with a neutron source in the center and later in the day without the source.[60]

The group worked for several weeks to improve their apparatus, while Christy sharpened his prediction of the critical mass.[61] With new data on cross sections, Christy made a quick correction to his original calculation of 600 g and revised the estimate to approximately 575 g. The group removed tubes in the tamper and filled holes used to insert a ^{238}U chamber into the center, thereby making the tamper shell more uniform. They found the critical mass of the uranium sulfate solution in the boiler to be 565 g.[62] Although theory and experiment arrived at an excellent agreement "somewhat fortuitously," this success helped to raise confidence in T-Division's ability to calculate critical mass.[63] Other studies in May measured the period of the reactor as a function of control rod setting and measured the effect of tamper materials on the critical mass.[64]

By June 1944, the Water Boiler had fulfilled its initial goals: it had provided a means of checking theoretical critical mass calculations, indicating how tamper materials affect critical mass, and it offered experience in assembling a supercritical system. It was time to address new problems. The group had already started work on hydride critical assemblies (Chapter 17). A vital project was the high-power Water Boiler, which would be constructed in part from dismantled equipment from the first Water Boiler. The power of the new reactor, 1 kW, would allow power operation and a high neutron flux, but would not contaminate the material so much that it could not be reclaimed. After June 1944, Water Boiler experiments would be designed to facilitate the transition to this new device.[65]

The Super

By February 1944, it was clear that making a Super would be far more difficult than originally thought, and research on the thermonuclear weapon began to receive less attention. Only Teller's theoretical group and Egon Bretscher's small experimental group that measured cross sections relevant to the Super continued to work on the problem. This work did not contribute to the main goals of Project Y, but it ultimately broke ground for the successful work on hydrogen bombs in the early 1950s by furnishing background information on cross sections for deuterium reactions and on the transfer of energy from a fission reaction to the surrounding medium.

In September 1943, Teller felt optimistic enough to ask the Governing Board to increase the level of the Super effort. Upward revision of the D-D and D-T cross sections implied that less material would be needed. As further justification, Teller cited indications that the Germans were going to use deuterium for similar purposes. However, the Governing

Board recommended that no more than one full-time person – Teller, Emil Konopinski, or Nicholas Metropolis – should work on the problem. Oppenheimer selected Konopinski, who worked on the Super for the remainder of the war.

In February 1944, Teller brought Stanislaw Ulam, Jane Roberg, Geoffrey Chew, and Harold and Mary Argo into his T-Division group. Mathematician Ulam began calculations on inverse Compton cooling of the system; Roberg calculated the ignition temperature of deuterium–tritium mixtures. That month, Teller officially informed the Governing Board that because of theoretical difficulties and the possibility that the Super would require tritium, "it appears that the development of the super may require longer than was originally anticipated." At the summer 1942 Berkeley conferences, Konopinski had suggested the need for tritium. Because of the scarcity of this material, tritium had been given little attention.[66] In February 1944, he suggested increasing the priority of the program.[67] But the board did not wish to expand Super work, because "the members of the board desired to produce something that would play a part in this war." However, they did not wish to cut it back entirely, as long as the research did not interfere with the more immediate program.[68] As a result, wartime research on the Super dropped in priority.[69] In the period between February and June 1944, Teller and Bethe frequently argued about the reduced status of Super work, which Teller opposed. After Oppenheimer separated Teller's group from T-Division in June 1944 (Chapter 8), Teller reported directly to Oppenheimer. A few months later, Teller's group would join the newly formed F-Division under Fermi. In this uncertain institutional arrangement, the preliminary research for the present hydrogen bomb proceeded with considerable difficulty. However, Teller never wavered in his dedication to the Super project, despite its dim prognosis for use during World War II and its limited support from the laboratory throughout Project Y.

11

Uranium and Plutonium: Early 1943 to August 1944

The story of the production of fissionable materials at Los Alamos is about the challenges of working with little-known, scarce substances under difficult experimental conditions, as well as the excitement of discoveries and unexpected turns in the course of all-out efforts to achieve practical results quickly. The interplay between the plutonium and uranium efforts within CM-Division reflects the wartime strategy of pairing complicated and straightforward tasks. In this way, personnel, equipment, and time could be focused on the most demanding problems. Thus, the relatively simple effort to produce uranium gun parts at Los Alamos complemented the more difficult effort to produce plutonium spheres, just as the relatively simpler gun program as a whole later complemented the more complex implosion program.[1] Those implementing the less intricate effort were under pressure to proceed rapidly and produce absolutely reliable results meeting all contingencies so that more of the group's resources could be diverted to the thornier problem. Consequently, the uranium program was remarkably fast-paced and rigorous. The need to make the most of resources and save time weighed especially on Joseph Kennedy, Arthur Wahl, and Cyril Stanley Smith in CM-Division, because they had to adjust to the changing requirements

* This chapter is based on a manuscript by Catherine Westfall. We are grateful to Gordon Baym, Les Redman, and Robert Penneman for their detailed technical editing of the chapter.

of the other divisions, for whom they provided support services, while at the same time working to achieve their own goals.

Although uranium chemistry and metallurgy had been investigated before the beginning of Project Y, the element provided a few surprises. Planning for a uranium hydride weapon yielded new insights. The division discovered the correct chemical formula of the pure compound, UH_3, and found that the hydride would have an unexpectedly high critical mass because it could be produced with a density far below theoretical estimates. With this finding, the uranium hydride weapon was doomed.

However, the chemistry and metallurgy of plutonium, an element discovered only two years before the start of Project Y, provided the greatest drama. The most daunting chemical task was to remove light element impurities. This stumbling block in the development of the plutonium gun was tackled by a number of workers, most notably the codiscoverers of the element: Glenn Seaborg at the Chicago Metallurgical Laboratory and Kennedy and Wahl at Los Alamos. Progress was spurred by heated competition, as both groups worked feverishly to produce their own purification schemes. By the time Wahl found ways to produce plutonium that met the necessary stringent purity requirements, however, the competition was moot. At just this point, Segrè's group discovered that too much ^{240}Pu was created in pile-produced plutonium to allow plutonium to be used in a gun. Ultra-pure plutonium was no longer needed. Because existing procedures could easily produce plutonium of a purity suitable for the implosion weapon, the crisis that shook the laboratory in August 1944 ironically lightened CM-Division's load.

Surprising results also emerged in plutonium metallurgy, the responsibility of Smith, head of the metallurgy group. Because too little plutonium was available at Los Alamos for extensive experimentation, Smith assumed that the substance would have the charactistics of uranium and used uranium as a stand-in for preliminary plutonium research. Later results showed this assumption to be invalid. Puzzling discrepancies in density measurements led to the realization that plutonium metal, the most complicated metal known to man, has six allotropic phases, more than any other metal. The melting points of ranium and plutonium also differ by hundreds of degrees. Understanding these characteristics was crucial to creating the combat sphere for the plutonium bomb.

Planning for Plutonium Metallurgy, Early 1943 to August 1943

In a December 1942 conference with Arthur Compton, Seaborg had determined that his Chicago plutonium research group, with the help of Berkeley, would be responsible for both "the purification and preparation of plutonium metal."[2] Thus, when planning the Los Alamos chemistry program in early 1943, Oppenheimer envisioned only a small staff of chemists to provide service chemistry for other divisions. Kennedy managed this effort, while Smith looked after metallurgy. Overlapping small-scale plutonium purification efforts were also under way at Ames and Berkeley, which hosted a vigorous program in developing for plutonium metallurgy the necessary "refractories" or "crucibles" (vessels capable of withstanding the action of adverse physical and chemical conditions, especially high temperatures and corrosive materials).[3]

By April 1943 the new laboratory had been assigned full responsibility for large-scale plutonium metallurgy, a task that could not be performed at the Chicago Metallurgical Laboratory, since it lacked the staff and the facilities. Seaborg and his staff were still eager to find a small-scale plutonium metal production method so they could make the first physical measurements of plutonium metal at Chicago. Among the properties to be measured was density, a vital component of critical mass calculations. As former Met Lab (and later Los Alamos) staff member Alan Florin recalls, "There was a sense of competition for making plutonium metal ... if we could get the metal first, it would be a feather in our cap." To accomplish the task, the group had to grapple with the inherent difficulties of working with the scarce, newly discovered element and produce results quickly or the task would certainly be assumed by Los Alamos metallurgists when grams of plutonium became available.[4]

By May, the Chicago group's position became less secure. Groves decided, on the Lewis review committee's recommendations of 10 May, to recruit a considerable number of chemists at Los Alamos so that plutonium could be purified there. This decision called into question the allocation of purification responsibilities. In the competition with Los Alamos, as Florin recalls, Seaborg's group decidedly had the edge at this juncture. Because du Pont engineers progressively took over the separation procedures throughout 1943, metal production and purification became the Chicago group's most important tasks.[5]

Luckily for Met Lab chemists, Oppenheimer wanted to keep the purification research performed at Los Alamos to a minimum. The CM-

Fig. 11.1. Button of plutonium used in fabricating the Fat Man. LA Photo, 1833.

Division at the laboratory was already heavily burdened by its service chemistry tasks. The work, which came under the leadership of Richard Dodson in fall 1943, included such major tasks as providing material for the Water Boiler and RaLa, as well as a stream of minor but time-consuming tasks such as preparing plutonium and uranium foils for physics experiments and producing BF_3 for ionization chambers.[6] In addition, the CM-Division was charged with providing fissionable, tamper, and case materials for the bomb, even though many crucial questions had not yet been answered concerning the fissionable materials to be used, the level of their purity, and the time that could be spent on research before sufficient material arrived for metallurgical procedures. Oppenheimer explained to Groves in late May 1943 that although "the headquarters and ultimate concentration" of purification work would be at the new laboratory, "every advantage will be taken of the laboratories which exist elsewhere and which are equipped to study this problem." With this provision in mind, Kennedy estimated in early May that the Los Alamos chemistry effort would require "about twenty-five chemists and twenty-five assistants."[7] The agreement struck was that when gram amounts of plutonium became available – by the earliest estimates, this would not occur until November – Los Alamos would focus on larger-scale chemical and metallurgical processes. A small microscale effort would be established at Los Alamos, but Berkeley and the Met Lab were encouraged to pursue vigorous plutonium purification research programs.[8]

To ensure that Los Alamos received the full benefit of the plutonium research, it was important to facilitate communication between Los Alamos and the other laboratories, particularly the Met Lab, where the bulk of the work was being performed. Groves, who took a hard line on compartmentalization in the wake of his argument with Oppenheimer over open colloquia, resisted such communication. On 17 June, the general issued a memorandum specifying that only authorized persons would be involved in the interlaboratory exchange of information. He listed the allowed and forbidden topics of discussion.[9] To circumvent such restrictions for plutonium research, Oppenheimer first recruited Charles A. Thomas, research director of Monsanto Chemical Company, to coordinate plutonium efforts among the laboratories and improve communication between the Met Lab and Los Alamos. Plans were soon under way for a monthly research status meeting in Chicago to be attended by representatives from groups working on plutonium chemistry and metallurgy.[10] In July, Thomas began attending Met Lab meetings, and by the next month he reported to Conant and Groves that interlaboratory monthly meetings on plutonium research would be established at Chicago.[11]

The Start of Metallurgy, April–October 1943

Because plutonium was still available only in microgram amounts, which were insufficient for metallurgical research, and since plutonium chemistry remained the province of the Met Lab, CM-Division immediately began an active program in uranium metallurgy.

The energetic British-born metallurgist, Smith, found a number of challenges awaiting him when he left the American Brass Company to head the Los Alamos metallurgy group in April 1943. He did not know for sure whether ^{235}U, ^{239}Pu, or some other substance would be used as the active material in the weapon, whether the form of the material would be a compound or a metal, whether alloys would be needed, or when the material would arrive. Although Frank Spedding at Ames had successfully developed foundry-scale procedures at the 25-pound scale and higher for producing the natural uranium metal (composed of both ^{235}U and ^{238}U) needed for the plutonium production piles, Smith could not use these procedures directly for the ^{235}U to be used in the weapon, for the highly fissionable isotope would achieve critical mass at the scale of only a few pounds. Also, it was scarce and had to be extremely pure to be used in the weapon.

Besides devising workable ^{235}U metal production procedures, Smith had to provide fissionable and nonfissionable material for various tests and experiments and also make the bomb core, tamper, and case. Realizing that the effort to create plutonium metal would consume most of the group's time and effort once sufficient quantities of the precious substance became available, he bore the considerable burden of providing reliable procedures and anticipating a variety of contingencies so that uranium metallurgy could proceed as smoothly as possible. The first item on Smith's agenda was recruitment, a difficult task because of the high wartime demand for metallurgists, who were needed to produce steel for guns and ships.[12] He also had to arrange for the acquisition, transport, and assembly of the necessary metallurgical equipment, including presses, large furnaces, and vacuum systems. Although this was standard equipment, acquisition and transport were slowed by wartime conditions. Despite these difficulties, enough equipment had been assembled by July for research to begin.[13]

Smith first tackled the production of uranium hydride, which was being considered for use in a bomb because of its high concentration of neutron-moderating hydrogen. Even before all the necessary equipment had arrived, the Ordnance Division requested pressed uranium hydride cubes for tests.[14] The research on uranium hydride began with an investigation of the characteristics and preparation of the compound. Anticipating bomb requirements, the group was particularly interested in producing a high-density, nonpyrophoric form of the substance.[15] To reduce pyrophoricity, the researchers coated hydride samples with a film of paraffin. To study the possibility of creating material having a density close to the theoretical value of 10, they made several small batches of uranium hydride powder in July and checked the rate of uranium hydride formation at various temperatures. They found a maximum rate of uranium hydride production at 225° C and noted that the compound did not form at temperatures of 450° C or greater. However even when a sample was subjected to "pressures of 200,000 lbs per square inch," it was possible to obtain only a density of 8. The lower density of uranium hydride in preparation increased its critical mass twofold and thereby provided further evidence that the material was unsuitable for use in a bomb.[16]

Although the practical goal of these studies was to produce fissionable material for a bomb, they yielded new insight into uranium hydride. Spedding, the first to make the compound, had identified its chemical formula as UH_4. By 31 July, Smith reported that when hydrogen absorption was measured in decomposition tests, the amount absorbed was nearly that of UH_3 every time. Independent tests conducted by chemist Morris Perlman in August confirmed that there are indeed three atoms of hydrogen per atom of uranium in the hydride.

Smith and his team of metallurgists shouldered most of the responsibility for uranium research, but the chemistry group also made important contributions. By fall 1943, chemist Lindsay Helmholtz was investigating various aqueous uranium compounds for use in the Water Boiler, the first critical assembly at Los Alamos; in addition, the chemistry group was determining density, a job within their expertise because the samples were so small. By investigating hydride samples produced in a steel container ("bomb"), Perlman and Samuel Weissman were able to confirm in October that uranium hydride had a density of 10.[17] That month, the Governing Board asked the metallurgy group to produce centimeter cubes of uranium hydride. These cubes would have to have high density, accurate dimensions, and adjustable hydrogen–uranium ratios. In addition, the group was asked to furnish material for nonaqueous critical assemblies and other tests.[18]

At the same time, the metallurgy group began developing techniques for producing ^{235}U metal. Whereas Spedding had developed an entirely satisfactory ton-scale production of natural uranium lugs for plutonium production reactors, Smith at Los Alamos had to worry about avoiding criticality hazards with highly enriched ^{235}U, maintaining chemical and isotopic purity of the material, and preventing the loss of even milligrams of the precious ^{235}U. In view of these considerations, plus the desire to provide absolutely reliable procedures under a variety of conditions, the group decided to look at more expensive and exotic methods of producing metal, such as thermal dissociation and atomic hydrogen reduction of uranium compounds, as well as by the more standard bomb reduction

Fig. 11.2. Metallurgist Cyril Stanley Smith co-head with Joseph Kennedy of the chemistry-metallurgy division. LA Photo, LAT 751.

method. This work gave them experience in providing uranium metal for experiments and weapon production. The work also promised to yield additional information. Guessing that uranium was metallurgically similar to plutonium, Smith planned to use the more abundant substance as a stand-in for plutonium.

Although Spedding's large-scale uranium metal production techniques had used stationary bombs for the smaller-scale and higher-purity reductions needed at Los Alamos, Smith's group had to devise a new bomb design, as well as develop suitable refractories, investigate materials to be used in the reaction, and develop techniques for each reduction scale. Determined to proceed in a rigorous, careful manner, the group planned experiments to test reductions over the range of 1–100 g. Fortunately, Smith had been able to recruit Richard Baker, who had a knack for devising practical techniques and, having worked with Spedding, had a thorough knowledge of the existing methods of producing uranium metal. By early 1943, Baker had successfuly adapted the standard Ames process for use at Los Alamos.[19] By fall, the metallurgy group had begun to work with a 50-kW vacuum furnace, casting projectiles for the Ordnance Division and searching for promising uranium alloys.

Plutonium Chemistry and Metallurgy Begin, Summer to Fall 1943

In the summer and fall of 1943, the CM-Division also began preliminary work on plutonium chemistry and metallurgy. The Division's chemists spent the summer setting up a full-scale chemistry laboratory and providing service to other groups. By fall, the effort had expanded to include purification research aimed primarily at providing analytical techniques. In this period, David Lipkin began devising an analytical method for detecting oxygen that would be sufficiently sensitive to test whether the strict purification standard had been achieved. A spectrographic laboratory was established under the direction of Weissman to facilitate further analytical work. In addition, Wahl began purification studies using plutonium obtained from uranium irradiated in the Berkeley 60-inch cyclotron, originally used in his plutonium separation studies.[20] In the fall of 1943, however, plutonium chemistry was a small effort at Los Alamos in comparison with Chicago; whereas Seaborg had thirteen people working on purification, Los Alamos had only six.[21] According to plan, Chicago took the dominant role in purification work before gram amounts of plutonium were available.[22]

By the end of the year, the Los Alamos purification group decided to abandon microchemical purification projects and work instead with the metallurgy group to devise procedures for the first large plutonium shipments scheduled to arrive early in 1944 from Clinton. By this time, the metallurgists had squarely tackled the challenge of large-scale plutonium metal production. While waiting for gram amounts of plutonium, Smith implemented a vigorous metal reduction program using uranium as a stand-in for plutonium. By the fall of 1943, almost 250 reductions had been made, so Smith's group was prepared to undertake large-scale plutonium metal production.

Efforts to produce plutonium metal on a microgram scale at the Met Lab proceeded with greater difficulty. In March 1943, Paul Kirk, who had come from Berkeley in January to lead metal production efforts in Seaborg's group, began working with microgram amounts of plutonium.[23] Throughout the spring and summer of 1943, however, Kirk and the others in Seaborg's group had trouble finding workable crucibles and making a solid pellet for density measurements. Moreover, they did not know whether they were producing plutonium metal or some other compounds, perhaps with impurities.[24] At this stage, the chemists were handicapped by their dependence on inherently inaccurate small-scale procedures and by the sparsity of information about plutonium; because they could not be sure of the content of the experimental solutions, they had difficulty making a creditable chemical identification of the end product. By August, Kirk reported that his group had encountered "serious difficulties" in their attempts to produce microgram amounts of plutonium metal, adding that unexpected differences between plutonium and uranium made uranium seem an inadequate stand-in.[25]

Kirk's difficulties underlined the need for more plutonium at the Met Lab. At just this time, the need for plutonium at Los Alamos was also increasing. The agenda for the newly launched physics program included, first, the measurement of the neutron number from the fissioning of ^{239}Pu (Chapter 5).[26] Oppenheimer began making special requests for the necessary plutonium. In June, Seaborg learned that Oppenheimer wanted to borrow 200–400 μg of plutonium for two weeks. After consulting with Seaborg, James Franck, then director of chemistry research at the Met Lab, decided they could lend 200 μg, half the amount recently delivered to Chicago. The decision prompted Chicago chemist Burris B. Cunningham to complain that the loan would leave the purification and metal production experiments in Chicago in short supply.[27] Seaborg had the chance to recover the controversial loan in person. After being ordered to take a vacation, he coyly retired to the Santa Fe area. Although he did spend some time relaxing, he also had the opportunity to confer with Los Alamos scientists (but not at Los Alamos). At a predawn meeting in late July, Robert Wilson returned the precious 200-μg sample, "which he was guarding with his personal Winchester ... deer-hunting rifle."[28]

Plutonium was not the only resource in short supply. Seaborg, feeling the pinch in manpower, began lobbying in May to get a higher pay scale for his workers, complaining that he had lost three men to his "competitors." In July, faced with the transfer of many staff members to Tennessee to help with the Clinton pile, Seaborg termed his manpower situation "critical." The NDRC policy against recruiting employees already committed to Manhattan Engineer District projects was breaking down. In response, Joyce Stearns, the Met Lab personnel manager, issued a memo insisting that employers be contacted in the course of recruitment, an admonishment prompted by the problem of "pirating of personnel" from one project to another.[29]

Interlaboratory Collaboration for Chemists and Metallurgists, Fall 1943

In fall 1943, through interlaboratory meetings, researchers began coordinating chemistry and metallurgy at Los Alamos, the Met Lab, and other MED facilities. The meetings were attended regularly by members of Seaborg's group, Berkeley representatives, and Kennedy and Smith. A special interlaboratory meeting at the Met Lab in September examined plutonium metallurgy, and the first official monthly interlaboratory meeting at Chicago held in October focused on plutonium purification and metallurgy.[30]

Disagreements over the division of small-scale metal production surfaced immediately. Although plutonium compounds had been identified, the Chicago group had not yet produced an unambiguous sample of plutonium metal in time for the September meeting. After hearing a report from Kirk on the status of metal production, Thomas suggested that further effort be deferred until larger quantities of plutonium were available. Kirk and Seaborg, however, persuaded Thomas that metal production studies should continue at Chicago.[31]

Determined to obtain a density measurement in the next few months both to expedite the war effort and scoop Los Alamos before gram amounts of plutonium were available there, the Met Lab chemists now took steps to procure additional expertise and material. In September, William H. Zachariasen, a prominent innovative X-ray crystallographer from Norway who had worked with X-ray diffraction since the 1920s, set up an X-ray powder crystallography laboratory at the Met Lab and almost immediately began providing crystal structure information on bismuth phosphate to aid in the separation efforts.[32] Zachariasen adapted X-ray powder crystallography into a unique tool for analyzing the contents of microgram samples of plutonium compounds. Although crystalline structure is usually determined in materials of known chemical identity and density, Zachariasen found he could use an alternative procedure in which he first made partial crystal structure identifications on microgram amounts of a compound. He used this as a starting point for determining chemical identity. He could then, by an iterative procedure, successively refine his determinations of crystal structure and chemical identity.[33]

After the September meeting, the Met Lab chemists thought of another use for Zachariasen's skills: if a large enough sample of plutonium metal could be obtained, perhaps X-ray powder crystallography could be used to determine its structure and density. Accordingly, Franck wrote Thomas urging him to let Kirk have more plutonium for metal production attempts, even if the material had to be taken from that allotted to Los Alamos. Through Thomas, Franck subsequently asked Los Alamos to relinquish 500 μg of plutonium, later explaining that he figured Zachariasen needed 100 μg of the metal for X-ray diffraction studies and that Kirk needed 500 μg of unpurified product to supply this amount. Although Oppenheimer agreed that density measurements should be made by Chicago chemists, not by the overworked Los Alamos group, 500 μg equaled almost the total supply allotted to the laboratory, whereas the Met Lab had almost 2 mg. As Oppenheimer noted in a 14 October meeting at Project Y, "the stocks of material" were "very much greater" at the Met Lab than at Los Alamos, and since Los Alamos could "ill afford" to process plutonium for others, Franck's request was denied. Franck apologized for the request and Zachariasen found a way to work with much smaller amounts. Nonetheless, the continuing problem of plutonium allocation increased tensions between the two laboratories.[34]

In November, the pile at Clinton went critical, after the group had solved a minor problem caused by improper agitation in one stage of the bismuth phosphate process and had found a way to can uranium slugs for irradiation.[35] From the beginning, the pile was clearly a success. Before the end of 1943, a few hundred micrograms of plutonium were sent to Chicago; production schedules indicated that by spring 1944, gram quantities would come to Los Alamos.[36]

The successful operation of the Clinton pile had many ramifications. As a testing ground for a full-scale production plant, it served as a warning that an all-out effort would have to be organized to put protective metal coverings on uranium slugs. Clinton also proved that Seaborg's group and du Pont had developed a successful industrial-scale method for separating plutonium from uranium and fission products, although both Seaborg and du Pont representatives knew that this method would have to be modified for the much larger scale Hanford processes.[37]

The First Burst of Progress in Uranium Metallurgy, Winter 1943–1944

The main pieces of metallurgical equipment were installed while efforts were being made to coordinate the difficult, time-consuming task of plutonium purification in the fall of 1943. During this period, the CM-Division defined the standard uranium metal production method and got a solid start in the development of the other chemical and metallurgical procedures needed to produce and fabricate the uranium for Project Y.

To provide uranium hydride cubes for criticality experiments (Chapter 17), the metallurgy group studied the kinetics of hydride formation and ways to coat the pyrophoric powder to make it inert.[38] By January, the metallurgy group was using polystyrene to impregnate UH_3 powder, producing a nonpyrophoric substance that was readily pressed, although it had a hydrogen-to-carbon ratio of only one. In March, Bacher announced that integral hydride experiments would start soon, and the group set up an apparatus for large-scale uranium hydride production and studied the effect of pressing variables on the density of plastic bonded compacts.

The metallurgy group also began implementing the careful, rigorous uranium metal production program that Smith and Baker had planned. A series of exploratory electrolytic experiments were performed. Stationary bomb efforts met purity and yield objectives. In addition, in January and February they initiated a series of tests on the 1-g scale using a graphite centrifuge and on the 10-g scale using a stationary bomb. The work with the stationary bomb pinned down optimum heating and firing procedures and demonstrated the superiority of the stationary bomb method, which became the standard uranium metal production procedure at Los Alamos.[39]

During this period, the CM-Division also began two new projects. The chemistry group began working on uranium recovery procedures and the metallurgy group developed uranium casting procedures. Recovery procedures were vital to collect the residues remaining in crucibles and slag. The uranium metal would undergo remelting and casting, both to remove impurities and to shape the material. The casting procedures were meant for unenriched uranium to be used in gun tests, because such parts could be cast directly into the desired shape. In contrast, active material parts required further fabrication to closer tolerances. Casting began in the last two months of 1943.[40]

The Difficulties of ^{235}U Procurement, Spring 1943 to Summer 1944

While the CM-Division was defining procedures for producing uranium metal and preparing for the challenges of plutonium purification and metallurgy, Oppenheimer worried about obtaining a sufficient quantity of ^{235}U. He had estimated at the April 1943 conferences that the electromagnetic separation method would produce 100 g of ^{235}U per day in 25 percent concentration by 1 January 1944, but this goal proved impossible. Despite a massive research effort by more than a thousand scientists at Kellex and Columbia University, a suitable barrier for the gaseous diffusion plant had not yet been found, and Groves decided to eliminate the upper parts of the gaseous diffusion cascade and use the product to feed the electromagnetic plant that was under construction at Oak Ridge that summer. When it appeared that the hope for producing sufficient ^{235}U lay with this plant, Lawrence was able to convince Groves to increase its size. But when the first oval isotope separation apparatus, or "racetrack," was completed in late October, a series of problems erupted, most notably an electrical shorting in the magnet coils. By early December, the plant's only racetrack had stopped working.[41]

In early 1944, a small amount of low-enrichment uranium became available for nuclear physics tests: 150 g were sent from Berkeley in early February, and the first small shipment from Oak Ridge came in early March. Nonetheless, delays at Oak Ridge inevitably caused problems for Oppenheimer. Whereas Bacher was told in November that he could expect uranium for the Water Boiler by 15 January, sufficient material had not arrived by February, whereupon a 1 April deadline was announced.[42]

Oppenheimer's problems grew worse in late 1943. Details of the first ^{235}U delivery were difficult to wrest from Groves, who wanted to keep information about Oak Ridge, including production schedules, highly secret. Arranging for the uranium specifications was also a headache. In January, when Oppenheimer instructed that the material be shipped, he included the impurity limitations and requested that the material be shipped according to enrichment level. Col. Kenneth Nichols, Groves's contact at Oak Ridge, insisted that such segregation would delay delivery by at least two months, although he promised that the enrichment would not drop below 10 percent, an arrangement that ultimately proved acceptable to the Governing Board. In April and May, the specifications were still being worked out between Los Alamos and Oak Ridge.[43]

Los Alamos metallurgists still had to wait for material. Although by May 1944 sufficient material was available for criticality experiments with the Water Boiler, the metallurgy group did not have ^{235}U to work with until August 1944. Concerned about the ^{235}U procurement, Oppenheimer urged Groves to consider a third method of isotope separation, thermal diffusion. Philip Abelson, who had helped Edwin McMillan discover neptunium in 1940, had designed a series of thermal diffusion columns with navy funding. Groves agreed, and in June 1944 the H. K. Ferguson Company of Cleveland signed a contract to build a thermal diffusion plant at Oak Ridge.[44]

Gearing Up for ^{235}U Metallurgy, Spring–Summer 1944

While waiting for the first large shipment of ^{235}U in the spring and summer of 1944, the CM-Division defined most of the methods necessary for the production of uranium hydride, implemented procedures for making uranium metal, and laid the groundwork for the recovery and fabrication of uranium. At this point, the chemistry and metallurgy groups were under considerable strain for two reasons: they had to do the uranium work at the same time that they were struggling to prepare proedures for plutonium, as explained in the next section, and they had to hurry, since they knew there would be little time to devote to uranium metallurgy once the plutonium chemistry and metallurgy efforts began in earnest.

In the spring of 1944, a useful procedure was defined for making plastic compacts with a high hydrogen-to-carbon ratio. By April, researchers had discovered that the newly introduced product, polyethylene (which had two hydrogen atoms for every carbon atom), was too difficult to dissolve to allow good impregnation. With polyethylene, refabrication of UH_{10} into UH_6 would require the complete reprocessing of UH_3. By July, Richard H. Kirby had devised a way to hydrogenate the double bonds in polystyrene. The resulting plastic, which had a high density and a hydrogen-to-carbon ratio of 1.75, made good UH_{10} compacts. It was also readily soluble and allowed easy removal of the plastic. By 1 August, 100 g of this plastic had been made, and by the next month, final fabrication methods had been developed and all dies and production equipment designed. In addition, Monsanto had been persuaded to produce hydrogenated polystyrene and delivered an initial order of three pounds.[45]

In the spring of 1944, the rigorous program on uranium metal production yielded procedures capable of producing high-quality uranium metal with minimal loss. Reductions were conducted on the scale of 1–10 g with UF_4 and UCl_3, both with stationary bombs and with the graphite centrifuge. In this period, the metallurgists also worked on the 200-g scale using unenriched fluorides from Oak Ridge. On July 1, UF_4 stationary bomb reductions on the 250-g scale produced well-formed buttons (disks of solidified metal, as recovered from the reactor) with a 99 percent yield. Stationary bomb reductions of UCl_3 on the 200-g scale also produced buttons with excellent 99.5 percent yields. However, the metallurgy group concluded that, although the reduction of UCl_3 could be used as a production method, it had some drawbacks – notably, the hydroscopic nature of UCl_3, which introduced undesirable oxygen and allowed the slag to penetrate the liner. Work also continued on reduction by electrolysis, on the 50-mg, 40-g, and 200-g scales. By summer, work focused on defining optimum conditions on the 200-g scale. In August, the group concluded that it was now "proven that uranium can be electrolytically reduced on production basis."[46]

In this period, the chemistry group defined the uranium purification and hydrofluorination procedures needed for uranium recovery.[47] Although the metallurgy group had little time to work on uranium remelting and casting because of the heavy burden of plutonium metallurgy, a fabrication group was started in July under the direction of Alan U. Seybolt. One of its first tasks was to investigate impurity behavior during 200-g uranium remelts. The group also took the first steps in devising the more sophisticated fabrication procedures needed to preserve the precious enriched product by making 25-g castings of unenriched material for investigations of optimal procedures for rolling ^{235}U. By August, preliminary work was also under way in rolling and cladding, hot-pressing, and forming uranium metal.

Pile-Produced Plutonium Reaches Los Alamos, November 1943 to April 1944

Organizational difficulties continued to plague plutonium efforts during the fall and winter of 1943–44. The operation of the Clinton pile in November 1943 opened new questions about the division of chemistry and metallurgy responsibilities. Participants at the February interlaboratory meeting agreed that a flowsheet should be put together each month to provide a concrete focus for discussions on the best plutonium purification scheme. Seaborg's summary of the meeting indicated that Met Lab work, which included three final purification procedures (precipitation, solvent extraction, and the use of volatile compounds), formed the main line of purification research. Kennedy had other ideas. This talented young chemist, whose genius for instrumentation had been crucial in the discovery of plutonium, had begun to resent the ambitious, publicity-prone Seaborg during their early collaboration in plutonium discovery. Kennedy had delegated purification responsibilities to Wahl, another plutonium codiscoverer, and amid rumors that top Los Alamos chemists had lobbied to exclude Seaborg from Los Alamos, Wahl devised his own purification flowsheet. Despite the efforts of Oppenheimer and Thomas to coordinate research, Wahl was not particularly interested in the Chicago work and drew instead on his substantial experience with plutonium chemistry at Berkeley, uranium stand-in work, and purification studies performed the previous fall at Los Alamos.[48]

While the chemistry responsibilities were being debated, members of Seaborg's group were still strugging to produce plutonium metal on a small scale, so that they could make an unambiguous density determination. In the last months of 1943, the group obtained confusing and disappointing results. One puzzle, which would not be unraveled until early the next year, was that sometimes a light-colored product was produced when the plutonium was converted to fluoride, whereas at other times a dark-colored product appeared, even though the chemical procedure seemed identical in both cases.[49] Problems with density measurements prompted even more concern. When the first unambiguous plutonium metal sample was finally produced by reducing PuF_4 with barium in November 1943, the density was measured first as 15.5 and then as 15 g/cm^3.[50] Because metallurgists, extrapolating from uranium, had expected the value to be around 19, it now appeared that more plutonium than originally anticipated was needed to make a bomb. Uncertain of their measuring techniques, they asked Zachariasen to carry out a measurement by X-ray diffraction. In December, he concluded that the material (incorrectly) thought to be plutonium metal had a density of 13. The Met Lab desperately hoped the sample was impure – if the density of metallic plutonium was really 13, the project would need two full-scale production plants to provide the plutonium needed for the weapon program.[51]

Chicago chemists faced this disheartening possibility just as a wave of anxiety was spreading through the Met Lab. The previous August,

Compton had voiced his fear that the Germans would be using radioactive weapons within six months. In December, in the midst of the first discouraging density measurements, Franck worried about the German announcement of a secret weapon, which he took to be a sure indication that a German atomic bomb was under construction. Compton, Henry Smyth, and Fermi anxiously met to discuss what should be done if a German weapon were used first.[52] Throughout the bomb project, fear provided a powerful incentive, and at this juncture, Met Lab chemists faced particularly grim prospects concerning the possibilities of failure as opposed to success.

The best hope at the beginning of 1944 for answering basic metallurgy questions was gram-scale procedures. Although the Chicago scientists were told at the January meeting to focus their efforts on further density measurements, Smith and his team at Los Alamos were ready to test large-scale procedures more likely to resolve the matter. Applying the knowledge gained from earlier stand-in work, they mapped out a metal reduction program, which they tested in early February with uranium.[53] When the first half-gram batch of plutonium reached Los Alamos later that month, the metallurgy group eagerly subjected it to three tested procedures: two metallothermic reductions using calcium in a stationary bomb and in a centrifuge, and fused-salt electrolysis.[54] No one was surprised when Los Alamos density measurements, taken by the capillary displacement method, showed values around 16. Further confirmation came from Zachariasen, who identified a characteristic face-centered cubic metal crystal structure from a Met Lab sample with a density also around 16.[55]

Suddenly, however, the situation changed. At the February interlaboratory meeting, Chicago chemists announced that Zachariasen's X-ray diffraction studies of their samples revealed two distinct crystal structures in the plutonium fluorides, PuF_3 and PuF_4, which had both been used indiscriminately to produce plutonium metal. This explained why the fluoride product was sometimes dark-colored and sometimes light-colored. At first, they were unaware that cylinders of hydrogen fluoride often contain hydrogen, which reduces PuF_4 to PuF_3. When PuF_4 was formed, the product was light; when PuF_3 was formed, the product was dark. Once they learned how to make the two compounds selectively on a routine basis, the chemists found it more efficient to make plutonium metal from PuF_4 than PuF_3, particularly in small-scale metal production; the difference in yield was not significant for gram-scale procedures.[56]

On 22 March 1944, the day after he heard of the Los Alamos density measurement, Chicago researcher Sherman Fried demonstrated to a colleague a technique for preparing and weighing a plutonium sample. To his great surprise, he obtained a density greater than 19. Assuming that he had made a mistake, Fried made several more density determinations using recently received plutonium samples. Each time, the density was close to 20. Further testing at Los Alamos confirmed these results, but no one could explain the unexpected change. In April, Smith told Kennedy that with the expected plutonium allotment of 3 g a month, his group would be able, with extensive recovery procedures planned to allow reuse, to run about fifty 1-g reductions by the end of July. Although the first series of metal reductions had produced only one success – Ted Magel produced some rather cokey plutonium in a 1-g run using PuF_4 reduced with lithium in a centrifuge – Smith felt confident that they would be able to develop a technique for producing good metal in three months.[57]

Purification Achieved and Plutonium Metal Unveiled, April 1944 to July 1944

Still unaware of the discovery of substantial spontaneous fission in pile-produced plutonium (Chapter 12), Chicago and Los Alamos researchers continued to vie for control over purification procedures. At the April interlaboratory meeting, Seaborg presented two purification schemes and Kennedy presented a quite different scheme prepared by Wahl. All procedures purified plutonium first with "wet chemistry" – through oxidation-reduction cycles – then converting the product to halide with "dry chemistry." However, both Chicago schemes used a solvent to extract Pu-VI and then either precipitated or extracted the Pu-IV.[58] In contrast, the Los Alamos scheme focused on precipitating Pu-VI, first as sodium-Pu-VI acetate, then used ether to extract Pu-VI, before reducing it to Pu-IV and converting that to the chloride. Furthermore, the Chicago procedure used tetrafluoride for dry conversion whereas the Los Alamos procedure used trichloride. Kennedy was confident of his group's methods because Wahl had already successfully tried the wet chemistry procedure on a 10-mg sample of plutonium in the new air-conditioned, dust-free chemistry facility, building D, completed in late 1943.[59]

Although members of the group felt that they should agree on a single purification scheme, they decided that all ideas should be entertained. Indeed, by this time, Los Alamos chemists were already successfully purifying plutonium from Clinton for the first 1/2-g and 1-g metal production experiments at Los Alamos.[60] In these experiments, the plutonium did not have to meet the strict weapon standard for light impurities. Nevertheless, Wahl confidently told Kennedy a few days before the April meeting that as long as care was taken to avoid contamination from reagents and the experimental environment, he could meet purity requirements through wet procedures, with only a few modifications to his current method.

Others were not so optimistic about the status of purification efforts. In February or March 1944, Compton called a special meeting between Chicago chemists and a representative from Clinton because of "the growing realization" at Chicago that achieving purification standards would be an "exceedingly difficult" task. Unlike Wahl, Chicago researchers did not think that necessary purification standards could be met, especially in light of the possibility of contamination. The group suggested that a mass spectrograph might be used for purification, but Thomas later vetoed this idea after preliminary work indicated that the method would take at least one year to develop.[61]

At the 17 April interlaboratory meeting, Seaborg summarized the current knowledge of plutonium metal. By then researchers knew that plutonium metal came in at least two forms, one with the face-centered cubic structure identified by Zachariasen with a density around 16, and the other, having a complicated, unidentified crystal structure and a density around 20. Spectroscopic analysis had shown samples of the latter to be extremely pure, and Seaborg theorized that lower densities were the result of impurities. He also gave the estimated melting point of plutonium as 950–1,000° C, a value close to that of uranium. By June, Los Alamos researchers had significantly corrected this view.[62]

Ten days after the April interlaboratory meeting, Seybolt reported results from a Los Alamos experiment aimed at exploring how plutonium metal reacted to heat. The metallurgists found five "temperature arrests" (i.e., temperatures at which the material would absorb heat without a temperature increase) between 137° and 580° C, the first occurring between 137° and 147° C.[63] Although this evidence suggested transitions between multiple phases of the metal, these results were so unusual that Los Alamos metallurgists remained highly skeptical of the apparent high number of phase transitions. They continued to test heat arrests, and when more material became available, they performed thermal expansion measurements. In one test, a plutonium sample was sealed with hexadecane in a pycnometer, a vessel of known volume that indicates the change in volume of a sample with heating. As Smith explains, the expansion was rapid "beyond all expectation." By the end of June, the metallurgists had "fully established a large density change near 135° C."[64]

Recognizing that there were at least two allotropic forms of plutonium, the metallurgists concluded that the room-temperature phase, the α phase, had a density around 20 and a complex crystal structure, and that the second phase, formed at about 135° C, had a density around 16 and a face-centered cubic crystal structure. Although then identified with the β phase of plutonium, the phase in fact seen was the face-centered cubic δ phase. The existence of allotropic forms in plutonium explained the puzzling difference in density measurements for plutonium metal. In the early experiments at Chicago and Los Alamos, impurities stabilized the low-density form. About the time Fried made his fateful measurement in March 1944, the plutonium being studied was purer, and thus the higher-density form could be created.[65] Los Alamos metallurgists were relieved to solve the density puzzle but worried that the large volume changes between transitions at such low temperatures would cause fabrication difficulties.[66]

Soon another unexpected characteristic of plutonium metal emerged. In May, while attempting metal reduction by electrolysis using tungsten cathodes, Morris Kolodney, who had both the temperament and skills necessary to maintain the meticulous control of experimental conditions required by this method, dissolved $PuCl_3$ in a fused salt bath. Although most metal reduction experiments were being run between 1,100 and 1,300° C, he ran the reduction at 650° C because he was using a pyrex glass container, which softens at high temperatures, and a salt bath with a low melting point. To his surprise, clean metal beads, much nicer than the usual cokey product, formed at this lower temperature. To his astonishment, further tests indicated that the melting point of plutonium was about 635° C, more than 400° C lower than previous estimates. Metallurgists had been led astray because they expected the melting point of plutonium to be as high as the known melting point of uranium, an expectation apparently confirmed when early small samples did not flow, but retained their shape at high temperatures. Actually, since plutonium is a highly reactive metal, it easily forms an oxide-nitride skin that envelops small samples in a stiff shell.[67]

This discovery of the low-temperature allotropic transformations and the low melting point of plutonium marked a giant step forward for Los Alamos scientists. Now that they understood that metal reductions were being run at excessive temperatures, they were able to formulate more effective procedures. Moreover, the quality of Kolodney's product quelled fears that it would be difficult to produce good, malleable plutonium. In addition, the understanding of allotropic transformations allowed the metallurgists to avoid phase transformation temperatures during fabrication.[68]

The biggest problem facing Kennedy now was plutonium purification, but Los Alamos chemists made considerable progress in this area in the spring and early summer of 1944. During this period, dry conversion efforts focused on trichloride studies, although studies were also made with tetrafluoride, trifluoride, and tribromide.[69] In the meantime, runs with larger amounts of plutonium had revealed serious problems with Wahl's original wet purification scheme: it caused too much uranium to be retained; produced plutonium nitrate end products, which caused splattering when converted to PuO_2; exposed workers to considerable toxicity when operated on the multigram scale; and reduced impurities insufficiently, probably because of contamination.[70]

To remedy these defects, Wahl devised an improved scheme, called the "A" process, which separated uranium more thoroughly, yielding a plutonium product, which then converted smoothly to PuO_2 without splattering. In early July, Kennedy judged that the new scheme was "good and safe." It had a yield of approximately 90 percent, and because the precipitates settled readily, they could be separated by decantation from their supernatants, and an enclosed apparatus could be used to reduce the chance of plutonium contamination and to protect workers.[71]

Chemists were uncertain about the extent of light-element impurities in Clinton plutonium because analytical techniques were unreliable for small samples, and they had not yet received enough material to risk a thorough analysis. In addition, although the analytical group had settled on the cupferron method of measuring a wide range of impurities in plutonium, the difficult task of measuring fluorine and oxygen was not yet complete, despite the progress on an ingenious scheme for oxygen analysis that grew out of Lipkin's efforts the previous fall. Nonetheless, Wahl felt that the scheme would produce plutonium meeting light-element purity requirements. If it did not, he was confident that he could devise the necessary additional purification procedures.[72]

At this point, Oppenheimer announced the discovery of spontaneous fission in pile-produced plutonium, and the problem of purification from light elements became moot. On 14 July, Oppenheimer told Kennedy that the lifting of plutonium purity requirements "was a virtual certainty." Four days later, Compton announced the disturbing news to Seaborg along with other Met Lab researchers.[73]

In light of the pressing need for trained people at other facilities, particularly Los Alamos, many scientists left Chicago in the summer of 1944. Like other group leaders, Seaborg lost several members. As plans for the Hanford production plant accelerated in mid-1944 and Clinton continued operation, Met Lab chemists and du Pont engineers remained tied up in developing and maintaining large-scale plutonium separation procedures. Although plutonium chemistry was still a part of Met Lab work, the focus of plutonium research and development shifted to Los Alamos.[74]

12

The Discovery of Spontaneous Fission in Plutonium and the Reorganization of Los Alamos

During the spring of 1944, Emilio Segrè's group in P-Division made the startling observation that the first samples of pile-produced ^{239}Pu had an unusually high spontaneous fission rate, with a neutron emission approximately five times that of cyclotron-produced ^{239}Pu. This finding confirmed the gnawing suspicions of Fermi, Segrè, Seaborg, and others that the neutron bath in the production piles at Clinton and Hanford might cause the formation of a significant quantity of ^{240}Pu, an as-yet-unobserved spontaneously fissioning isotope of plutonium. However, the alarmingly high rate of the spontaneous fission was unexpected. This rate increased the neutron background enough to make it highly probable for a gun-assembled gadget to predetonate and thus undermined the plutonium gun program.

Determined not to lose the heavy investment made in plutonium production, Groves forced the laboratory to change course. The primary technical objective shifted from developing a gun weapon to developing a plutonium implosion assembly. Within days after Oppenheimer officially announced the spontaneous fission discovery, the laboratory reorganized its work force to focus on implosion. Two new divisions were established – X (Explosives) under Kistiakowsky, and G (Gadget)

* This chapter is based on a draft by Lillian Hoddeson to which Paul Henriksen contributed the section on the summer 1944 reorganization. The technical content was edited by Gordon Baym and Les Redman. An early draft was edited by Richard Hewlett.

under Bacher. Most of the groups in these new divisions were moved out of the earlier Research and Ordnance Engineering divisions. Unfortunately, at this point experiments in the implosion diagnostic program were indicating that an implosion weapon would be extremely difficult, if not impossible, to achieve.

Spontaneous Fission Studies at Berkeley, 1942–1943

One of the first experiments to search for spontaneous fission was conducted in the late 1930s by Willard F. Libby, but it was not successful.[1] Shortly afterward, Niels Bohr and John Wheeler touched on the theory of spontaneous fission in their comprehensive treatment of nuclear fission in 1940.[2] Fermi and Arthur Compton certainly worried about the possibility of spontaneous fission in designing the first atomic pile, because the added neutron background from spontaneous fission would influence pile functioning.[3] The first to observe spontaneous fission were K. Petrzhak and G. N. Flerov, in their Leningrad uranium experiments in 1940.[4]

Like several other wartime Los Alamos developments, the spontaneous fission program grew out of nuclear studies at the Radiation Laboratory of the University of California in Berkeley. Starting in late 1941 or early 1942, Segrè and several of his Berkeley graduate students – Owen Chamberlain, George Farwell, Gustave Linenberger, and Clyde Wiegand – conducted pioneering spontaneous fission studies. Operating within the Berkeley tradition of cooperative research established by Ernest Lawrence, Segrè's group collaborated with Berkeley chemists Joseph Kennedy and Arthur Wahl; both would later join Project Y.[5]

Segrè's spontaneous fission group also worked closely with another Berkeley group that employed many of the same techniques and staff. This group measured nuclear properties, such as α activity and the neutron-induced fission rate. Both groups used very thin layers of material and needed extremely stable amplifiers for their counting of fission events. Among other problems, this nuclear physics team analyzed the isotopic content of different mixtures of uranium produced in Lawrence's mass spectrograph, the Calutron. Segrè and Kennedy were in both groups, as was Wiegand.[6]

Using ionization chambers to detect fission fragments, the Segrè team explored spontaneous fission in a variety of elements, including tuballoy (the naturally occurring mix of ^{234}U, ^{235}U, and ^{238}U) and plutonium.

Their extremely small samples of precious plutonium (of the order of micrograms) were made by deuteron bombardment of ^{238}U in Berkeley's "Crocker" 60-inch cyclotron. They had to make thin layers of material, particularly the uranium – to ensure that the short-range fission fragments would enter the ionization chamber and register full-size pulses – while avoiding false counts caused by the coincidence of several α-particle pulses.[7]

Delicate experimental problems arose requiring the detectors to be cutting edge technologies. For one thing, the electronics had to be highly sensitive, as well as fast and versatile.[8] The experimenters worried constantly about detecting false counts or missing real ones. To avoid electrically induced false counts, the apparatus was entirely battery-operated. When more counts were seen during daytime hours than at night, the group initially thought that the difference was caused by cellos in rooms nearby resonating with the apparatus.[9] They eventually traced the difference to the light in the room; photoelectrons were needed to begin the electron multiplication avalanche that registered fissions in their gas-filled cold-cathode detector. Simply leaving a flashlight on at night inside the apparatus equalized the day and night counts.[10] They also tried to avoid the false counts caused by the coincidence of several α-particle pulses.[11] On 24 June 1943, the group established an upper limit for spontaneous fission of 5 fissions/kg sec in such plutonium, or 18 fissions/g hr.[12]

Spontaneous Fission Studies Begin at Los Alamos

Several researchers had suggested that a mass 240 isotope of plutonium would be produced in nuclear piles even before Los Alamos established its scientific program. The argument drew on an analogy with uranium. When thermal neutrons interact with ^{235}U, two competing processes take place: fission and neutron capture, which lead to the formation of ^{236}U through the emission of gamma rays.[13] Plutonium-239 was expected to behave analogously, since ^{239}Pu, like ^{235}U, is fissionable with thermal neutrons.[14] Neutron capture with the emission of a photon could presumably cause the formation of ^{240}Pu in proportion to the square of the neutron irradiation in the pile. The new isotope, having an even number of neutrons and protons, would be likely to fission spontaneously.[15] At the Chicago Metallurgical Laboratory, Seaborg recorded in his diary on 18 March 1943:

> The possibility of forming 94-238, 94-240, 93-237, 95-240 and other isotopes in a high power chain-reacting pile ... a cross-section one percent of the fission cross-section would result in enough 94-240 to complicate the purity problem if it is a short-lived α emitter .... If the spontaneous fission rate of 94-240 is high, e.g., a half-life of less than 10^{10} years, it might be serious. Another possibility is an appreciable n,γ reaction with 93-239, which could lead to the formation of 94-240 by β emission."[16]

Fermi, having assembled the first chain-reacting pile at Chicago in 1942, and for many years concerned with the creation of new elements through neutron capture, was – like Wahl, Kennedy, and Segrè – suitably placed to conceive of the formation of ^{240}Pu. Although these (and probably other) researchers saw that ^{240}Pu might be an impurity in pile-produced plutonium, no one appears to have suspected the magnitude – that the spontaneous fission rate of pile-produced plutonium would be a million times that of ^{235}U. The expected neutron emission rate would not have affected the gun program; rather, the overriding concern was about neutrons resulting from the interaction of α particles emitted by the plutonium with light-element impurities in the fissile materials. That spontaneously emitted neutrons might overwhelm those from light-element interactions to an extent that could produce predetonation could not be determined experimentally until the neutron fluxes in the plutonium-producing piles were actually measured.[17]

At the April 1943 Los Alamos conference, predetonation of the atomic bomb was one topic of discussion. Chemists both at Los Alamos and the Chicago Metallurgical laboratories planned to cope with the α,n reactions by "super purification," a program directed toward preparing highly pure material (Chapter 11). In the scientists' opinion, the discovery of a significant level of spontaneous fission would make the costly purification effort superfluous, but they judged such spontaneous fission unlikely. Nevertheless, for completeness in the study, Oppenheimer asked Segrè, who was attending the conference as a visitor, to look into the spontaneous fission issue. He invited Segrè to move the Berkeley spontaneous fission program to Los Alamos.

Segrè immediately began to prepare for this new assignment, assembling materials and his small group of young physicists – Chamberlain, at this time almost twenty-three, with a B.A. from Dartmouth College in 1941; Farwell, age twenty-three, with a B.S. from Harvard University in 1941; Linenberger, almost twenty-three, with a B.A. from Rice University in 1941; and Wiegand, twenty-eight, with an A.B. from Willamette

Fig. 12.1. Experimental physicist Emilio Segrè was head of the nuclear physics group responsible for exploring spontaneous fission. LA Photo, 1582.

College in 1940 and a commercial radio operator license from the federal government in 1933. Farwell recalls that in mid-June 1943, "we all packed up, bags and counters, detectors, electronics and all, and went off to Los Alamos ... where we became Segrè's group." The equipment was shipped to New Mexico in an Allied moving van, in which Linenberger rode "shotgun." Segrè and Farwell flew to New Mexico in a commercial DC-3, arriving in Santa Fe on 18 June. Initially the group set up its apparatus in temporary quarters in the technical area of the laboratory.[18]

For their long-term experiments, however, they searched for an undisturbed site, one with "peace and quiet from electrical and audible disturbances, and shielding from cosmic rays." Farwell recalls exploring "caves at the bases of various cliffs, looking for a spot that might be easy to dig into and make a laboratory .... We ended up being allowed to put our experiment in a Forest Service cabin, an old log cabin in Pajarito Canyon, a site located 14 miles from the main technical area by

Fig. 12.2. United States Forest Service cabin in Pajarito Canyon used by the spontaneous fission group of the physics research division. Photo courtesy of George Farwell.

the route traveled," which was shielded from the radiation background and other activities of the laboratory.[19] Segrè recalls: "It was a most poetic place ... we went there by jeep every day. There was a bed in it, somebody occasionally slept there."[20]

While Segrè and his group were traveling to Los Alamos from Berkeley, a report from Paris arrived via the grapevine of Frédéric Joliot, Pierre Auger, Arthur Compton, Richard Tolman, and Oppenheimer. The report described Joliot's discovery of spontaneous neutron emission from polonium and lent further support to the decision to explore spontaneous fission. At the Governing Board Meeting on 17 June 1943, Oppenheimer read from Tolman's report on his recent visit to Chicago: "in connection with the purification problem, Dr. Compton wishes me to call your attention to a report, coming through Ogier [*sic* Auger], that Joliot has found a spontaneous emission of neutrons from polonium, as well as the production of neutrons through the (α,n) reactions with impurities. Compton guesses that there might be one spontaneous neutron to 10^7 α particles from polonium and is having the matter looked into at Chicago." If true, this effect would rule out the polonium–beryllium initiator intended to begin the nuclear reaction, because of the accompanying spontaneous neutron background.[21] Of significance for the spontaneous fission studies of plutonium was Joliot's speculation, recorded by Tolman, "that perhaps in general a small spontaneous neutron emission might be expected to accompany α emission, and Fermi is inclined to agree on the basis of his ideas as to nuclear rearrangements."[22] The results therefore raised the serious possibility that plutonium, itself an α emitter, might create a threatening neutron background by fissioning spontaneously.

As David Hawkins, who attended this meeting, recalls, Joliot's results were not trusted at Los Alamos: "Because the difficulties of purifying polonium were already well known at Los Alamos, it was generally believed that Joliot must have overestimated the purity of his material, and that his neutrons were really from the (α,n) reaction of light element impurities."[23] Bethe commented at the meeting: "The reaction might easily be due to impurities .... A neutron could be expected to follow the α decay if the nucleus were sufficiently excited." Nevertheless, the " 'Joliot effect,' if real, might materially affect the program of the Laboratory,"[24] and the decision was reconfirmed to look carefully into whether plutonium exhibited spontaneous neutron emission. Oppenheimer stated at the meeting that Segrè "should be given facilities and guards in Pajarito Canyon and go into all the spontaneous fission questions."[25]

Early Studies at Pajarito and the Effect of Cosmic Rays

Recreating the Berkeley procedure, Segrè's group had set up its battery-powered α-particle-counting apparatus in Pajarito Canyon by the first of August 1943. By the middle of that month, the team was studying samples of polonium, protactinium, ionium (thorium-230), ^{234}U, enriched ^{235}U, impure ^{238}U, and ^{239}Pu. The precious ^{239}Pu – five 20-μg samples – had been made in the Berkeley 60-inch cyclotron. Later in the program, increased sample sizes made it necessary to construct new ionization chambers and new amplifiers.[26]

Chamberlain, Farwell, and Wiegand improved on the nitrogen-filled ionization chambers they had designed in Berkeley to replace the earlier air-filled chambers. Roughly 5 inches in diameter and a foot tall, the chambers were filled with boron trifluoride and installed in 20-inch cubes of paraffin. To make the best use of the nitrogen-filled chambers, the group built amplifiers with better stability and frequency characteristics, as well as less attenuation, than in the Berkeley experiments. Wiegand designed most of the electronics, then based on the 6AK5 vacuum tube, the stand-by of that time. Six-volt automobile batteries, arranged in racks, powered the filaments, while arrays of 45-V B batteries supplied 90 V for the plates and screens.[27] Wiegand recalls that Fermi, a close associate of Segrè (the two had worked together in Rome), took great interest in this experiment, often sitting with him as he developed the electronics. They "looked in the scope and we'd change the resistors and just make little changes as I was trying to make the pulses go faster and faster."[28] Fermi would later play a critical role in confirming the experimental results.

By 19 August, the group was collecting its first data.[29] Six weeks later, the chambers were regularly counting ^{238}U, ^{235}U, ^{234}U, and ^{239}Pu, as well as thorium. To reduce error, they systematically interchanged the samples between the chambers.[30] However, with the very small samples available thus far, the group was not yet able to determine the actual spontaneous fission rate of ^{239}Pu.[31] The rate of ^{240}Pu would not be measured for some time to come.

To determine the spontaneous fission rate of an element, the group had not only to observe and count fission fragments, but also to determine ν, the average number of neutrons emitted per fission, a parameter that thus far had only been estimated as approximately equal to that for slow-neutron fission. Combining coincidence chamber measurements on 90 μg of plutonium with data obtained by the German researchers W. Maurer and H. Pose, Farwell and Chamberlain obtained a value of $\nu = 2.3$ for spontaneous fission of ^{239}Pu, almost the same as that obtained in neutron-induced fission.[32] In a concurrent but separate determination, Wiegand and Segrè reported $\nu = 2.37 \pm 0.3$ for ^{239}Pu. They also measured the distribution of the fission fragments and the ionization produced by fission fragments of various sizes.[33] The group concurrently studied the spontaneous fission of uranium, using separated uranium isotopes prepared at Berkeley.

Within a month after these experiments began at Los Alamos, Segrè's group noticed that the fission rates for ^{238}U at Berkeley and Los Alamos were in agreement, but that the Los Alamos rate for ^{235}U was substantially higher than the Berkeley rate. They concluded by November 1943 that the discrepancy arose from cosmic rays, which are more numerous at Los Alamos, owing to the laboratory's high altitude. Cosmic-ray neutrons were inducing fission in the uranium samples, an effect showing up as false spontaneous fission counts.[34] Because ^{235}U showed the altitude effect and ^{238}U did not, and because the rate of spontaneous fission greatly exceeded the rate of induced fission by cosmic-ray neutrons, they learned that spontaneous fission, which causes far greater neutron production than change in altitude, occurs in ^{238}U and not in ^{235}U. Thus, the serendipitous change in altitude revealed that there was no need for concern about predetonation due to spontaneous fission in ^{235}U, a finding of practical importance for the uranium gun program.

During the last three months of 1943, the group focused on improving the scope of the experiments. For example, as explained in their report on 15 October, the group hoped to make the rate of collection on plutonium "essentially ten times as fast," first, by preparing "a thin foil of approximately 40 μg (as opposed to the 20-μg foils previously used) and, second, by adding 100 μg of material from the recent Berkeley irradiation." With these improvements, they "hoped that the actual spontaneous fission value [could] be obtained in the course of two or three months."[35]

The 20-μg samples of Berkeley plutonium remained in the ionization chambers, slowly registering counts. By 31 January 1944, these samples had registered six fission events in 80,000 μg hr of observation during August through December 1943, roughly six counts in five months.[36] The five chambers in operation over this five-month period were running approximately twenty-two hours a day throughout the week.[37] To shield the chambers used in both the plutonium and uranium studies from cosmic-ray neutrons, they added a boron compound to the ionization chambers.[38] The shielded samples produced only one count for 40,000 μg observation hours, or three counts in five months – an extremely low data collection rate, with correspondingly poor statistics.[39]

Tests on Pile-Produced Plutonium

All the work done between April 1943 and April 1944 was in preparation for testing pile-produced plutonium, which was not yet available at Los Alamos for spontaneous fission studies. In the fall of 1943, Los Alamos expected to receive 50 g of pile-produced plutonium from X-site – Clinton, at Oak Ridge – around 1 April 1944, to be followed at least three months later by "main production" samples from Hanford.

In the intervening winter, while the group was awaiting the pile-produced material, Fermi suggested a decisive experiment for measuring the spontaneous fission rate of such plutonium. Recognizing the need to amplify the effect from the spontaneously fissioning component, Fermi argued that if pile-produced plutonium showed more spontaneous fission than cyclotron-produced plutonium, owing to irradiation by the neutron bath in the pile, then samples sent through the pile again would show proportionately more spontaneous fission. If ^{240}Pu were produced through radiative neutron capture in the pile, its contribution to the spontaneous fission of a sample of pile-produced ^{239}Pu would be proportional to the square of the total "irradiation" in the pile. In a re-irradiated sample, this contribution would be proportional to the fourth power of the total irradiation.[40] The sample would have to be of a sufficient size before the quadratic pile-exposure term could be detected. Fermi proposed that the test be made on approximately 50 mg of Clinton-produced plutonium.

Accordingly, on 25 January 1944 Samuel Allison asked Compton to arrange to retain and re-irradiate at Clinton 50 mg of the first 1-g shipment of plutonium going to Los Alamos. "This will enable us to study the new atomic species produced by such an irradiation which will, of course, take place in our Hanford piles. In particular, we may be able to detect 94-240. Fermi tells me that Oppenheimer agrees that the experiment is worthy of being carried out."[41]

The first sign that plutonium might not work in a gun assembly was seen in data taken early in March on cyclotron-produced plutonium. By this time, only four counts had been recorded from the boron-shielded Berkeley samples, a number that translated into a spontaneous fission rate of 40 fissions/g hr. This rate would have been just tolerable in the plutonium gun, but a rate several times that was dangerous. What concerned the scientists was that the uncertainty of the data was so high that, as Oppenheimer wrote to Kennedy and Captain Parsons, "a fission rate twice that indicated is certainly compatible with the data."[42] As it turned out, every subsequent measurement confirmed the borderline safe result of 40 fissions/g hr and also improved the statistics. Of the eight counts registered by 12 April, the group estimated that at least seven were real. At this point, the error, although down from the 50 percent level of 6 March, was still a huge ± 30 percent.

Great care had to be taken to avoid accidental counts from any source whatever. They worried most about "α pileups" and a secondary effect from α particles emitted by the plutonium leading to the production of neutrons by (α,n) reactions, which in turn caused what appeared to be spontaneous fissions, while actually being neutron-induced. To check whether these effects were taking place, Segrè proposed using samples of two or more different sizes and determining whether they gave the same answer. Larger samples would give larger amounts of α particles and thus more likelihood of α pileup or (α,n) reactions leading to neutron-induced fission. When the check was made, it showed virtually identical results for the different samples, confirming the lack of an effect from α pileups or secondary neutrons. Other more obvious sources of error, from which the experimenters were careful to protect their delicate apparatus, included lightning, automobiles, and people.[43]

The reality of the spontaneous fission problem hit within days after the group received its first eight samples of production plutonium. The first sample was placed in the detection chambers on 5 April. Seven more were added on 12 April. By 15 April, startling, if tentative, results were out: during the first *three days* of observing, the pile-produced product was showing five times as much spontaneous fission as the plutonium from Berkeley![44] Aware that this unwelcome result was based on only eight counts, Segrè very cautiously reported on 15 April:

> Reliable data with the modified apparatus is not yet available because the samples have not been accurately calibrated, and the working conditions of all the amplifiers have not been checked. It will also be necessary to investigate by using 49 samples of different sizes whether there is any secondary effect like the production of neutrons by the αs of 49 which makes the apparent spontaneous fission rate dependent upon the amount of 49 used. With all these reservations, which may produce a drastic change in the result, we give a tentative result of 200 f/g hour.[45]

The group chose not to circulate this worrisome result widely, claiming experimental uncertainties and poor statistics. They could not yet identify ^{240}Pu, because the samples were still too small to allow them to observe the quadratic irradiation term.[46] Despite their efforts to keep the results quiet, news of the spontaneous fission in plutonium soon traveled through the laboratory, troubling certain staff members. Wiegand and Farwell recall that Edwin McMillan, in charge of the threatened gun program, drove to Pajarito Canyon to, as Farwell recalls, "see this thing first hand. And when he saw the traces ... he believed it. Wiegand recalls McMillan being impressed by "even the very first count."[47]

Given the small number of counts and great experimental uncertainty, why were both McMillan and the Segrè group already alarmed? The reason lay in the experimental methodology: the recording technique was so designed that each count could be trusted to a high degree of probability to represent a real fission. Their Esterline–Angus chart recorder was an accurate clock that registered the precise time of each recorded pulse. A specific time correlation in the counts would have suggested an external time-related factor, but they found the fission pulses occurring randomly. To determine whether any individual pulse was full height, and therefore corresponded to a fission fragment, they calibrated the chambers using ^{235}U and used radium–beryllium sources to generate neutron-induced fissions. And to check whether "α particle pileup" was occurring, they did another calibration experiment in which they introduced increasing amounts of plutonium into the chambers and determined how many α particles could be added before such αs produced a false fission. Knowing this number, they could adjust their detection system so that at least a month would elapse before α pileups could simulate a fission.[48]

Many more counts from the Clinton plutonium were registered in the following weeks. By 5 May, the group had seen 66 counts, and by 9 May, 40 more, giving a total of 106 counts registered in 0.406 g hr of observation between 15 April and 9 May. These data therefore indicated a spontaneous fission rate for Clinton plutonium of 106/.406 = 261 fissions per g hr, compared with 40–50 per g hr from the Berkeley plutonium (with rather poor statistics on both).[49] Counting continued until 24 July, by which time the boron-shielded early production plutonium yielded 131 counts, which corresponded to 180 f/g hr ±16, as opposed to 50 f/g hr for the cyclotron-produced plutonium.[50] It was becoming clearer each day that production plutonium contains a strong spontaneous fissioning component.

Bacher Brings the News of Spontaneous Fission to Chicago

At first the Chicago Met Lab was reluctant to accept the results. An experience of P-Division leader Bacher (to whom Segrè reported) in early June 1944 offers an illustration. As the Los Alamos liaison since late 1943, Bacher had been a regular visitor to the Chicago Metallurgical Laboratory, which was then spearheading the plutonium purification program. For his visit between 31 May and 8 June 1944, Bacher told Met Lab Director Arthur Compton that he wished to report on the Segrè group's recent results. Bacher assumed that the Met Lab had heard about the findings, for Los Alamos had reported them earlier to Groves, "with the request that he pass them on to the Chicago group," whose work they obviously affected. However, Bacher recalls that Compton, obviously hearing the news for the first time, "went just as white as that sheet of paper." After remarking, "This will just cause great troubles in our laboratory," Compton added, "I'm not sure that should be reported here [Chicago]. I'm sure Groves would be very much upset."[51]

Bacher felt strongly that his report fell "within the limits of what he [Groves] told me I should report down here," and he told Compton that he intended "to report it unless told not to." So Bacher telephoned Groves, who, even at this late date, asked "Do you think that needs to be reported to them?" Bacher replied, "Of course .... It's a fundamental fact of the material they're working on." Only then did Groves realize that holding back the results could actually impede the work of the Manhattan Project. Groves authorized Bacher to deliver his report but asked him to limit the audience to only some half dozen. Bacher followed this authorization, but reflected recently: "I'm sure it didn't take more than a half an hour to get all over the laboratory .... This just meant trouble ahead."[52]

Given the vast importance of the spontaneous fission results to the Chicago effort, why had Fermi, then still on the Chicago staff, failed to notify his colleagues that their painstaking purification efforts were likely in vain? Segrè proposed in a recent interview that Fermi was simply "a very tight-lipped person," or that the results were possibly not then prominent in his mind.[53] But it is hard to believe that in the context of wartime pressure Fermi could have been indifferent to whether his colleagues were wasting time. Reflecting further on Fermi's role in the discovery of ^{240}Pu, Segrè suggested that Fermi, like Groves, might consciously or unconsciously have been wishing the results would go away, so that in the hurried atmosphere of the time, he forgot about the problem.[54]

Confirmation and Impact of Plutonium 240

Oppenheimer presented the evidence for spontaneous fission to the laboratory in a colloquium on 4 July 1944 (clearly not a lab holiday that year). Captain Ralph Carlisle Smith wrote in his colloquium notes that day, "Plutonium 240 and its spontaneous fission ... might limit the use of plutonium in the gun method because of high neutron background resulting therefrom."[55] Conant, who was at Los Alamos between 2 and 6 July, wrote in his work sheets of 4 July 1944: "There is a very real possibility that it may prove impossible to use the gun method with '49' because of spontaneous neutron emission. This will be clear in about a month. The material from pile appears to differ from material from cyclotron."[56]

Attention turned to implosion. In his 4 July report, Conant remarked: "There is little doubt that an implosion method can be made to work for both '25' and '49'. The question is how efficient can such bombs be made to be and how fast can a moderately efficient bomb be developed .... Eventually it may be possible to design an implosion bomb with such speeds of assembly and such high compressions that very great efficiencies result (much greater than with the gun assembly). But at present one must be content with lower compressions and low efficiencies."[57] Yet implosion was the laboratory's only real hope for using the expensive production effort at Hanford.

Meanwhile, in the first weeks of July, Segrè's group carried out Fermi's test on re-irradiated plutonium.[58] Preliminary data from a 25-mg sample, introduced into the ionization chambers on 9 July 1944, were available by 11 July 1944.[59] Oppenheimer teletyped Conant: "The Pajarito Counting Rate of the re-irradiated sample is increased by an amount corresponding to its increased radiation period .... No conclusive proof exists that the substance giving these counts is four ten [^{240}Pu], but the a priori probability is very high and the chemical operations performed on the material obviously did not separate it from four nine."[60] The results from more re-irradiated samples, added to the counting apparatus on 11, 12, and 13 July, confirmed the earlier evidence. Farwell recalls that every one of the re-irradiated samples "started off like a string of firecrackers."[61]

In the heat of the crisis, gun researchers considered other shooting possibilities, for example, faster guns. But these ideas were soon given up as impractical (Chapter 13). Similarly, the chemists and physicists pondered the problem of separating out the ^{240}Pu, possibly using Lawrence's electromagnetic method. They concluded, as Oppenheimer explained to Groves on 14 July 1944, "My opinion is that this is not a job which can be developed within any reasonable time scale but that it should be referred to [Lawrence] for more expert consideration."[62] The separation idea was considered in detail on 17 July, at a meeting in Chicago that included Conant, Oppenheimer, Charles Thomas, Compton, Groves, Colonel Nichols, and Fermi. All agreed that to separate out the ^{240}Pu, although possible by electromagnetic methods, "would postpone the weapon indefinitely."[63] John Manley later reflected, "One could have separated out those bad plutonium isotopes from the good ones, but that would have meant duplicating everything that had been done for uranium isotope separation – all those big plants – and there was just no time to do that."[64] Manley summed up, "The choice was to junk the whole discovery of the chain reaction that produced plutonium, and all of the investment in time and effort of the Hanford plant, unless somebody could come up with a way of assembling the plutonium material into a weapon that would explode."[65]

At the Chicago meeting, Conant stressed authorizing Los Alamos to build a low-efficiency implosion bomb of only a few hundred tons of TNT based on a mixture of both plutonium and uranium. Such a weapon would be easier and quicker to build than a more efficient one and would also have less neutron background than a pure plutonium bomb. Conant felt that after building and testing the low-efficiency device, the laboratory could then turn "with less nervousness" to building a more efficient gadget.[66] However, Oppenheimer opted for the plutonium implosion device.

The report Oppenheimer sent Groves on 18 July 1944 explained the laboratory's dilemma:

> Samples have been investigated whose neutron radiation has varied over a range of 50. The fission rate of the sample appears to be proportional to the number of neutrons which have previously passed over the material. This strongly suggests that there is an ingredient in the samples formed by the neutrons, and responsible for the fission .... It is known that when thermal neutrons pass through plutonium about 1 in 3 of the neutrons captured leads to the formation of the isotope 94-240, which has as the code name 410 .... There is an overwhelming presumption at the present time that it is for practical purposes stable. It is natural to ascribe the spontaneous fission observed in irradiated samples of plutonium to the presence of this isotope .... We have investigated briefly the possibility of an electromagnetic separation .... It is our opinion that this method is in principle a possible one but that the necessary developments involved are in no way compatible with present ideas of schedule.[67]

He therefore proposed that Los Alamos focus on developing an implosion bomb: "In the light of the above facts, it appears reasonable to discontinue the intensive effort to achieve higher purity for plutonium and to concentrate attention on methods of assembly which do not require a low neutron background for their success. At the present time the method to which an over-riding priority must be assigned is the method of implosion." But he advocated caution:

> Since the results outlined above are new and since there is a possibility that the interpretation placed on them may not be completely correct, it was agreed in our discussion that although the discontinuance of the purification and neutron-less assembly program should be started immediately, it should be so conducted that at any time within the next month a return to these programs can be made without loss of more than a month's time. In particular, no essential personnel or installations should be permanently lost to the project within that period.[68]

Two days later, Oppenheimer announced to the Administrative Board that, "Essentially all work on the 49 gun program and the extreme purification of 49 should be stopped immediately." He added: "All possible priority should be given to the implosion program. At the same time, nothing essential to the twenty-five-gun program should be left undone."[69]

The next day, Thomas, coordinator of the Manhattan Project's purification effort, supported Oppenheimer's recommendation in a poignant statement to Groves:

> At Hanford levels, the neutron emission from this isotope would be several hundred times that resulting from the chemical impurities even when the specifications for chemical purity of 49 are met. It therefore seems unnecessary for the military objective to continue the program on final purification. This information comes at a time when we have progressed to a point where we feel confident that the chemical purification of 49 can be accomplished. As a matter of fact, Y has produced small batches of metal almost entirely free of many of the objectionable impurities. Following the decisions reached in the above meeting, I am proceeding with the demobilization of my staff which was handling the coordination and general direction of the chemistry, purification and final metallurgy of 49.[70]

Conant scribbled across Thomas's earlier 13 June optimistic report on light element purification of plutonium, "All to no avail, alas!"[71] Meanwhile, as the plans for redirecting the plutonium effort were being cast, definitive results from Fermi's test on the re-irradiated plutonium arrived. The first run, which began on 6 July, produced 154 counts by 24 July, in a period dramatically short compared with those in which counts had been recorded for Berkeley plutonium, or even for Clinton plutonium that was not re-irradiated.[72] Farwell recalls, "We began to see 3, 4, 5, or even 10 or more counts in a single day on a single counter, and that was like lightning," compared to the earlier rates of one a month for Berkeley plutonium, or 1 to 4 counts a day for production plutonium. The 154 counts converted to 1,580 ± 120 f/g hr, in comparison with 50 f/g hr for Berkeley material and 180 f/g hr for production plutonium. The results were now indisputable.[73] At the colloquium on 25 July 1944, R. C. Smith made this note: "Colloquium. Oppenheimer presided, told of abandonment of gun method for 49. No need of purity for implosion."[74]

Although Segrè's group was virtually certain that ^{240}Pu explained their results, they carefully looked into all other possibilities, in particular the presence in the ^{239}Pu of elements 92, 93.[75] The chemists soon ruled out the existence of other elements, either, as Kennedy suggested, by using X-ray analysis or, as Wahl suggested, by studying samples having various isotopes removed. Between July and September 1944, Farwell analyzed samples with various "Wahl fractions," for example, one from which Wahl had chemically removed elements 92 and 93. The subtractions made no difference in the counting.[76] While ^{238}Pu could conceivably still be making a small contribution, its presence could not explain the bulk of the data.[77] The only remaining explanation of the observed spontaneous fission was the new isotope ^{240}Pu.

The group began a quantitative analysis of the fraction of ^{240}Pu in the pile-produced ^{239}Pu samples. In reactor-produced plutonium, they expected the fraction of ^{240}Pu in ^{239}Pu to be a constant times the fraction of ^{239}Pu in ^{238}U. The estimated fractions Farwell noted in his data book on 24 July were 0.6 parts per million (ppm) for the Berkeley material, 50 ppm for the Clinton plutonium, and 1,000 ppm for the re-irradiated Clinton plutonium – rough numbers calculated from the best-known values of the reactor flux and ^{239}Pu radiative capture cross section.[78] They carefully checked these fractions over the next year and determined the constant to be about 71. Farwell's doctoral thesis showed that the number calculated from the theory of production agreed with that measured from spontaneous fission, clinching the identification of ^{240}Pu. Segrè's group also further refined its study of spontaneous fission, while the Los Alamos Laboratory was carrying out its ultimately successful program to develop the implosion bomb, improving its measurement of the nuclear parameters of the observed spontaneous fission, for example, the number of neutrons emitted per spontaneous fission, the spectrum of the spontaneous fission fragments, and the half-life of ^{240}Pu. The group established the spontaneous fission rate of ^{240}Pu as 1.6×10^6 f/g hr, and that of ^{239}Pu as 40 f/g hr.[79]

Reorganization of the Laboratory

Fortunately, by the time the ^{240}Pu crisis hit the laboratory, the nuclear physics program was advanced enough for Oppenheimer to continue it with a fraction of the original staff, and the uranium gun program had also by and large solved its main problems. Oppenheimer was thus able to refocus the laboratory's program on implosion.[80]

Changes in Group Structure in August 1944

The reorganization was remarkably rapid. Discussed first in an Administrative Board meeting on 20 July, the new alignments were in place within two weeks.[81] The reorganization transferred technical control over implosion from Parsons to Bacher and Kistiakowsky, although interdivisional questions and broad policy questions were still to be cleared through Parsons. McMillan retained technical control of the gun.[82] At this time, the laboratory added two associate directors: Fermi, to oversee the research and theoretical divisions and all nuclear physics problems; and Parsons, to direct the division responsible for ordnance, assembly, delivery, and engineering.[83] Two assistant directors were also added: D. A. Shane, who handled personnel, and Dana Mitchell, who administered procurement. With shop groups now under the direction of Earl Long, the methods used in V Shop could be extended to C Shop.

The new Gadget (G) Division's objective was to conduct experiments on the critical assembly of active materials, study the hydrodynamics of implosion, and design and procure the implosion tamper and active core. This division combined portions of the earlier P- and E-Divisions. Only one G-Division group, Electronics, had existed as a unit before the reorganization; the others were created by subdividing existing groups or by elevating research projects to the status of groups.[84]

The new Explosives (X) Division was commissioned to design the explosive components of the implosion bomb, develop methods of detonating high explosives, improve the quality of castings and lens systems, and provide explosive charges for implosion studies. This division was created from the implosion groups that had evolved within E-Division. Neddermeyer's former group E-5, Implosion Experimentation, became X-1, under Norris Bradbury (with some members moved to the X-ray section of G-Division).[85] The distinct teams that had made up E-5, each studying implosion by a particular diagnostic technique, became sections of X-1, which, according to Kistiakowsky, were not designated groups because their leaders were too young to be put on the same administrative level as Luis Alvarez, McMillan, and Edward Teller.[86] Group E-9, under Kenneth Bainbridge, High Explosive Development, became X-2, Development, Engineering, and Tests. Both before and after the reorganization, this group focused on designing and testing full-scale high-explosive assemblies. Eventually, X-2 would engineer the Trinity test. Group E-10, which operated the explosives production facility (S-Site), became X-3, with no change in its duties, but with a new group leader, Army Captain Jerome Ackerman.

The remaining groups in E-Division, all dealing either with the gun or with bomb delivery, formed the new Ordnance (O) Division, which remained under Parsons and was responsible for the final weapon design and delivery. The main tasks for this newly composed engineering division were to coordinate lab activities with those of outside military groups conducting flight tests at Wendover, Utah, or overseeing transport to Tinian, and to ensure timely production of gun parts.

The original P-Division groups that studied the production and measurement of neutrons made up the new Research (R) Division, headed by Robert Wilson, who recalls that he reluctantly accepted the position on the understanding that he would do as little administration as possible. (He claims he carried all his administrative papers in the inside band of his Stetson.) The other former P-Division groups that dealt with the Water Boiler, detectors, and electronics were moved into either the new Fermi (F) or the G-Division.

F-Division (named after Fermi) consolidated several miscellaneous research projects, including the Water Boiler, Teller's group on Super theory, and Egon Bretscher's group on Super experimentation. By the time Fermi was able to join the Los Alamos staff in September 1944, the most important division assignments were already in place. Yet Fermi's senior standing and talents qualified him for a top position. Oppenheimer's compromise was to place him at the head his own division.

The laboratory-wide reorganization of summer 1944 had little effect on the organization of T-Division. The division was still divided into five groups: T-1, implosion under Rudolf Peierls; T-2, diffusion theory under Robert Serber: T-3, efficiency under Victor Weisskopf; T-4, diffusion problems under Richard Feynman; and T-5, computation under Donald Flanders. Group T-6, under Eldred Nelson and Stanley Frankel, was added in September to handle the IBM machines. Group T-7, under Joseph Hirschfelder, would be formed in November to deal with blast. In May 1945 an eighth group would be formed under George Placzek to deal with the development of a "combined" plutonium and uranium gadget.[87] However, CM-Division's work load was reduced when the plutonium purification program ended and approximately 7 percent of the chemistry staff was sent to other divisions or assigned to other jobs, one of which was to fabricate implosion initiators.[88]

As before, multidisciplinary collaboration was designed into th new organization. T, CM, and R offered service for the two implosion divisions, G and X, and for the gun division, O. T-Division continued the liaison program that Bethe had begun at the start of Project Y, in which particular theorists were assigned to help with specific experimental projects in other divisions.[89]

The August 1944 reorganization was drastic. Although the laboratory had initially been organized around scientific and engineering tasks, henceforth the organizing principle was whether work applied to implosion or the gun program. In general, group and division leaders were more affected by the reorganization than individual group members, who in many cases kept doing the same thing even though the name of their division had changed. The August 1944 reorganization was the largest of a sequence of reorganizations of a laboratory whose structure was by nature ephemeral; experiments and responsibilities changed overnight as the priorities that the war gave to the program shifted.[90]

Panels and Boards after the Reorganization

Project Y was an important early model for today's committee-run big science. In the early days of the laboratory, the Governing Board and Coordinating Council were Oppenheimer's principal advisory panels. Members of the Governing Board, the division leaders, and heads of administrative departments met almost weekly to consider issues related to both the technical work and community life. Just before the formal reorganization in August 1944, Oppenheimer divided the board into a technical and an administrative section, because he felt that too much of the board's time was spent on administrative matters at the expense of technical problems.[91]

Although the Administrative Board continued to meet throughout the war, the Technical Board met only a few times before it was disbanded, its role carried out by a number of boards and committees created to deal with specific issues. The first such specific technical committee was the Intermediate Scheduling Conference, under Parsons. Created in August 1944, this committee coordinated aspects of the final packaging of the gun and implosion weapons for testing and delivery to their final destinations. By March 1945, the conference had evolved into a purely administrative group with technical problems taken over by the Weapons Committee.[92] The Technical and Scheduling Conference was an interdivisional coordinating group started in December 1944 to help with the scheduling and coordination of specific projects, particularly those using active material. This group also coordinated shop time.[93] In March 1945 the Cowpuncher Committee was formed to "ride herd" on the implosion program (Chapter 16). Allison chaired both the Cowpuncher Committee and the Technical and Scheduling Conference. In addition, several smaller groups coordinated more specific sections of the increasingly complicated effort. For a short time in early 1945, combat delivery of the bombs was the province of the Weapons Committee, which was chaired by Norman Ramsey and answered to Parsons.[94] The Detonator Committee (Alvarez, Bainbridge, and Col. R. W. Lockridge) was formed in October 1944 to deal with the external procurement of electric detonators. The Initiator Committee (Bethe, Robert Christy, Fermi, and Bohr) gave advice on implosion initiator design from February 1945. An unnamed committee of Bacher, Fermi, and Wilson scheduled the use of ^{235}U in experiments to ensure that this precious material was used to full advantage.

The story of the discovery of spontaneous fission in plutonium and the dramatic impact of this finding on the wartime Los Alamos laboratory underscores the important role played by scientific research in this historic institution. In this episode, a purely physical effect, detectable only with great difficulty by the most sophisticated electronic devices then available, caused a large-scale restructuring of the laboratory's program.[95] Although members of the gun program did their best to explore means for overcoming the predetonation problem, none of the ideas suggested were practical.[96] The unpleasant fact that implosion was still unproved, and possibly unworkable, caused McMillan to make the pithy observation in mid-July 1944 that "one may have to accept an appreciable fizzle probability in preference to not having a [plutonium] gadget at all." However, Groves rejected this position and the civilian–military laboratory realigned its scientific troops for a massive attack on implosion.

13

Building the Uranium Bomb: August 1944 to July 1945

After abandoning the plutonium gun, Oppenheimer streamlined the uranium gun program. The remaining work, all experimental and straightforward, was consolidated in one ordnance group under Lt. Comdr. A. Francis Birch. Birch's group completed and tested the uranium gadget design by February 1945. Meanwhile, metallurgists decided that ^{235}U was strong enough to be used in its natural state, so that alloy research could be discontinued. They developed suitable large-scale metal reduction and fabrication techniques. Polonium production for the initiators was on schedule.

As ^{235}U production slowly increased and the laboratory made arrangements for fabricating gun parts, a quiet but dramatic change was occurring in the gun program. Before March 1945, this program had centered on perfecting a reliable method of assembling ^{235}U. Now Birch had to turn the gadget into a bomb that could be delivered by an airplane. The bomb had to contain the gun gadget, offer protection from antiaircraft fire, house all the components (including antenna, fuzing, and circuitry), and provide a reliable, stable flight. The design of the gadget could still be changed slightly, after being frozen in February. But once the drop

* This chapter is based on a manuscript by Roger A. Meade, to which Westfall contributed the sections on metallurgy and initiator development. We are grateful to Robert Penneman for his detailed editing of the technical content of the metallurgy and initiator sections.

tests began, extensive changes could not be made without jeopardizing Groves's July deadline.

Conversion to the Uranium Gun

With the cancellation of the plutonium gun in July 1944, the need to develop a multifaceted high-velocity gun suddenly vanished. Fifteen months of intensive effort had gone into the physical design of the plutonium gun's components, into interior ballistics, and the mechanical properties of plutonium. With this work ended, the remaining work on a ^{235}U assembly could all be located in one group – the gun group, O-1, of the Ordnance Division. This group was led by Birch.[1]

Birch had a mature, unflappable personality that made the gun group one of the laboratory's most smoothly operating groups.[2] A Harvard geophysicist, Birch had an extensive background in physics, electronics, and mechanical design. At Los Alamos he had been a member of the instrumentation group headed by Kenneth Bainbridge, with whom he had worked at MIT. The MIT experience served him well, since most of the gun work from August 1944 involved testing and instrumentation. Also assigned to the new gun group was Lawrence Langer, a physicist trained at New York University, who came to Los Alamos in November 1943. As the alternate group leader, Langer served as the group's chief physicist and completed the development of the gun initiator. His close work on the gun since 1943 with Edwin McMillan brought continuity to the new group. He also acted as the Ordnance Division liaison with the shops and proved invaluable in expediting gun-gadget machine work, at a time when shop work held up many development schedules.[3]

The uranium gun offered Los Alamos its only certain hope for a wartime combat weapon. Elaborate and thorough testing turned this hope into a reality.[4] Langer commented recently that much of his work on the gun gadget was surveillance testing to make certain that all contingencies were considered and tested for.[5] As Birch recalled immediately after the war, the uncertainties surrounding implosion led his group to formulate extremely conservative specifications for the gun bomb, including field operations. Faced with the need to produce a combat weapon guaranteed to work, Birch and his group could be grateful that ^{235}U had a very low spontaneous fission background.

Metallurgy and Initiator Development for the Gun

In August 1944, when the first large shipment of ^{235}U was sent to Los Alamos from the Oak Ridge electromagnetic plant, prospects looked dim for procuring enough ^{235}U in time for a wartime weapon. Construction of the thermal diffusion plant had begun, but it would not be in operation until the end of the year. Although the gaseous diffusion plant was also under construction in Oak Ridge, the researchers were having difficulty producing a newly designed nickel barrier on an industrial scale. Progress with the electromagnetic plant was also disappointing. Although four separation "racetracks" were in operation, they had yielded only a fraction of the expected material. A fifth track was experiencing minor operating problems, and newly built tracks were beset with insulator failures.[6]

By January 1945, however, ^{235}U production was increasing steadily. The electromagnetic plant was operating with all its racetracks, and the gaseous diffusion plant had begun isotope separation with a workable barrier in place. By 1 March, the thermal diffusion plant was also in operation, producing slightly enriched material that was fed into the electromagnetic plant for further enrichment.[7]

By the time of the first large shipment of ^{235}U from Oak Ridge, workable reduction and fabrication schemes had been devised using natural uranium, with the result that the Oak Ridge tetrafluoride product could be reduced to metal on the 225-g scale, recast, and rolled. Recovery procedures were also working, so material could be successfully recovered from bomb reduction, purified, hydrofluorinated, and returned to the metal reduction group. The success in processing and recovering ^{235}U was reassuring in light of the uncertainty as to whether a plutonium weapon could be built and the final confirmation that the uranium weapon would use ^{235}U metal. Los Alamos researchers could at least feel confident that ^{235}U metallurgy would pose few problems.

From summer 1944 to early 1945, the uranium effort was streamlined. With the success in bomb reduction, the electrolysis effort was discontinued. Similarly, upon discovering that unalloyed uranium had the required strength, both Los Alamos and MIT were able to stop the alloy projects. Attention now focused on refining recovery procedures, gaining experience with processing larger batches of ^{235}U, finding optimal procedures and crucibles for reduction, and developing fabrication procedures for weapon components. Since large amounts of ^{235}U were not yet available, work continued with natural uranium, both to gain experience with larger-scale processing and to provide the uranium metal.

After the first ^{235}U run, the recovery group experimented with a simplified ether extraction process for slag and refractory linings. When they found in December that excessive uranium was retained with liner solutions, unless the ether was redistilled and reacidified, a 2.5-inch column with an ether still and reacidifier column was installed in Building M. In the next month, they recovered refractory liners using the column. In January 1945, they found that the stripped solution contained only about 0.1 mg of uranium per liter. By this time, the group, now gearing up for the massive task of recovering uranium from UH_{10} plastic cubes (Chapter 17), felt confident that equally satisfactory results could be obtained if slags were dissolved by acid in a saturated $Al(NO_3)_3$ solution and then extracted with ether. In December, the group also found that extraction proved useful for purifying some of the special materials that came through recovery. Although they decided to retain the precipitation of uranyl oxalate as the final purification step, a subgroup was established to handle ether extraction in these cases.

In the fall of 1944, the metallurgists made a number of ^{235}U metal reduction runs on the scale of 100–200 g to explore the possibility of handling larger quantities of the fissionable material.[8] In the process, they discovered that the first Oak Ridge tetrafluoride product had a low bulk density. Eric Jette complained that this characteristic, which resulted in increased charge volume, decreased processing efficiency, because the apparatus was designed for a set charge volume. As he explained in a report of October 1944, "When the bulk density is low, the charge volume is increased. Our apparatus is designed ... to make a certain number of reductions per day. If the volume of charge mixture per unit weight of metal exceeds that planned for in our bombs and refractory liners, we have only two alternatives, (1) redesign the equipment or (2) make more reductions." By January 1945, Oppenheimer had talked Oak Ridge representatives into promising a higher bulk density. Jette also believed he could devise a 500-g-scale reduction method that could efficiently handle low densities. Indeed, in February 1945, Richard D. Baker and his group made twenty reductions on the 360-g scale with the Oak Ridge product and twenty-seven reductions on the 500-g scale with enriched ^{235}U fluoride from the recovery group.

In the fall, progressively larger runs were also made with unenriched uranium to test MgO refractories from different sources and experiment with various mixtures. In November they discovered that although liners from the Los Alamos group and those from MIT using Vitrafrax MgO worked well, some liners from MIT using MgO from General Electric were causing trouble. Since the Vitrafrax material was no longer available, testing began on the General Electric MgO liners. Despite this complication, in December the first 500- and 1,000-g reductions were run, both with a 99 percent yield. In February, Baker and his co-workers were able to provide 3 kg of biscuit metal for the fabrication group with runs on the 500- to 1,000-g scale having an average yield of 99.8 percent.

During this period, the fabrication group, under Alan Seybolt, also worked to fabricate ^{235}U metal. By the time ^{235}U metal was available to the group in September, a method for bottom pouring from an upper crucible into a lower mold had been developed for unenriched uranium metal that regularly produced satisfactory castings using MgO or graphite molds. Since the metallurgists needed to process smaller amounts of ^{235}U than unenriched uranium, both to conserve the fissionable material and to avoid processing close to criticality, special furnaces of limited size were needed for ^{235}U fabrication.

As the remelting scale increased in the fall and winter, the group had trouble finding a good crucible for ^{235}U remelts. In October, they hot-pressed several ^{235}U hemispheres for use by physicists. In November, the group obtained a new vacuum furnace, which aided in investment molding. By the next month, a 5-kg furnace was under design for still castings. After a round of practice investment castings in January, the crucible problems were solved, and in February ^{235}U spheres for criticality work with outside diameters up to 3.5 inches were completed without difficulty. They made the spheres and accessory pieces by melting buttons in crucibles (MgO for the spheres, BeO for the pieces) using resistance heating, casting into investment molds, and machining to remove risers and shape the pieces to final dimensions. The next step was to devise an efficient casting method for the larger pieces needed for the gun weapon. Investment molds of the larger size needed for weapon parts could not be made free of disqualifying defects. Centrifugal casting was tried, for it did not require an excessive amount of material (such as in risers) and thus reduced the difficulties regarding nuclear safety, with the result that more of the limited amount of ^{235}U allowed per casting could go into the final finished piece.[9]

Meanwhile, the chemists were making considerable progress in producing polonium in the form needed to fabricate the initiator. When the foils from Dayton reached Los Alamos, they dissolved the polonium from the platinum and plated it. The resulting foils, which were used to form the initiator, had to meet neutron emission requirements. To measure the exact polonium content of the foils, the chemists first had to improve their assay devices, a task complicated by the element's tendency to migrate. They constructed a vacuum demultiplier to measure larger samples and a gas target apparatus to measure polonium on curved surfaces. Later a calorimeter was used; measurement of the heat from alpha decay indicated the polonium content.[10]

By summer 1944, Allan Ayers had devised a design for the initiator of a gun-type weapon, and Dayton and Los Alamos chemists worked to reduce neutron emissions to levels acceptable to the physicists designing the weapon. According to Joseph Kennedy, this background was "low enough to permit their use" in the uranium gun-type weapon. In the same month, Oppenheimer wrote to praise Charles Thomas and his Monsanto colleague James H. Lum for their part in the effort, saying that some Monsanto foils could be used "without any further concern."[11]

The Gun-Gadget Test Program

The caliber of the ^{235}U gun had been decided by the time of the reorganization. Carlton Green, chief gun designer for the navy's Bureau of Ordnance, based his design on Hirschfelder's interior ballistics calculations. While the Naval Gun Factory worked on the uranium gun, Birch and his group focused on developing the target, target insert, and projectile.

Birch's test program used all the tools at his disposal. The 20-mm gun continued as the primary testing device. Other guns tested target designs that looked promising after being tested at the 20-mm scale. These other guns gave Birch and his group a feeling for how an assembly would behave at larger scales without having to fabricate a full-scale assembly. Birch salvaged the high-pressure gun originally designed for the plutonium gadget and used it to test recoil effects of the muzzle fittings. The operating pressures and velocities of this gun provided a substantial margin of safety. Birch reported much success by late September. The recoil tests on the high-pressure gun proved the adequacy of the muzzle fittings. The new primer fittings gave faster ignition than the older radial design. The proposed target design performed well, although some designs involving new materials were scheduled to be tested at the 20-mm scale. The proposed initiator design also looked good.

Although the design of the gun and its components seemed workable, McMillan, the person responsible for the entire gun program, asked Robert Serber to clarify a number of matters before proceeding with the final development. For example, McMillan needed to know how the yield might be affected by a tamper; he also asked for Serber's analyis on how the presence of initiators and the ring system might affect mechanical assembly.

In October, Birch's interest turned to drop testing to check the performance of the gadget, including the flight performance of ballistic shapes, fuzes, firing circuits, suspension, and lug; and most important, to check the behavior of the projectile when the gadget hit the ground. Birch also wanted to guarantee the reliability of the gadget by accumulating experience in simulated combat and to give key personnel experience in every aspect of handling the gadget in preparation for the final, overseas operation. In November, Birch's group routinely achieved consistent performance in assembly. By January 1945, the gun group routinely attained assembly of all parts.

Groves upset the steady pace of the gun program in December when he asked Oppenheimer to give higher priority to the gun program than "to his corpulent competitor" (Fat Man, the implosion weapon). Groves wanted all research and development on the gun completed by 1 July 1945, the point at which sufficient uranium and plutonium would be available for the combat weapons. For the first time, he assigned a definite completion time for the project. Consequently, the Technical and Scheduling Conference devoted its entire next two meetings to examining the status of the gun program. At its 28 December meeting, the conference examined a program for critical mass measurements of ^{235}U spheres. Fermi believed that finding the critical mass should be the overriding priority. Feynman suggested calculating the size of the mass that could be placed in the gun itself. Meeting next on New Year's Day, the conference reviewed the entire gun program in light of Oppenheimer's view that Groves wanted the gun gadget ready by 1 July 1945, to ensure that the laboratory would not be "holding up the military application of the weapon." Production of ^{235}U was apparently ahead of schedule. Reporting to the conference, Birch summarized his current program of target shots and drop tests, which were being conducted independently of each other. The target tests were designed to demonstrate the consistency and quality of the material assembly. The drop tests, conducted with models of varying degrees of likeness to the final gadget, tested the release mechanism, stability of flight, fuze performance, ballistic behavior, and in some tests, whether assembly took place from the shock of ground impact.[12]

Although confident that his program could deliver a combat gadget by July, Birch pointed out problems that needed to be solved to meet Groves's schedule. Gun deliveries, particularly for drop tests, were behind schedule and the exact critical mass of ^{235}U remained "subject to considerable uncertainty," a fact already noted by Fermi. Although Birch felt that last-minute changes, particularly in critical mass estimates, could invalidate much of the current testing, he believed that a gadget could be ready by 1 July.[13]

The Augmented Gun Gadget

No one knew for certain how many critical masses could be placed in the uranium gun, or what the isotopic purity of the available active material would be. Accordingly, Oppenheimer, Parsons, Birch, McMillan, and Joseph Hirschfelder discussed the possibility of modifying the current design to use different quantities of active material. Birch believed that it would be much easier to design a new gun using the current target and projectile geometries rather than modify either the target or the projectile. The problems of developing a new gun would be primarily logistical, not technical. However, all design work and testing, whether for a new gun or for modifications to the existing model, would compete with the ^{235}U gun work under way and could possibly delay combat use.

Responding to an inquiry from Oppenheimer, Parsons suggested establishing a "certain ritual" to investigate a new gun gadget. T-Division could then evaluate any new designs. He emphasized the need to order such a gadget from the Naval Ordnance Factory immediately, if the gun was to be ready by the fall of 1945, a goal suggested at the meeting on the augmented gun gadget. Although Oppenheimer approved nonengineering studies, on the assumption that they would give a better idea of what would be required, he declined to pursue the methodical study proposed by Parsons.

Hirschfelder prepared an interior ballistics appraisal. He believed that the powder used in the original gun would be more than adequate for a new model, but that final decisions regarding powder should be subject to projectile weight. As the coordinator of the gun program, McMillan recommended to Parsons on 23 November that any decision regarding a new gun should be based on Hirschfelder's parameters. Although

McMillan and Hirschfelder had shown that a new gun was possible, Birch was unconvinced. He encouraged in-depth study of development time, the quantity and enrichment of active material expected, and the relative military value of each weapon.

On 7 December 1944, Oppenheimer gave formal approval to the design of a new gun. The approval reflected "our best present guess of what is required."[14] The basic design took shape that December. On 18 December 1944, Parsons authorized site selection for testing and the use of two existing, unused guns for experimental firings. The 20-mm gun, by this time no longer being used for the Little Boy program, saw new life in testing target and projectile components for the augmented gadget. Tests with the 20-mm gun conducted throughout January and February 1945 proved the design of the new gadget.

Despite his decision to proceed with an augmented gadget, Oppenheimer still had doubts about the weapon. He wanted to know whether the new gadget would work before investing too much effort in it. He asked Robert Wilson to test the subcriticality of the target and projectile assemblies using slow-neutron simulation. Specifically, Oppenheimer wanted to know "what if any distribution of [active] material will remain subcritical."

Contracts for gun components were let in February 1945. So that no company would have the complete design, three independent plants built parts: the Naval Gun Factory in Washington, D.C., built the gun and breech; the Naval Ordnance Plant at Centerline, Michigan, made the target case and a few other parts; and the Expert Tool and Die Company, Detroit, Michigan, constructed the bomb tail fairing and various mounting brackets. A number of smaller components such as the projectile and fuze elements were either manufactured or modified at Los Alamos.

After much discussion with Hirschfelder, Parsons placed the order for both the expansion equipment and three tubes on 23 April 1945.[15] The pace of the war, however, made the decision moot. On 28 July, the Naval Gun Factory was ordered to stop all work on project guns. Four months after Hiroshima, in December, Navy Capt. Ralph A. Larkin, who succeeded Parsons at the laboratory, asked the new laboratory director, Norris Bradbury, what to do about the guns and components at the factory. Bradbury replied: "Keep drawings at Navy Yard. Junk anything else."[16]

Preparations for Hiroshima

By February 1945, work on Little Boy had progressed so well that the design was evident. Birch began making plans to ship the gadget to the yet unselected military base, where it would be readied for combat. Birch proposed two methods of shipping the active material: shipping parts separately to the final assembly point, or shipping the active material in two special containers to the final assembly point, where each would be placed in its respective place just before combat delivery.[17] Although he favored the first method of shipment, Birch recognized that the second offered several advantages. First, a number of target cases could be stockpiled at the assembly point. Any losses or damage to such units would be offset by sheer numbers. Finally, if the active material was sent separately, security would be enhanced, because only the active material would have to be guarded carefully during transit. Birch ultimately recommended the second method.[18] The two containers were shipped to Tinian welded to the deck of the USS *Indianapolis* with a twenty-four-hour armed guard.

Once Birch decided on the method of shipment, Marshall Holloway went to work on safety, proposing several tests to model the nuclear behavior of the active material that could result from handling, an accident, or combat delivery.[19] Holloway suggested tests to determine how safe it would be to handle active material under the stress of being shipped long distances. These tests included immersing the active material in water, placing it on a simulated work area, and arranging many people in close proximity to it. Holloway noted that immersion in water should give an idea of the danger to personnel and eliminate the danger of exposing personnel to active material. A fourth possible hazard, fire, could not be tested. Intense heat could ignite ^{235}U and thus form uranium oxides, which in turn would create a diffuse mass that might support a chain reaction because the changed chemistry, from metal to compoundes, would provide moderation for neutrons. (The decrease in neutron energy would lead to a decrease in critical mass.) This last hazard was referred to T- and CM-Division for study.

Holloway recommended two tests for imitating the hazards of assembling active material. In the first, a target and projectile would actually be assembled with several people in close proximity. This test would give hands-on experience in handling ^{235}U metal. In the second test, target and projectile pieces would be inserted into the gun under conditions thought to be similar to those at the advance base. Six experiments were designed to test possible catastrophes during combat delivery. Of these, the sixth was designed to test the critical separation distance between the target and projectile. Tests at Wendover had shown that the impact of the gadget hitting the ground could dislodge the projectile and drive it into the target. Noting that "war is a violent business," and that there is a possibility that the projectile could become accidentally dislodged and explode prematurely, Holloway proposed measuring this critical distance and perhaps designing a safety feature that would keep the projectile from seating in the target in the event of an accident.[20]

With these preparations, the uranium gun program neared completion. Although the elimination of the plutonium gun had simplified the gun program, greater urgency had been placed on the uranium gun since the implosion weapon was uncertain until July 1945. Birch kept the program on a steady course, analyzing all aspects of the gun program and never deviating from his mission of giving the laboratory a gun of conservative design guaranteed to operate in combat.

The work that remained in preparing the final combat gun weapon was to find the appropriate ballistic shape and to develop a fuze that would detonate the gadget at approximately 1,000 feet over the target. The first task consisted of building and testing models. Several ballistic shapes were tried. The earliest of these, designed and built even before the gun type was selected, was constructed of lead weights and sheet metal and had approximately the same dimensions and weight as the final combat unit. The design was flexible because initially the purpose was to test the many possible combinations of weight and centers of gravity. Birch used the sixteen units constructed in this series, each identified by its drawing number, to test fuzing, suspension rigging within B-29 bombers, and drop ballistics. More sophisticated models, closely resembling the exterior appearance of the combat weapon, soon came into production. Containing a dummy gun, they gave Birch an opportunity to simulate the weight and weight distribution of the combat unit.

If the fuze either failed or malfunctioned and prematurely detonated the bomb, both the mission and the strike force would be destroyed. Were the fuze to fail, the resulting dud and active material might be recovered by the enemy. Any fuze that failed even 1 percent of the time on average was unacceptable. Fuzing was also important because Oppenheimer and others believed that the bomb had to be detonated at a considerable altitude over any target. No fuze had yet been developed to operate in this way. Until Trinity, the detonation altitude remained an unknown, since it could only be calculated from the expected size of the explosion, which had to be extrapolated from the current theory of blast damage.

The job of developing the fuzing system for the gun fell to physicist Robert Brode.[21] Parsons noted in 1945 that Brode accomplished this work "with such smoothness and thoroughness and obvious soundness, that this difficult feature of the fission bomb has caused no real worry." Brode pursued fuze development along three lines: barometric switches, electronic circuitry, and clock switches. All were problematic. The barometric switches, although simple mechanical devices, were unproved in falling bombs. Electronic circuitry, although identical to the circuits found in proximity fuzes and radio altimeters, had not yet been developed specifically as long-range detonating devices. The third alternative, clocks, was pursued as a backup system. Clocks proved difficult to work with because they required resetting each time they were used, which increased their probability of failure. Most of Brode's efforts were concentrated on developing and testing electronic designs.

To help respond to the reliability requirement and altitude consideration, Brode developed a "philosophy of fuzing" in line with the strategy of using overlapping approaches. The philosophy called for backups for every component and circuit. When Brode put a fuze in a bomb, he expected it to work. He developed and supervised a program of statistical testing and constantly reminded his people of the importance of their jobs. Brode's approach to fuzing matched Birch's program of overdeveloping the gun gadget, as well as Langer's surveillance testing of initiators.

The actual development of the electronic circuitry was assigned to Dr. Horace Crane of the University of Michigan, under a contract let by Section T of the OSRD. By early 1944, Crane had an amplitude-modulated radio proximity device ready for testing. It was capable of operating at an altitude of 150 feet. Before the device was tested, however, T-Division revised its estimate of the required altitude to 3,000 feet, which was far too high for Crane's device to work. The device remained unusable even when the firing height was subsequently lowered to approximately 1,000 feet.

Brode immediately redirected electronic work to two other devices: frequency-modulated and pulse-operated units. Again through Section T, Brode contracted out the actual development. Norden Laboratories, the contractor, had done much of the work on a frequency-modulated device under an earlier navy contract. They used this work to build a model nicknamed "Archie." The Archie device proved exceedingly efficient and soon became the primary electronic fuzing circuit. Because the Archie device became available so quickly, all fuze work beginning in mid-1944 centered on perfecting and testing all three parts of the fuze: barometric switches, Archie circuits, and clock switches. The entire fuze mechanism consisted of a radar relay, barometric switches, and a series of clocks arranged in parallel. Four Archie units, triggered to act at a prescribed altitude, supplied current to a network of relays. These relays fed the current to a group of barometric switches designed to begin operating in parallel at an altitude of 5,000 feet. These switches, in turn, connected a 30-V potential to a bank of clocks, also designed to work in parallel. The clocks fed the 30-V potential to a high-voltage relay that, in turn, fed current through a high-voltage condenser to the detonation line.

After the Trinity detonation and its confirmation of efficiency estimates, Oppenheimer directed Brode to increase the firing height from the last estimate of 1,000 feet to 1,850 feet. Oppenheimer also directed that an additional 180 feet of altitude be added to compensate for any intrinsic delays caused by instrumentation. Acknowledging that a precise setting was not possible, Oppenheimer told Brode that detonation could take place within 100 feet of the desired 1,850 feet.

The shift in emphasis from the gadget as a gun to the gadget as a bomb raised two concerns. Since the reliability of Little Boy remained somewhat uncertain, it became important to know the fate of the projectile if the gadget did not detonate. A special circuit known as an informer was designed to provide information about the projectile in case it turned out to be a "dud" – that is, in case the main fuze failed or the projectile did not accelerate properly and meet the target. Birch recommended a switch, activated by impact, that would either extinguish or modulate a radio transmission signal that would be monitored from a second aircraft. He asked Brode to begin work on such a switch and to consider first testing the ability of the switch to withstand the violence of the bomb hitting ground and the resulting impact of the projectile.

The informers never worked well, even with their simple circuit. The distance between transmitter and receiver, the orientation of the transmitter to the receiver, and the level of background electrical noise surrounding all of the equipment caused recurring problems, which made it difficult to collect useful information. Although new, more powerful informers were being developed, they would not be ready in time for the first combat drops. Given the unreliability of the informer circuit and the high probability that ground impact alone would cause an explosion, Brode asked Oppenheimer to eliminate the informer from Little Boy. Both Parsons and Birch agreed with Brode's assessment, and on 17 July, Oppenheimer granted permission to eliminate the informer.

The second concern was how to prevent a dud. Whereas the informer circuit was designed to analyze such an event, a backup fuze, known as an inertial impact switch, seemed the best way to prevent such a failure. From the beginning, this switch had caused problems. During high-altitude drops, the switch failed completely. It only worked on drops conducted below 6,000 feet, far below the altitude of a combat drop. Summing up test results, Birch noted that the switch could be used in combat only if a target was reached, the bomb was released from an altitude below 13,000 feet over water or below 6,000 feet over land, and the main fuze failed. Although the switch would detonate the bomb in this instance, the explosion would kill the aircrews. Since an experienced aircrew would know well in advance whether or not proper altitude could be reached, Birch recommended that the switch be eliminated from Little Boy, and Brode seconded Birch's recommendation on 17 July. Brode pointed out that the switch itself could cause the main firing circuit to fail. Consequently, on 18 July, Oppenheimer granted permission to omit the switch. Although personally wanting the switch included, he could not overlook its many problems, particularly the hazard to the primary firing circuit. Like the informer, the inertial switch never proved reliable and in the end it was judged unnecessary.

By early May, Little Boy was ready for combat. Component tests of the initiator, projectile, target, and propellant had proved satisfactory, as had system tests of the blind target configuration and simulated combat drops. But as the time for the combat mission drew closer, Birch became concerned about the safety of the gadget if the strike plane crashed, either during take-off or landing. A crash would in both instances start fires, which could ignite the bomb's powder charge and cause a nuclear detonation. Birch was most concerned about a crash on take-off. Although in general few accidents had ever occurred on take-off, if it happened to a nuclear-armed plane, the result could be devastating. The Army Air Forces group selected and trained to fly the combat missions, the 509th, had witnessed a take-off crash in which the high-octane aviation fuel of a B-29 exploded into a huge fireball and ignited the plane's load of incendiary bombs.[22] Birch's solution was to equip Little Boy with a removable breech plug that would allow the crew to load the gadget in the airplane in a safe condition and then arm the bomb in flight. The bomb, assembled without its powder charge, would be harmless during take-off. Once the strike plane took off and achieved the necessary altitude, the plug would be removed, the powder inserted, and the plug replaced. In the event of an emergency or an aborted mission, the process could be reversed and the powder returned to its special external cannister.

Uranium Metallurgy and Initiator Development for the Gun

Throughout the spring and summer of 1945, ^{235}U production progressed slowly but steadily. When the final racetracks began operating at the electromagnetic plant in April, Oak Ridge finally had all three plants running at full capacity. In that month, Oppenheimer transmitted the proposed uranium production schedule from Nichols to Smith and Bacher. The schedule predicted that by early June, sufficient ^{235}U might be available to fabricate gun components.[23]

By this time, the production plans were well defined at Oak Ridge. In September 1944, Groves, Nichols, and a committee of representatives from each isotope separation project had carefully considered losses, capacities, and construction schedules to plan the flow of feed through the three isotope separation plants so that the greatest amount of material with the highest enrichment could be produced as quickly as possible. Once the thermal diffusion plant went into full operation (in March 1945), it was able to produce feed for the electromagnetic plant, which began receiving feed from the gaseous diffusion plant in April. During April, the gaseous diffusion plant achieved a higher enrichment of material, and the configuration for producing ^{235}U for the first weapon was set: the thermal diffusion plant fed material to the gaseous diffusion plant, which fed material to the electromagnetic plant for the final stages of enrichment.[24]

Although ^{235}U production surged in May, by the end of the month it was clear that there would still not be enough shipments to complete the fabrication of gun components in June. Samuel Allison and Smith now estimated that the necessary metal for the gun weapon would be available on 1 August. To allow the cube experiments to continue until the last possible moment, Smith and Allison carefully coordinated metallurgical schedules with those for the rest of the laboratory.[25]

While the metallurgists awaited ^{235}U shipments in the spring and summer of 1945, metal production and fabrication proceeded smoothly.

The metallurgy group reaped the benefit of previous efforts to devise efficient ^{235}U metallurgical techniques and prepare for contingencies. ^{235}U metal production procedures were complicated by the fact that the active material had to be reprocessed quickly from ^{235}U hydride cubes. Because of the efficient reprocessing procedures, however, Clifford Garner's chemists were ready to reprocess the impure metal "on a three shift per day basis" to prepare the necessary fluoride for reduction to ^{235}U metal. As Kennedy and Smith noted, Richard Baker could reduce enriched material "as a matter of course" with the stationary bomb method he devised in 1943. In March, Baker achieved "an astonishing record" in preserving the precious metal during reduction, with an average yield of 99.96 percent. Smith later summarized Baker's overall ^{235}U reduction efforts as "incredibly efficient."

Metal fabrication procedures were implemented without trouble. After Seybolt's group had tested casting procedures with unenriched material, Smith and Kennedy reported in June that "processes for casting and machining have been completely developed." Centifugal casting, an established metallurgical practice for pieces of simple geometry, produced castings of good physical properties and high density, held up less material, and required less machining for finishing the parts. The group fabricated gun components in the following manner: biscuits of metal were received from Baker's group and pressed to the required size with a hydraulic press enclosed in a box to prevent metal loss. Broken biscuits were then placed in a MgO crucible, which was covered and placed in a small furnace arranged for bottom pouring. Then the metal was heated and kept molten, with a maximum temperature of about 1,350° C, then cooled to 1,270° C. At this temperature, the metal was cast into a mold made with a steel ring inserted between magnesia top and bottom plates. After approximately two hours, the furnace was opened and castings were removed.

In June, Smith supervised test castings of the first ^{235}U by I. C. Schoonover's group. In early July, Smith estimated that the final target assembly would be ready for delivery by 27 July and reported proudly, "15 castings were made without failure." Schoonover's group in fact completed the ^{235}U gun assembly on 24 July.

During the spring and early summer of 1945, progress also was made on the initiator for the gun-type weapon. As in the implosion initiator effort, neutron background was a prime concern. The chemists worked with the ordnance group to ensure that neutron emission would not cause predetonation in the gun-type device. Once the initiators met this standard, extensive tests were made to make sure the low neutron level was maintained. Twenty initiator units were checked for several months "without suffering any serious rise in neutron background." Other tests checked whether neutron backgrounds increased under combat conditions. In each case, no change in neutron emission was observed. As a final check, measuring equipment was transported overseas along with approximately forty initiators. When sixteen of these were checked on 28 July, all neutron measurements were identical to those made in the United States.

Fabrication of the Active Material

At the end of May, all that remained to ready the gun gadget for combat was to fabricate the projectile and target insert. A three-step process was used to make sure the insert would contain ^{235}U of the highest possible isotopic purity. Because the metal on hand did not have the expected purity and because the insert needed to be safety-tested, the plan actually called for the target insert to be fabricated twice and the projectile once. First, the CM group would select the purest metal available. Then they would fabricate the insert and conduct safety tests and paraffin tests of criticality. After these tests, the insert would be returned to CM-Division for purification and refabrication into the projectile. The projectile would then be safety-tested and made ready for shipment to the 509th's headquarters on Tinian Island. Finally, after sufficient material arrived, the target insert for the combat bomb would be fabricated. No additional safety tests were required. Under this procedure, all fabrication would be completed by mid-July.

The first ^{235}U projectile component was completed on 15 June and the target insert was completed on 24 July and tested on 25 July. The projectile, along with the combat ballistic case, was then shipped to Tinian on the *Indianapolis*. Birch accompanied the target insert, which was flown to Tinian. On 30 July, the insert, projectile, and initiators were assembled in the bomb. All other components were added on 31 July. The powder was placed in a special container for insertion once the plane was safely airborne.

Before the gadget was sent into combat, an effort was made to estimate how the unit would behave, both before and during assembly – how the yield would be affected by the compression of the core during impact, what the absolute probability was that the bomb would detonate, and what the probability was of predetonation and a fizzle. Calculations by Leonard Schiff showed that the expected yield of Little Boy would be the equivalent of 13,400 tons of TNT. Establishing the absolute probability that the gadget would detonate at any given point in time seemed critical to understanding performance. Schiff calculated the probabilities for five time intervals.

In the early days of the project, the gun had been the laboratory's primary device, but soon most of its effort was diverted to the plutonium gun. When the plutonium gun was abandoned, the uranium gun again took center stage, for until implosion was perfected, the uranium device was the only certain means by which the laboratory could fulfill its mission of producing a fission bomb. Later, as the prospects for the implosion grew more optimistic, the uranium bomb fell back into relative obscurity. However, the gadget's shepherds – Parsons, Birch, and Brode – kept working and ultimately brought the weapon to Tinian. When Birch's Tinian crew assembled Little Boy, the least controversial bomb of the project became a combat weapon. The uranium gun did not require proof firing, because of the relative simplicity of the nuclear assembly design, because of the unstinting development work of Parsons, Birch, and Brode, and, what is most significant, because the Trinity test confirmed that a nuclear fission explosion was indeed feasible.[26]

14

Exploring the Plutonium Implosion Weapon: August 1944 to February 1945

After August 1944, the implosion program began to engulf a growing fraction of the laboratory's personnel. Considerable research during the fall and following winter focused on experimental diagnostics. Seven parallel experimental programs – RaLa, betatron, magnetic, and electric pin studies, in addition to the original X-ray, photographic, and terminal observations – brought the most current techniques of experimental physics to bear on implosion problems. The foremost tasks were to determine the collapse time, compression, and symmetry, and to assess different explosives and explosive system designs. Informed trial and error was the approach most frequently taken in these simultaneous lines of experimental inquiry, since theoretical understanding was still incomplete. While each program offered its particular advantages, many efforts overlapped, adding modest confirmation to the amassing body of understanding. Despite lingering uncertainties about the feasibility of an implosion weapon, the six months following the August 1944 reorganization saw the central research question of the laboratory shift from "Can the implosion be built?" to "How can the implosion weapon be made?"

* This chapter is based on a draft by Lillian Hoddeson, to which Catherine Westfall contributed the section on plutonium metallurgy. We are grateful to Gordon Baym and Les Redman for their detailed editing of the technical content of the chapter, and to Robert Penneman for contributing to the section on the RaLa method and for editing the plutonium metallurgy section.

Much work also remained before plutonium components could be produced. In establishing a plutonium production system in the limited time available, the chemists and metallurgists, like the physicists, often relied on empirical methods guided by intuition, since little theory was available and methodical procedures were extremely time-consuming.[1]

Experimental Diagnostics

The history of the seven-pronged experimental program to study implosion during World War II is one of painstaking progress, with few highlights or definitive measurements, many ambiguous steps, and numerous failures. G-Division groups worked on the RaLa, pin, magnetic, betatron, and counter X-ray methods. Within X-Division, group X-1, which had absorbed most of the former group E-5's program, worked on the photographic, flash X-ray, and terminal observations method. The original three diagnostics – X-ray, photographic, and terminal observations – continued to gather useful data in the four months following the reorganization. Although outstripped in power by the newer methods, these older diagnostics yielded results helpful in developing the electric detonator, initiator, and explosive lens.

The RaLa Method

The principal advantage of the RaLa method, being examined in G-Division under the direction of Bruno Rossi, was its capability of determining the compression of the imploded metal as a function of time. Since plutonium was not yet available, materials having mechanical properties similar to plutonium had to be substituted. Uranium had the disadvantage of being a strong γ-ray absorber. Metals of lower opacity – such as iron, copper, or cadmium – were more successful in RaLa test shots. Most of the shots employed cadmium.[2]

The radiolanthanum came from Oak Ridge; the first large shipment of more than 300 curies arrived at Los Alamos in mid-September.[3] Unfortunately, because of impurities from the shipping container, this source yielded only 40 curies of the desired lanthanum-140. After separation, the Los Alamos chemists packed the lanthanum in a lead container and trucked it to the test site, where it was removed by remote handling and inserted into the center of the shell.

The first RaLa shot, with an iron "pit mockup" surrounded by Comp B, and collapsed onto the rather weak source of 40 curies, was fired on

22 September.[4] In one of the army tanks that Luis Alvarez had procured (Chapter 8), oscilloscopes for the four banks of eight parallel-connected ionization chambers recorded the data.[5] The test confirmed that the ionization chambers were performing well and also provided a time of collapse. But the source was too weak to provide conclusive information on the symmetry and compression. One happy result was that the contamination of the firing site was far less than expected; for this reason, the group gave up its plans to work in a mobile tank laboratory. It was possible to work at a permanent RaLa firing site.

The mildly encouraging results from the first shot stimulated recommendations from theorists about further RaLa experiments. On 3 October 1944, Bethe suggested that the lab (1) select "the size of the gadget ... so as to give a fairly considerable but not too great attenuation of the γ-rays in the assembled state"; (2) choose the mockup material so that the time it would stay together after firing, the "sitting time," would be "long enough to be measurable, i.e., long compared with the resolution time of about 2 to 3 microseconds," a condition that ruled out tuballoy "except possibly in some very special arrangement"; (3) choose the compressibility of the material to maximize the compression without spalling – "the only materials which be can used with some confidence in the standard geometry are steel and copper"; (4) use an amount of explosive corresponding "as much as possible ... to specifications for the actual gadget"; (5) use, if possible, a tamper material that would permit researchers to take a magnetic record simultaneously (i.e., copper rather than iron). He also recommended particular dimensions and various materials.

Slowly but systematically, the RaLa program contributed valuable information about implosion as the strengths of radiation, materials, size, and nature of explosives were varied. The second shot, fired on 4 October, was similar to the first, but used a much stronger source, about 130 curies. In addition, a copper liner was used instead of an iron one.[6] The third shot was fired on 14 October with a somewhat smaller quantity of RaLa. Neither shot showed evidence of compression, but in making these shots the group learned how to carry out the experiment.

Possibly distressed at having failed to observe compression in the first shots, Rossi and Hans Staub prepared a memorandum listing "conditions that must be fulfilled in order that reliable data may be obtained." They proposed experiments to investigate the "influence of the nature of the HE," the "influence of the mode of initiation," and the "influence of the nature and thickness of tampers," as well as a number of

Fig. 14.1. "Remote handling" of a kilocurie source of radiolanthanum for RaLa testing. LA Photo, 26406.

auxiliary programs, geared, for example, to designing "circuits for the investigation of the simultaneity of detonation" and to investigating the electric detonator, "as soon as Alvarez' group produces a satisfactory" one.[7] Oppenheimer agreed to "all elements of the program" but decided not to try out electric detonators "in so costly an experiment as the RaLa until a little further experience had been obtained."

Meanwhile, in late September theoretician Robert Christy suggested using a solid rather than hollow core (described under "Theoretical Studies," Chapter 15), expecting in this way to reduce significantly the jets and spalling expected in the standard arrangement. Rossi was immediately enthusiastic about exploring Christy's suggestion, and at the RaLa planning committee meeting on 5 October he urged Oppenheimer to authorize RaLa shots using solid spheres. The group concluded that

25 October would be "the earliest date for solid spheres."[8] This date, as well as the notion that the group could obtain rapid definitive results on the Christy concept, proved overly optimistic.

The first RaLa shot using a solid sphere was actually fired early in December, but the results were erratic. Before that, the group had seen "some evidence of compression." However, the shot fired on 14 December was a milestone, for it showed "definite evidence of compression," a result Bacher considered "quite encouraging in the limiting case in which the asymmetries of the implosion play the smallest role." Christy then immediately examined the theoretical implications of these first solid RaLa shots, attempting "to find a reasonable way of calculating what went on such that the observed [γ-ray] transmission curve could be reproduced." He also embarked on a program to estimate efficiencies for various initiators.

The first two shots using electric detonators and solid cores were fired on 7 and 14 February 1945. They showed a great improvement in quality and caused a definite, if modest, turn in the implosion program. By providing the first observations of sizable compression, they confirmed significant improvement was possible when electric detonators were used in place of Primacord systems. After these shots, electric detonators were used in all RaLa tests and the results essentially settled the design for the Trinity gadget. By the end of February, Rossi could quote Oppenheimer as having said in a meeting that month, "Now we have our bomb."[9]

The Pin Method

The pin method gave vital information on asymmetries in spherically convergent arrangements. Early in August, Oppenheimer asked Darol Froman, a member of John Manley's P-Division group, to join Edwin McMillan and cohead a group devoted to electric and magnetic methods. Froman accepted, but mentioned that he did not favor the joint leadership. He proposed instead that two separate groups be set up, one to explore electric and the other magnetic methods. Accepting this suggestion, Oppenheimer assigned responsibility for the electric pin method to the G-Division's group G-8 under Froman.

Early in September, Froman wrote a careful description of the pin method: "to determine the shape and dimensions of the liner of a tamper as a function of time during the process of implosion .... The principle ... is to determine by means of either electrical pickup or direct connection the positions of various points on the inside surface of the liner as a function of time." He proposed that the electrical record be taken by an oscilloscope having an accurately timed sweep. Although not in principle limited by the scale of the assembly, the method could not work on a complete sphere, since space was needed for the electrical wires. But the method could be used together with the magnetic one, providing an independent calibration.[10]

Since the pin method was based entirely on familiar technology, it could potentially yield useful data quickly. William Higinbotham's electronics group promptly constructed the necessary electronics, about which Froman wrote gratefully, "After hauling [it] several miles in a truck we simply plugged it in, and every part has been operating perfectly since .... I think it quite significant that in its two and one-half days of use we have already obtained data of some value to the project."

By early October, pin experiments were already "quite encouraging," and the group was planning "to investigate the early phases of acceleration around a configuration of lenses." Within only a few weeks, first measurements were on hand; a 3-inch Pentolite cylinder propelling a 1/8-inch steel plate had been studied "with satisfactory contacts observed from conductor points placed at 1 and 2 mm from the plate." In its use of trial and error, Froman's program resembled all the other diagnostics. He planned "to shoot many identical shots with many contact points to test the consistency of the method, to measure the variation of starting time and velocity across the plates and to try the method on hemispheres." By 1 November, the group was comparing the use of Pentolite versus Torpex in the acceleration of steel plates, finding "very little difference." Furthermore, tests of small HE lenses coincided with studies using Marley's rotating camera. The method was particularly useful in providing accurate values of the velocity of different points on the imploded object. By mid-February, electric detonators were being tested on the pin method in hemispherical shots.

The Magnetic Method

Soon after the reorganization, McMillan became head of the G-Division group (G-3) that hoped to obtain good measurements of timing and total compression using the magnetic method. Unlike the other diagnostics, the magnetic method appeared capable of working on full-scale shots. The method was based on measuring the induced voltage created in a pickup coil as the imploding metallic sphere collapsed in a static magnetic field.

The full program was divided into four sections of G-3: one under J. L. Fowler on South Mesa Extension, a second under J. R. Wieneke of G-2 at P-site, a third under Henry A. Fairbank of G-6 in Bayo Canyon, and a fourth under Jack Smith of X-1, in Building 2, Anchor Ranch.[11] The last three sections hoped to combine the magnetic method with X-ray and RaLa shots, using the magnetic method to time precisely when motion of the metal parts of the core began. In mid-October, however, the work was "still confined almost entirely to Fowler's section owing to the lack of personnel." Although preliminary results in September on small spheres of uranium oxide tamper and copper liner indicated "some compression," many further months of work would show that it would be extremely difficult to achieve useful data by this approach.

Throughout the fall, the magnetic experiments were troubled by a spurious electromagnetic background. In an effort to reduce the background, the laboratory constructed a separate proving ground for magnetic shots in Pajarito Canyon, a site completed by 15 December and put into immediate operation under the leadership of Edward Creutz.[12] Interference problems also plagued attempts to coordinate measurements using magnetic and other diagnostics, as the circuits of the individual methods "appeared to interfere" with each other.

Kistiakowsky hoped to use the magnetic method to study jets. He wrote to McMillan: "We are extremely poor in methods for proving that collapse of spheres in its late stages is not completely messed up by jets of the type observed by X-1 on very small spheres and on cylinders. In particular, the RaLa experiments may give us information on compression but will give only very indirect information on mixing, i.e., the jets, and then only if combined with successful tests by the betatron method." Because Fowler's work had shown "that a strong electro-magnetic signal is sent out when a permanent magnet is struck very violently, as by a detonation wave," he suggested levitating a magnet placed in the center of a sphere. "You may then be able to get a sudden signal when this magnet is struck .... If no jets are present this will be the instant when the inner cavity is closed up. If jets are there the instant will be earlier .... By comparing the times thus obtained with, for instance, the RaLa records it may be possible to determine whether jets were present in the RaLa collapse or not." In a second memorandum that day on the same subject, he mentioned a method suggested by Norris Bradbury for calibrating the experiment. He added yet another piece to the same memorandum on the following day: "It doesn't matter how you do the experiment and whether you will find out anything. Just do it quickly." In fact, by mid-January comparative results between the magnetic and pin methods were offering quite specific information about the jets. By mid-February, the magnetic group was regularly firing shots.

At about the same time, the group began to employ yet another parallel approach, the magnetic "pin-loop method," to study timing in small lenses. In this method, the detonation suddenly ionizes the material between two pins, which thus become electrically connected to create a closed electrical loop. But this alternative method turned out to have little advantage over the ordinary electric pin method.

The Betatron Method

The betatron group, jointly headed by Neddermeyer and Kerst, was formed within G-Division (G-5) to conduct experiments analogous to the flash X-ray method, but using high-energy γ radiation obtained from accelerating electrons in a betatron. The plan was to record the radiation – after it passed through the imploding assembly – by a cloud chamber, rather than by ion chambers or electronic counters. The joint leadership of the group represented unusual expertise: Neddermeyer had been a pioneer in developing the cloud chamber into a useful tool for high-energy physics, and Donald Kerst had invented the betatron at the University of Illinois.

In mid-August, they set to paper a first detailed description of the experimental setup. Their plan called for two reinforced-concrete flatiron buildings, or "bunkers," pointed at each other, with the betatron erected in one bunker and the cloud chamber in the other. The explosion would occur in between. The concrete was several feet thick, to protect the betatron and cloud chamber from the blast. The distance between betatron and cloud chamber was about 40 feet. The γ rays emerging from the betatron were to pass out of the betatron building through special aluminum "nose pieces" placed over the exit holes, and all the delicate equipment was to be placed on shock mountings.[13]

Both the betatron and pin methods depended for their timing on the fast-sweep oscilloscopes recently developed in the wartime radar program. Until these devices came along, oscilloscope sweep times were inadequate for monitoring implosion, being in the range of microseconds rather than the required tenths of microseconds. In Neddermeyer and

Kerst's planned betatron program, the fast-sweep oscilloscopes would be housed in the cloud-chamber room.

The program could not go forward, however, without an appropriate betatron. Oppenheimer, who was actively engaged in planning the betatron experiments in the latter part of August, stepped in with maneuvers that illustrate the extraordinary network of cooperation between the Manhattan Project and institutions all over the country in that wartime period. Within one week after receiving the Neddermeyer–Kerst memo on the betatron method, Oppenheimer had located the only existing suitable betatron in the United States and had arranged for its transfer to Los Alamos. Teletyping Conant that tests by Neddermeyer and Kerst on timing pulses and on the sharpness of collimation of the betatron "lead us to believe that this instrument may be very valuable in large scale studies on our main program," he requested that one of the betatrons being constructed under an NDRC contract be diverted to Los Alamos by 15 October. He inquired about the logistics of such a diversion.[14] Conant replied that the Allis Chalmers Company was making such betatrons, and that thus far only a single one had been completed, a 20-MV unit made for the Rock Island Arsenal.[15] This machine was just then at the University of Illinois in Urbana for testing and tuning. The next day, Oppenheimer teletyped back to Washington: "We should like to have this machine diverted to site Y. According to Conant's information Allis Chalmers can complete fabrication of unit within two months of receipt of order."[16] Later that day, Oppenheimer received the go-ahead from Edward Moreland, executive officer of the NDRC, for negotiating with Army Ordnance to obtain ownership of the betatron.[17] Four days later, the Washington Liaison Office teletyped back that it would be eight weeks before the betatron was fully tested.

The betatron group's progress in the fall of 1944 also illustrates the speed with which high-priority projects developed at Los Alamos. In early September, while waiting for the machine to arrive, Neddermeyer and Kerst composed a memorandum to Bacher covering basic issues concerning the experiment, such as the number of tracks expected in the cloud chamber, the testing of the betatron in Illinois, the preliminary schedule, and personnel estimates. The memorandum, as well as the earlier one of 18 August, was then thrashed about at a meeting on 7 September, which included Oppenheimer, Parsons, Bethe, Weisskopf, Kerst, Bradbury, W. A. Stevens, Neddermeyer, Kistiakowsky, Alvarez, Rossi, Parratt, and Bacher.[18] Rossi commented that two-dimensional symmetry would be examined. Neddermeyer pointed out that one could make three simultaneous observations, one using the cloud chamber, a second using an ion chamber (to test compression at the center), and a third using counters. Oppenheimer questioned whether the compression would also be tested with RaLa, but felt that the symmetry would be well determined.[19] Bethe commented that the method might observe the dodecahedral structure seen in X-ray shots. On the issue of schedule, Bacher correctly estimated "that results would be forthcoming in the 4th month." Kistiakowsky, whose division was then grappling with the problem of lens design, pointed out that "the betatron method would be very important if it could look into the Baronal lenses," an experiment Bethe calculated to be feasible. In closing the meeting, Oppenheimer "stated that the program is likely to be started."[20]

Within a few days, Kerst and Neddermeyer had assessed the personnel needs of the project, and Kerst had also set down details of the betatron procurement, indicating "that the unit with associated equipment should be here without fail by November 10, and that it should be equipped according to specifications supplied by me." Two days later, he itemized the required personnel, man-days of miscellaneous service, and the details of transportation, procurement, and space.[21] On the same day, Tolman, now vice-chairman of the NDRC, informed Groves of the installation plans.[22] Oppenheimer also notified Groves of his wish to proceed with the betatron experiment.[23] Then, on 3 October, the Army Ordnance Department formally relinquished its claim on the betatron.[24] By early October, Los Alamos had begun to prepare for installation, estimating the job could be completed in only fifty to sixty days, during which time the cloud chamber would be constructed and the light source tested. The buildings at the future location of the betatron, K-Site, were targeted for completion on 15 December.

Kerst and Neddermayer led the group jointly; Kerst was responsible for the betatron, Neddermeyer for the cloud chamber. The group included William Ogle on the betatron; John Streib, C. F. Sayre, and Allison on the cloud chamber; Donald Mueller responsible for construction; and Melvin Tamarelly for shop work. Construction began immediately. Meanwhile, the Urbana tests showed the machine to be operating well, "with an output of 70 R/min which is approximately three times the intensity on which preliminary estimates were made." By 1 November, arrangements had been made to transport the 6-ton betatron in early or mid-December.

The cloud chamber was "functioning satisfactorily" by mid-November. After a ten-day delay, because the contractor had failed to meet the tight construction schedule, installation of the betatron began on 16 December. One month later, on 15 January, the group took their first betatron pictures of implosion, using a steel mockup with a gold core. Because of the low initial intensity of the betatron, too few tracks could be seen in the region representing the tamper; further experiments at higher intensity and better alignment were planned to examine "how serious these difficulties may be." By mid-February, the betatron was operating at approximately 50 R/min, and was expected "to contribute valuable information to the implosion study" by the end of the month.

The X-ray Method

The X-ray program gradually shifted its focus from studying gross features of implosion to jets, detonators, and lenses. In this period, the effort fell into two main sections – group X-1, under Kenneth Greisen, which employed small-scale flash X-rays; and G-2, under Lyman Parratt, which aimed at large-scale studies using the counter X-ray method.

Although the continuing construction at Anchor Ranch often prevented the firing of test X-ray shots, Greisen's group systematically examined jets and asymmetries, focusing initially on shots made with spheres filled with gases, such as hydrogen (at 200 atm), argon (at 200 atm), or xenon (at 50, 20, and 10 atm), liquids, such as methyl iodide, and solids such as paraffin. Greisen's program gradually refocused its attention on the detonation problem, to which Greisen would make major contributions in the pre-Trinity months. The group measured the spread in detonation times and the average time between the detonation and the X-ray flash. They attempted to reduce time spreads by reducing Primacord lengths and inserting tetryl pellets at the branches to improve communication of the detonation. For this program, Ernest Titterton, of the British Mission, designed a circuit for mixing the pulses from pairs of pins placed in Primacords and for displaying the mixed pulse on an oscilloscope to check detonation velocity variations. Such pins could also be used to observe the position of a detonation front in a lens. To measure the time spread among detonations and show whether or not all the detonators had fired, Titterton developed "the informer," a device consisting of a set of switches that produced an electrical signal when each detonator went off.[25] During December and January, the group conducted extensive tests of electric bridgewire detonators, and of explosive lenses. They made a concerted effort to learn which explosives worked best.[26]

Meanwhile, Parratt's group worked at P-Site on developing the counter X-ray method, which was to work in conjunction with the magnetic method, exploring small-scale lenses. This program, which James Tuck oversaw, studied an enormous number of jets and spalls in different materials in a variety of configurations. By early October, Parratt was reporting "appreciable compression" for certain lens systems. Although the lack of available lenses for tests was "seriously holding up this program," the group observed definite compression in November and produced a series of interesting shots of Comp B slabs on iron plates showing both jets and spalling.

Progress slowed in December owing to the collapse of a lead shielding structure, which wrecked the counters inside. After repairs, the group further explored the structure of jets formed in slabs of explosives set on a steel plate, comparing their results with those of Walter Koski in the flash photography program. Despite their expectations, they saw jets in two-point detonated slabs when the detonation waves moved directly against each other. When the waves hit obliquely, they could see a structure in the jets, which appeared to be hollow and to consist, at least in their early stages, of a series of laminæ. Similar experiments in January examined multiple spalling in slab shots with 1/4-inch plates, and single spalls in slab shots with 1/8-inch plates. These spalls increased in number reproducibly, as the wave angle moved toward normal incidence on the plate. Many further slab studies of jets and spalls used steel plates and very small intersection angles. Jetting remained strong down to about 10°.

Theoreticians helped experimentalists in the X-ray program. For example, Peierls suggested the ingenious "heap-of-disks" experiment, in which a pile of metal disks, initially fixed in contact with the high explosive, is dispersed by impact from the detonation wave. By taking X-ray photographs of the displacement of each disk, Parratt's group observed behavioral differences between Comp B, Torpex, and Pentolite. The differences in displacement of individual disks also gave information about the detonation, the wave profile, and the reaction. The different experimental groups were encouraged to collaborate as well as compete. Parrat's group put a great deal of effort into coordinating its work with the magnetic studies.

As the X-ray groups grappled with the technical difficulties and intrinsic limitations of the counter and flash X-ray methods, the betatron, pin, magnetic, and RaLa efforts began gathering useful data. By March

1945, these more powerful diagnostics had superseded the X-ray methods in implosion study.

Flash Photography and Rotating Prism Camera

Photographic methods were incapable of studying the implosion of solid spheres, because a hole large enough for light to pass through would be disruptive. Thus, as the program shifted from hollow to solid cores, the subject of photographic studies shifted from general implosions to lenses, initiators, and detonators.

Koski's section in X-Division (X-1C) continued the flash photographic study of imploding cylindrical shells, with the aim of pinning down some detailed features of jets. The group also weighed the advantages and disadvantages of various explosives and explosive arrangements. During August, for example, they found that short liners placed between the explosive and metal cylinder increased the asymmetry. They also found that Torpex gave smaller jets than Comp B and Pentolite.[27] Although they eventually obtained three fairly symmetric implosions of high velocity, the results were not reproducible and many lens studies in August proved "failures as far as symmetry is concerned." Further cylindrical shots throughout the fall and winter of 1944 used many different metals, different numbers of detonation points, and various explosives (including loose tetryl, Comp B, Pentolite, Comp C, Torpex, cast TNT, Comp B-tetryl).

Koski's group tried out a number of mechanical devices to smooth out the jets, with only partial success; for example, they placed gaps, Baronal pads, or barriers made of layers of sheet lead and steel screen in the system. When the detonation waves hit the metal head-on or with a glancing blow during the experiments, the jets disappeared. Klaus Fuchs, who arrived at Los Alamos on 14 August 1944 as a member of the British Mission, analyzed Koski's photographs of jets theoretically and concluded that the important factors giving rise to both the jets and spalls were the interaction of detonation waves arising from different detonation points, the interaction between shock waves in the liner, and the rarefaction propagating from the free surface after the shock reached it. How much of Koski's work Fuchs communicated to the Soviet Union is not certain. It was learned later that David Greenglass, a G.I. who worked as a machinist cutting lenses for Koski's group part of the time, passed at least one of Koski's sketches on to Harry Gold, a member of the Soviet spy ring. Fuchs's theory was able to correlate the size of the jets with the timing imprecision in the firing of neighboring detonation points. Koski's detailed studies of the causes and nature of the jets would be applied subsequently in developing the modulated implosion initiator, a device based on the jetting phenomenon (see Chapter 15, The Implosion Initiator).

Another section (X-1D), under Morris Patapoff and Joseph Hoffman, used the rotating drum and mirror cameras to study the same problems that Koski's group addressed, with the emphasis on detonation timing, however. Several months of work confirmed the finding by Koski's section, namely, that increasing the number of detonation points did not improve symmetry. Thereafter, the section changed course; in late December, at the request of Parsons, Hoffman reorganized the rotating camera team to focus on the explosive lens, directing this work together with W. G. Marley.

Terminal Observations

The terminal observations program in X-Division, under Henry Linschitz and Walter Kauzmann (section X-1B), became increasingly concerned with explosive lenses. The major accomplishment in this period, aside from developing a methodology for studying lenses, was the discovery of the problem of "velocity aberrations," which refers to the differences between the predicted and observed velocities in the explosives. The division continued to study this problem thoughout the fall of 1944 but made little progress. The full explanation was not found in the war period.

The typical two-dimensional lens experiment performed by members of the terminal observation program, according to the recollections of group member Lilli Hornig, began by outlining the shape of a lens on a steel plate, roughly 8 × 12 inches in size, as a guide to where the fast and slow explosive components of the lens would be placed. The explosives were poured into "what looked like cookie molds," and a detonator was placed at the top. After detonation, the wave traveling through the lens left an imprint on the steel plate, from which the jet shape could be measured with great precision.[28] Hornig recalls empirically trying many permutations in the hope of discovering "the combination that had the right ratio of refractive indices so that we would get the right ratio of wave propagation .... There was no theoretical calculation that would give the appropriate lens shape that one could then rely on. Another part of the problem was simply quality control; nobody had cast explosives to the kinds of precision that we needed."[29]

Plutonium Metallurgy

While the physicists explored implosion as a mechanical phenomenon, the chemists developed ways of treating the large quantities of plutonium expected from Hanford. Although in early August, Cyril Smith argued that stringent purity requirements still had to be met, because of the effect of impurities on physical properties, corrosion resistance, fabrication, and alloy development, the standards he set were ten to several hundred times less strict than those required before the plutonium-240 crisis.[30] Arthur Wahl's current purification scheme seemed too rigorous in light of the new specifications, but he decided to play it safe and continue at full speed on the development of process "A," the "wet chemistry" scheme he devised in summer 1944 (Chapter 11). In August, when the process was tested in a 1-g enclosed apparatus, it showed a satisfying 95 percent yield. The group also worked to simplify the design, so as to decrease the volume of residue solutions in anticipation of larger-scale processing. By 1 September, they had built an 8-g closed purification apparatus, both to further check the efficiency of the process and to furnish plutonium to the metallurgy group. The researchers were gratified when the device produced an excellent product with 98 percent yield in October and used the 8-g design as a guide for the design and construction of the 160-g apparatus chosen for standard production-scale purification.[31] By January, Wahl had several reasons to celebrate. In December, they understood process A well enough to send specifications to Hanford, and by January they were ready to run the 160-g apparatus.[32]

Although plutonium trichloride proved adequate for the 8- to 10-g metal reduction runs in the summer of 1944, the chemists reevaluated dry chemistry procedures when the purity requirements were relaxed. Without strict limitations on fluorine in the purified plutonium, the highly hydroscopic chloride seemed a much less attractive choice for dry conversion than fluoride, especially because the fluoride produced superior yields. By the end of the summer, Eric Jette reported that the use of plutonium tetrafluoride allowed a reduction to metal on the 500-mg scale with satisfactory purity and "exceptionally high" yields. Consequently, the work on other preparations ceased.[33]

Through the fall, the dry chemistry team, led by Clifford Garner and Iral B. Johns, labored to improve the new dry conversion process (later referred to as an extension of process "A") to match the escalating quantities of plutonium being processed. The dry conversion group studied the process in multigram runs that provided plutonium fluoride for the metallurgists while also conducting experiments on the 200-mg scale. To ensure the optimum temperature for obtaining the maximum hydrofluorination rate of PuO_2, a special apparatus continuously weighed the Pu compound during treatment.[34]

By 1 January 1945, the first large-scale dry conversion unit was completed and tested successfully on the 30-g scale. The unit processed plutonium in several steps. When the oxalate slurry arrived from wet chemistry, it was mechanically transferred to two platinum-rhodium boats and loaded into a unit for dry ignition. The resulting plutonium dioxide was weighed and transferred to the hydrofluorination furnace, where it was converted to the fluoride. The fluoride was weighed, to check the extent of conversion, and placed in bottles by a lucite apparatus operated inside a dry box. The bottles were then transferred to metallurgy groups. Although they had not quite achieved their goal of reducing hydrofluorination time to less than twenty-four hours, the dry conversion team was ready with an apparatus to accommodate the first 160-g run in February.[35]

Progress also came quickly for the metallurgists after August 1944. Now that they had enough plutonium to devise large-scale procedures, the metal reduction group worked in the summer and fall of 1944 to establish a large-scale metal reduction process, all the while struggling to find adequate crucibles. At the same time, other metallurgists plunged farther into the new realm of plutonium metallurgy to develop fabrication methods. The metal reduction group stopped working on electrolytic methods, because it was very difficult to create a molten salt bath with the proper viscosity and solvent power for the production of metal. In comparison with bomb reductions, the electrolytic method produced consistently lower yields in May and June 1944.[36]

Although the first successful gram-scale reduction was accomplished with the centrifuge method, by mid-July attention focused on the stationary bomb method.[37] Before the summer of 1944, the centrifuge method had two advantages over the stationary bomb method. First, it was optimal for reduction on the scale of 50 mg to 1 g, since reductions on this scale tend to produce a large number of single droplets that easily become coated with refractory material.[38] Second, the stationary bomb method, unlike the centrifuge method, was handicapped by mistaken estimates of the melting point of plutonium, since it depended on precise control of thermal conditions, which differ with the amount of material being reduced. However, once the true melting point of plutonium was known and the scale of reductions increased to more than a gram, Baker's stationary bomb reduction consistently produced excellent yields of collected metal and supplanted the centrifuge technique.[39]

To determine optimum stationary bomb conditions for each scale of operations, Baker and his co-workers had to establish whether supplementary heat was needed to allow proper separation of the metal and slag. This heat could be obtained by adding external heat or by introducing a "booster" material that would react to produce the heat during the primary reduction. They also had to ensure that the metal product stayed molten long enough and that the slag was low enough in viscosity to allow metal agglomeration. Although they could do nothing about the melting point of plutonium, they could modify the slag. In addition to achieving the optimal temperatures for firing procedures, they had to determine the best liner refractory material that would retain reaction products and not contaminate the metal product.[40] Throughout the summer and fall of 1944, the metallurgists ran many metal reductions to investigate these factors. Since fabrication studies were also being pursued intensively by this time, they produced metal for these efforts while refining their procedures. In addition to gathering information from the larger runs, which had increased to the 20- to 25-g scale by the end of the year, the stationary bomb group also performed 500-mg plutonium reductions and reductions on scales of 500 mg to 120 g with cerium, which proved to be a more suitable stand-in than uranium, once the true melting point of plutonium was known.[41]

The group suffered some setbacks. For example, the first 8-g reduction, known as the "jinxed batch" because of a series of purification and recovery mishaps, was a total failure. Nonetheless, by December they had devised a metal reduction scheme that consistently produced good metal with yields greater than 96 percent on the 6- to 10-g scale, as well as on the 20- to 25-g scale.[42] A December 1944 report explained that the fluoride was mixed with calcium, in excess of the stoichiometric ratio by about 25 percent, and often some iodine and further calcium were added. This mixture was placed in a liner composed of magnesium oxide, contained in a steel bomb filled with argon, and heated by high-frequency induction, to set off the reaction. For 25-g charges, liner temperatures were set at 450° C; smaller charges required a higher temperature.[43]

Although the development of analytical equipment and methods came to a halt after the work on the plutonium gun stopped, and thus researchers had no way of measuring accurately some light-element impurities, they felt that the purity of the metal emerging from their purification and production procedures was "within the range" of the first strict tolerance limits for most light elements. Since these limits had been greatly relaxed, they were confident that the metal would "meet all reasonable requirements" for producing plutonium components for the implosion weapon. With an efficient technique that produced metal of the necessary purity, the stationary bomb group was ready by the beginning of 1945 to handle the first samples of Hanford plutonium.[44]

Throughout the summer and fall, Los Alamos metallurgists paid increasing attention to fabrication methods as they worked against the clock to develop this final stage in the production process. The first step in fabrication was remelting, which further purified the metal and combined separate buttons into a single, larger coherent piece. Metallurgists also used remelting methods to make ingots, which were then mechanically shaped, and to make simple shapes directly by casting in an appropriate mold.[45] The first successful remelt of plutonium came in April 1944, and throughout the summer and fall, they conducted a vigorous program to design optimal remelting procedures, the most important problems being the choice of an appropriate crucible (see below) and the development of protective coatings. By 1 December, 15- to 20-g lots of plutonium were successfully remelted.

During the summer and fall, they also developed fabrication techniques. Los Alamos metallurgist Claire Balke, who came to the project from the Fansteel Metallurgical Company with an unusual expertise in the use of high-strength steels for metallurgical operations, had developed hot-pressing techniques for the difficult task of working beryllium oxide and tungsten carbide. Thus, the metallurgists were able to hot work plutonium soon after gram quantities became available.[46] By November, work was under way on rolling and extrusion. The problems of fabricating final shapes and surface protection were identified as the "principal unsolved chemical and metallurgical problems" in the production of plutonium assemblies.[47]

Hoping to avoid using the brittle α phase of plutonium, which was stable at room temperature, the metallurgists initiated alloy research by the end of 1944. This alloy "scouting work" had prompted "keen interest in several alloy systems, in particular the plutonium rich U-Pu and Al-Pu systems." When they found that some alloy forms indeed suspended transformation of more malleable phases to the brittle α phase, they checked with the physics groups about the effect of alloys on the amount of critical mass needed for the bomb and obtained the limits for adding alloying material.

To develop successful fabrication processes, the metallurgists had to deepen their understanding of plutonium's physical properties. Besides studying allotropic transformations, melting point, and density, a group under Jette measured the coefficient of thermal expansion, electrical conductivity, and deformation under pressure. In this effort, Frank Schnettler proved particularly talented in making the most difficult measurements.[48] Although they first assumed that the β-phase was the most malleable one and received confirmation that they could use β-phase metal or alloyed β-phase metal for the Trinity and combat devices, by November they had realized that the most malleable phase was the γ, not the β phase.[49]

By the end of the year, Los Alamos knew that the α phase had an orthorhombic structure and a density of 19, and they understood that the δ, not the β phase, had a face-centered cubic crystal structure and a density of 16, whereas the β phase had a density of almost 18. The complicated crystal structure of this phase was still unsolved. In reviewing wartime progress, postwar plutonium metallurgy expert Fred W. Schonfeld concluded that Los Alamos metallurgists "probably roughed out more than sixty percent of the field," thus making a "very impressive" contribution to the understanding of the basic characteristics of plutonium metal. This contribution is all the more impressive because metallurgists were forced to use the hurried approach demanded by wartime conditions.[50]

The knowledge gained about the properties of plutonium metal allowed the metallurgists to make the following conclusions about the fabrication of plutonium metal by the end of 1944. It was possible to roll or press β-state metal at about 150° C into thin sheets or foils without serious cracks. Practically, however, it was better to use the even greater plasticity of the γ phase by heating the metal to about 250° C. The researchers recognized that hot oil was a heating medium that also protected the metal against oxidation and protected operators from oxide or metal dust. From a toxicological point of view, pressing in hard steel dies, the preferred shaping method, proved to be successful on small but not too thin pieces. Rolling, with the help of improvised equipment, produced high scrap, owing to warping and cracking, because it was not possible to achieve close temperature control and the metal was often cooled below the phase-transformation temperature. By early 1945, Los Alamos metallurgists were ready to fabricate material from the 160-g run.[51]

Continued successes in metallurgy also depended on research on refractory materials for crucible manufacture. Considerable work was clearly needed on refractory material because of the special problems caused by the extreme reactivity of plutonium and uranium, as well as the fact that refractory requirements differed for reduction, remelting, and casting.[52] Before gram amounts of plutonium were available, Los Alamos metallurgists conducted preliminary research on the reaction of various crucible materials when subjected to micro- and milligram quantities of plutonium and to larger quantities of stand-in materials. When they became deeply involved with metal reduction experiments, some of this effort was delegated. On 7 March 1944, they decided to create a central refractory laboratory at MIT under the supervision of F. H. Norton and the technical direction of John Chipman.[53]

Before the refractory laboratory could produce crucibles, Los Alamos needed suitable containers for early plutonium metal reductions. Fortunately, Wendell Latimer had formed a small group at Berkeley in early 1943 to investigate potential crucibles for plutonium metal processing. The group was supervised by E. D. Eastman and included Leo Brewer, a young Ph.D. in chemistry, and LeRoy Bromley, Paul Gilles, and Norman Lofgren, all Berkeley undergraduates. As Brewer recalls, their assignment was to "predict the worst properties that plutonium might have," on the basis of the current knowledge of plutonium and models for predicting chemical properties. From these studies they concluded that cerium monosulfide, a compound not as yet prepared, was needed to ensure "that plutonium metal could be cast and fabricated and still maintain the desired purity." Eastman's group designed and built the necessary equipment, and by spring 1943, when the first gram-scale quantities of plutonium arrived in Los Alamos, they were able to send a small supply of these exotic crucibles to Los Alamos. Although difficult and costly to make, because of the many failures in production, these crucibles worked well for containing the highly reactive metal and produced metal of the necessary purity.[54]

By summer, various crucibles were being produced and ceramic crucible research was under way both at Los Alamos, under Balke, and at the MIT refractory laboratory. Such ceramic research broke new ground. For example, although clay customarily had been used almost exclusively as a bonding agent for crucibles, retaining high-purity uranium and plutonium after reduction and remelting was nearly impossible in the presence of significant amounts of silica in the container. An additional complicating factor was that those designing refractories had to worry not only about the chemical characteristics of the material used, but also about its physical characteristics. For example, if a crucible was too sensitive to thermal shock and cracked on firing, it was useless, no matter how desirable its chemical composition. Because of the special metallurgy needs at Los Alamos, crucibles had to be manufactured from entirely new materials. Research proceeded on a variety of oxide, sulfide, and nitride refractories.[55]

By 22 December, several types of refractories were looking "promising" in the metallurgy of plutonium.[56] Although the metallurgists had not had enough experience with casting procedures to judge the best material for this purpose, cerium sulfide emerged as the first choice for remelting. Magnesium oxide had been identified as the best refractory for liners. Magnesium oxide crucibles were much easier to produce, and worked well at firing temperatures as high as 1,000° C, but they reacted with the calcium reductant, causing contamination at the even higher temperatures first thought necessary for plutonium production. Because researchers therefore assumed that cerium sulfide crucibles were necessary, the MIT laboratory was assigned the task of producing cerium sulfide crucibles large enough for production-scale remelting.[57]

After tests during the summer of 1944 indicated that plutonium corroded much more easily than uranium, two groups – one under the direction of Samuel I. Weissman, the other under the direction of Alan Seybolt – began testing coatings to protect both the metal and those working with it. By the end of 1944, attempts to coat plutonium with evaporated aluminum and electrodeposited silver and zinc had failed, and a new round of testing was planned.[58]

Another crucial part of the plutonium production system was recovery. In the fall of 1944, while Wahl experimented with the 160-g apparatus, most of the plutonium for metallurgical efforts was purified by the recovery group. By then, this versatile group had amassed considerable experience in providing usable material to fuel the chemistry/metallurgy effort. After demonstrating his skill in recovering small amounts of material, Frank Pittman developed and instituted procedures aimed at recovering plutonium from "any element or group of elements," assuming it "was in any possible form and mixed with any possible type of material."[59] With such procedures, the group scraped together material for the first 8-g metal reduction (the "jinxed batch") in July 1944. They recovered the material for this run three times. Because of faulty glassware and a broken centrifuge, it was spilled twice in recovery and everything in the room had to go through recovery, including the floor. The ill-fated material was burned because of a malfunctioning furnace controller in dry conversion and had to be recovered a third time. Researchers were not terribly surprised when the 8-g reduction subsequently failed, as a result of oxygen contamination in the starting material.

A measure of the value of plutonium can be inferred from the incredible pains taken in the recovery effort. Besides assuming purification tasks, on occasion the recovery group acted as conservators of plutonium. Almost every procedure using plutonium was followed by a recovery attempt. The recovery group developed specific procedures to recover material from purification, remelting, and casting processes, as well as other metallurgical procedures. In addition, they developed special methods for plutonium recovery from such nonroutine materials as analytical solutions and cutting oil. The group also recovered plutonium from paper, rags, and paper tissues. Figure 14.2 shows the range of procedures that the recovery group devised for the recovery and purification of plutonium.[60] By June, plutonium losses had been reduced to approximately 1 percent of the quantity processed, and researchers hoped to reduce this figure further by an order of magnitude.

Recovery work was not only difficult, it was dangerous. Although the laboratory introduced standardized procedures and enclosures to protect workers in the purification, dry chemistry, and metallurgy efforts, most recovery work had to be performed in open systems because of its diverse nature. Thus, recovery group members had a much higher exposure to plutonium than other researchers. By the time the plutonium for the Trinity and Nagasaki devices had been processed, almost half of the experienced workers had been removed permanently from plutonium processing, owing to the alarming levels of plutonium in their urine.

Special safety precautions were taken to reduce the hazards of handling plutonium. After the experience of the 8-g disaster, and in anticipation of increasing difficulties and dangers with increased amounts of plutonium, researchers carefully planned processing facilities and procedures. By September, plans were under way for a special "hot" area containing closed cubicles, free of furniture except for a large hood, with washable walls. To increase the chance of recovery in the event of spills, all outlets, from sink, hood, and floor, led to reservoirs. The laboratory decided that workers would wear face masks and protective clothing, which would be specially laundered after use. Researchers were

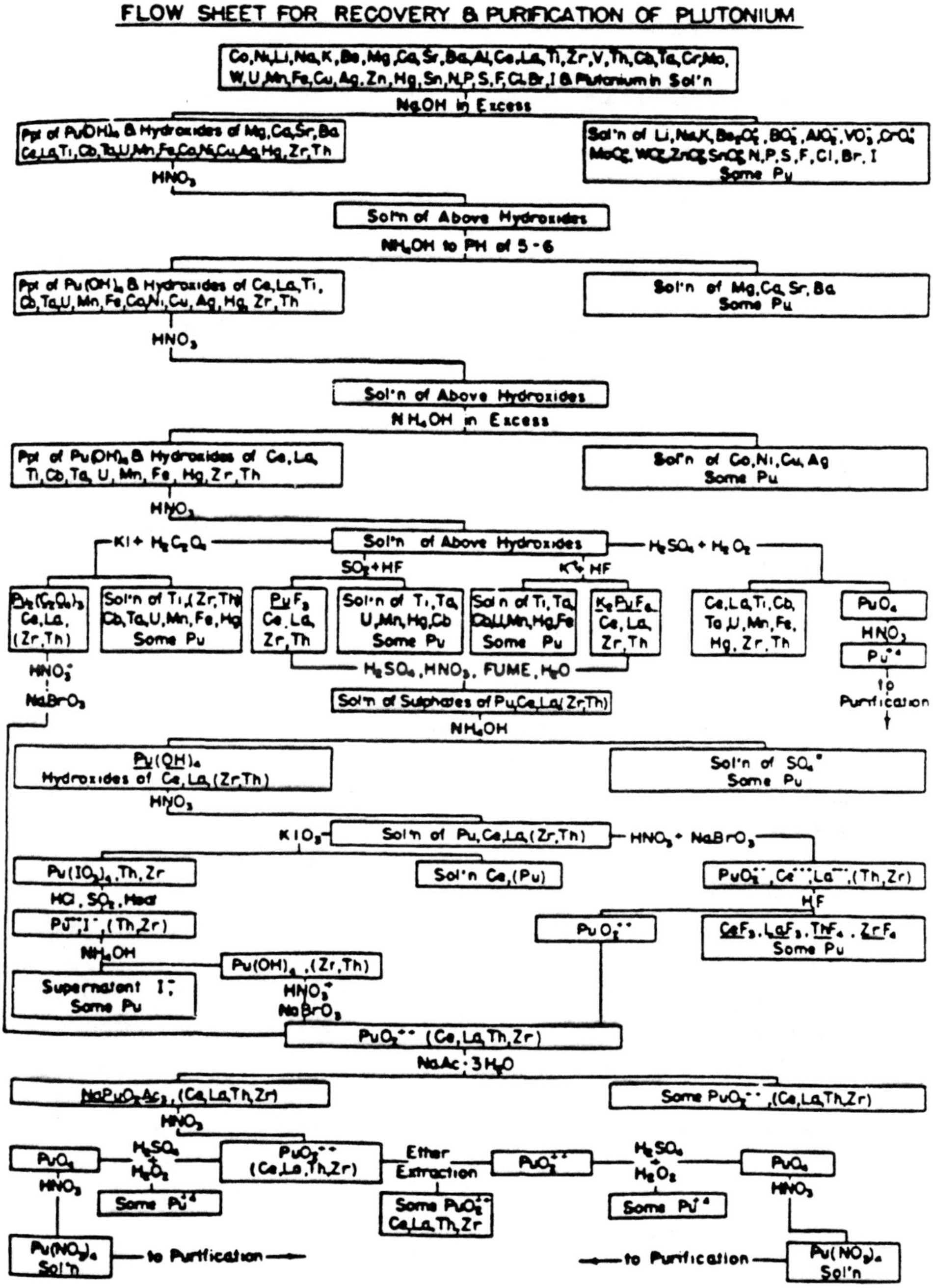

Fig. 14.2. Flow sheet for recovery and purification of plutonium.

also concerned about unintentional accumulation of a critical mass. In November, distribution of plutonium and ^{235}U was carefully controlled. A special group handled all plutonium and ^{235}U transfers under the direction of a central switcher.[61]

The roof of C Shop, a few hundred yards west of D Building, caught fire the night of 18 January 1945. Although the fire was not serious, Kennedy worried that if a fire started in D Building, the toxic plutonium fumes might endanger the entire community. Despite the heavy work load, researchers organized a twenty-four-hour fire watch, which included holidays and weekends, and Kennedy enlisted Thomas's help to lobby for the construction of "a fire-proof factory" farther away from the community to handle all routine plutonium processing.[62] They agreed that the current system sufficed for building a few weapons, but a safer, more carefully designed facility was needed for the future to ensure the safety of the community and workers. By early February, Groves had authorized the construction.[63] During the last tense months of the war, the chemistry/metallurgy group made plans for this new facility. which came to be called DP-Site.[64]

Early in 1945, the wartime plutonium production process was nearing a milestone. Although Wahl was not able to start the 160-g run on 1 January 1945, because the necessary Clinton shipment was about half that expected, by 2 January more material arrived and the wet chemistry part of 160-g run began. It took twenty-four hours, about as long as estimated, and provided a 95 percent yield. More than 97 percent of this was converted to PuF_4 in dry conversion and reduced to plutonium metal. The resulting material was then remelted, reduced, and hot-pressed into two 0.90-inch-diameter hemispheres weighing 60 g each. The researchers concluded that the purity of the metal was "very good" by current standards and "surprisingly close to the old tolerance level."[65] The spheres were then given to the Research Division for neutron multiplication studies, which were crucial in confirming critical-mass estimates.[66]

With the success of the 160-g run, the next crucial step was the first Hanford delivery. By this time, many at Los Alamos worried about when plutonium would arrive and whether it would provide adequate starting material for the Los Alamos production system. The first Hanford pile had started operating in late September 1944, but within hours it had shut itself down, as a result of poisoning by ^{135}Xe, a fission product that captured enough neutrons to prevent the propagation of a chain reaction as subsequent tests indicated. As Los Alamos waited anxiously in November and December, Fermi and Zinn diagnosed the poisoning problem. The problem was solved by adding uranium slugs to the extra channels proved by the conservative du Pont engineers, thus increasing reactivity to override the poisoning. By late December, two piles were operating at full loading, and plutonium production had begun.[67]

Plans for receiving the precious substance had already been made. On 6 December, Kennedy and Wahl submitted requests detailing the specifications they hoped for in plutonium shipments. On the same day, the chemists met with Oppenheimer, Tolman, Lauritsen and Colonel Peterson and decided that Peterson would ask that Kennedy be allowed to visit Hanford to discuss chemical specifications. On 5 January, Oppenheimer told Kennedy to prepare for a visit on 18 and 19 January, but on 12 January the laboratory director announced that the visit had been "indefinitely postponed."

On the last day of January, Oppenheimer announced that the first Hanford plutonium shipment, an 80-g batch, was expected around 7 February. After this date, Los Alamos would depend for plutonium on the Hanford facility. The last Clinton shipment of 33 g would arrive a few days later. Oppenheimer was not optimistic about the ease of interacting with Hanford. He had been told that the quality of the material was "somewhat low," but Hanford officials refused to provide further details. In addition, a 26 January letter from Peterson, written in response to the Los Alamos specifications, was noncommittal about whether specifications could be met. In the absence of a direct channel of communication, Oppenheimer felt that details about the shipments would have to be worked out after the material arrived and that therefore they "should anticipate some difficulty in the immediate future."[68]

Oppenheimer's pessimism was unwarranted. By mid-February, arrangements had been made for Kennedy to visit the Washington site. He met with Hanford officials on 19 February and although "du Pont did not want to discuss specs," Kennedy was "amazed to learn that specs – perhaps based on our table – were probably possible." When the first shipment arrived, Los Alamos chemists judged it favorably, for the most part. On the negative side, they pointed out that it arrived as a solid and therefore had to be dissolved; furthermore, it was assayed at less than the expected 80 g. They also worried that the silica content would necessitate extra filtering. Nonetheless, the material was "very satisfactory," much better, in fact, than Wahl had expected. The 95 percent purification specifications had been met, and the gamma activity was lower than expected.[69]

As time went on, this generally favorable judgment held, although Los Alamos researchers also recognized that the successful operation of the processing system depended on the continued high quality of plutonium from Hanford, and that they had little influence on Hanford plutonium production. Despite their concern, by February the metallurgists concluded that Hanford plutonium was as pure after remelting as that received from Clinton and reacted the same to hot pressing. Wahl had some trouble later in the spring with unpredictable oxidation states and impurity levels, particularly excess silicon and tin compounds, but still was able to produce "an excellent product" with yields "only a few percent below the expected 95 to 98%." Although Wahl also continued to take exception to Hanford assays, complaining in June that results from the Los Alamos radioassay group led by Rebecca Bradford indicated that the laboratory had received 5 percent less in its shipments than Hanford claimed, in general there were few problems with the Hanford product.[70] Purification standards were largely met, uranium content was minimal, and γ activity remained low. The Met Lab and du Pont had accomplished their task well. The group used several methods for assaying plutonium. Perhaps the most important was α-particle counting, the only method that worked on the tracer scale, which required careful measuring with micropipettes and evaporating under controlled conditions onto counting disks.[71]

15

Finding the Implosion Design: August 1944 to February 1945

The accelerated implosion effort, which began in August 1944, made rapid progress. By October 1944, James Conant was giving a lensed implosion device a 50–50 chance of working on schedule, if all went smoothly, for a test at Trinity on 1 May 1945 and a "3:1" chance for a test on 1 July. But he added, "In my opinion, the probabilities of success by the gun method (Mark 1) within the next year are very much greater than by the implosion method. Indeed the gun method seems as nearly certain as any untried new procedure can be."[1]

Overcoming asymmetries remained the outstanding technical problem of the implosion program. By mid-fall 1944, two experimental strands of the implosion program were converging on this problem: research on the explosive lens and on the electric detonator. In addition, in T-Division, Robert Christy put forth a conservative proposal for overcoming the asymmetry: try to implode a solid sphere rather than a spherical shell of active material. However, calculations indicated that the "Christy gadget" was intrinsically far less efficient than the hollow weapon, and that such a device would require a modulated initiator to activate the explosion at the most favorable moment. The call for the development of the implosion initiator added another thorny problem to the program.

* This chapter is based on a manuscript by Lillian Hoddeson, to which Catherine Westfall contributed pieces on polonium incorporated in the initiator section, Les Redman contributed a section on explosives, and Paul Henriksen a section on acquiring the Trinity site. We are grateful to Gordon Baym and Redman for their detailed editing of the technical content of this chapter.

As the time approached when sizable quantities of plutonium would become available, gross design features had to be frozen in order to begin final bomb production. On 28 February 1945, Oppenheimer and Groves fixed provisionally on the Christy gadget, with electric detonators and explosive lenses made of Comp B and Baratol. The practicability of this design hinged on the future development of a modulated initiator, and on lenses and detonators yet to be designed for the final weapon.

Implosion work from August 1944 to February 1945 was increasingly dictated by detailed design questions – for example, how to ensure that multipoint detonation would be simultaneous, what the assembly time would be, the amount of compression that could be achieved, whether bridgewire or spark-gap detonators should be used, whether an explosive or electronic switch should fire the detonator circuits, whether the active material should be in the form of a spherical shell or solid sphere, whether the design should include lenses to focus the detonation wave, what explosives the lenses should be made of, and whether a modulated initiator should or could be designed to enhance the efficiency of a solid sphere implosion. Trial and error was the dominant mode of attack, especially in the development of the lens and detonator, both of which were as yet poorly understood by the scientists. In theoretical study, numerical analysis and iteration often replaced full analytical solutions, because the basics of shock hydrodynamics were not yet known. Nor were enough data available for developing detailed quantitative treatments. In 1944, there existed analytic solutions for shocks in perfect gases, but few understood shocks in other materials, and there were no data for nuclear materials, particularly in the temperature-pressure-energy regions of interest surrounding the implosion.

Explosives: Development of Lenses

The most challenging explosives problem in the fall of 1944 was how to develop adequate-quality lenses – both for the full-scale implosion systems to be tested at Trinity and used on Nagasaki, and for the extensive research program. Two of X-Division's three groups in the summer of 1944 were involved with lenses: X-1, under Bradbury, an implosion research group; and X-3, under Capt. Jerome Ackerman, essentially an in-house explosives factory for casting lenses and Comp B charges at S-Site.[2]

The division considered two aspects of lens-making: developing their intrinsic design, and selecting and developing appropriate fast and slow lens components. Kistiakowsky shaped the program with characteristic spirit and verve.

Lens Problems

By July 1944, von Neumann and others had established the shape of the lens. The number of lenses to be used would depend on both their reliability and complexity of construction. Members of T-Division, particularly Rudolf Peierls, worked out a first theory of explosive lenses, which unfortunately could not be applied directly to actual systems. The main obstacle to developing such a theory was the fact that the velocities of detonation depended on the shape of the system; thus, in contrast to situations in optical systems, the propagation was both nonlocal and nonlinear. One drastic consequence was that the velocities did not scale simply with the geometry and dimensions of the explosive.

The fundamental problem was that detonation waves, which the lenses were required to manipulate precisely, had been studied very little, and their behavior in lens configurations proved not to be a direct extension of their stand-alone behavior. For example, the time taken by the detonation front to travel in a particular HE from one point to another was typically measured in a "rate stick," a sample of explosive a few inches across and many inches long.[3] However, the propagation of the detonation wave proved to be quite dependent on "edge effects": the detonation front diverged spherically as it progressed through the HE, and when it reached the stick's surface, energy left the stick, decreasing the velocity locally. In their experiments, the members of X-Division found that when the detonation in the rate stick was initiated by a large booster that produced a less divergent wave, the observed detonation velocity increased.

Another large problem was that the refractive index could only be determined by iteration. The slow HE generally behaved faster than its stick velocity indicated. To measure the effective detonation velocity of the slow explosive in a lens, one had to (1) make a rough guess of the detonation velocity, (2) design and procure a mold to make a lens, and (3) fire it and observe the emergent wave shape. Only then could the designer know to what extent his original guess at the slow component velocity had been too fast or too slow. The iterative process was slow and frustrating. Responding to such problems in the experimental study of lenses by terminal observations, Linschitz summarized in March 1945,

Fig. 15.1. Original explosives bunker at S-Site, as it appeared in 1990. LA Photo, 87-372 10.

"The rate studies carried out by X-1B during the past few months have shown the lens situation to be so complex that it is clear that an empirical approach to the pattern of lens design will be the only feasible one for a long time to come."

Because of the difficulties in making three-dimensional lenses, including a shortage of molds, the diagnostic program tested only two-dimensional (or planar) lenses in the early experiments. The two-dimensional lens was easy to make – the fast HE could be cast to shape, or, as in the early work, prepared by hand from a cast sheet – and its performance was easy to diagnose. However, such lenses gave cylindrically, rather than spherically, convergent waves.

The lack of lenses for both the diagnostic program and the actual bombs remained one of the vexing challenges for Project Y. There were

Fig. 15.2. Explosives casting building at S-Site. LA Photo, 3179.

never enough trained workers or space in buildings to deal with the variety and quality of lens demands; few technically trained people had either the experience or the inclination for HE work. The S-Site staff was about 90 percent SEDs. In addition, the original estimates of the number of lenses equired for the experimental program were too low. As a consequence, there was considerable pressure on the understaffed lens production facility. Kistiakowsky explained in a later history of the lens-making effort that in more than eighteen months of lens making during World War II, "well over twenty thousand castings were delivered to the firing sites, while the number of castings rejected because of poor quality or destroyed for other reasons is several times this figure."[4]

Lens Making

Out of wartime necessity, lenses were made in the simplest molds corresponding to the shape geometrically required – a pentagon or hexagon. Various spherical inserts and caps were used with the molds to provide the desired spherical surfaces on the tops and bottoms of the cast lenses or inner charges. Although straightforward in principle, mold design, carried out by section A of group X-2, under Robert W. Henderson, and later by group X-4, was in fact an extremely complex enterprise. Throughout the war period, the exact design needed for the molds eluded the Los Alamos researchers. Only in the pre-Trinity months would X-Division researchers succeed in matching the slow explosive velocity to the mold, and only then by changing the composition of the slow explosive after the mold was already on hand and castings had been tested.

It was particularly difficult to cast the high explosives accurately and avoid cracks, bubbles, and other imperfections. Cooling cycles had to be long to minimize thermal stress cracks. Castings had to be wrapped in insulation before being transported between buildings. The molds were made to the nominal size of the HE block. Scaled versions of lens assemblies were not confined, but merely supported as necessary with holding fixtures.[5]

Casting technology developed slowly and painfully at Los Alamos, by a succession of reasonable steps, that consistently failed to give completely satisfactory results. Eventually, the problems were overcome by two sections of X-3. From the beginning of X-Division, a research and development section under Lt. J. D. Hopper worked on methods of fabricating high-explosive components. In November 1944, four chemists from the Met Lab under David Gurinsky formed the nucleus of a "special problems" unit, whose principal assignment was casting technology; their work became the basis for much of the casting procedure used in production.[6] During this period, the unit tried many tactics, such as adding pellets of Comp B to slurry in the mold; pouring Comp B in increments, both horizontal and vertical; varying the cooling cycles for the castings; increasing the size of risers and heating them; and casting into evacuated molds.

The casting of the HE differed from conventional foundry work, because the material being cast was a two-phase slurry consisting of a dense solid phase dispersed in molten TNT. The denser solids tended to settle and the result was a nonuniform material. Heat flowed through the material poorly, yet the heat of fusion, which was released on solidification, was quite great. Thinner sections, or regions with short paths to cooler regions, would tend to solidify completely before thicker sections did, thus closing off flow from the riser and leaving low-density regions or voids. An indispensable program under M/Sgt. Gerold Tenney in Section X-1E examined the integrity of the castings. Large components of Baratol were opaque to the 250-keV X rays employed, and when such components came into use, radium became the radiation source. The numbers of charges examined grew rapidly; a record was achieved in December when 2,266 charges were examined, as opposed to 1,576 the previous month.

Developing Explosives for the Lenses

Throughout the fall and winter of 1944, considerable effort went into developing the fast and slow explosives to be used in the lenses. The fast component was selected in a relatively efficient process. At the beginning of the period, Kistiakowsky felt that only Torpex, among the fast explosives that could be cast, would support a stable detonation wave. During the fall, the unit examined PTX-2 (Picatinny ternary explosive 2); this material proved to be somewhat unstable under the conditions encountered in production casting, and was thus discarded. Later in the winter, PTX-2 enjoyed use under less severe conditions. Comp B, which was obtained from Holston Ordnance Works, varied in the quality of such features as particle size and viscosity of melts, until Los Alamos insisted on standardization, after which successful casting procedures became possible. Although some scaled lenses continued to be made with Pentolite, PTX-2, and Torpex in late 1944, Comp B became established as the fast HE, both in the lens and inner charge. It was relatively safe and stable, and relatively easy to cast and make homogeneous.

In contrast, selection of the slow component was a diffuse, exploratory program, which took more than half a year. When X-Division was formed in August 1944, Kistiakowsky expected to use the barium nitrate explosive Baronal as a cast slow component, unless "by 1 Oct a slow castable component is discovered whose velocity is well below 5000 m/sec." In that case, he planned to discard the Baronal and use the new material "for all future lens design and manufacture." In the summer of 1944, Eugene Eyster's group at ERL developed a procedure for preparing Baratol in a form that could be readily cast.[7]

The Nonlens Gadget

Not everyone at the laboratory agreed with Kistiakowksy's decision to pursue the lensed gadget rather than a nonlensed one. The fact that the lens program was diverting resources and a large number of workers from other projects detracted strongly from its appeal. Kistiakowsky, Peierls, and Bethe were among those who insisted that lenses were both feasible to design and essential to obtaining the desired wave shape. Parsons, considerably more skeptical than the others about lenses, felt that vigorous research on nonlensed systems should be continued.[8] Whether to continue research on a nonlens gadget was the central topic of a 2 November meeting called by Oppenheimer, but the discussion was inconclusive.[9] Oppenheimer therefore authorized the continuation of "work on both lens and non-lens problems," a situation he nevertheless acknowledged to be "undesirable."[10]

The prospect of focusing on lensed systems perturbed Parsons, who wrote Oppenheimer on 19 February 1945, "It is difficult in cold blood to look for an adequate tested lens implosion gadget in 1945." He added that

> the non-lens implosion gadget as a limited objective (June–September, 1945) I believe could be engineered if there is good luck at every turn and if the philosophy is kept straight. It would be tragic to prune or suppress work now underway here which is materially contributing to the solution of the lens or non-lens implosion problem .... It is obvious from the above that I feel that the possible Summer 1945 gadget is a non-lens model, and that lens research toward a better model can and should continue here.[11]

At the decisive 28 February meeting, however, the argument put forth by Parsons was rejected. In retrospect, were lenses the best approach to the wartime implosion gadget? Theoreticians then working at Los Alamos, including Bethe and Peierls, were almost certain that lenses offered the only approach to a successful implosion.[12] But the crucial decision in February to use lenses in the Trinity gadget was a gamble, because not many lenses had yet been used in the diagnostic program. At the time of the August reorganization, the preparation of spherically convergent lenses was only beginning. The first shots of two-lensed systems in October, performed in the X-ray program, gave only somewhat faster rates of collapse and relatively small compressions, while the difficulties of making molds and casting the lenses were painfully evident. By 1 March 1945, the production of effective small lenses suggested that this capability could be extended readily to the production of full-scale systems, but this projection turned out to be too optimistic.

The Detonator

Alvarez and Lawrence Johnston demonstrated the feasibility of making simultaneous electric detonators in May 1944. However, two important problems had to be overcome before an electric detonation system could be practical. First, the detonators had to be rugged, reliable, and safe. Second the firing circuit had to be able to survive both the cold of a high-altitude flight and the severe vibration of a falling bomb, with a rapid (low-inductance) switch to initiate the condenser discharge. Logically, these problems should have been solved sequentially, but with the tight deadline for Project Y, it was imperative to develop the firing circuit well before the final detonator design was in hand. This was only one of the difficulties confronting the detonator program in the pressured final year of Project Y.

Detonator research and development in the fall of 1944 was based in group G-7 under Alvarez, and section X-2C under Lewis Fussell, a man remembered by his colleagues as a "solid engineering type." The Alvarez group bore primary responsibility for the detonators. Aided by a large staff of SEDs and Hispanic women from nearby towns, they examined every aspect of both bridgewire and spark-gap detonators – including composition of the wire, explosives and insulation, the voltage versus the capacity of the energy source, the circuit impedance, the wire size, and the methods of preparing and loading the detonators with explosive.[13]

Fussell's group focused on the firing circuits, the "X-Units." For immediate use in experiments and drop tests, they developed "crash" sets – small-scale firing units, hurriedly constructed by hand. Group members Donald Hornig and Keith Henderson, who worked on designing the spark-gap switch and condensers, were quite frustrated at first, for there was "very little to go on in designing the circuit."[14] Between August 1944 and February 1945, they examined three types of switches for the firing circuits: a mechanical, an explosive, and an electronic spark-gap switch. The relatively simple mechanical switch, initially used in experimental work and included in what was called the "Model I" X-unit, was abandoned in late fall, because it could not fire all detonators simultaneously. The explosive switch, in which the detonation of an explosive charge caused electrical contact by breaking through a dielectric layer separating two metal disks and driving these plates together, was used extensively in the experimental program. However, being totally self-destructive, the explosive switch could not be tested before actual use and was for this reason also given up in late fall as impractical for use in the final bomb.

Hornig had been motivated to work on electronic switches even before X-Division was formed and before any actual switch work was done at Los Alamos. Shortly before the August reorganization, he attended a seminar in which Oppenheimer laid out what he then knew of the difficulties with the explosive switch. In response, Hornig wrote Oppenheimer a memorandum suggesting that Los Alamos explore a triggered spark-gap switch – the type eventually chosen – based on discharging banks of condensers, the discharge triggered by charging a third "probe" electrode to high voltage, of the order of 15 kV. This spark-gap switch, Hornig felt, could be made to fire in a fraction of a microsecond. In contrast to the explosive switch system, where a single switch set off the entire system, the spark-gap system had identical rapidly operating switches for each condenser bank. As Hornig recalls, "more or less immediately ... Oppenheimer put us in business ... to explore the utility of triggered spark gap switches."

By fall 1944, the Los Alamos program of manufacturing experimental detonators was being aided by a manufacturing effort based in Pasadena, at the California Institutute of Technology (Caltech). Physicists Charles Lauritsen and William Fowler, as well as Lauritsen's son Thomas, were responsible for Caltech's involvement in building accurate airplane-fired rockets for the navy to use in the Pacific. After they completed the rocket mission, Charles Lauritsen asked Oppenheimer whether they might be able to help with the Los Alamos program. Oppenheimer invited him to visit the laboratory. A fruitful collaboration between Caltech and Los Alamos began after Lauritsen learned from Alvarez about the problem of manufacturing detonators.[15] Lauritsen put Alvarez in touch with manufacturers in the Los Angeles area who could make the bridgewire detonator bodies. Alvarez recalls ordering thousands. Subsequently, the Caltech group would carry on a substantial portion of the research and development program on detonators.[16]

By late September, Alvarez and Johnston were proudly reporting the small time differences of bridgewire detonators. At the time, they were estimating that bridgewire detonators would be ready for use at Los Alamos in about one month. Hoping to "keep firing voltage and energy to minimum values," Alvarez, Robert Alldredge, and Johnston were also researching the explosive switch "to fire large numbers of wire detonators in completely independent circuits."

Although the detonators promised good timing, the currents in the detonators were almost impossible to measure because of electrical problems. Since a detailed knowledge of the detonator current was needed to develop the firing circuits, Fussell's group faced the impossible task of developing a device without knowledge of its most fundamental parameter. Johnston commented, "I advised Mr. Fussell sometime ago that this was likely to be the case, and that hence his electrical detonating circuits would have to be developed largely by trial and error, by actually using them to fire detonators and measuring the timing." Characteristically, the implosion program had to rely on the cut-and-try approach.

Difficulties arose between the Alvarez detonator group and Fussell's firing circuit section, owing to a mismatch between the problems of developing the detonator and its firing circuit. The current to set off the detonators had to be low enough to allow the firing circuits to fire many detonators at the same time. But Alvarez's group was not optimistic that a current low enough to meet this requirement could reliably fire the bridgewire detonators. Spark-gap detonators, which fired at a lower voltage than bridgewire detonators, were regarded as a possible solution, although they were considered unsafe, because lead azide, on which they were based, was an extremely sensitive compound.[17] Whereas the bridgewire device was set off by exploding a wire embedded in explosives, the spark-gap detonator was set off by an applied voltage causing a spark to jump between two pieces of metal placed a small distance apart in the explosive. Since not enough manpower had been assigned to the detonator program, it was not clear that either group could meet the schedules for the drop tests of the bomb prototypes then planned for October or November 1944 at Wendover, Utah. In his report for September, Fussell said the situation was "considerably blacker now than it was a month ago." With no knowledge yet of "how many detonators are to be used per gadget, nor what their electrical characteristics will be," Fussell feared that firing devices for the detonators would not be available in February.

Kistiakowsky attempted to mediate between the Alvarez and Fussell groups. On 29 September, Kistiakowsky wrote that both "Alvarez and Fussell have agreed to put intensive effort into improving their products, Alvarez attempting to obtain detonators requiring lower voltages and energies and Fussell developing switches which will handle larger currents and stand off higher voltages and, therefore, suitable for greater numbers of present type detonators." New designs of lead azide spark-gap detonators promised to make the traditional mechanical safety gates unnecessary.

During September, Bainbridge proposed that the Raytheon Company, in Waltham, Massachusetts, be asked to produce the firing units for the upcoming drop tests at Kingman. Both Fussell and Bainbridge were familiar with this company, which had done much work for the MIT Radiation Laboratory. Raytheon, as Bainbridge explained to Kistiakowsky, had "proved in the past that they can work rapidly as the concern is not too small or too big and they have one of the best procurement organizations in the country. I believe that Raytheon might cut the time of manufacture of a large number of units, from three or more months to two or better." At the 26 September detonator meeting, those present agreed that Raytheon was "a suitable firm to be approached for the purpose of manufacturing firing ciruits for the drop tests at Kingman," and, furthermore, that two types of firing circuits should be made. They hoped "that two preliminary sets will be available here for test about 25 November." By the end of October, negotiations were under way to have Raytheon construct 100 sets in a crash program.

Early in November, Kistiakowsky phased the Hercules Company out of the bridgewire program, since it had made little progress in producing the agreed-on 2,000 bridgewire detonators. However, Hercules was to continue working on spark-gap detonators, which it would load at a "starting rate of 1000 per day with an increase to 2000 per day within two weeks." Los Alamos also examined British electric detonators, ordering several hundred from Imperial Chemical Industries.

In an early October memorandum, Kistiakowsky speculated to Bainbridge, that spark-gap detonators "will turn out to be as reproducible as the bridgewire type. If that is the case, the only advantage of bridgewire detonators will be a slight added safety because of absence of primary explosives .... My own bet at present is on spark gap detonators. I believe that their use can be made adequately safe."

At this point, the issue of which switch to use in the firing circuits was also unsettled. Although Kistiakowsky strongly favored the explosive switch, Hornig boldly objected, arguing strongly on 9 October against the explosive switch and in favor of the spark-gap switch that he and his co-workers had been exploring for the past several months. The simple solution he offered to the mismatch between detonators and firing circuits was to use multiple switches. Hornig realized that the essential problem was that too many units were being put in parallel at a given voltage. That resulted in slower timing. Following Hornig's suggestions, two switches were placed in parallel on each condenser bank.

Administrative problems were still hampering cooperation between the G- and X-Division groups working on the detonator program. Kistiakowsky joked on 18 October, "I believe it [the program] can be put effectively in operation only if there is a strong AMG [Allied Military Government] with a military force in temporary occupation of the entire disputed territory." He listed the areas over which G-7 should have jurisdiction – essentially

> the entire research program on detonators and firing circuits ... detonators, both bridgewire and spark gap type, specifications of experimental types of detonators ... supply of finished and semi-finished detonators ... research on firing circuits ... construction of firing circuits, 'crash sets,' determination of ... physical and other performance data on experimental detonators.

X-2 was to have "complete jurisdiction of and responsibility for the engineering and construction of detonators and firing circuits for the gadget." In this mock military memorandum, he added: "All exchanges of information and requests between Group X-2 and G-7 will be in writing with copies to Bacher and Kistiakowsky. In case of disputes the final authority in all matters pertaining to firing circuits rests with Bacher; final authority on matters pertaining to detonators rests with Kistiakowsky." That month, Oppenheimer appointed a detonator committee composed of Alvarez, Bainbridge, and Lt. Col. R. W. Lockridge, to "determine technically and administratively all questions connected with the external procurement of electric detonators." At a meeting on 25 October, Kistiakowsky proposed a plan "for handling the research and engineering aspects of the detonator program," a plan accepted in principle by those at the meeting. They specified that G-7 would handle the procurement and fabrication of special research detonators and would initiate and supervise a contract for the manufacture of detonators to meet experimental needs. They noted that by 1 November G-7 was expecting to have a final design for "an experimental model spark gap detonator." The responsibility for developing the crash sets was transferred from X-2 to G-7, with X-2 retaining "jurisdiction of and responsibility for the engineering and construction and test of detonators and firing circuits for the gadget and prototype models of the gadget initiating system." They also agreed that the Raytheon units, as well as other units involving many detonators, would be tested at Los Alamos on South Mesa, but that Primacord testing and other timing studies would remain at Two-

Mile Mesa. To help alleviate G-7's serious understaffing problem, this group received a new staff member in late November, Edward Lofgren, who had previously worked with Ernest Lawrence on the electromagnetic method.[18]

Alvarez wanted the tests by group X-2 conducted on South Mesa in order to use a rotating mirror camera available there around 15 November, which could study many detonators simultaneously and provide an individual record of the timing of each detonation. By mid-February, the rotating mirror would fully replace the oscillograph method of timing detonators.[19]

Through November and December, the research on detonators was beset by delays and poor manufacturing quality; Bacher reported that "neither bridgewire nor spark gap detonators seem sufficiently reliable at the present time." In mid-December, they chose the spark-gap switch "on the basis of practical engineering considerations"; operationally, this switch seemed barely different from the electrically untestable explosive switch.

By mid-January, group G-7 could boast "a very considerable improvement in the simultaneity and reliability of both wire and spark detonators." Most of the earlier problems of both the spark-gap and bridgewire types were traced to poor control of the explosive. The group began making "strong efforts" to bring electric detonators into the experimental program, "setting up to detonate charges at four field sites. First priority is the Ra-La, together with associated experimental shots." This new emphasis caused detonator production to increase. As reported in mid-January, "Wire detonator shipments to us from Caltech have reached approximately 300 per day, with promise of achieving 1000 per day in one to two weeks," while the shipments of spark detonators to Hercules from Caltech "have reached approximately 150 per day." Group G-7 estimated that its production section "can load and prepare 200 detonators per day. Their eventual capacity is expected to be 400 per day." Detonator production was becoming a factory effort.

Meanwhile, Los Alamos worried about whether Raytheon would produce the firing circuits on schedule. They learned with alarm that Raytheon had waited until 27 November to place its condenser order with the Sprague company, "and then only by telephone to be followed by letter confirmation." Furthermore, "Sprague claimed that engineering on this order would take from three to four weeks." As would become clear later in the program, this worry was well founded. However, the situation seemed to be getting little attention at this time.

Detonator and switch performance seemed adequate by the end of February. Two successful RaLa shots had been fired using electric detonators (see Chapter 14). Hornig reported that "out of approximately 200 detonators [of the hand-tamped type] fired there were *no failures* .... Yesterday a shot, using the Raytheon mechanical switch, was fired on 40 of the machine pressed detonators .... It is fairly evident by now that the new detonators live up to what was claimed for them in respect to simultaneity."

Theoretical Studies: the Christy Gadget

After the August 1944 reorganization of the laboratory, the implosion calculations in T-Division remained the province of group T-1 under Peierls, which received considerable assistance from group T-6, the numerical calculations group under Eldred Nelson and Stanley Frankel. The calculations of an ideal spherical implosion (described in Chapter 8) were virtually complete by this time, and the T-1 group's emphasis "shifted more to the actual implosion as against the ideal implosion." Thus, the group became more concerned with the discrepancies between experimental test shots and theoretical calculations of quantities such as densities and velocities, both of which were predicted to be higher than actually found. IBM studies of implosion proceeded slowly in this period, with only six problems completed in seven months.[20] The group also worked on theories for jets and spalling, equations of state for matter at high densities, temperature dependence, the stability of the shock waves, and a modulated neutron initiator. In January 1945, Feynman joined group T-6, as did Metropolis. As a consultant, Feynman changed procedures and greatly speeded up the progress of implosion calculations. He was soon named group leader. Frankel moved into Teller's group in F-Division.

At the time of the reorganization, Christy, a member of T-1, was working on the problem of asymmetric implosions.[21] In exploring the idea of inserting a "pusher" into the implosion assembly to replace a portion of the high explosives as a possible means of smoothing out jets and avoiding the Taylor instability, Christy opened a line of study that would lead him to the all-important "Christy gadget." This conservative, solid-core implosion assembly reduced the asymmetry and jets by brute force, but to meet efficiency requirements, it included an initiator that would begin the reaction near the optimal moment. Oppenheimer and his division leaders assumed that an appropriate initiator could be developed and decided in February 1945 to proceed with the Christy design for the combat weapon on a tentative basis. With the successful development of the modulated initiator, this decision became firm on 1 May 1945.

As the theoretical program for the Trinity gadget entered its final period, the already close relationship between theory and experiment at Los Alamos grew even closer. From the beginning of Project Y, Bethe had insisted that members of T-Division work with the experimental groups. In January 1945, Bethe formalized the policy by assigning theorists to specific experimental groups.[22] The precedent set by such successful imposed relationships between theory and experiment in physics would have a lasting impact on postwar research.[23]

The Implosion Initiator

The initiator problem was undertaken both by T-Division and the initiator group in G-Division, G-10, under the direction of Charles Critchfield, who had worked earlier on the gun. Although research on the gun initiators had little direct influence on implosion, the work was of indirect benefit to the program by establishing radiochemical techniques and a polonium supply line. Two approaches to the modulated initiator were explored – one based on the (α,n) reaction of ^{210}Po and ^{9}Be, the other on the (γ, n) reaction on ^{9}Be, as suggested by Alvarez. In the (α,n) initiator, arrival of the shock wave from the high explosive at the center of the assembly caused mechanical mixing that allowed α particles to impinge on the beryllium and yield neutrons.[24]

It was still unclear in late December, however, whether a practical initiator could be made for the Christy gadget. Bacher recalls that just before Christmas Oppenheimer asked group G-10 to report by 1 February 1945 on whether they could make the initiator. Preliminary research had given them some confidence, but Bacher cautiously replied, "I haven't any idea." At this point, he recalled recently, "the group really went at it hammer and tongs." Several researchers were added and Bacher "gave them just about top priority to go ahead and dig into this."[25] The Los Alamos Laboratory's capability to amass an all-out attack on this high-priority problem would pay off handsomely.

The job was intrinsically difficult. Realistic experiments were impossible; a definitive test would have required using the initiator in a full-scale gadget with active material. The work was reduced to implosion tests on a small scale, similar to those made in the terminal observation program, with subsequent recovery of the material and counting of the neutrons. By January, they had selected the (α,n) reaction for the initiator. Rubby Sherr, Bethe, and others had proposed several possible designs. On hearing news of this progress, Oppenheimer asked Bacher: "Can't you advance that date when you'll tell me that you can do this?" Bacher agreed. As he recalls, he told Oppenheimer on 31 January "that we thought we could make one."[26]

Oppenheimer now took several steps on his own to advance the initiator work. On 9 February, he formed an advisory committee on implosion initiators composed of Bethe, Christy, and Fermi.[27] This committee, which typically met weekly, kept in close contact with G-10, with the members of T-Division studying the initiator, and with the radiochemists who were preparing the initiator material.[28]

Oppenheimer also authorized a new firing site for initiator experiments in Sandia Canyon. As in every other sector of the implosion program, empirical tests of large numbers of small-scale models were undertaken. This approach, as Bacher described in February 1945, provided a way for testing "many of the suggested types of initiators under conditions less violent than actual, but nevertheless approaching them." By mid-February, Bacher was able to report encouraging results by group G-10, which received help from Koski's photographic group in X-Division in the analysis of recovered metal initiator components.

As the implosion initiator program expanded in 1945, the laboratory became concerned about meeting its polonium needs. Project Y had in 1943 arranged for a steady supply of polonium from the Monsanto Company, which agreed to prepare the element on small platinum foils (Chapter 7). But the implosion initiator program required much more polonium than anticipated; the material polonium was needed for testing all the initiator designs. Accordingly, Oppenheimer sent an emergency telegram to Thomas at the end of December 1944 asking that shipments be increased to 20 curies per month.[29] Fortunately, Monsanto was able to meet the increased polonium requirements. In February 1945, Richard Dodson and Charles Thomas revised the schedule for polonium shipments, calling for 100 curies a month by June and 500 curies a month by December, amounts needed for the implosion initiator designs being pursued at the height of design work.[30] In June, as the laboratory was making the final preparations for Trinity and the combat devices,

Monsanto agreed, after a special request by Allison, to ship 35 curies of polonium per week.[31]

Acquiring the Trinity Site

Oppenheimer, Kenneth Bainbridge, Robert Henderson, Maj. William Stevens, and Maj. Peer de Silva constituted the site selection committee for the Trinity test. The site had to be flat, so that the blast wave could spread unhindered; hard, so that the containment vessel ("Jumbo") could be carted in without trouble; arid, so that storms and haze would not interfere; and secluded – far from population centers, but not too far from Los Alamos, to allow equipment and people to be brought in quickly. A less urgent consideration was the number of persons who would have to be displaced. Secretary of the Interior Harold Ickes would have to approve the movement of any Indians to accommodate the testing. He was known to be a prickly individual to deal with.[32] Eventually Bainbridge and the committee narrowed the choice down to two sites: the Jornada del Muerto Valley in the Alamogordo Bombing Range in New Mexico and the desert training area near Rice, California.[33] They settled on the Jornada del Muerto region. After obtaining tentative approval from Oppenheimer and Groves, Bainbridge and Parsons approached Gen. Uzal G. Ent, commanding general of the Second Air Force, which governed the bombing range. In early September 1944, he consented to the use of the area.[34] Bainbridge determined the exact location in consultation with Comdt. William O. Eareckson, who provided aerial photographs of a site in the northwest corner of the bombing range.

Rapid construction then began. Oppenheimer approved the plans on 27 October 1944, and Groves did so five days later.[35] Planning the layout and overseeing construction were jobs that fell to John Williams, who had performed such roles in the construction of Los Alamos. An Army Corps of Engineers group, led by Stevens, designed the camp, but Capt. Samuel P. Davalos oversaw the actual construction by soldiers, military police (MPs), and itinerant workers.[36] Lt. Howard C. Bush, head of the MP detachment assigned to guard the base camp, saw to the day-to-day operation and was in charge also of the site administration and general morale after the scientists arrived. Groves's attempt to keep the Los Alamos connection secret failed, because most of the workers recognized Bush, whom they had seen at Los Alamos.[37]

The Corps of Engineers constructed cheap roads by compressing the desert dirt and rocks. They built shelters designed by R. W. Carlson and George Reynolds for instruments and personnel 10,000 yards from Ground Zero to the north, west, and south. David L. Anderson was in charge of gathering earth samples for chemical analysis and obtained meteorological equipment to begin compiling weather information on the site.[38] Others prepared samples of cables to determine how they would stand up to the electrical load and desert weather; the test would require satisfactory operation of many miles of test cables. Phillip Moon of the British Mission carried some of the supervisory load.

The observation and instrument stations for the Trinity Project encompassed a circular area just over 100 square miles within an 18-by-24-mile rectangle inside the air base. At the center stood a 100-foot metal tower to hold the gadget. The nearest buildings were the McDonald Ranch House and a few ancillary buildings about 2 miles to the southeast. The Trinity scientists used the Ranch House, named after the previous owner, the McDonald family, as a laboratory to test the nuclear activity of bomb parts and as the final assembly building for the core. The main observation and instrumentation stations were just over 5 miles from the steel tower, to the north, west, and south. From these stations the scientists would later watch as the final signals were sent by radio and wire to the device on the tower.[39] The Base Camp, where the scientists and military ate and slept while on Trinity duty, was erected 9.5 miles south of Ground Zero and consisted of a cluster of bunkhouses, a mess hall, and utility buildings.[40] The experiments and measuring devices were scattered throughout the space between the tower and the 10,000-yard stations; most of them were clustered within 1 mile of the tower.

Freezing the Design

In the last weeks of 1944, informed outsiders were still very skeptical about the possible success of the implosion weapon effort. Conant thought that the "gun method seems sure but needs a little pushing ... 1 August earliest date." As for the fast implosion using a shell assembly, he noted, "Difficulties still enormous." He confessed, "My own bets are very much against it." The Christy gadget, he felt, "looks possible," but "without a 'modulated source' which will be difficult to arrange the efficiency is very low (less than 1%?)." He added that "the electrical detonators and the accuracy of detonation points still very much a point at issue."[41]

Despite this bleak prognosis, the time had come to curtail the exploratory program and fix on a particular implosion design, for by late February, Hanford was estimating that enough plutonium would be on hand for an implosion gadget early in June. Four decisions had to be made immediately: (1) whether to use a hollow or solid sphere, (2) whether to use explosive lenses, (3) whether a modulated initiator could be made, and (4) how to divide the work between Los Alamos and other participating laboratories.

Accordingly, on 28 February 1945, a decisive meeting held in Oppenheimer's office settled these issues. Present were Groves, Conant, Tolman, Kistiakowsky, Bethe, Oppenheimer, and Charles Lauritsen. This group decided to concentrate all further work on the conservative, solid-core, lensed Christy gadget, using a modulated initiator and electric detonators. The composition of the lenses was specified, with allowance for fine-tuning their composition, since it was possible that the estimated velocity, which was based on small-scale models, could be incorrect for the full-scale device; this adjustment would indeed prove necessary. Research on other slow component materials was to be stopped. The kind of plutonium and the number of lenses and detonators was settled. The group also decided that the nonlens program would be taken over by Caltech.[42] A second meeting on the same subject, on 1 March, formalized these decisions in a series of resolutions, which were subsequently included in the introduction to the first Cowpuncher Committee's Report (Chapter 16).[43] This document also specified particular jobs to be done by X-, G-, and CM-divisions through June 1945.

A detailed schedule for implementing the historic February resolutions was established immediately: by 2 April, full-scale lens molds were to be delivered for full-scale casting; on 15 April, the problem of the timing of multipoint electric detonation would be solved; by 25 April, hemisphere shots for the observation of the emergent wave front were to be ready; and between 15 March and 15 April, detonators were to be in routine production. By 15 April, large-scale lens production for engineering tests would begin. Between 15 April and 1 May, a full-scale test, by the magnetic method, would be made. And between 15 May and 15 June, full-scale plutonium spheres were to be made and tested for their degree of criticality. By 4 June, the fabrication of lenses for the Trinity test was to be in process. Finally, by 4 July, fabrication and assembly of the Trinity spheres was to begin.[44]

Once the design of the implosion and gun weapons was fixed, Project Y shifted from research to development and testing.[45] Although many parts – including lenses, HE blocks, and pit – were made at Los Alamos, outside companies, such as Raytheon, played an increasing role after March 1945. However, it was the laboratory's responsibility to ensure that necessary raw materials were in the hands of the manufacturers and that parts were actually made. New groups formed to handle the testing, delivery, and manufacturing.

The group and division structure shifted to reflect the new focus. For example, during February and March 1945 group X-2 (Development, Engineering and Tests) split into four sections. Engineering problems were localized in section X-2A, which became X-2 (Engineering Design). High-explosives development became X-2B, which was subsequently absorbed into X-6 (Assembly and Assembly Testing). X-2C (Test Measurements) became X-5 (Detonating Circuits) and X-1A (Flash X-ray Photography). X-1, which had been the "vanguard of field research on implosion," stopped working on flash photography of cylindrical charges and flash X-ray studies of implosion and switched to reflection stereophotography of imploding hemispheres and also began developing an explosive switch for the electronic informer. Trinity test preparations, previously overseen by X-Division, were given division status, with Bainbridge responsible for Trinity planning.[46]

By March, the delivery effort in O-Division had evolved into a division known as Project A. The number working on both delivery and Trinity problems increased. R-Division worked extensively on Trinity measurements beginning in April 1945.

The tremendous increase in the amounts of ^{239}Pu and ^{235}U that were available made it even more important to monitor the presence of those isotopes. Thus CM-12 (Health Instruments, Decontamination) was created in April by splitting off from CM-1 those staff members who had been doing health instrument and decontamination work. The requirements imposed on RaLa production became more stringent, justifying a separate group. Part of the radiochemistry group (CM-4) was assigned to work on RaLa chemistry (CM-14) or to prepare sources and purify the increased amount of polonium necessary for initiator testing (CM-15). CM-5 (Uranium and Plutonium Purification) split to form separate uranium and plutonium purification groups. A month later, group CM-10, which had been working on Jumbo design (Chapter 18), was dissolved, for Jumbo work had been abandoned.

The general coordination committees – the Technical and Administrative Board and Coordinating Council – were supplemented by smaller committees focused on particular aspects of the weapon development: the Cowpuncher Committee, the Weapons Committee, the Intermediate Scheduling Conference, the Technical and Scheduling Conference, and the Initiator Advisory Board. With these alignments in place, the implosion program now rapidly turned from research to the development of a combat weapon.

16

Building the Implosion Gadget: March 1945 to July 1945

After the implosion design was frozen at the end of February 1945, the program shifted its emphasis from research to constructing actual bomb components, including explosive lenses, detonators, initiators, and the plutonium hemispheres. Kistiakowsky reported on X-Division's work in April: "One can now state with a reasonable degree of assurance that all major research and design gambles involved in the freeze of the program of the X-Division have been won. Progress is more and more determined by the rate of supply of manufactured items." By May, he concluded, "The activities in X-Division have lost all semblance to research and have become so largely production and inspection and testing that their brief summary here seems impractical" (at which point his division progress reports stopped abruptly). However, most of the crucial components of the implosion gadget remained problematic, almost to the time of the Trinity test, and most underwent last-minute change.

* This chapter is based on a manuscript by Lillian Hoddeson, to which Catherine Westfall contributed sections on making the plutonium hemispheres and on polonium development, and Les Redman contributed most of the section on explosive lenses. We are grateful to Gordon Baym and Redman for their detailed editing of the technical content of the entire chapter and to Robert Penneman for editing the technical content of the sections on plutonium, polonium, initiator, and RaLa.

The Cowpuncher Committee

Oppenheimer launched the final phase of Project Y's implosion effort with his appointment on 1 March 1945 of the powerful Cowpuncher Committee to "ride herd" on the program. Besides Oppenheimer, the members included Bethe, Kistiakowsky, Parsons, Bacher, Samuel Allison, Cyril Smith, and Kenneth Bainbridge. The committee oversaw eight major programs: (1) fabrication and inspection of explosive lenses; (2) design and construction of electric detonators and detonator circuits; (3) diagnostic tests to determine timing, compression, and symmetry; (4) research in chemistry and metallurgy; (5) study of the critical mass and time constant of the plutonium nuclear explosion; (6) design of the inner metal parts of the implosion assembly; (7) coordination of the Trinity program; and (8) assignment of shop priorities. At weekly meetings, this committee relentlessly defined and redefined the assignments to individual groups, while constantly adjusting scheduled milestones.

At the first Cowpuncher meeting, held on 3 March, the members prepared an outline of the coordinated program. "Maximum pressure" was put on the initiator, whose feasibility was to be determined by 1 May 1945. Other high-priority items included the electric detonators, the Raytheon firing units, the handling of plutonium shipments from Hanford, lens testing, RaLa, magnetic and betatron shots, and the procurement of lens molds.[1]

The Initiator

The initiator problem boiled down to increasing the number of neutrons in the system very rapidly at a particular time, to increase the efficiency of the gadget by starting the chain reaction during the few microseconds when the system was fully compressed. The (α,n) reaction of beryllium with α particles emitted by polonium was the most promising method for producing the neutrons. Attention focused on how best to mix the polonium and beryllium to initiate the reaction, and how to keep the two substances separate until the neutrons were needed.[2] A large number of designs were suggested; by March, only six remained in the running.

Since the program required a substantial quantity of polonium, the polonium effort grew rapidly during the spring of 1945. Most of the burden for coordinating the polonium work fell to Richard Dodson, for by the time the initiator had become a vital research effort, Charles Thomas of Monsanto Corporation had resigned as coordinator.[3] In March, a new group, CM-15, was established to handle the burgeoning polonium effort. In the same month, Oppenheimer told Kennedy that construction of the building needed to conduct polonium research would have priority over construction of the new plutonium plant.

Among those working in G-Division on the difficult initiator experiments were Charles Critchfield (group head), Milo Sampson, Sidney Barnes, and Rubby Sherr. Their principal concern was to study recovered implosion remains. But unlike the terminal observations program to study the implosion, in which recovered fragments could provide substantial information, initiator tests required complete recovery of a nearly intact initiator. The efficiency of neutron production could be determined only by counting the total number of neutrons emitted. It was not possible to determine precisely how fast the neutrons had been formed, or to measure the number emitted in the first microseconds after the initiator was hit by the shock wave. The initiator studies also employed the counter X-ray method, flash photography, and the pin method, but none of the approaches proved satisfactory.

Given the inconclusiveness of the experiments, the choice of initiator design hinged strongly on theory, which included the calculation of efficiency. Klaus Fuchs, who had been working on the theory of jets, collaborated with Bethe, Paul Stein, and Robert Christy in developing a rudimentary theory of the Urchin, a favorite design.[4] Several elder physics statesmen also lent support to the initiator program. Having "meditated" on the initiator during March, I. I. Rabi concluded that the present initiator program was "well conceived and that we are being sufficiently cautious in our investigation." Although Fermi remained skeptical about the initiator, Rabi assessed, "it would be most astounding if we were unable to make a reasonably satisfactory initiator."

Bacher was disturbed by Fermi's mistrust of the initiator. He recalls that Fermi would come "every second day or so ... with a new reason why the initiator wouldn't work." To allay Fermi's concerns, Bacher proposed in the spring of 1945 that Oppenheimer ask Bohr to assess the initiator program. When Bohr and his teenage son Aage told Fermi that the device promised to work, Bacher reflects, "it made a lot of difference ... because ... it was somebody from the outside."[5]

By mid-April, the Los Alamos scientists had in general come around to believing that a modulated implosion initiator was technically feasible. Bacher noted in his diary, "The more satisfactory condition of the initiator is best indicated by the fact that Fermi forgot to come [to the 14 April initiator committee meeting]." At the 24 April meeting of the Initiator Advisory Board, Bethe, Christy, Critchfield, Fermi, Sherr, Bacher, and Wilson agreed unanimously "that such an initiator can be made." This agreement, Bacher told Critchfield on the following day, "was much greater than we had any right to hope for." The board members went on to eliminate certain models and rank others. Bacher ruled "that any type of initiator still in after May 1st should immediately be designed as an actual initiator."

Bethe, Fermi, and Christy attempted to settle the initiator design later that day. Christy and Bethe argued for the Urchin, whereas Fermi favored one of the other designs. On the following day, Bethe and Bacher decided that the Urchin should be assigned top priority. The formal unanimous decision of the Initiator Advisory Board on 30 April was that "a modulated initiator is feasible and should be incorporated in the first design of the gadget." The board recommended "putting the Urchin first .... Fermi dissented from the majority report, but apparently did not feel very strongly about it .... Since there is a general feeling that any one of these initiators will work, the decision from one to another is rather hard to make, and probably not on very sound grounds." Years later, Bethe reflected: "Voting against Fermi was a very hazardous matter, and so I felt somewhat nervous ... especially during the night of the Trinity test .... If it didn't work it was my fault mainly."[6]

Attention then turned to the details of fabricating Urchin initiators, for example, how to deposit polonium on various metals as a means of separating the polonium from the beryllium. The fabrication task required Los Alamos chemists and metallurgists to collaborate with G-Division physicists. On 4 May, Kennedy sketched out the responsibilities of particular sections of the Chemistry and Metallurgy Division. Dodson was responsible for coordinating CM-Division's initiator activities. Smith was to be liaison with the beryllium metal suppliers.[7] Tensions mounted between the chemistry and physics leaders over the sensitive issue of reducing the neutron background. Kennedy's group was already severely taxed with last-minute tasks for the plutonium and uranium efforts. Strained as well by initiator demands, he noted in mid-May that his group had a "virtual 100% Urchin program."

By 1 June, Kennedy's group was able to report that, because of the purity of Monsanto foils and the success of fabrication procedures, all the initiators met neutron background standards. The first complete Urchin unit was ready soon afterward.[8] Despite last-minute excitement when the first full-scale Urchin nearly dropped down a drainpipe during canning, Johns was able to show the device to the anxiously awaiting

Fig. 16.1. Covered passage over Los Alamos main road. Building P, housing the experimental physics division, is on the left. Building T, housing the theoretical physics division, is on the right, partially hidden by the bridge. LA Photo, 3736.

Kennedy and Oppenheimer on 21 June. By the end of the month, another implosion initiator had been completed, and Kennedy reported that both full-scale initiators "had a neutron background well below tolerances." They sent the completed Urchins for acceptance testing to G-Division, where they were dropped, vibrated, exposed to water vapor, and checked for leaks. In each case, neutron emission levels remained the same. When one initiator was opened, the group found no evidence of the redistribution of polonium.

Explosive Lenses

In the final phase of the implosion program, X-Division focused on producing explosive lenses according to the 28 February 1945 design specifications.[9] Again, many problems had to be solved by trial and error. For example, the design freeze took into account the fact that observed velocities in the components differed from calculated ones by specifying that if the ratio of velocities suggested by small-scale tests proved unworkable for full-size lenses, then the composition of the slow component, the Baratol, could be adjusted. Such an adjustment proved necessary, in part because mold procurement was very slow, especially for full-scale charges. Not until early June was there an adequate supply of full-scale molds, and, indeed, the ratio of velocities corresponding to these full-scale molds was not able to produce a proper wave shape.[10] The patch-up solution was to alter the physical composition of the Baratol.

Although this adjustment solved the wave-shape problem, it made new casting procedures necessary, to ensure the production of quality lenses. Casting to the stringent Los Alamos requirements on dimensions, uniformity of density, freedom from cracks and voids, and chemical composition was at best a "black art." As the lenses were scaled up to full size, new physical problems arose, such as cracking. The materials were weak and brittle, and the large and heavy full-scale charges were awkward to handle. There were many minor accidents, such as corners breaking. One of the most frightening new concerns was that an explosion might occur while people were handling or machining a piece.[11] The S-Site laboratory sections and casting room learned how to produce acceptable results using empirical, iterative procedures. John Russell recalls that to prevent throwing away too many of the precious lenses, he and Kistiakowsky worked together immediately before Trinity to rescue defective lenses. Like dentists, they drilled holes directly into cavities in the HE. Then they used a steam-heated pot to melt small amounts of explosive, which they used to make repairs. Although the procedure could not completely fix the composition and contour, it was far "better than just going ahead and shutting your eyes and saying, well I hope that busted-off corner doesn't hurt anything."[12]

Detonators

Detonator development in the six months before Trinity was almost as frustrating as lens making. Detonation reliability was still questionable until weeks before Trinity. The production of both detonators and firing circuits consistently lagged behind schedule.

During March, Kenneth Greisen's X-1 group terminated its small-scale flash X-ray studies and turned its entire attention to detonator research and development, supplementing the work of the G-division detonator group, G-7, which that month was placed under Edward Lofgren's leadership. While Greisen's and Lofgren's groups focused on the actual detonators, Lewis Fussell's section of Robert Henderson's X-Division group, X-2C, continued work on the firing circuit, collaborating with Leon Fisher's section of G-7, which focused on spark-gap switches. Spark-gap switches were looking better than ever in mid-March; Fisher reported that in a shot using eighteen detonators and Donald Hornig's triggered spark-gap switch, "there were no failures," the timing was adequate, and "the gap did blow up. The amount of energy we discharged through the gap was twice what Hornig expects to pass through the gap." The bridgewire program depended heavily on manufacturing help from the Caltech group under Charles Lauritsen. Henderson's group at Los Alamos and Lauritsen's group in California shared responsibilities in their mutual effort to make 1,000 detonator assemblies per week.

March saw the beginning of the program centered on the "handlebar," a bridgewire detonator system that appeared less likely to fail. However, even after the handlebar program entered production, G-Division continued extensive research on other kinds of detonators. Spark-gap detonator research had been dropped for the time being, despite promising results; but a month later Lofgren told Bacher, "Unless there is serious objection we revive the azide spark work."

By this time, detonators were being used in several diagnostic programs, in particular, wherever multiple-lens systems were used. But the detonators used in such research were not the most advanced models. The best detonators and lenses had to be saved for the full-scale implosions.

In April, when Greisen's group assumed responsibility for testing the handlebar detonators, the failure rate was still too high.[13] In preparing a detonator testing schedule, Bradbury took into account the fact that the detonators were not yet working well. This schedule provides another illustration of the pervasive impact of military procedures on the Los

Alamos research program. Attempting to allow for all contingencies, Bradbury announced that all detonators produced

> would be devoted to timing tests on the ground at Y in an effort to produce a satisfactory detonator. Such a program will be continued until 300 have been tested and satisfactory results reached. If satisfactory results have not been obtained at this time, the tests will be continued on this basis until they are. When a reasonably satisfactory detonator is reached, 50% of the production will be time tested and 50% will be available for detonator switch combination tests .... Subsequent tests will be devoted 50% to timing tests and 50% to field tests or, if it seems more satisfactory and the production is sufficient, 33% timing, 33% local field tests, and 33% field tests elsewhere.

Referring to the shortage of staff, he pointed out that present production was being kept up only by "taking men off other high priority jobs." The cause was "the fact that the mechanical parts coming from Cal-Tech are in an unacceptable condition. The same statement applies to production from our local shops." He suggested "commandeering" seven persons from X and G Divisions, adding to G-7 "3 more men of only modest or mediocre ability," and "4 more men or girls" to help them to "increase their production to 75 detonators per day." He added, "In view of the fact that Cal-Tech has never met their promised production quotas, it appears highly desirable to make our own production as much as possible." Lofgren was also concerned that the Caltech contribution was more than three weeks behind in its scheduled production of 1,000 detonators per week. He complained that those Caltech did make required "additional mechanical operations by us because of incompleteness and poor fit." He grumbled that the upcoming tests "will be made with detonators about which there can be no assurance of even moderately good timing." He saw the basic problem – a result of the incredible time press of the project – as "the premature diversion of detonators from the problem of their own development," and he cautioned that this practice would "further delay an already late program."

Like countless other efforts at Los Alamos, the detonator program resorted to overlapping development as a means for meeting military deadlines. Thus, while research on the handlebar detonators went on, work also continued on another bridgewire model, as well as on spark-gap detonators, which managed to survive in the program despite their lower priority and the various decisions to discontinue research on them. At the same time, the filling of detonators with explosive at South Mesa was going well, and the laboratory estimated that soon 50 detonators a day would be loaded.

At this time, Fussell's X-5 group was devoting its energies to preparing for drop tests scheduled at the end of May at Wendover field in Utah, and to bringing the Raytheon units, whose design had been frozen in late March, into production. Raytheon had begun its work on detonator firing circuits with a unit that contained a mechanical switch. In March, Los Alamos changed the design to a unit that instead contained a spark-gap switch. Two prototypes of the latter model had been built and shipped to Raytheon, arriving at a moment when Raytheon was only beginning production of the now outdated original model. The company had not yet delivered a sufficient number of these for use in the diagnostic program.

Raytheon was never given a reasonable explanation of why the change was made to the Model II units, and the company was slow to alter its assembly line to meet the new assignment.[14] Not until 2 May did a few of the first Raytheon Model II units arrive at Los Alamos, after passing all their electrical and mechanical tests. One of them fell off the truck carrying them to South Mesa and rolled some 50 feet into a canyon, providing fortuitously, as Hornig recalls, "the ultimate shake test – it worked fine after it was pulled back up." Another of the Model IIs had a defective switch. The rest worked well.

Through the spring of 1945, thoughts turned to the dangers of accidentally setting off the detonators during bomb-drop tests. Parsons, Oppenheimer, Kistiakowsky, Bainbridge, Lauritsen, and others decided in late April that it would be wise to have some expert check the design "on the spot for possible errors." They arranged for R. L. Grauman, a fuze and detonator expert of the Naval Ordnance Laboratory, to assess the detonator safety situation. At a meeting held on 1 May 1945, Grauman judged the handlebar detonators to be mechanically safe and "the best under the present circumstances" but recommended a series of tests for "rough handling." He also specified that "particular attention must be paid to the design in order to avoid any possible disturbance of explosive contents." Parsons ordered "jolt and jumble tests" for the bridgewire detonators.

On 8 May, a new group, X-7, was formed under Greisen to handle the large-scale testing and manufacture of the handlebar detonators, and also to conduct detailed research into detonator timing. Greisen's section X-1A dissolved, and most of its members moved into X-7. On 12 May, Lofgren nervously turned the newly frozen specifications over to Greisen, cautioning him that since the physics of the detonators was still not fully understood, "we cannot claim that the following specifications represent an optimum." But he remained confident that acceptable models could be made. Lofgren's group now turned its entire attention to the spark detonators, working in collaboration with Caltech. The pressure was on Greisen's group to provide detonators, not only for the experimental program, but also for the bombs to be used against Japan. On 23 May, Bradbury asked Greisen for a group of detonators to use in tests at "Destination," with an order that they leave Los Alamos about 5 June.

The arrangements for detonator testing at Trinity were made two weeks later. Fussell, whose group would be responsible for the Trinity testing and the firing of detonator circuits, asked Hornig, James Buchanan, and Sgt. R. Brown to construct and install a modified Model II Raytheon set. At this time, the schedule called for the Trinity test on 20 July, with various rehearsals before. Kistiakowsky asked Fussell to set aside one Raytheon unit to be "made shippable to Trinity by 1 July, with two further units so prepared by 7 July."

Unfortunately, just as these requests were being issued for detonators at both Destination and Trinity, Greisen was noting that "the present status of these detonators is not very satisfactory." In particular, the insulation material used in lead-in wires often cracked and led to short-circuiting. A different, artificial molded material was needed. But, as Greisen wrote Oppenheimer on 6 June:

> facilities for the molding are limited to a few small companies in the east. Our requests for prompt deliveries from these companies are being met with unwillingness to cooperate on the part of those companies. In particular, the one company which has begun to turn out cores (Brillhart Co., Great Neck, Long Island) could double their output (which otherwise will be insufficient) by employing *one* extra person on night shift; but this they are unwilling to do. The general attitude seems to be that they do not want war jobs any more, and are directing their efforts towards post-war work instead. Cannot some pressure be brought to bear against such resistance?

Available documents do not indicate Oppenheimer's response to this memorandum, but they suggest that at least a few of the molded cores were ordered almost immediately for testing about two weeks later. One month before Trinity, Los Alamos was still struggling to produce the implosion detonators. In a memorandum to Lofgren on 11 June, Greisen bargained for as much time as possible – he hoped for twenty more days – to make the decision as to what kind should be used and outlined a contingency plan. Given the uncertainties surrounding the detonators, group G-7, collaborating with X-7, invested a little more time in its alternate bridgewire program. In mid-June, the plan was to make 3,000 alternate assemblies, but then the group found that the lead azide detonators might be unsafe, for they could be set off "with the static charge obtained by a man."

On 21 June, with the clock to Trinity ticking away and the detonator problem still unresolved, Oppenheimer set up a detonator committee composed of Fussell, Greisen, Lofgren, and Frank Oppenheimer to coordinate all components of the detonator program. Handlebar production was increased to more than 500 detonators a week. The committee decided to relieve Caltech of its responsibilities on the primary detonator model, so that it could focus on the alternate bridgewire type.

Despite the uncertainties and difficulties of detonator development, plans to install and test detonators at Trinity pushed forward. On 9 July, in a memorandum on the Trinity "hot run," Bradbury listed the various detonator jobs to be carried out at Trinity. By 5 p.m. on 14 July, he projected, the gadget would be "complete." On Saturday, 15 July, "Look for rabbit's feet and four leafed clovers. Should we have the Chaplain down there?" On Monday, 16 July at 0400: "Bang."

Diagnostics

Through the spring and early summer, the original three diagnostic techniques – X-ray, photographic, and terminal observations – continued to study lenses, detonators, and the initiator, while the RaLa, pin, betatron, and magnetic methods provided increasing evidence of compression and symmetry, as well as assurance that the velocities would be sufficiently high.

Walter Koski's flash photography group abandoned its study of cylindrical charges and focused on jet formation in small-scale hemisphere and slab shots, initiator studies, and lens shots. Since lens production was very slow, Koski would often make his own two-dimensional lens molds and lenses. The mirror camera was used extensively in testing lenses, as well as detonators. Various special problems were examined, such as those suggested by Peierls in March, which included edge effects, cracks, and "the straightening of a shock wave in Aluminum as it traverses different thicknesses of the metal."

Terminal observations measured the time it took for a detonation wave to travel from one point of a lens to another, and to determine the temperature in the implosion by observing possible melting of selected materials at the implosion center in recovered cores.[15] The X-ray program worked principally on the initiator. As mentioned, the flash X-ray program was discontinued in March 1945. The counter X-ray method, only then beginning to work, was also dropped, since the same data could be gathered using the betatron method.[16] The counter X-ray group redirected its attention to the initiator, using photographic methods.

The month of March saw little RaLa testing because of an insufficient supply of radiolanthanum. One solid shot, fired on 3 March, used the combined remains of shipments four and five, but nevertheless proved inconclusive because of source weakness. An important milestone was reached on 1 April when explosive lenses were used for the first time in a RaLa shot, but electromagnetic disturbances destroyed the data's usefulness. However, a subsequent shot, using the same RaLa batch, showed definite compression of tuballoy, a density increase, and reasonable acceleration. A few further RaLa shots, using the remains of shipments four and five, gave evidence of compression in aluminum.

In the spring, because of the difficulties of separating the lanthanum from the barium and the dangerously high radiation levels of larger sources, Rod Spence and Norman Gross developed a new chemical procedure for carrying out separation of the lanthanum through precipitation of lanthanum hydroxide at a distance of 90 feet. Although the accuracy of RaLa work was still improving thoughout this period, by June the main pre-Trinity task of the RaLa experiment – to aid in settling the design for the Trinity gadget – had been accomplished, and prominent members of the group had turned their attention to Trinity work. Rossi, for instance, was attempting to measure "α," the e-folding time for neutron population growth.

In these pre-Trinity months, the pin method group (G-8) focused on studies of lens assemblies and on measurements of the velocity at various points inside test models. They made a number of shock velocity measurements in nonmetals, "in an attempt to find a material which would be a suitable replacement for aluminum in the magnetic test of the first full scale implosion." The pin group also studied "initial material velocities at positions corresponding to points far inside the gadget and under conditions which are suitable mainly for study of initiators." In late March, results from small-scale lenses and tuballoy showed velocities that agreed well with the theoretical calculations of Bethe and Christy. By mid-April, the pin method experiments were seen as "very encouraging, both from the way the observed velocities can be repeated and from the actual magnitude of the velocities obtained," which were "in extraordinarily good agreement with that calculated by the theorists." Shock velocity experiments continued in various metals, as did measurements of surface initial velocities "to get some idea of the violence of shock in the various implosion tests."

With the improvements in both the betatron and cloud chamber, the precision of the betatron method increased. By mid-March, the betatron group had seen definite evidence of compression in both tuballoy and aluminum. They improved their measurements of compression and radii by integrating the oscilloscope trace. By mid-March, they were able to obtain average densities to 5 percent accuracy. Uranium was used sparingly in these tests because it was pyrophoric; the shots often set off forest fires. The handful of uranium shots that were carried out did show, by mid-May, definite reduction in diameter, in excellent agreement with theory.

Throughout the spring, the magnetic method group devoted much effort to testing background, particularly to studies of electrostatic and magnetic "hash." One of their more serious problems – which occurred particularly with uranium – was that the explosion caused disturbances that would be registered by the pickup coil. The reason for these difficulties was still a mystery in mid-May. However, one excellent magnetic record taken in mid-June on a small-scale shot showed clearly the motion of the aluminum pusher and the shock-wave reflection.

During this time, preparations were moving along for the full-scale magnetic test at Pajarito Canyon – the only full-scale nonnuclear test shot. This test occurred on 14 July, two days before Trinity. To the dismay of all concerned, the initial interpretation raised grave doubts about whether the implosion could succeed at all, for the test seemed to indicate such a slow velocity that it cast doubt on the bomb's effectiveness. Fortunately, Bethe uncovered an error in the analysis. After taking Bethe's correction into account, the group concluded the test had demonstrated good detonator functioning and good symmetry.[17]

Fabricating the Plutonium Hemispheres

The arrival of Hanford plutonium early in February signaled the beginning of full-scale wet and dry plutonium purification (Chapter 15). The wet and dry purification groups focused on making the "A" process, developed by Arthur Wahl in summer 1944, in an effort to produce an adequate supply of plutonium.[18]

Although the A process generally worked well for wet purification, Wahl had to overcome certain problems with the apparatus. As he recently explained, "The folly of developing the process with too little attention to engineering principles made itself felt, and it was necessary to make innumerable modifications in the apparatus during the six-months construction and testing period which followed."[19] In the first run, for example, metal parts in the apparatus became corroded and some precipitates did not settle properly. When enough material was available in March, they made two more 160-g runs. At this point, the researchers had trouble "with the peptizing of the final oxalate precipitate" and with the H_2O_2 left in solution from recovery operations. To resolve these problems, they ran eighteen 8-g tests.

The program also suffered from a severe shortage of workers, a problem that affected much of the laboratory when Los Alamos switched from research to production but was particularly acute in the chemistry–metallurgy effort, since original staffing estimates had been far too low.[20] In the case of wet purification, three 160-g purification units were ready in May but could not be run simultaneously until June owing to a shortage of operators.

Despite these problems, an adequate quantity of plutonium was successfully processed through wet purification in June and July of 1945. By the end of July, a new, more efficient purification method, the "B" process, had been developed, but by then the plutonium for Trinity and the combat sphere had been produced almost entirely with the A process.[21]

In the last frenzied months before the first plutonium components were made, the dry purification group also improved upon their process, making it possible to hydrofluorinate the necessary plutonium. The group's goal now was to reduce the hydrofluorination time. In April, they tried direct hydrofluorination of the oxalate on a 128-g and a 158-g batch. Although the decomposing oxalate flew around inside the reactor and the bulk density of one of the fluorides was low, yields of 94 and 96 percent, respectively, were obtained in 3.5 hours. Although six hydrofluorinator units were in operating condition in May, the equipment required constant repair, and until June the group was too short-staffed to handle both processing and maintenance. In May, the group faced the additional frustration of having two runs ruined by "accidental misconnection of controlling thermocouples." By this time, however, it was clear that a new oxalate ignition cycle and HF procedures had solved the problem of excessive hydrofluorination times. In June and July, while processing plutonium for the Trinity and Nagasaki devices, the dry conversion group was able to process a sufficient amount of plutonium in six to eleven hours of hydrofluorination with 87–100 percent yields.

At the same time that the wet and dry chemistry groups were trying to maintain the flow of purified plutonium, the metallurgists were struggling to produce plutonium hemispheres that would meet the exacting requirements of the Christy design within the deadline. Seeing that Richard Baker's stationary bomb method, devised for 6- to 25-g runs, continued to work well as quantities rose to 300 g, attention focused on the work of E. F. Hammel and his co-workers, who were improving the process of large-scale remelting. In February, the metallurgists also began tests of "the minimum remelting conditions needed to give satisfactory metal."[22]

By March, when a new, safer, all-metal apparatus had been installed for larger remelts, the electrochemists had revealed that plutonium metal could be produced at lower temperatures. But they worried about its purity. In April, with neutron multiplication measurements on hand, taken on metal pieces produced by remelting experiments, they were able to evaluate metal purity more precisely. By this time, the new metal vacuum system had successfully remelted four 70-g lots and four 140-g lots with 98 percent yields. Ultimately, Los Alamos produced almost all the necessary refractories for wartime production of both plutonium and uranium. The MIT laboratory had been able to produce few acceptable CeS crucibles before the decision was made to use refractories of MgO, the material that became the obvious choice after the true, unexpected low melting point of plutonium was known. Smith acknowledged in an apologetic letter to John Chipman that this turn of events "must be very painful" for the MIT staff that had spent so much time, effort, and expense on developing these CeS crucibles, and so much trouble obtaining the material to make the MgO crucibles.[23]

In addition, the metallurgists were seeking reliable methods for fabricating larger hemispheres, which could also be used for the multiplication studies.[24] To employ plutonium's δ phase, the metallurgists had to stabilize it. One possibility was to find an alloy that would depress the δ–γ transformation to a temperature close to room temperature, so that the metal could be worked and retain its shape. By April, Eric Jette and his group had demonstrated that a 3 atomic percent gallium alloy would retain the δ phase indefinitely at temperatures well below ambient.[25] Smith felt that the stability of the δ phase had been established adequately for it to be used in weapon components, and Jette agreed. However the time dependence of the phase stability of the plutonium alloy was not yet completely established. Jette worried that the metal might change phase between the time the alloy was prepared and then used. That might amount to catastrophic changes in dimension that would ruin the weapon. Smith considered the risk worth taking. He wanted to proceed with an alloy. He discussed the problem with Oppenheimer after dinner at the house of his friend Edith Warner, a resident of nearby Otowi who sometimes served meals to Los Alamos scientists. Oppenheimer, who "had an incredible ability to understand fine points at the bench level and relate them to the larger objectives of the laboratory," accepted Smith's evaluation. Now the metallurgists could define the final fabrication plans, although the metallurgy troubles were by no means over.[26]

At a 23 June meeting of the Cowpuncher Committee, Smith was asked to complete the combat sphere as soon as possible, along with the components for the Trinity device. Although Kennedy privately decided that they could finish the combat spheres around 21 July, Smith and Garner gave Allison a 23 July date, adding that the proposed deadline allowed "no contingency for disaster in any stage of the processing." They urged that exact information on Hanford shipments be relayed "as far in advance as possible." The Trinity hemispheres were completed and delivered on 2 July 1945. After this, the metallurgists worked feverishly to produce components for the first combat unit. A mockup of the Trinity device was first assembled and checked for criticality in Omega Site, on July. Although it was one of the first holidays the laboratory had celebrated for a long time, Holloway, Smith, Bacher, and several others worked all that day, without the benefit of the usual support staff. Bacher recalls "getting somebody from someplace to get me into the safes so that I could get some gold foil and platinum foil."[27]

Problems continued to plague the group, even at the eleventh hour. When they received the Hanford schedules in early July, it looked as though the required plutonium would not arrive in time to meet the 23 July deadline for delivery of the combat weapon. When a 30 July date was suggested, Groves objected strenuously. Under this pressure, Kennedy and Oppenheimer decided to include in the combat sphere an amount of plutonium previously set aside for stability and nuclear tests. Assuming that Hanford shipments continued on schedule through 23 July, they would have sufficient plutonium. However, the Hanford shipment slipped another two days. To compensate, the deadline was pushed up to 26 July. They now had no extra plutonium to compensate for the difference if Hanford experienced further delays. The following week, yet another problem emerged. A new last-minute design was formulated to solve a severe jetting problem that had been recognized just before the planned delivery of the combat weapon sphere on 24 July.[28] Unfortunately, one of the hemispheres for Nagasaki had to be remelted and recast because it was underweight. The strategy of redundancy paid off in this instance. Provision had been made to produce three hemispheres, so two usable ones remained.[29]

Despite the last-minute problems, the combat hemispheres were successfully fabricated with the same technique used for the Trinity hemispheres.[30] They were completed on 23 July and shipped 26 July. To the satisfaction of the researchers in the chemistry and metallurgy group, analytical tests revealed that the α,n yields from the combat hemispheres were well within weapon purity requirements.

Theoretical Work

Three substantial theoretical problems remained unresolved at the beginning of March: (1) researchers had not yet been able to develop a theory for the modulated initiator, (2) they still had to calculate the effect that a temperature increase in the materials struck by the incoming shock wave would have on compression at the center of the implosion, and (3) they still did not understand the stability of shock waves. The initiator theory developed by Fuchs, Bethe, and Critchfield by mid-April had enabled the lab staff to design and test the device successfully. Temperature problems were considered by Joseph Keller and others. Peierls and A. E. Roberts examined the stability of shocks. T-Division also worked continuously on clearing up discrepancies between experimental and theoretical results. In June, Bethe felt that the most pressing remaining problem was initiator timing.

The IBM implosion hydrodynamics studies had been moving very slowly, until March 1945, when Feynman was moved from T-4 to T-6 to take over the project. Whereas up to that point only eleven problems had been solved, by the end of April, Feynman's group had worked out five further problems. Feynman tells how he and others on the staff would often fix the IBM machines to save time, and also how Frankel eventually suffered from the "computer disease," a preoccupation with using the computer itself rather than running designated problems on it. This disease interfered so much with his functioning in the implosion program that Bethe in March 1945 asked Feynman to take over the computation group. At this point, progress on implosion became notably faster, in no small measure because Feynman had explained to the people on the computation staff how their work fitted into the program as a whole.[31]

Throughout this last phase of implosion work, T-Division turned increasingly to calculations related to the Trinity experiments. Efficiency calculations by Weisskopf's group T-5 were refined considerably. According to new estimates in June, the Trinity explosion would be comparable to the blast from 4,000–13,000 tons of TNT (Chapter 17).

Serious concern about predetonation voiced by members of T-Division during May and June led Divisions CM and G (with which T-Division had worked closely in late spring) to make significant last-minute changes in design of the implosion assembly.

Last-Minute Assembly Tests

In the spring of 1945, a final program of last-minute tests on the assembly of the two combat weapons was established in G-Division. That part of the program devoted to the implosion weapon was a response to Kistiakowsky's concerns about the status of the Fat Man at a meeting with Bacher in early March 1945. Kistiakowsky worried that while A. Francis Birch was overseeing assembly of the gun weapon in the field, no one was responsible for physical assembly of the implosion bomb. Bacher agreed that an engineer with responsibilities analogous to Birch's was needed for the implosion assembly.

The Cowpuncher Committee considered the matter at the end of the month and made G-Division responsible for designing and carrying out the implosion pit assembly. Early in April, Bacher assigned this engineering task to two members of the critical assembly group, Holloway and Morrison. Called the "G-Engineers," they were responsible for the final "readiness" of the implosion bomb, including the procurement, fabrication, and testing of all components from the inside of the high explosive to the outside of the initiator at the center of the bomb. In line with the Los Alamos emphasis on exacting empirical studies, these two physicists performed a series of mechanical and nuclear tests on both the ^{239}Pu and initiator before inserting them into the pit at Trinity.

Next, Bacher and the others pondered how to organize the pit team. They wondered whether to put a senior staff member in charge of "instruments" to work with Holloway and Morrison. "In any event," Bacher thought it "desirable to get someone in electronics specifically assigned to the electronic instruments for this work." On 18 April, he asked William Higinbotham to contribute a member of G-4 to the pit team. Higinbotham agreed to let Bacher discuss the new assignment with Boyce McDaniel. Raemer Schreiber had earlier asked to transfer to G-1. Both accepted the pit team assignment.[32]

Later, at the Trinity site, where Slotin and Morrison arrived with the plutonium and the initiators on 12 July 1945, the G-Engineers repeated a number of the same kind of critical assembly tests and measurements that they had made earlier at Los Alamos.

In May 1945, in the period that the G-Engineers were being formed, Kistiakowsky put forth his innovative "trap-door" suggestion for the pit, which simplified both shipping and combat operations. Except for an inner charge, two explosive lenses, and the "capsule" containing the active material, the bomb could now be assembled at Los Alamos rather than in the Pacific. During shipping, plaster plugs replaced the high-explosive blocks. In the field, the crews could insert the missing inner charge and HE lens blocks after the capsule containing the active material was inserted. The sphere would then be ready for its armor case. All implosion assemblies made after 4 August 1945 were trap-door models.[33] Kistiakowsky, Holloway, and Morrison worked out the relatively simple changes in the sphere that would allow the active material to be placed inside the cylindrical tuballoy section that formed the capsule.[34]

On the Eve of Trinity

By early July, the Trinity program was in its final stage. On 6 July, the uranium reflector was machined and ready to be assembled in the Trinity device. On 10 July, the best lens castings were selected for the Trinity shot; the next best castings were assigned to the full-scale magnetic shot. Assembly of the Trinity device, without its plutonium center, occurred on 11–12 July at Los Alamos. The plan was to insert the plutonium core at Trinity through the trap door. At 12 a.m. on 13 July, the assembly left Los Alamos by truck for the test site. The plutonium center was inserted on the afternoon of 13 July, and in the evening, the full bomb was hoisted to the top of the tower. Until the full-scale magnetic test on 14 July, all diagnostic work on implosion had been done on small-scale models, and it was by no means certain that the full-scale test would go as planned. On the eve of the historic test, as several hundred researchers engaged in last-minute activities, one question stood in the minds of all involved: Would the implosion gadget work?

17

Critical Assemblies and Nuclear Physics: August 1944 to July 1945

Through mid-1945 the Los Alamos Laboratory continued to pursue a strategy of overlapping approaches in its programs to test and refine critical mass and perform other calculations affecting bomb design and deployment. These activities included work in R-, G-, and F-Divisions with critical and subcritical assemblies, nuclear constant and other measurements by R-Division, and theoretical fine-tuning by T-Division, along with a backburner effort on the Super in F-Division. In addition, new projects were begun after mid-1944, among them a series of tests on critical assembly by G-Division, work on a high-power Water Boiler (nicknamed Hypo) to be used by F-Division for critical mass measurements, and the development of a spectacular assembly suggested by Frisch, which, unlike other experimental assemblies, went critical with prompt neutrons alone.

Besides diverting resources from implosion development and preparations for the Trinity test, these projects were sometimes quite dangerous and often risked the loss of precious fissionable materials. Nonetheless, they were mounted with considerable enthusiasm, and characteristically subjected to empirical testing. As had been the case in early nuclear constant measurements by P-Division and in the metallurgy program,

* This chapter is based on a manuscript by Paul Henriksen and Catherine Westfall. We are grateful to Gordon Baym for his detailed editing of the technical content of the theory section, and to Leslie Redman for a technical review of the remaining sections.

researchers attempted to compensate for the lack of ^{239}Pu by working with ^{235}U, in hopes that the results could be extrapolated to reveal the properties of the heavier element.

Continuation of the Critical Mass Studies

Short of the real explosion, there was no way to determine precisely the extent of supercriticality achieved in the gun- or implosion-assembled bomb, or to measure other chain reaction properties, such as the neutron population growth rate. However, model experiments with subcritical, near-critical, and critical assemblies could check theoretical estimates. Aqueous solutions, hydride, metal spheres, hemispheres, and building blocks such as cubes and rods were used in these experiments. Tampers, too, were often used, although tamped systems were harder to calculate than bare ones. The ultimate goal of this line of research, which began with the Water Boiler program in the first year of the project, was to check predictions by measuring neutron multiplication in ^{235}U and ^{239}Pu metal critical assemblies. Although the laboratory did not have enough fissionable material for such measurements, members of R-, F-, and G-Divisions were determined to obtain as much information as possible with the material at hand.

The first subcritical metal assembly, which was suggested in September 1944 by Alfred Hanson, Robert Serber, and John Williams, used part of the laboratory's meager supply of ^{235}U to make a series of small metal spheres (built up from about twenty smaller pieces). The inverse multiplication of neutrons in these spheres was plotted against the mass of the subcritical assembly and extrapolated to find the critical mass.

R-Division's subsequent program on subcritical metallic ^{235}U spheres was divided into three sections: Robert Wilson (who also led the division) headed a group at the cyclotron, John Manley led a group at the Cockcroft–Walton, and Williams headed a group at the Van de Graaff. In measuring neutron multiplication, they placed a mock fission source in the ^{235}U sphere, sometimes set inside a tamper.[1] They tried to interpret the effects of the tamper on the assembly, count the neutrons emerging from the assembly accurately without affecting the geometry of the sphere and tamper, and count the neutrons over a wide range of energies. These experiments would provide evidence that α, the capture-to-fission ratio, was small.[2] The quantity measured in the untamped sphere experiments was the number of neutrons in excess of the minimal number required for the fission process, ν-1-α. This quantity entered most of the calculations of criticality.

The experiments began with a small-scale model, a 1.5-inch-diameter sphere of uranium enriched to 70 percent ^{235}U and having a source of extra neutrons inside.[3] The mock fission sources had to mimic the fission neutron spectrum, a problem solved to within 1 percent by using a polonium–beryllium–boron–fluorine mixture. The emerging neutrons had to be detected with equal sensitivity over their range of energies.

Hanson's group used the long counter, which was uniformly sensitive to neutrons in the energy range from a few keV to 2 MeV, to measure the neutron output from the 1.5-inch sphere and then from 73 percent ^{235}U spheres, 2, 2.5, 3.5, and 4.5 inches in diameter.[4] They determined the fission cross section and multiplication by measuring the induced activity in foils of ^{235}U placed between the hemispheres. The multiplication for the tamped sphere had to be calculated from the bare sphere multiplication and the "tamper effect," that is, the ratio T of the number of fissions in the tamped sphere to the number in the untamped sphere.[5] Hanson's measurements agreed surprisingly well with the predicted multiplication.[6] These estimates of the tamped critical mass, considered accurate to 5 percent, allowed laboratory staff to design various parts of the bomb and bomb casing months before the active material existed.

Other forms of sphere experiments were carried out. Using the same spheres as Hanson had used, for example, Thoma Snyder and William Woodward used ^{235}U and ^{238}U fission chambers to measure the multiplication of neutrons that the cyclotron directed onto the outside of the spheres.[7] Fermi and L. D. P. King of group F-2 and Herbert Anderson of group F-4 used the same spheres, but different detectors, to measure the critical mass.[8] A large number of similar metal-sphere experiments were conducted. In November, for example, R-Division examined a 2.5-inch untamped sphere with a mock fission source.[9] After some technical debate, they then decided to construct a 3.5-inch and then a 4.5-inch sphere.[10]

In the meantime, while waiting for ^{235}U metal, G-1 members made a moderated critical assembly using $UH_{10}C_4$ cubes. The group planned to compare the critical mass for beryllium oxide, iron, and tungsten carbide tampers and also made a nearly critical assembly using lead.

Recasting the ^{235}U metal for these experiments proved to be a challenge for the CM-Division metallurgy group. They prepared uranium hydride by treating clean uranium metal from the spheres with hydrogen.

Since hydride formation is exothermic, the temperature had to be carefully controlled at about 225° C. The resulting hydride powder was then slurried in a water-free benzene solution of the desired plastic. Monsanto Chemical Company eventually provided the necessary 50 pounds of hydrogenated polystyrene. After it was mixed to a consistency of enamel paint, the paste was dried in thin layers on stainless-steel trays and then ground to make a uniform powder. Next, the powder was pressed into a die, heated to 100° to 150° C, pressed, and then cooled under pressure.

From November 1944 to January 1945, the metallurgy group handled some fifty lots of material, fabricating 1,350 individual $UH_{10}C_4$ cubes from 12 kg of beta-stage ^{235}U. Because roughly ten fabrication operations were required for each lot of material, extreme care was taken to prevent any loss of material. As Joseph Burke, the metallurgist in charge of the operation later recalled, after being used by the physicists in a variety of critical assemblies, the cubes were given to Edward Wichers, a skillful analytical chemist recruited from the Bureau of Standards, "who added to them all wiping tissues, empty bottles and scrapings from dirty dies and converted it all back to UF_4." A 1945 report stated proudly "that the total loss over all operations – hydriding, plastic bonding, pressing, many critical mass experiments and finally recovery – was 12.0 gm, less than 0.01% of the total amount handled."[11]

Group G-1 took the uranium hydride cubes from the metallurgy group and set about determining the critical size of a hydride assembly. The addition of ^{235}U hydride cubes made from R-Division's small ^{235}U sphere to existing uranium hydride cubes allowed the group to achieve a critical assembly of UH_{10} and hydrogenous plastic cubes in a beryllium oxide tamper on 28 November 1944. As more uranium arrived, the pure plastic cubes were replaced, until the first critical assembly, with a 1:10 ratio of uranium to hydrogen, was achieved on 11 December. Other work on the hydride assembly included measuring the neutron spectrum and distribution and the fission rate at various positions within the reacting assembly. G-1 also investigated the capabilities of several different tampers; however, they all required a greater critical mass than the beryllium oxide. Successful as these experiments were, at the next Technical and Scheduling Conference Oppenheimer emphasized that the hydride program, even though interesting, would not be allowed to delay the ^{235}U metal critical mass program, which would provide important information for the plutonium weapon, as well as check calculations for the gun design.

Accordingly, preparations were promptly made to take further measurements after 12 March 1945, when sufficient ^{235}U had been accumulated to make a metal critical mass. CM-Division metallurgists took the first step, processing the material into 4.5-inch hemispheres. G-Division physicists then began the experiment. Although the sphere proved to be slightly less than critical, the physicists placed a layer of rods and blocks of ^{235}U between the hemispheres to achieve a critical mass.

In this way, smaller pieces of ^{235}U allowed the program to be more flexible. When the results from successively larger assemblies of "pseudospheres" (i.e., configurations of pieces of material arranged to approximate a spherical shape) were compared with extrapolations made from the sphere measurements, researchers gained information not only on the approach to criticality for the gun weapon, but also on important details for the ^{239}Pu program. Unlike ^{235}U, which was separated into two subcritical parts in the gun weapon, ^{239}Pu had to be assembled in one subcritical piece. Without accurate knowledge of the critical mass, the laboratory risked reaching criticality while the bomb was being assembled. Because the laboratory had accumulated insufficient ^{239}Pu to assemble a metal critical mass, or even to create metal spheres for testing that were close to criticality, it had to make do for the time being with information obtained from a ^{235}U metal critical assembly.

The multiplication measurements on the 4.5-inch sphere were completed on 1 April, and on 4 April the group came to within 1 percent of achieving a critical assembly with a combination of ^{235}U metal hemispheres and cubes. They did not go further with that arrangement, for Holloway and Bacher recommended that the spheres be refabricated. The metallurgy group then refabricated all the material into hemispheres. One week later, a milestone was reached: the assemblies group created the laboratory's first metal critical assembly with the new ^{235}U metal hemispheres. Although calculation of the critical mass of the sphere proved nontrivial and required many corrections, the predictions differed from the experimental values by only 3 percent. The measurements of ν-1-α, σ_f, and critical mass were completed in April; some quantities remained uncertain.

The plutonium critical mass program began in January 1945, limited at every stage by the pressure of time and the lack of ^{239}Pu. The first sphere was 0.90 inch in diameter. R-Division made the first measurements by bombarding it with neutrons and measuring the multiplication, as they had for the ^{235}U spheres.[12] Because there was not enough ^{239}Pu to make the thin foils to be activated by neutrons, a ^{235}U catcher had to be substituted, and thus the accuracy of the multiplication experiments was reduced.

Originally, G-Division had planned to construct a 1.5-inch and then a 2.5-inch plutonium sphere as soon as the material became available. They then planned to make cubes and proceed to the critical mass in the small steps that had proved effective for the ^{235}U. The metallurgy group proposed making it into octants, which could be increased in size gradually until the critical mass was reached. Owing to time constraints, however, the 1.5- and 2-inch-diameter spheres were bypassed in favor of proceeding directly to a 2.75-inch sphere. The metallurgy group admitted that the die construction for the octants was more difficult than expected, and the cubes could not be fabricated, because the smaller critical mass of the ^{239}Pu would require smaller cubes for pseudospheres, and these could not be prepared in time for the experiments. Fermi suggested making just a few hemispheres and no intermediate sizes, but Frisch, Morrison, and Holloway favored experimentation with more hemispheres. They finally compromised and made a few intermediate hemispheres. At their 15 March meeting, Frisch, Morrison, and Holloway also decided to measure the critical mass of a ^{239}Pu solution in water with a beryllium tamper. This solution was created in April and became the first plutonium critical assembly.[13] G-Division never had the opportunity to create a critical ^{239}Pu metal assembly because of time pressures and the scarcity of material, but group members made a limited series of experiments with subcritical assemblies, which allowed them to estimate the critical mass for the implosion weapon.

The critical mass studies that successfully checked bomb design calculations were sometimes risky and inconvenient. One scare occurred in January 1945, when the experiments with tamped UH_{10} cubes indicated that the measured critical masses were considerably higher than had been predicted by T-Division: 2.1 kg had been the predicted critical mass of UH_{10} cubes in a BeO tamper, whereas G-1's experiments showed that 3.2 kg were needed. A careful reinvestigation of the arrangement suggested that at least part of the problem lay in the experimental apparatus. G-1 members then tried everything they could think of to eliminate extraneous matters from the tests, such as external counters, air gaps, and some of the control and safety devices. However, the problem persisted into May, postponing the fabrication of gun parts until the end of the month, when the cubes were recalled to allow staff to prepare for gun component fabrication (Chapter 13).

It was perhaps better for the health and safety of the hydride critical

Fig. 17.1. Beryllium block configuration used to determine the critical masses of uranium and plutonium. LA Photo, 1986.

assemblies group that the experiments ended when the ^{235}U was converted to metal. The hydride group studied some assemblies of hydride cubes so close to critical that just the added tamping of someone sitting on top made the assembly supercritical. (They would hop off just as the assembly went critical.) In other experiments, the cubes, hemispheres, and tampers were manipulated by hand, a practice that continued (in defiance of the rules G-Division had established regarding critical assemblies) until the accidents after the war that took the lives of Harry Daghlian and Louis Slotin. Daghlian was working alone at night on 21 August 1945 with a plutonium sphere and heavy tamper blocks. The final block slipped out of his grasp and caused a runaway chain reaction. On 21 May 1946, Slotin was showing Alvin Graves how to make criticality measurements, using the same plutonium sphere Daghlian had used. Slotin was gradually decreasing the separation between hemispherical beryllium reflectors with a screwdriver when the tool slipped and the reflector closed completely. Again, a runaway chain reaction ensued. In each instance, thermal expansion brought the excursion to a halt immediately after supercriticality was attained, but the two men received lethal doses of neutrons and gamma rays.[14]

Continuation of the Experimental Physics Program

Sphere multiplication was only one of the topics investigated by R-Division in the last year of the project. Until February 1945, when the division focused almost exclusively on assisting with Trinity measurements, R-Division conducted a number of experiments designed to improve nuclear constant data and provide checks on bomb design.[15]

Measurements of the ratio between the neutron number induced by fast fission and that induced by slow fission had been made by the Williams, Wilson, and Manley groups. All three found that the neutron number changed little with incident neutron energy. By late 1944, Williams's group had measured a ratio of 0.98 for ^{235}U and 1.01 for ^{239}Pu.[16] Williams's and Wilson's groups remeasured important cross sections, including those for ^{239}Pu, which had not been extensively studied. Williams's group found that new measurements of ^{235}U and ^{239}Pu fission cross sections confirmed earlier results.[17] Wilson's group measured both absorption and fission cross sections as a function of energy. The greatest difficulty in directly measuring fission cross sections for ^{239}Pu was finding a plutonium fission chamber with the sensitivity necessary to discriminate fission pulses from the large background of α particles.[18] Using small samples of ^{239}Pu (around 30 mg) and a neutron energy range from 0.01 to 200 eV, they confirmed the existence of a resonance at 0.3 eV, and reported two poorly resolved resonances at 12 and 100 eV. The fission cross section at thermal energies was measured as 708 barns, a result close to that expected from relative measurements and also impressively close to the modern value of 741 barns. Measurements of the absorption cross section, which were made with both metal and PuO_2 absorbers, confirmed the existence of resonances at 0.3 and 12 eV.[19]

Wilson's group made an interesting observation during its measurements of the fissionable isotope ^{233}U: "the number of neutrons per fission" of the isotope was "a few percent higher" than that of ^{235}U. In addition, beyond 150 keV, the ^{233}U fission cross was 2.1 times that of ^{235}U.[20]

In 1945, R-Division mounted a number of experiments aimed at simulating selected properties of the bomb. Two notable examples were the gun-model experiment and its measurement of the fast multiplication rate as a function of the mass of the fissionable material. Using a mockup of the gun and the pulsed beam from the cyclotron to bombard the target with slow neutrons (to simulate the scattering and absorbing properties of active material and tamper), Wilson's group measured how much active material could be safely placed in the target of the weapon and then determined the assembly's multiplication rate as the projectile moved toward the target. In March, they measured the half-life for slow neutron decay in the various configurations of the solid tamper model. By April, they had determined how much active material could be put in the target and projectile and when criticality would be reached as the target and projectile came together.[21]

They began measuring the neutron multiplication rate as a function of the mass of fissionable material in May 1945. They used two methods here, one of which drew on a technique devised by Bruno Rossi. In Rossi's technique, the first neutron triggered counting of subsequent neutrons as a function of time. The other method employed the fast modulation equipment at the cyclotron. The cyclotron neutron pulse started the chain reaction, and the decay of the burst was measured as a function of time. These experiments determined the change in decay time as small changes occurred in the degree of criticality of the system. By June, the group completed numerous measurements with both methods, using ^{235}U and ^{239}Pu tamped spheres, and obtained the neutron multiplication rate of the critical material as a function of the mass.[22]

Continuation of the Theory Program

In the last year of the project, theorists conducted the first detailed investigations of damage. When Hirschfelder's group was formed in November 1944, Bethe outlined a number of topics requiring further study, including the formation and expansion of shock waves, the motion and disintegration of the ball of fire, and special hydrodynamical problems associated with damage, such as the effect of rain and fog. By January

1945, Hirschfelder and British physicist William J. Penney had gathered a great deal of data from Britain on the structural damage caused by German high-explosive bombs. These data proved extremely useful in the group's further calculations, and by the next month it had developed a hypothetical "history" of the explosion of a nuclear weapon with the explosive power of 10,000 tons of TNT. As Bethe summarized, by this time they had "obtained a fairly complete picture of the development of the blast wave in air and the accompanying radiation phenomena."

The evolving damage estimates helped in planning for Trinity measurements (and in planning for the protection of personnel). They also gave a clearer picture of the damage that would be caused by the two bombs to be used in combat. As Bethe later summarized, the theorists found that nuclear explosions differ from high-explosive blasts of the same energy in the following ways: (1) pressure is higher at small distances; (2) pressure is lower at large distances; and (3) radiation is emitted, and thus heat effects are greater. In a later study by Penney, Serber, and Ens. G. T. Reynolds of the blast effects at Nagasaki and Hiroshima, Penney concluded, "The heights of burst were correctly chosen with regard to the type of destruction it was desired to cause," and they also kept fallout to a minimum.

The height of burst was set at 1,850 feet, on the basis of the Trinity experiments. Los Alamos researchers knew that radioactive contamination is severe only when the fireball touches the ground, because dirt is sucked up into the blast column and is vaporized. Fission products from the detonation then condense on the dirt and fall to the ground near the blast, creating hazardous contamination. When the burst takes place at a higher altitude, the fireball does not pick up dirt, and the fission products are carried higher into the atmosphere and dispersed over a wider geographical area, with the result that fallout is rendered harmless.[23]

Theorists were eager to check and improve their calculations from late 1944 to mid-1945, even though such work was unlikely to affect the bomb designs, which were both frozen in February 1945. For example, the data from the sphere multiplication experiments prompted refinements in the calculations of neutron diffusion. Noteworthy was a clever approximate method developed by Serber in December 1944 that separated the neutron diffusion problem into two parts, core and tamper. The two problems were related by the "conservation law" at the boundary between the two subsystems, which stated that the rate at which neutrons are generated in the core had to be equal to the rate at which they flowed into the tamper. With this scheme, as Bethe has explained, "it is possible to determine critical masses by the use of two graphs which can be slid over each other by amounts indicating the ratio of mean free path in tamper and core." The method also simplified calculations of the rate of multiplication.[24]

T-Division had less success in comparing estimates of the hydride critical mass with R-Division experiments in January and February 1945. Hugh Richards's improved fission spectrum measurements agreed more closely with theory than did earlier measurements. But despite the efforts of Feynman and his group, the discrepancies between theory and experiment for the fission spectrum remained large enough to suggest that the theoretical methods were inadequate. However, the "perfect agreement" in the fall of 1944 between neutron multiplication measurements of the first ^{235}U sphere experiments with the calculations of Serber's group indicated that some aspects of the theory did work well. Ultimately, they concluded that the theory was "fairly accurate" and that most of the discrepancy arose from nuclear constant inaccuracies. Nevertheless, "a very confused situation" surfaced in April 1945 when the theorists experienced "considerable difficulty" interpreting time-scale measurements made by groups R-1 and G-1. Bethe speculated that the problems could have been in the fission spectrum data employed in the calculations, the assumed absorption in the tamper, or the transport cross section in the tamper. The point became moot three months later when the Trinity data became available.[25]

From summer 1944 to summer 1945, T-Division continued to improve its understanding of efficiency calculations, including those for the Trinity gadget. On the eve of Trinity, Bethe reported the "best estimate" of efficiency for the test. However, inferring the actual efficiency of the Trinity gadget proved to be more complex than anticipated; in particular, it was important to distinguish the final blast efficiency from the efficiency of the initial nuclear reactions.

The Super

Although the Governing Board had in February 1944 judged the Super would not be developed in time for use during the war, research on the weapon continued as a low-priority effort in the second year of Project Y. Most of the experimental studies were performed in group F-3 by Egon Bretscher, Anthony P. French, and Michael J. Poole. The work began in October 1944 with the design and building of a low-voltage arc accelerator for measuring cross sections as a function of energy, a project that occupied group F-3 for the rest of the war. In April 1945, as installation of the system neared completion, the group planned to measure the variation with deuteron energy of the cross section of the reaction-yielding protons. They also planned to obtain the branching ratio of the two competing D–D reactions,

$$D + D = H + T,$$

and

$$D + D = n + {}^{3}He,$$

by observing the relative numbers of H (protons) and ^{3}He (α particles) from these reactions. In May, they determined the variation of proton yield with change in the deuteron energy.

They also calculated the cross section for an energy of 10 to 40 kV. But the results were inconclusive. By July, they were ready to measure the variation of the number of α particles with bombarding energy from the deuterium–tritium reaction,

$$D + T = n + {}^{4}He.$$

The first experiments determined the number of α particles emitted from the T + D reaction as a function of particle energy of the accelerated tritons. In a separate run, with deuterium as the particles bombarding the deuterium target, they determined the number of protons under geometrically identical conditions. These data, together with the energy loss per centimeter in the target for each particle, yielded the ratio of the cross sections for the two reactions.

Meanwhile, theoretical work in group F-1 finally laid to rest the question of whether the hydrogen bomb might ignite the atmosphere.[26] Up to the end of the war, the main task of Group F-1 was to conduct a series of relatively primitive calculations connected with various Super concepts. Although no significant breakthroughs on the Super came during World War II, Teller remained enthusiastic about its future importance.

The Dragon Experiment

Of the three new programs introduced in the second year of Project Y, The "Dragon," or "drop," experiment was the most spectacular. It emerged from a suggestion by Otto Frisch in memos to Oppenheimer on 17 and 24 October 1944 proposing the dropping of a ^{235}U hydride slug through a just subcritical assembly, making it supercritical for prompt neutrons for an extremely short time, like a nuclear weapon.[27] Since the assembly was supercritical for such a short time, the heat and radioactivity would not build up enough to prevent workers from handling the material. Given the daring nature of the experiment, Frisch was surprised to learn that the Coordinating Council judged the experiment worth pursuing. At that meeting, Feynman compared it to "tickling the dragon's tail."[28]

Frisch's group began to design the Dragon apparatus later that month. Charles Baker worked on the optical system to measure the velocity of the falling slug, and Joseph Rotblat and James Hughes worked on the detection equipment. Two weeks later, the electronics for measuring the slug velocity was on hand and four different detectors of strong neutron pulses had been built.

As the year drew to a close, the experiment gained priority and was assigned more manpower, because the hydride would soon have to be returned to the metallurgists. The equipment, some of it simplified from the original plans, was ready by mid-December. During the next few weeks, the uranium hydride was prepared and positioned, and on 18 January 1945 the apparatus succeeded in producing the world's first chain reaction using prompt neutrons. By dropping the slug several times in quick succession to increase the system's neutron background, the scientists were able to increase the neutron output of the apparatus, because in strengthening the source they raised the overall neutron population for the same multiplication factor. By 21 January, this phase of the experiment was complete.

From 28 January through 1 February the assembly was used again, with more active material and with a cadmium sheet between tamper and core to reduce the return of thermal neutrons from the tamper. Most of the information about the assembly was obtained during this period, which ended when two-thirds of the UH_{10} was returned to CM-Division for conversion to metal. Together with additional new material from Oak Ridge, the metal was used to make 3.5-inch and 4.5-inch spheres for criticality sphere experiments later in 1945. For a short period, the division used a third assembly, moderated to UH_{30} by polyethylene blocks, to measure delayed neutrons and gamma rays.[29]

The size of the burst was controlled by the number of neutrons present when multiplication occurred; three drops only seconds apart gave the largest yield attempted. In a final burst, the heat blistered and swelled the plastic to such an extent that criticality could not be obtained again. In one drop, Frisch's group measured an energy production of

20 MW/sec when the temperature rose 6° C in 0.003 sec. The burst of neutrons and dramatic temperature rise gave "very direct evidence of a nuclear explosion nipped in the bud" and a much-needed confidence boost to the scientists, for the results of the assembly experiments agreed with their theoretical predictions.

The High-Power Water Boiler

Another new project in the second year of Project Y was the high-power Water Boiler, which got its start after the low-power Water Boiler went critical in June 1944. At this point, Fermi and Bacher advocated the construction of a higher-power Water Boiler to operate at 1 kW. They argued that the reactor could be built quickly and easily with parts from the original Water Boiler. In addition, it would provide a new, more powerful neutron source for neutron multiplication experiments. They also proposed that the reactor be used to model weapon conditions, a program that would be particularly important if large-scale tests later revealed unexpected problems. Groves initially resisted but soon deferred to Oppenheimer, who had made up his mind that Hypo was worth building.

With Groves's approval, the plans for Hypo began. The original Water Boiler was dismantled and converted into a high-quality neutron source during October, November, and December 1944.[30] Even though Hypo was deemed a worthwhile project, it had relatively low priority and consequently had difficulty getting adequate personnel from the laboratory's small labor pool. Obtaining materials was also difficult, but fortunately for the timely operation of Hypo, both active and inactive material was on hand in barely sufficient quantities.

G-Division also took on a new program of safety studies concerned with ^{235}U in both metal and hydride and with plutonium metal. Because water was such a good moderator, and because the bombing run in Japan would be conducted mainly over water, the division needed to learn what would happen if the bomb was inadvertently dropped in the water, or, for that matter, if the airplane carrying the weapon crashed on land. Holloway's straightforward approach in testing the effect of water on the criticality was simply to immerse the gun parts in a large tank of water and take measurements. Although Oppenheimer considered the procedure inelegant, Bacher and Holloway convinced him that it was reliable. By March 1945, G-1 had determined that hemispheres immersed in water showed little additional multiplication, but, if the material were divided into pieces, the moderation effect was greater and could, for sufficiently fine division, produce a supercritical state. In some of their tests, the group also placed the ^{235}U in a mockup of the gun model. There was no time for mockup tests with plutonium in water, although the group made one simple experiment. They immersed 5.2 kg of ^{239}Pu metal with a surface area of 450 cm^2 in water and found no measurable multiplication.

The wide range of relevant but unessential projects conducted in the final year of Project Y produced no startling new information. Nonetheless, these projects were useful. The information gathered provided extra insurance against last-minute problems, but perhaps even more important, the absense of negative findings suggested that the laboratory was on the right course for creating a workable atomic weapon.

18

The Test at Trinity: January 1944 to July 1945

Just before dawn on 16 July 1945, the area selected for the Trinity test – the desolate Jornada del Muerto region of New Mexico – no longer swarmed with activity, as it had in the past several weeks. The thunderstorms that had worried Groves and Oppenheimer through the night had stopped. The scientists, who had worked almost nonstop in preparing for the first atomic bomb test, waited tensely for the test to begin.

Arranging their apparatuses around the gadget – ionization chambers, seismographs, motion picture cameras, and other devices – they prepared to record physical aspects of the explosion: light, heat, neutrons, gamma rays, and other features. The data would indicate what to expect of combat atomic bombs and how to achieve the most destruction. But even the most careful preparations could not guarantee a successful test, because the weather had to be just right to prevent heavy fallout from reaching populated areas. Completing the test on schedule became of paramount importance when President Harry S. Truman announced that he would meet with Churchill and Stalin at Potsdam on 16 July 1945.

* This chapter is based on a draft by Paul W. Henriksen. We are grateful to Kenneth Bainbridge for his detailed editing of an earlier version of this chapter.

The Experimental Program

Because only a limited number of measurements could be taken at Trinity, the ones to be selected became a critical topic of discussion. A panel consisting of Fussell, Moon, Bernard Waldman, and Victor Weisskopf was assembled to evaluate proposals. Data were needed on both the performance and the effects of the weapon. Especially important were shocks, both the air blast, which would determine the height of their combat burst, and ground shock. Other crucial effects were visual, thermal, and nuclear radiation. Devices for measuring blast and ground shock could be readily derived from known systems. Optical observations were adequate for general purposes, such as recording what the explosion looked like, as well as special purposes, such as following the growth of the fireball and the intensity and incendiary effects of the thermal and visible regions of the spectrum. The nuclear radiation of the expected intensity, coming first from the reacting device and shortly afterward from the fission products, would be a novel experience. However, accurate extrapolation could be made from laboratory situations.

Diagnostic instrumentation was less certain. Many questions would be of interest: How close together did the detonators fire? What was the time between detonator firing and the first appearance of γ rays from the nuclear reaction (HE transit time)? What was the speed of the nuclear reaction (α)? And what was the energy release (yield)? Yield information could be derived from the growth rate of the fireball or from efficiency. Herbert Anderson suggested that the percentage of material fissioned (efficiency) might be measured by radiochemistry, by comparing fission products with residual plutonium in samples taken from the crater.

Diagnostic experiments were divided into three broad categories established at a conference in Oppenheimer's office on 23 December 1944: "essential, desirable, and unnecessary." Essential experiments included those designed to determine the efficiency of the reaction, the pressure of the blast wave, and the time spread in the firing of the detonators. "Desirable" ones – those judged to be important, but not of such high priority as to interfere with the work on the detonation system, either by shop personnel or by the scientists – included photographic and spectrographic analysis of the fireball. Such records would have been of inestimable value if the explosion "fizzled." As it turned out, they provided a visual record for posterity and an objective measurement of the size and duration of the fireball. A second group of desirable experiments

Fig. 18.1. Base camp, completed in December 1944, was the main service area for the Trinity Test in July 1945. LA Photo, TR49.

would measure the earth's motion during the explosion. These were to provide evidence should lawsuits be brought against the laboratory for blast damage.[1] All other experiments were deemed unnecessary.

Test Selection Process

The Trinity test was intended to resemble the explosion over enemy territory, and nearly all monitoring devices and experiments were designed to measure aspects of the bomb without affecting its operation. According to Moon, who helped design some of the monitoring experiments, the Trinity test was seen as a calibration for the entire series of combat weapons. Although the scientists by and large adhered to this stricture, some differences were allowed. For example, the gadget had an "informer" switch at each detonator. The goal of similarity was superseded in this casè by the need to measure the simultaneity of the detonators, one of the essential experiments.[2]

The scientists were prolific at devising tests. Individual experiments had to be carefully scutinized by the review committee because of limited available shop time. Weisskopf, on the evaluation team, kept Bainbridge up to date on predictions of the explosive power, radioactivity, and other nuclear products. In selecting experiments, the review committee drew on advice from Fermi in 1944 on nuclear physics experiments, Penney in 1945 on blast problems, and Joseph O. Hirschfelder in 1945 on fallout expectations.[3]

Simple experiments that did not occupy a great deal of shop time took precedence over other experiments; monetary cost was less of a concern. Although repetition was avoided in the more complicated experiments, some overlap was allowed for the simpler ones. Bainbridge's policy was that the test should not be delayed by any experiment, no matter how important it seemed at the time.[4]

Desirable Measurements

Seismographs and other devices measured earth motion during the explosion. Fussell's committee doubted the need for such tests, and seismology expert L. Don Leet of Harvard, who was asked to make seismograph measurements during the test, agreed. It did not seem possible that the explosion could shake the ground enough for earth motion to show up plainly on a seismograph record taken far from the explosion. There was no question, however, that the measurements would be made, for they were relatively simple and would not use any shop time or manpower. Furthermore, Groves was anxious to have accurate earth motion records because he feared the explosion might be noticed in neighboring towns and might prompt lawsuits when the Manhattan Project became public knowledge. Subsequently, ground shock measurements were made both near (800, 1,500, and 10,000 yards) and far (50 to 100 miles) from the gadget.

A variety of instruments measured the earth motion. At Penney's suggestion, the Corps of Engineers erected stakes at accurately measured distances from Ground Zero. After the 100-ton test (see below) and after Trinity, the positions of the stakes could be remeasured so as to determine the permanent displacement of the ground near the explosion. The yield could be estimated from these data. Another measurement used the geophone, a device for transforming ground vibrations into electric signals.[5] The most extensive measurements were made with seismographs. Tests at the 100-ton explosion confirmed Leet's conclusion that any explosion with less than 50,000 tons of TNT would not cause damage from ground vibrations. In fact, the test suggested that the radius for damage would probably be on the order of 1,000 yards.[6] Seismograph measurements were made on site at N 9000, Tularosa, Carrizozo, San Antonio, and at the base camp.[7]

The primary purpose of the photography effort was to film the explosion from the moment it began until the radioactive cloud was out of visual range. A good photographic record would be useful for both spectrographic and yield analysis if the device worked properly. It would be even more important if the detonation was imperfect, because it might indicate the nature of the asymmetry. However, such photography presented a challenge. Fastax cameras, exposing up to 10,000 frames a second, had to be started and stopped with great accuracy if they were to take photographs continuously during, but not before or after, the explosion. The first few crucial microseconds were especially difficult to pinpoint. Furthermore, the wavelength and intensity of the light could not be predicted with certainty, and the amount of light would be constantly changing.

Trinity photography was the responsibility of Julian Mack and Berlyn Brixner.[8] Of the more than fifty cameras used by the photography group, most took motion pictures in black and white or in color. However, different stages of the explosion required different film speeds, lenses, and exposures, and no one knew the amount or kind of light that would be emitted during the explosion. Fastax cameras taking 10,000 frames per second were used to record minute details of the beginning of the explosion. Spectrographic cameras monitored light wavelengths emitted by the ball of fire. Pinhole cameras recorded gamma rays.[9] The photography group did not rely solely on their professional photographers. They distributed an ample supply of hand-held movie cameras to the scientists and military personnel observing the test. Some of the best color photos of the explosion were taken in this manner by Jack Aeby.

Trinity photography was further complicated by the radiation, heat, and shock wave, which would damage cameras and film placed near Ground Zero. Fastax cameras placed only 800 yards from the blast were protected with a steel and lead glass bunker designed by Brixner. They were mounted on a sled that could be pulled out of the contaminated area by a chain attached to the lead-lined tank sent in to obtain soil samples. Fastax cameras, mounted in the bunker and activated by an electrical signal from the timing circuits, would exhaust their film supplies in several hundredths of a second.[10]

Other kinds of cameras also recorded the explosion. To obtain the spectrum over the first 1/100th of a second, a rotating drum spectrograph was placed at a 10,000-yard station. A guided, slow-recording, low-power spectrograph was set up to follow the ball of fire.[11] By measuring the intensity and spectral composition of the light from the blast, scientists could determine the temperature of the ball of fire. Its energy could be calculated from the dimensions read from the film. Pinhole cameras photographed the gamma rays released in the explosion, in the capture of prompt neutrons in the high explosive or from the decaying fission fragments. The resulting photographs revealed the expansion of the high explosive.

Essential Measurements

One of the essential factors to measure was the simultaneity of the detonators. The degree necessary in an efficient implosion was unknown.[12] Oscilloscopes received signals from switches connected to the detonators; automatic cameras photographed the traces. It was also essential to measure the time interval between the detonation of the high explosive and the beginning of the chain reaction. This interval would reveal whether the nuclear reaction began prematurely or was started by the initiator.[13]

The time for the neutron population to increase by the factor e (the base of the natural logarithms), was designated α. This indicated how fast the fission neutrons were multiplying.[14] The Bethe–Feynman theory for the efficiency maintained that α should be nearly constant until the thermal expansion enlarged the fissionable core enough to make the system no longer supercritical. Robert Wilson and his group were given the task of measuring α. He treated α as a variable and measured it at different times during the explosion. His group measured the γ-ray flux instead of the neutrons themselves, because the γ-ray flux would be more intense than the neutron pulse and the γ rays would not be delayed by the tamper and HE in leaving the bomb. Electron multiplier tubes measured the γ rays given off in the explosion; the tubes had the capacity to reset themselves so quickly that pulses arriving at slightly different times could be distinguished. The Wilson group performed two experiments. In one, they fed the signal from one set of electron multiplier tubes into an oscilloscope whose signal was inversely proportional to α. The signal was stored in a charge collection box built into the face of the oscilloscope. The second experiment used the detector from the first experiment and a similar one placed at a different distance from the bomb. An electronic timer measured the time difference between pulses from the detectors and generated a signal proportional to the time difference. From this signal the group determined α.[15]

Bruno Rossi proposed an alternative means of measuring α. Although Rossi was not officially part of Wilson's group, Wilson welcomed his help and provided him with precious space for his Trinity experiment. Rossi's method also measured γ rays, but he used ionization chambers to record their flux and feed a voltage directly to an oscilloscope.[16] Here Rossi ran into difficulties because the available ionization chambers could not record individual pulses quickly enough to distinguish between closely spaced bursts. Furthermore, the pulses would be arriving so fast that it would be almost impossible to start the oscilloscope at just the right moment. Rossi ingeniously circumvented both problems. He avoided the problem of starting the linear sweep at the right moment by using a sine wave in place of the sweep. He then calculated α from the wavelike trace. However, Rossi had a difficult time convincing his fellow scientists, especially Hans Staub, that he could make the method work.

Rossi and Wilson solved one other problem. It was important for the ionization chamber and electron multiplier tubes to be close to the gadget, but the oscilloscopes had to be far from the blast to avoid destruction. They designed a line to transit the pulse without diminishing it. Drawing on advice from E. M. Purcell, transmission line expert from the MIT Rad Lab, Rossi used a hollow copper tube 3 inches in diameter with internal cylinders of copper of decreasing radius tapering off from the end near the blast. The central part of the line was a thin copper wire. The device allowed Rossi to increase the pulse at the oscilloscope end without using an amplifier.[17] The closely spaced metal cylinders also acted as the electrodes on the chamber.

The energy released by the bomb was measured in several ways, which turned out to be a fortunate circumstance because some of the experiments did not work properly. Moon suggested determining the number of fissions in the explosion by measuring the number and intensity of the γ rays emitted. He also proposed measuring the time at which the nuclear reaction occurred, estimating its duration by the same method. He was confident that the prompt and delayed γ rays could be separated. Ionization from the prompt γ rays would be measured by ionization chambers. Neutron fluence (time-integrated flux) measurements would be made by activating gold foils exposed to the blast. The ionization from the delayed γ rays could be measured by suitable devices within 10 or 20 miles of the gadget. Knowing the number and energy of the γ rays, one could derive the number of fissions and calculate the efficiency and yield of the bomb.

Emilio Segrè and Group R-4 had the task of detecting γ rays.[18] Segrè's equipment consisted of an ionization chamber, a multiple amplifier, and a Heiland recorder designed to measure the ionization produced by the γ rays from 0.01 sec after the explosion until the shock wave reached the equipment and destroyed it. The delicate part of this operation was deciding where to place the chambers. If too close to Ground Zero, they would be destroyed before they could transmit any data. If too distant, the γ rays would be absorbed before they reached the apparatus.[19] One chamber was placed on the ground 550 m from Ground Zero. Another was placed at the same distance, but attached to a barrage balloon floating near the level of the bomb. The research group hoped the airborne chamber would not be affected too soon by earth thrown into the air by the explosion.[20]

Besides spewing shock waves, light, heat, and γ rays, the Trinity explosion also released neutrons in abundance. Their energies and distribution provided another means of calculating the yield, but they were more difficult to measure accurately since they were more likely to be degraded in energy or absorbed before reaching the measuring devices.[21] Moon, Waldman, Fussell, and Weisskopf thought they could estimate the number of neutrons absorbed by the tamper to within a factor of two, but even this poor accuracy was highly uncertain.

Moon originally planned to pinpoint the beginning of the nuclear reaction by measuring the neutron flux, but he soon realized that the velocities of the neutrons would vary a great deal and thus make it difficult to deduce the time variation of the reaction from the neutron flux.[22] Gamma rays would be better indicators of the time variation because they traveled uniformly at the speed of light. They were not perfect either, for they would also be released from reactions other than the main chain reaction. Moon began to reconsider neutrons and in February 1945 decided they should be measured. Examining several options, he settled on the gold foil experiment, which survived the selection process because of its simplicity. Gold foils were placed in protective tubes and scattered around the bomb. Slow neutrons from the bomb induced short-term radioactivity in the gold.[23] The gold foils, which were scattered from 300 to 1,000 m around Ground Zero, measured the number of

Fig. 18.2. Lead-lined tank with trap door underneath, used to recover soil samples from the Trinity site. LA Photo, J10F129-12.

neutrons per square centimeter.[24] Two other experiments measured the neutron activity. One gave a time-differentiated neutron record using cellophane films passing rapidly between two ^{235}U plates to catch fission fragments.[25] The other used a sulphur detector with a threshold of 3 MeV to measure high-energy neutrons.[26]

The problem of estimating the efficiency of the explosion by measuring fission fragments left in the soil was considered at a 20 July 1944 meeting. Bainbridge realized that a direct examination of the soil might provide the most accurate measurement for the yield of the bomb, an opinion supported by Moon in February 1945.[27] A fission fragment experiment was approved and assigned to Columbia University graduate student Herbert Anderson, who had come to Los Alamos in November 1944. Anderson proposed a method of separating the plutonium and fission products from the soil. The analysis was accomplished with the aid of Nathan Sugarman from the Chicago Met Lab.[28]

The most important Trinity measurements were concerned with the bomb's destructive power. Since the principal goal was to achieve maximum blast wave energy from the least material, the laboratory placed the greatest emphasis on measuring the energy in the blast wave. It used a variety of techniques for this purpose. This energy, which was thought to be a reliable source of information on the total energy release, would have impact on the bomb's use in combat. However, it was not clear how much of the total energy would go into the blast wave and how much into heat and light.

One device for measuring blast wave pressure had been in existence for some time at ordnance proving grounds: a pressure gauge based on a piezoelectric quartz crystal, the electrical characteristics of which changed in response to pressure.[29] Waldman and Alvarez developed another device for use in the combat drops, and also at Trinity. They suggested that the only quantity that could be measured accurately from 20 miles away during combat use of the bomb was the change in pressure generated by the blast wave. Neutrons and γ rays would be too attenuated to be useful. In this method, a pair of small beryllium–copper diaphragm microphones recorded the pressure peak following rarefaction from the explosion.[30]

One of the more sophisticated methods for measuring the blast wave energy was the excess velocity method, which consisted of making a precise measurement of the velocity of sound at the site of the explosion and then comparing it with the velocity of the blast wave. The difference between the two velocities could be used to calculate the energy of the blast. Barschall, G. Martin, and others conducted the experiment.

In June 1945, Manley suggested using spring-loaded piston gauges to measure the peak pressure from the blast, for according to theory the pistons would all move the same distance when hit by the same blast wave, and the final position of the piston would indicate the pressure. These gauges would also be impervious to electrical disturbances and could act as backups to the electrical methods. Other mechanical gauges consisted of aluminum foil diaphragms and water-filled pistons that would squirt their water when hit by a blast wave. Like the cameras that were set to photograph various intensities of light, they were set to record a variety of pressures ranging from 2.5 to 150 psi. In the spring of 1944, Parsons suggested a number of gauge methods, familiar to him from his tenure at the Dahlgren Naval Proving Ground.[31]

By placing a measuring device as close as possible to Ground Zero and recovering it after the blast, William Marley and Frederick Reines measured a different aspect of the blast pressure: the maximum pressure exerted by the bomb. G. I. Taylor suggested copper ball and cylinder gauges to measure the pressure generated in a gun barrel during firing.

These gauges consisted of hollow copper spheres and cylinders placed under pistons in rugged iron cylinders. The pressure from the blast was directly proportional to the compression of the piston and ball. Reines buried the gauges in the ground within 300 feet of Ground Zero, having the top of the iron housing flush with the top of the ground.

Aluminum diaphragm box gauges were among the simplest devices used at Trinity to measure blast pressure, and among the first suggested. The gauges operated on the principle that the blast pressure would break a diaphragm covering a large opening more easily than one over a small opening. A series of boxes with holes of different sizes covered by the same diaphragms, calibrated, and placed at various distances from Ground Zero, would provide an inexpensive and accurate measure of the pressure. Similar gauges had been used at the Aberdeen Proving Ground by Robert Sachs.[32] The greatest difficulty with these gauges was finding uniform diaphragms to cover the holes. Metal foil proved to be the most uniform thin covering.[33]

The third mechanical gauge was a water-filled tube connected to a piston. The water flowed out of the tube when the blast wave created a pressure difference between the ends of the tube. A stylus connected to the piston made scratches on a smoked glass disc, thus recording the flow rate. Knowing the flow rate, scientists could calculate the peak pressure, impulse, and duration of the positive phase of the blast. Theodore Jorgensen prepared a number of these gauges and placed twelve of them 350 to 2,000 yards from Ground Zero.[34]

The 100-Ton Test

The first chance to test many of these experiments under explosion conditions came on 7 May 1945 in the rehearsal for Trinity known as the 100-ton test. In the summer of 1944, Bainbridge decided to stage such a rehearsal by setting off a large pile of conventional explosives with a small amount of radioactive material added, so that its dispersion could be measured. Because explosions of more than a few tons of TNT have different characteristics than smaller ones, the size of the explosive was chosen to produce a blast from 100 tons of TNT, much larger than any previously measured explosion. The "100-ton test" would allow the group to check instruments and facilities, calibrate gauges, and uncover weaknesses in their plans and organization. Oppenheimer at first opposed the idea, but Bainbridge convinced him of the need for such a test.[35]

Fig. 18.3. The "100-ton test" – tower with 100 tons of explosive detonated ten weeks before Trinity to calibrate instruments. LA Photo, TR 216.

On a wooden platform some 800 yards southeast of Ground Zero, the scientists and military detachment stacked box after wooden box of high explosive until approximately 100 tons were in the pile. An irradiated slug from the Hanford pile was dissolved and poured into flexible tubing threaded through the HE. On 7 May, the pile was detonated. Although the results of many of the experiments at the 100-ton test were of little value, simply going through the procedures was a valuable exercise.

Anderson's group tested his method of measuring the blast efficiency. They learned how much of the radiation would be deposited over the area near the blast and whether the radioactive particles were deposited on the surface of the ground, mixed with the rubble, or dispersed into the air. They learned how well the fission products and the plutonium could be separated. On a more fundamental level, they satisfactorily tested the ability of an army tank to approach Ground Zero through the explosion rubble closely enough to gather the earth sample and leave without overexposing those inside to radiation. The blast compressed and blew the surrounding earth into a saucer-shaped crater, expelling about 40 percent of the dirt. Some 2 percent of the activity of the dissolved radioactive material was deposited in the crater out to a distance of 450 feet from the center.[36]

The piezoelectric gauges worked perfectly. They showed that the 108 (actual) tons of TNT in the pile exploded with the energy of 108 tons of TNT.[37] However, Alvarez and Waldman's condenser gauges measured peak pressure and impulse nearly 40 percent lower than the piezo gauges. Clearly, they needed more work before Trinity.[38] On the other hand, the aluminum diaphragm gauges agreed well with the piezo gauge data.[39]

The most serious problem that the 100-ton test turned up was a spurious electric signal generated by an unknown source. This electrical interference set the 100 tons off 0.25 sec early and destroyed data for the monitoring experiments that depended on a precise timing signal.[40] One of the mundane, but vital, lessons learned from the test was that the mess hall would have to be enlarged and cars serviced better to keep them from breaking down. Cars traveling on the dirt roads kicked up enough dust to ruin some experiments. As a result, the army began constructing 25 miles of temporary roads that lasted almost precisely the three months for which they were designed. As many cables as possible were elevated or buried to keep them from being damaged. The scientists found communication facilities inadequate and often out of order.[41]

As a precaution, the Trinity team decided not to allow any experiments to be installed within four weeks of the test date, so that two weeks could be used to set up the last experiments, leaving two weeks for final tune-ups and rehearsals.[42]

Preparing the Gadget

Predicting the Weather

The date of the Trinity test depended both on the readiness of components and on the weather. In the early months of 1945, gadget parts promised to be ready in June or July. The question was, when would the weather conditions be appropriate? Haze, dust, and mirage effects would interfere with photographic measurements; overcast skies would make flying more difficult for the airplanes that would drop the instruments. Thunderstorms would wreak havoc with the barrage balloons. Winds had to be favorable to keep the radioactive cloud away from inhabited areas to the east and north. Each Trinity group was asked to specify the best weather conditions for their experiment, and a weather group under meteorologist Jack Hubbard tried to find a date to match the requirements.

Hubbard's first task, in April 1945, was to choose a date for the 100-ton test. By 20 April, weathermen and machines were positioned at Trinity and the crew had begun making observations. The results were supplemented by other weather data supplied by Northwest Airlines, Caltech, the Army Air Force ten-day weather forecast, and upper-air maps from Kirtland Field and the Alamogordo Air Base.[43] Hubbard identified 7 May and 27 April as the days within the range of 20 April to 10 May when optimum conditions could be expected. The laboratory settled on 7 May as a day when weather conditions would be compatible with safe operation. Hubbard's predictions were borne out and the test took place under good weather conditions.

Next, Hubbard had to determine when temperature, humidity up to 1,000 feet, and velocity and direction of wind at all levels would be optimal for the Trinity test. He had to determine which day would most closely meet the detailed weather requirements of each group.

Meeting the weather needs of all groups proved impossible. The pit assembly team's request for humidity below 89 percent and Anderson's for no rain after the shot were easy to meet in the desert. But the groups had to compromise on wind needs. Manley requested calm air for his blast gauges. Holloway and Morrison of the pit assembly group also wanted little or no wind, to avoid dust in the air at the base of the tower. In contrast, Bainbridge asked for 10- to 15-mph winds to carry the cloud away from Ground Zero and to help disperse it. Bainbridge would later settle for calm winds at the surface, with more wind aloft.

Optimum winds at Trinity would draw the radioactive cloud away from the nearby towns and break it up as rapidly as possible. A northwest to southwest wind was judged best, with slightly south of west the most favorable. Fortunately, winds from this direction were the driest, making the sky less hospitable for thunderstorms that could concentrate the radioactivity in rain drops. The greater the change in wind direction per change in altitude, the more dispersion would result. However, no one was sure how high the cloud would go. Another helpful weather condition would be an inversion layer over nearby towns to prevent material from raining down.

Bainbridge boiled all the requests down to a final group of eleven. Four concerned moisture, and were easily met in a desert area: visibility greater than 45 miles; humidity less than 75 percent at all levels; clear skies; no precipitation in a 35-mile radius within twelve hours after the operation and no thunderstorms in the region 30 to 75 miles east within four hours after the operation. Four others specified optimum winds and inversion layers: no inversion layers of more than 1° at elevations below 10,000 feet above the terrain; ground inversion at about 1,500 feet and rather thick at the inversion point to prevent mirage effects; winds aloft westerly, running between 6° and 22.5° south of west and between 6° and 34.5° north of west, with the central zone of 12° avoided unless no inversions occurred between 2,000 and 18,000 feet; velocity at 10,000 feet no less than 30 mph, but surface wind calm (less than 3 mph), with air movement below the inversion 2–12 mph from west southwest through west northwest. Hubbard's job was to find such conditions on a particular day at the preferred time of several hours before dawn, and have his prediction come true! After procuring precise weather-measuring equipment, he projected that the dates that would meet "nearly every specification of the various groups" were 18 to 21 July, with 12 to 14 July second best.

When Bainbridge telephoned the information to Washington, however, he was told that Stalin, Churchill, and Truman had agreed to meet at Potsdam on Monday, 16 July, and that Truman wanted the Trinity results before the conference. Although Hubbard told Bainbridge that 16 July was a bad choice for a test date because thunderstorms were expected, he soon realized that political factors would cause the test to occur on that day despite the weather. Bainbridge tried to get an earlier date, but all parts of the bomb could not be ready before 16 July.[44]

Although Hubbard was opposed to the 16th, at Groves's weather conference that afternoon at the McDonald Ranch House he agreed the shot could be made then with some sacrifice to the experiments.[45] Groves broke off the discussion and took Oppenheimer into an adjoining office for a private discussion. They postponed the decision until the next weather conference at 2:00 a.m., the original detonation time for the test.[46] At 2:00 a.m. Groves, Farrell, Colonel Yates, Colonel Holzman, Oppenheimer, Tolman, and Hubbard again conferred at the McDonald Ranch House. Hubbard recommended postponing the test until 5:30

a.m., when the thunderstorms would be dissipated by the first rays of the sun. Groves apparently wanted to postpone even longer.[47]

The winds came around to the desired direction. By 4:45 a.m. it was clear that the radiation would not present any grave immediate danger to the site. The wind structure moved the bulk of the cloud to the northwest and northeast, thereby dispersing it. At the time of the explosion, the overall weather conditions satisfied fewer than half the optimum conditions for the test. The sky was clear to the east and over Ground Zero and south 10,000, but overcast to the west. Visibility was greater than 60 miles; 45 miles had been considered sufficient. The surface wind from the east southeast was 3–6 mph below 500 feet, which was almost satisfactory for the scientists who wanted calm conditions. However, they could not have the desired inversion layer at 1,500 feet; they had only two slight inversion layers at 100 and 500 feet and a third such layer at 17,000 feet.[48] Nor could they get dry conditions within a 30-mile radius within twelve hours of the test. However, the most important test condition was satisfied: the rain had stopped by 5:30 a.m.

Assembling the Trinity Gadget

Gadget assembly and delivery to Trinity had begun on 3 July, with partial assembly of a mockup at Los Alamos. The mockup was driven to Trinity, where it was assembled, but without active material and explosive lenses. It was returned to Los Alamos without incident. When the explosive lenses were ready on 7 July, they were added to the gadget, which was then loaded onto a truck and driven over rough roads for eight hours, without damage. Lenses for another gadget arrived at Los Alamos on 10 July. Kistiakowsky and Bradbury personally examined them for chips and cracks, setting aside the best charges for Trinity and leaving the rest for the full-scale magnetic test of the gadget at Pajarito Canyon without active material.[49] The HE was assembled at V Site and driven to Trinity with SED Alvin D. Van Vessem watching over it.[50]

One component of the Trinity array was the impressive metal device Jumbo, 214 tons of iron and steel, fashioned into a hollow cylinder having dome ends, 25 feet long and 12 feet in diameter. The planned containment vessel, although earlier an integral part of Trinity planning, had been downgraded in importance by the time of the test because the rate of plutonium production at Hanford was higher than expected.[51]

After Oppenheimer and Groves decided to proceed with the design

Fig. 18.4. The containment vessel called Jumbo, built for possible use at Trinity to recover precious plutonium. LA Photo, TR 17.

and construction of the containment vessel in February 1944, Los Alamos began consulting with foundries about the possibility of making a steel sphere, 13 to 15 feet in diameter, to withstand a pressure of 60,000 psi.[52] The search for a sphere manufacturer narrowed to three of the largest steel companies in the United States: Jones and Laughlin, Bethlehem Steel, and the General Engineering and Foundry Company, although none could guarantee success.[53] Group E-9 concluded, more optimistically than the steel companies, that a 150-ton steel sphere could contain the explosive force of 2 tons of high explosive and also be transported by rail.[54] The conflicting conclusions led to the provisional Los Alamos decision on 23 May 1944 to abandon construction of the spherical container then known as Jumbo, while continuing tests with scaled-down "Jumbinos."[55]

Several weeks later, Oppenheimer received promising news on Jumbo. Carlson, from Bainbridge's group, designed a cylindrical "Jumbo 2," which was easier to fabricate but still difficult to transport. Although still skeptical, Oppenheimer let Carlson obtain estimates from railroads and steel companies on a cylindrical container, 12 feet in diameter, 28 feet long, weighing 180 tons.[56] Carlson found that the Babcock and Wilcox Corporation in Barberton, Ohio, a manufacturer of boilers for the navy, had built a somewhat similar container for the oil industry and was willing to attempt the construction of the huge containment vessel.[57] Oppenheimer and Groves consented to building Jumbo sometime in the summer of 1944.[58] Carlson, Robert W. Henderson, and the Babcock and Wilcox Corporation settled on the final form of the container: a cylinder with hemispherical heads.[59]

By March 1945, Oppenheimer had become confident enough that sufficient plutonium would be procured for a second test, should the first one fail to rule that the full-scale test would not be done in Jumbo. However, the plan was to erect the vessel 800 yards from Ground Zero, so that it could stand ready to contain the next full-scale or partial-scale test.[60]

Jumbo was never to be used as a containment vessel in the test, and this engineering marvel was simply to be a 214-ton object placed in the path of the shock wave.[61] Never popular with the Los Alamos scientists, because it symbolized the nadir of the scientists' confidence in creating a nuclear explosion, Jumbo had helped to guarantee Groves's acceptance of the test at the time of planning Trinity. It is generally believed that Jumbo would have contained a fizzled Fat Man explosion, had it been used.[62]

Assembly of the gadget began at 1300 hours on Friday, 13 July; a date Kistiakowsky chose in the hope that it would bring the test luck. Bradbury led the assembly. The G-Engineers assembled the pit, with Holloway responsible for inserting the active material at the base of the tower. Bacher served as adviser. Louis Slotin and Harry Daghlian monitored the assembly for excess radiation. Roger Warner coordinated assembly of the high explosives and detonators, with help from Henderson, Henry Linschitz, Schaffer, T/3 Leo Jercinovic, Arthur B. Machen, Van Vessem, and Edward J. Lofgren. Kenneth Greisen, J. C. Anderson, T/3 Vincent Caleca, and Robert W. Williams verified that the special switches and circuits, inserted to check the simultaneity of the detona-

Fig. 18.5. The Trinity gadget on the day before its detonation at Trinity, with physicist Norris Bradbury. LA Photo, TR 311.

tors, were properly installed.[63] They brought the plutonium hemispheres to the McDonald Ranch House for final preparation. At 1518 hours on 13 July, when the plug containing the remainder of the critical assembly arrived at Ground Zero, they began placing active material in the HE assembly and closing the bomb. As expected, the radioactivity from the plutonium rose steadily as the plutonium sphere was assembled. The plug was turned over to the G-Engineers, who were poised to place it in the HE assembly.[64]

They first removed the brass alignment plug, which had replaced the active material when the high-explosive shell was constructed. The plug was a snug fit, but they soon removed it with the point of a wrecking bar and pliers.[65] As Holloway and Morrison lowered the plug into position, others kept track of the number of radiation counts coming from the assembly. The insertion proceeded smoothly for a few more seconds, until the plug stuck in the opening and, for "several frantic minutes,"

Fig. 18.6. The Trinity device being delivered to the shot tower LA Photo, TR 310.

would go no further. Holloway and others realized that the plug had expanded slightly from its own internal heat and the warmth of the desert sun. The HE assembly had been shaded and hence was cooler than the plug. As Holloway described it,

> We knew damn well it should have [fit in the hole] because the uranium had been in [the assembly] before the thing ever went down south. So I thought a minute and I believe I was the one who suggested, 'look, just let it stick there for a few minutes, and the heat will be conducted away by the rest of the pit,' and in less than a minute it just fell in and that crisis was over.

Kistiakowsky and the HE group stepped in to insert the inner charge and lens block. The men watching kept track of the neutron background and also had to be careful that the charge did not bump or scrape the charges already in place, even though the charges had been carefully

Fig. 18.7. Improvised lean-to sheltering workers at the Trinity site. LA Photo, 387.

padded with felt and paper. Luckily, the neutron count did not change even after the explosive lens was placed on top of the inner charge.

The assembly of active material and high explosives was finished at 1745 hours on 13 July. Once the polar cap was fitted securely into the sphere, the scientists hoisted the gadget to the top of the tower so that the final stage of the assembly could begin the next day. As the bomb was being hoisted, they left nothing to chance and piled up a truckload of mattresses under the bomb, should it fall.[66]

Explosion

At 10:00 p.m. on the evening of 15 July, the arming party, which included Bainbridge, Joseph McKibben, and Kistiakowsky, went back to the base of the tower, where they would wait near the locked steel box containing relays and switches. On their way, they stopped at the South 10,000-yard station to inspect the timing switches. Donald Hornig had just left his guardpost near the bomb on the top of the tower. McKibben lay down under the tower for a brief nap, to be awakened by Bainbridge to start arming the bomb.[67] Once Hubbard had issued his prediction that the weather would be satisfactory at 5:30 a.m. and Oppenheimer and the others had agreed that would be the test time, Bainbridge unlocked the box at the base of the tower and closed the four toggle switches on the lines to West 900.[68] They and McKibben went to the West 900 station to close the last timing switches. Bainbridge returned to the base of the tower one last time to close the last switch in the arming circuit. Now the bomb could be detonated from the South 10,000-yard station. Bainbridge's last act was to turn on a string of lights to guide the B-29s from Kirtland air base to Ground Zero during the test. Unfortunately, the thunderstorms raging through the area at that time prevented the bombers from taking their position. Finally, Bainbridge returned to South 10,000 at 5:00 a.m. While Oppenheimer and the others paced about nervously, or applied suntan lotion, Bainbridge unlocked the master switches and had McKibben start the automatic timing sequence with 20 minutes and 15 seconds left until 5:30. With 45 seconds left, McKibben activated the motorized drum that automatically turned on the rest of the electronic signals to the test equipment. With a few seconds left, Titterton's electronic timer began sending signals to the data-gathering experiments. Bainbridge stepped out of the shelter and waited for the last few seconds to tick away in the pre-atomic dawn.[69]

Most of the eyewitnesses remembered the intense color of the fiery cloud – the purples and reds of the smoke and the dazzling white light preceding them. They remembered the deep rumbling boom and the substantial air blast felt many miles from the explosion. Others remembered the shape of the cloud, first like a raspberry, then like a mushroom, or trunk of an elephant, or goblet.[70] Victor Weisskopf spoke for many people when he said "When the explosion went off, I was first dazzled by this indirect light which was much stronger than I anticipated, and I was not able to concentrate upon the view through the dark glasses and missed, therefore, the first stages of the implosion."[71] Maurice M. Shapiro had a more personal reaction: "At the time of the initial flash of light my eyes were not protected, and I was momentarily blinded .... After a couple of seconds I regained sufficient sight to see the entire sky (in the direction of Trinity) aglow with an orange hue."[72]

Some of the observers tried to estimate the power of the blast. Von Neumann's guess was at least 5,000 tons, probably more.[73] Fermi made a slightly more accurate estimate, based on a simple experiment he per-

Fig. 18.8. Trinity fireball, several seconds after detonation. LA Photo. 65 3994.

formed while watching. He tore a small sheet of paper into pieces, which he dropped in the still desert morning as the blast wave passed his location. The blast moved them about 2.5 feet, indicating by Fermi's calculation that the explosion had been equivalent to about 10,000 tons of TNT.[74] Fermi's impressive seat-of-the-pants estimate was low by only a factor of two, according to the latest estimates of the bomb's power: 20,000 to 22,000 tons of TNT.[75]

Photographic Measurements

The intense radiation from the explosion fogged the photographic film, even in the heavily shielded cameras set near Ground Zero. However, the prints were clear enough to allow Mack's team to measure the size of the ball of fire at various stages during the first 50 milliseconds.[76] The fireball was fairly symmetric, except for a few blisters and spikes shooting ahead of the sphere, until it struck the ground 0.65 $\pm$ 0.05 milliseconds after detonation. Three milliseconds after the detonation, an "irregular line of demarcation," appeared at the bottom of the ball "below which the surface was appreciably brighter than above." The line traveled to the top of the ball during the next 8 milliseconds. The shock wave-front was clearly visible for the first 0.10 sec. By accident, it remained visible even longer. One of the barrage balloons was tethered near the tower on a long metal cable. The thermal radiation vaporized the cable, but its particles were still visible on the film, indicating the path of the shock wave.

According to Mack's report,

> The ball of fire grew ever more slowly to a radius of about 300 meters, until the dust cloud growing out of the skirt almost enveloped it. The top of the ball started to rise again at 2 seconds. At 3.5 seconds a minimum horizontal diameter, or neck, appeared one-third of the way up the skirt, and the portion of the skirt above the neck formed a vortex ring. The neck narrowed, and the ring and the fast-growing pile of matter above it rose as a new cloud of smoke, carrying a convection stem of dust up behind it. A boundary within the cloud, between the ring and the upper part, persisted for at least 22 seconds. The stem appeared twisted like a left-handed screw. The cloud of smoke, surrounded by a faint purple haze, rose with its top travelling at 57 m/s, at least until the top reached 1.5 km. The later history of the cloud was not quantitatively recorded.[77]

Aftermath

The crater was a shallow depression, six feet deep at the center and covering a circle with a 250-feet radius.[78] On 30 and 31 July, Bainbridge led a group of six people into the crater area to recover test apparatus and films of the blast. Because the ground was still radioactive, team members took turns getting out of the car, so that no one would get a heavy dose. Others returned to the site to measure the γ-ray activity at various points. Their findings showed that the radiation was most intense at a radius of about 30 yards, "over an incomplete ring of greyish material. The distribution as a whole was notably unsymmetrical, the intensity being greater towards the north." The bulk of the γ-ray activity came from the surface of the soil, but the amount of trinitite in the sample greatly influenced the results. Where there was trinitite, the activity was higher.[79]

Local fallout was heavy enough to affect some animals on nearby ranches, but apparently no people.[80] After the Hiroshima blast, airborne surveys of the west coast of the United States were made to detect fission fragments from both the Trinity and Hiroshima blasts. The planes flew from Wendover air base in Utah to Bakersfield in California, up the west coast to Alaska and back to Wendover via Hanford, Washington. Using tissue paper filters mounted in B-29s, they found no activity from Trinity and only a small amount from the Hiroshima blast.[81]

Trinitite, the thin layer of glassy fused earth in the crater sometimes called atomsite, was one of the long-lasting effects of the bomb, and one of the most popular souvenirs from the test. Some of the material was swept out of the crater and fell from the air as perfect spheres within 1,000 yards of the explosion. Most of the trinitite was green, colored by iron in the sand. Copper produced red and yellow pieces. The radioactivity varied from piece to piece. Although not intensely radioactive, the pieces could cause radiation burns when worn in jewelry next to the skin.[82]

Results of Experiments

The most important result of the Trinity test was that the implosion worked, with sufficient efficiency. The yield and size of the fireball allowed the delivery group to fix the explosion height for the Hiroshima and Nagasaki bombs at 1,850 feet. The experiments collected a vast array of data, despite the extra-intense radiation that spoiled some results.

Most of the experiments performed as expected. Not all provided useful data, however, since the yield was almost three times larger than predicted. The blast-measuring devices performed quite well, but most of the γ-ray measuring devices were overloaded by the γ-ray flux, which was much larger than expected. This occurred because extrapolations from chemical explosives were somewhat inaccurate (because chemical explosives put more of their total energy output into the blast wave, whereas nuclear explosives put less into blast and more into thermal radiation, as well as a minor amount into neutrons, gamma rays, and fission products). Because the laboratory had anticipated the lower proportion of energy in the blast wave, but not the greater yield, the blast-measuring devices could cope with the output of the blast, whereas the γ-ray measuring devices had too much to measure. Even so, the piezoelectric blast gauges were thrown off scale and no records were obtained.[83] The higher-than-anticipated γ radiation also fogged the motion picture films slightly.

The blast damaged Segrè's γ-ray yield experiments.[84] The airborne meter, unprotected by the ground, was destroyed before it could transmit to the recorders.[85] The ground chamber fared better, but the great amount of radiation overloaded the meters in the first few seconds. Still, from about 10 to 20 sec after the blast, Segrè's team obtained reliable readings.[86] Unfortunately, few neutron detectors survived the blast. Of the three cellophane cameras, only the 600-m station survived long enough to give a "moving picture" of the neutrons.[87] Only two of the eight sulphur threshold detectors were recovered; they recorded high-energy neutron fluence at 200 m.[88] Seven of the gold foils were recovered and yielded the total number of neutrons per unit area.[89]

The storm conditions just before the blast had little effect on the experiments, with one exception. Waldman and Alvarez were poised to fly over Ground Zero just before the test and drop gauges by parachute from two Army Air Force B-29s. After the blast, these gauges were to radio data back to the planes. Unfortunately, the bad weather that delayed the test also kept the B-29s from moving into position. Alvarez maintains that shortly before the scheduled takeoff Oppenheimer asked him not to fly over Ground Zero because the test would be too dangerous. Although furious, Alvarez deferred to Oppenheimer. The gauges were dropped at Hiroshima, however, and contributed useful data.[90]

The seismographs provided exactly the information Groves wanted. They detected a tremor at the North 9000 station and at San Antonio 28 miles away. The maximum motion at San Antonio was less than enough to produce a small crack in the wall of a house. Other seismographs at Tucson, El Paso, and Denver, alerted by the Manhattan Project to the possibility of a tremor, showed no tremor. The scientists correctly predicted that the damage radius would be on the order of 1,000 yards.[91]

Data on the mechanical operation of the bomb were ambiguous. The intense γ rays ruined the measurements of detonator simultaneity.[92] However, the fact that the explosion was successful indicated the detonations were close enough to being simultaneous. Sutton measured the time interval between the firing of the Raytheon condenser unit and the appearance of the γ rays. Because the explosion was as large or larger than expected, it was logical to assume that the initiator had played a part in the detonation. If the explosion had been much smaller, this measurement would have determined whether the explosion had been initiated properly.[93]

In the α measurements, both groups were able to show that α was very close to the theoretical prediction and that the α value implied a more efficient explosion than anticipated. Rossi measured α for a single neutron generation.[94] Wilson found the upper limit of α from the single ionization chamber and the oscilloscope. Calibration traces from previous tests were used, because the calibration trace failed to show up on the screen with the trace from the explosion. The two-chamber electronic timer method gave a value for α. The simplicity of Rossi's method made it the method of choice during subsequent weapon tests.

Radiochemical analysis of the amount of plutonium fissioned allowed scientists to calculate efficiency and yield. From samples collected by the shielded tanks, they calculated that the bomb had been more efficient than predicted, with a yield of 18,600 tons of TNT, quite close to the currently accepted value of 20 to 22 kilotons. This method turned out to be the most accurate means of determining the efficiency of a nuclear explosion and was used for many years after. T-Division's predictions, based on conservative values provided by Serber's group, was only between 5 and 10 kilotons.[95]

The excess-velocity blast-yield measurement worked better at Trinity than during the 100-ton test and provided among the most accurate measurements of the blast pressure, which agreed fairly well with the other measurements.[96] Fifty of the aluminum diaphragm box gauges designed to read the peak pressure of the blast were scattered over the desert around the test site. They measured the yield at 9,900 ± 1,000 tons of TNT.[97]

Not all the blast gauges worked, however. The crusher gauges positioned directly under the bomb were destroyed by the blast pressure multiplied by its reflection from the ground. No gauge within 200 feet of Ground Zero survived. The remaining five ball gauges measured pressures ranging from nearly 5 tons per square inch at 208 feet to slightly more than 1 ton per square inch at 327 feet. The water-filled pistons also did not work well. Jorgensen and Rubby Sherr found they had functioned properly for only a small range of impulses. Thus Jorgensen guessed a range of impulses corresponding to 100 to 5,000 tons of TNT and set the pistons accordingly. Eight of the twelve pistons were recovered, but only four gave a record of the blast, and only one gave a reasonable result of about 10,000 tons of TNT.[98]

The results from Trinity led to some modifications in the bomb and to a slightly different design for the core of the Nagasaki bomb. Several proposals emerged concerning the combat use of the bombing program. Some suggested taking advantage of the greater-than-expected release of radiant energy as a visual weapon[99] and using the bomb at the proper time to create a thunderstorm.

The Trinity experiment launched the American program to build nuclear weapons. One immediate result was forecast by Oppenheimer in a teletype on 19 July to Groves: he proposed using the large amount of ^{235}U from Little Boy to make composite cores with plutonium and enriched uranium. Although Groves replied that "factors beyond our control" dictate proceeding according to existing schedules," the composite core was the next core design to go into stockpile many months later.[100]

19

Delivery: June 1943 to August 1945

After the Trinity test, Los Alamos could complete its "delivery" program to provide combat weapons – the program code-named Project Alberta (or Project A). The engineering tasks of the program had included choosing suitable airplanes, training the crew, designing a ballistically stable outer shell and tail, ensuring the bomb's safety from electronic interference by the enemy, and evaluating fuzes. The last phase of the program was bombing Hiroshima and Nagasaki.

Delivery Activities in 1943

The delivery program began in October 1943 with the establishment in the Ordnance Division of group E-7, "integration of design and delivery," made up of Norman F. Ramsey, Jr., the group leader, Sheldon Dike, and Bernard Waldman. Personable and outgoing, Ramsey was the son of an army general and trained as a molecular beams physicist at Columbia University, who had worked under I. I. Rabi. As a consultant in the field of microwave radar for the secretary of war, Ramsey was highly valued by Stimson's assistant, Edward Bowles. To bring Ramsey to Los Alamos, Groves arranged a compromise in which Ramsey officially

* This chapter is based on a manuscript by Paul W. Henriksen, to which Catherine Westfall added material on the decision to drop the bombs. An early draft was edited by Richard Hewlett.

remained on Bowles's staff while he served on permanent loan to Los Alamos.[1]

Ramsey's first tasks were to survey the Army Air Forces' stock of airplanes and determine the sizes and shapes of bombs they could carry.[2] To drop the long plutonium gun weapon, Project Y needed an airplane with a bomb bay at least 17 feet long and 23 inches in diameter. (The shape of the uranium gun was not a problem because it was much shorter.)[3] This length requirement left few airplanes from which to choose. The B-29, the largest and longest-range bomber in the American fleet, had a long enough bomb bay, assuming that the front and rear bays were joined. The plane was 99 feet from nose to tail. Its chief attraction was its two bomb bays, each 150 inches long and 64 inches wide.[4]

The British had a candidate that could have been modified for bomb use, the Lancaster. Capable of carrying the largest British bomb (the 12,000-pound Grand Slam blockbuster), the Lancaster's back bay was roughly as long as the plutonium gun bomb, but had a smaller diameter. Its major defect was its nationality: no American military man wanted the atomic bomb delivered by a foreign plane, even if it belonged to America's closest ally. The Lancaster's threat to American pride actually helped secure the cooperation of the military. When faced with the shortage of B-29s, Groves had only to mention that the British would be happy to make a few Lancasters available, and the B-29s were soon forthcoming.[5]

Although Los Alamos scientists tended to think that the mechanics of delivery would be simple to work out, and that only a rudimentary testing program would be needed, Parsons, from his years of experience in naval ordnance and radar programs, realized that the delivery of the exotic weapon would require a great deal of aircraft preparation. From the first efforts at Los Alamos, the delivery program spread to Dahlgren Naval Proving Ground in Virginia for scale-model tests; to Muroc Army Air Base in California for preliminary B-29 testing and ballistic data-gathering missions; to Utah's Wendover Army Air Base for the training of the main bombing crew and tests of the fuzing, detonators, and ballistics; and finally to Tinian island in the Pacific for practice bombing runs and the missions themselves.[6]

Dahlgren Naval Proving Ground: Scale Models

Together with H. H. Arnold, commanding general of the Army Air Forces, Parsons arranged for a drop test program under Ramsey's supervision to begin on 13 August 1943, at Dahlgren. Ramsey used a makeshift scale model of the currently conceived gun bomb, a standard 500-pound, 23-inch diameter bomb, cut in half, with the front and back joined by a length of pipe 14 inches in diameter.[7] Unfortunately, this "sewer-pipe" bomb fell in a flat spin.[8] However, some adjustments of the fin and a change in the center of gravity improved its stability. Tests on the sewer-pipe models continued through the end of 1943. In December, the first proximity fuzes from the University of Michigan were added to the test program.[9]

Seeing the problems at Dahlgren, Ramsey concluded that a full-scale delivery program was needed. Plans for the testing began after von Neumann, Ramsey, and Parsons selected the dimensions for the plutonium guns (17 feet by 23 inches in diameter) and for the implosion bomb (just over 9 feet long by 59 inches in diameter).[10] Los Alamos placed orders for the ballistic shells with industries in Detroit.[11]

Designing a case for the implosion bomb was complicated because of Fat Man's unwieldy dimensions. Ramsey and von Neumann decided the bomb would be roughly spherical. The largest sphere that would fit easily through the 64-inch bomb bay door in the B-29 would have a 59-inch diameter.[12] The Bureau of Standards bomb group designed the tail assembly.[13] A 23/59th-scale Fat Man model was developed by early 1944.

Arnold's office gave the modification of the B-29s top priority; the work began on 29 November 1943 at Wright Field, in Dayton, Ohio, home of the army's Air Materiel Command.[14] The crews changed the bomb bay doors, installed a frame, bracing, and release mechanism for Fat Man; sway bracing and a release mechanism for Thin Man; and special wiring for the fuze experiments. The release mechanism was a hook used for towing gliders.[15] The modified aircraft and full-scale dummy bombs were ready in February.

Muroc Army Air Base: Full-Scale Testing

On 3 March 1944, tests began at a new location, Muroc Dry Lake in California. The purpose was to check the fuzing equipment, stability, and ballistic characteristics of the bombs: the fieldwork facilities and suitability of the aircraft.[16] Ramsey recalls that Muroc Dry Lake "was supposed to have the largest number of days with clear weather and visibility to 30,000 ft of any place in the country." Unfortunately, "it did nothing but rain. Muroc Dry Lake was under about two feet of water. Tests that were supposed to take a week took about two months."[17]

High-speed photographers recorded the bomb drops at Muroc Dry Lake from the moment a bomb left the airplane until Fat Man splattered or Thin Man buried itself in the ground. SCR 584 radar was used to track the planes. Los Alamos photography group leader Julian Mack invented a camera to fit on the radar's tracking parabola. He assigned Berlyn Brixner the task of photographing the bombs. Having difficulty following the bomb during its entire flight, Brixner acquired the aiming mechanism of a machine gun mount and fastened two cameras to it. One camera was pointed at the ground to record the impact while the other followed the descent.[18] Navy photographers, assigned by Muroc, aimed their cameras by hand with good results.

In one harrowing test, Brixner mired his truck with the camera mount on it in the middle of the target on the test range. Rather than call off the test until he could free the truck, he radioed in to go ahead, having decided to photograph the bomb drop from near the bullseye. He judged this relatively safe, since they had never yet hit the target. Furthermore, he judged that he would have roughly a minute to run out of the way if it appeared that the target would be hit. The airplane dropped the bomb and through the camera's telephoto lens he could see the bomb's head, apparently heading right for him. A few seconds later, he knew he was safe when he began to see the tail of the bomb.[19]

The initial Muroc tests showed that few parts of the bomb were working properly: the fuzes were unreliable and Fat Man was unstable in flight. The B-29s were supposed to be able to fly to 35,000 feet, but they often overheated at that altitude. Other problems arose when the airplane took off on the water-covered runways. For example, the landing gear sprayed water on the release mechanism through cracks in the bomb bay doors. This water froze at the 30,000-foot altitude where the drops began.[20] Most critically, the bomb release mechanism failed to work properly. The testers believed that once the switch to release the bomb had been thrown, the bomb's weight would pull the release open. But they learned that the release jaws would actually have to be pulled apart, and thus there could be multisecond hangups in the bomb drops.[21]

During the drop tests, Ramsey's group uncovered an important defect of conventional bombs, which were also dropped during the tests as points of reference for the atomic bomb models. Most of them missed their target by a wide margin. The reason, revealed by the high-speed photographs, was that their tail fins folded up under the pressure, causing them to fall unpredictably. In reporting this finding to the army, Parsons believed that he was finally contributing to World War II on the combat level. Unfortunately, Los Alamos censors intercepted the report and the army did not learn about the problem from Parsons. The air force made the discovery independently a year later.[22]

The implosion model proved problematic. The spherical case and armor shell surrounding one model required more than 1,500 bolts to join them together. And still the parts didn't fit properly, nor did the bolt holes align. Furthermore, assembly took too long. A subsequent model with 95 percent fewer bolts had an inner shell composed of a central belt of three segments with polar caps. The new case allowed the armored shell to be attached after the detonators, fuze components, and wiring had been installed and checked. The outer case was redesigned to an ellipsoidal shape, which was much easier to bolt together. The fuzing and detonating circuitry fitted between the sphere and the ellipsoid.[23]

Another problem with Fat Man was that its tail caused it to wobble in flight. They tried changing from a circular to a square tail (59 inches on a side), but the wobble persisted. As a last resort, Ramsey acted on the suggestion of bombardier Capt. David Semple and had steel plates welded into the tail assembly at a 45° angle to the tail, forming a crude parachute. "To everyone's surprise," as Ramsey later reported, this modification succeeded, "the bomb being completely stable in its flight and the ballistic coefficient being improved rather than decreased as anticipated."[24] With its almost ideal ballistic shape, Little Boy presented no further problems to the drop team. Between June and October the Delivery group worked to replace the ballistic models with more realistic ones.[25]

The problems identified by the drop tests at Muroc – including the need for release hooks, for baffle plates in the tail fins of the Fat Man model, and for a complete set of ballistic tables – confirmed what Parsons had said about an all-out testing program being important.[26]

Wendover Air Base: Crew and Component Testing

On 11 August 1944, the Army Air Forces recommended starting the training of combat crews and freezing the ballistic shapes of the models, so that a new lot of B-29s could be modified. By October, the modified aircraft were available, and the air crews had been selected.[27] Parsons, in collaboration with Gen. Uzal G. Ent of the Second Air Force, chose an unused air base in Utah – Wendover, code-named Kingman, or sometimes W-47 – close to the Utah–Nevada border.[28] The Second Air Force, under General Ent, provided the air crews and support facilities. The 509th Composite Group, under Col. Paul Tibbets, became the combat unit responsible for dropping the bomb. The First Ordnance Squadron Special, commanded by Capt. Charles Begg, in charge of ordnance for the 509th, was authorized to assist the scientific teams from Los Alamos in assembling bomb components at Wendover and loading them in the B-29s. The scientists carried out most of the dangerous or complicated tasks, such as handling the high explosives. Many of the test units dropped were filled with concrete to simulate the weight of the high-explosive-filled bomb.[29]

Project Y personnel controlled the flow of bomb parts to Wendover and scheduled the tests. Although Parsons was responsible for the Wendover project, he spent little time there. Ramsey, as Parsons's technical deputy, spent a day or two at Wendover each week in the early spring of 1945. Most of the duties were assumed by Navy Comdr. Frederick Ashworth, who was put in charge of operations and made an alternate to Parsons in November 1944. He also took over Ramsey's Wendover duties. Overall coordination was left to Ashworth, who tended to give the scientists free rein in their testing programs. Parsons intended that Project Y personnel at Wendover would do most of the work on the more experimental models; enlisted men would make the inert assemblies. A similar policy would hold later at Tinian. By 1945, the enlisted ordnance personnel were doing most of the routine assembly, with only two or three Project Y representatives present for a test program of over 20 units per month.[30] The number of tests at Wendover increased steadily until August 1945. At first, only the fuzing and delivery groups were involved, but soon the gun, high-explosives, detonator, and ballistic groups were included in the effort.[31]

Group members from Los Alamos often traveled to Wendover for short periods to oversee segments of their group's test program. The goal was to have the bomb explode consistently at the determined height. Norris Bradbury led the high-explosive and mechanical assembly teams, with Roger Warner and George Galloway as his deputies. They coordinated production of the high-explosive spheres and assembled the armored shells of the weapons. Detonators were the responsibility of Lewis Fussell's group. Robert Brode led the fuzing team, with Edward Doll as his deputy. Dike led the aircraft ordnance team. Francis Birch was in charge of the gun team and Maurice Shapiro of ballistic measurements.

The scale of the testing program was somewhat controversial. Ramsey felt that only a minimum number of the practice bombs should be produced and tested, since no active material would be exploded in these prototypes. As he argued, all the testing in the world would not effectively prove the final bomb. If a practice bomb exploded at the correct height, that merely showed that the fuzing system had worked. Reliability would come from careful testing of the component parts and their assembly. However, Parsons, always the military man, preferred a large number of test drops, so as to check thoroughly the reliability of components. His viewpoint prevailed, and almost one drop each day was the rule during June and July of 1945. The program continued to progress without serious problems, although secrecy made procurement difficult.

In February 1945, a number of Little Boy drop tests were also made in California near the Naval Ordnance Test Station at Inyokern. The soil at Salton Sea in California had unfortunately failed to stop the plummeting bombs from burying themselves in the ground. Units needed to be recovered to check the seating of the projectile in the target. On the theory that the sandy ground at Inyokern would provide more stopping power, drop tests were moved there in late winter. However, it took several days of digging with earth-moving equipment to recover the first Little Boy dropped at Inyokern.[32] In addition, the water table was too high, and some units were never recovered because water kept filling the hole. Little Boy drops then shifted back to Wendover.[33]

The tests at Wendover and Inyokern were designed to yield information about true air speed and the time of fall. Both were needed to determine how much time the B-29 would have to make its escape and to indicate how smoothly the bomb fell through the air. Information on ballistic coefficients was required to make bombing tables for the bomb sites. Rotation, yaw, and striking velocity indicated how well the bomb was falling.

Another item of interest in the bombing tests was the fuze mechanism for detonating the bomb at a prescribed height. The primary component was a radar device that closed a relay at a preset altitude. One of the cruder methods of testing fuze behavior used smoke puffs produced when fuzes activated the detonators. They set off a series of small explosive charges that had been placed in the tail of the gadget. Although useful when other methods failed, the smoke puffs were not as reliable as the components being tested, and the method was discontinued. Another procedure used four informers in each bomb; their signals were modified by the actions of clocks and pressure switches on the bomb. Oscilloscopes on the ground monitored the signals. Continuously moving film cameras, specially designed to fail less often than conventional cameras, photographed the oscilloscope readouts.[34] Even though the Wendover tests were not as well coordinated and documented as Parsons would have liked, by July 1945 they were yielding useful data confirming that overall the components worked reliably.[35]

One aspect of the program had to do with the final assembly of both Little Boy and Fat Man. At that time assembly was being carried out in the field by scientists. In the case of Fat Man, the model having more than 1,500 bolts to connect the spherical case and armor shell was clearly unsuitable. The next design was much easier to assemble, but the sections could not be held together tightly enough by the much smaller number of bolts. A third model had a five-section equatorial belt, with flanges on the five sections drilled completely through to accommodate aircraft bolts. This model was easier to assemble and produced the necessary pressure on the HE blocks. Although basic changes stopped with this design, refinements continued until the end of the war.

The implosion case was only part of the bomb. Arming, fuzing, and firing components had to be mounted in the space between the sphere and armor. Finding space for all components was not a serious problem, but the frequent changes made in the circuits meant repositioning and rechecking clearances. Improvements in the outer covers proceeded along with improvements in the sphere. The cover could not be completely prefabricated because the 1,100 bolts had to be matched to the cover and hand fitted after the sphere and cover were put together. Neither could the tail cone be premade, since it had to be joined by drilling and tapping the sphere at the point of tangency. In the second model, mild steel ellipsoidal covers replaced the case and tail cone, and a tension union joined the ellipsoidal covers and sphere. The mild steel of which the ellipsoids were made had to be replaced with homogeneous armor. But when the armor segments were formed, heat treating warped the segments. These case-forming problems remained unresolved until almost the end of the program.

Assembly of the HE blocks also created a persistent problem. As late as May 1945, assembly was still not a routine operation. Merely eliminating masking tape from the edges of the blocks changed the alignment and left gaps between blocks. Cloth webbing was used to hold some blocks in place while the rest were assembled. Wooden dowels, inserted in the detonator holes, kept them aligned. These problems were not eliminated until Kistiakowsky conceived of the trap door assembly.

Shipping bomb parts and assembly buildings to Tinian was itself a major undertaking. In March 1945, Warner and Galloway began forming a shipping catalog that was made obsolete in May by the trap door. Site Y representatives traveled to Mare Island Navy Yard to examine shipping crates and provide information on parts to be packed by the Mare Island crews. The first batch shipment was completed at Wendover on 30 April 1945 to test the shipping procedures. The second batch shipment (a modifications building) left Wendover on 31 May 1945.

Tinian

By June or July of 1944, discussions about a Pacific combat base for the weapon assembly were under way at Los Alamos. The scientists felt that the assembly should be handled by Project Y personnel, with possible help from the enlisted ordnance personnel. Indeed, the first assembly team was composed of Los Alamos personnel supplemented by the ordnance squadron of the 509th. Duties were to be turned over to the military after the bomb assembly had become routine. However, the war did not last long enough for this to occur.[36]

In December 1944, Manhattan Project and Army Air Forces officials conferred on a base location. The air force suggested an island in the Marianas group (principally Saipan, Guam, and Tinian) in bombing range of Japan. In mid-January, Parsons and Ashworth met with air force officers to discuss the organization and problems of such a base.[37]

Serious problems in the conception of the delivery operation emerged in a subsequent conference held in January 1945 between Project Y and the air force. One problem was that the only realistic plans for the overseas assembly base thus far were designed for the B-29 squadron and for B-29 problems. Another was the lack of communication between Project Y and the air force. None of the parties seemed to understand the complexity of carrying out the delivery at the advance base. Parsons described the meeting: "when each one was struck by this understanding; the symptoms were practically those of horror .... The morning conference ended in an uproar, with several Lt. Colonels fighting for the floor to explain how difficult our problem was."[38]

In February 1945, Ashworth, acting on behalf of Los Alamos, informed the Pacific Fleet of the nature of the bomb project. Ashworth carried the message to Adm. Chester Nimitz, commander in chief, Pacific Ocean Areas, on Guam.[39] His other task was to visit and evaluate Tinian Island. He had already decided that Guam's harbor was too busy to handle shipments from Los Alamos quickly; Guam also had no construction crews to build the required facilities. Tinian, in contrast, seemed able to support the project since it had an operating B-29 airfield. In consultation with Groves, Ashworth selected Tinian. He reserved space for the 509th operations, staked out an area for the bomb assembly buildings, and chose a location for the loading pits.[40]

In setting up their laboratories on Tinian, Los Alamos personnel experienced problems similar to those encountered in setting up Project Y. Col. Elmer E. Kirkpatrick went to Tinian in April to oversee the construction of the bomb assembly buildings. They included standard quonset huts, general-use buildings, larger assembly buildings, and even larger warehouse and ordnance administration buildings. One of the buildings was equipped with a powerful air conditioner. Warehouses were to store components of the bombs shipped from Los Alamos. However, the June 1945 acceptance of the trap-door plan for assembling the Fat Man bomb made the many cases of bomb parts shipped to Tinian unnecessary. (There were enough for forty-five or fifty bombs; however, there was never enough active material on Tinian for more than the two combat bombs.)[41]

Tools and specialized equipment had to be shipped secretly, with the additional complication that the shipments had to pass through military channels and a port of embarkation. These constraints were intolerable, since it was never certain what would be needed, and last-minute changes were certain. Anticipating this difficulty, Ramsey included in the Table of Organization for Project A, one "bomb assembly kit," which included anything from screwdrivers to quonset huts judged possibly useful in last-minute changes.[42] Each group prepared a detailed list of the needed equipment, including several spares of each item, and the number of such kits was multiplied by three. The "kit" idea worked. And when any problems were encountered, Parsons used "Silverplate," the code name assigned to all atomic bomb–related activities within the military, to keep material moving to Tinian.[43] This name required instant cooperation from all military personnel.

In May, the shipments of "kit" materials began. Rough handling and less than top-priority treatment resulted in delays and damage. Shipping to Tinian was plagued by the same kind of catch-22 situations that plagued Los Alamos: because the shipments were secret, the shipping clerks did not know that the crates were important and did not give them special treatment. Shipping personnel had to see that the ships were loaded in a certain order, so that materials for Tinian would be unloaded first in the normal course of unloading. Kirkpatrick sent these shipments, called "Bowery," or "Bronx," or "Red Ball," directly to Tinian instead of taking the normal path to Guam, which would have involved a two-week delay, possible damage, or even loss, because of extra handling in Guam.[44]

By 9 May 1945, D. M. Dennison, who had worked for Parsons, had developed the general procedures for bombing Japan.[45] His report noted that, "in the bombing missions which have been flown heretofore, the crew is more valuable than the aircraft, and the aircraft more valuable than the bomb load. In the present case the bomb is far more valuable than the aircraft." (The relative value of bomb and crews was not discussed.) The report went into detail on the weather over Japan and suggested a range of six days for the bombing. Radar bombing runs were to be made only as a last resort, since far less accuracy could be achieved with the radar bomb site than with visual bombing. The report also stressed the importance of dress rehearsals over Japanese territory. The test drops of bombs filled with tons of high explosive but no active material came to be called the "pumpkin" missions. By June 1945, the advance base was nearing completion, the training missions were in full swing, and the basic bombing technique had been worked out. What was lacking was the active material for the bomb and the personnel from Site Y to assemble it.

The 509th squadron remained at Wendover through June 1945, practicing bombing techniques and preparing for the move to Tinian. Tibbets completed an inspection in the first part of July, but this was probably only a ceremonial tour, since the space had already been selected by Ashworth and in a few days construction under Kirkpatrick's direction would be complete. Tibbets knew about the atomic nature of the bomb, having been briefed by Ramsey at the time he was selected to lead the squadron.[46]

Preparations for Bombing

When Truman assumed the presidency after Roosevelt's death on 12 April 1945, he inherited the assumption that the atomic bomb would be built and used as soon as possible. His top advisers, including Secretary of War Henry L. Stimson, stressed that the bomb would also make the Soviet Union more tractable and thus help to guarantee postwar peace. This view was challenged by Niels Bohr, who had warned Roosevelt in August 1944 of the dangers of a postwar arms race, and by Leo Szilard, who began agitating in May 1945 for an organized protest by scientists against the bomb. A particularly coherent argument against the moral, diplomatic, and political advisability of dropping the bomb was given in a document prepared by a Metallurgical Lab committee chaired by James Franck and transmitted to Stimson. However, even the Scientific Panel of the S-1 Committee, which included Fermi, Compton, Lawrence, and Oppenheimer, gave little credence to their argument. Instead the committee reported to Stimson on 16 June 1945 that they could propose "no acceptable alternative to direct military use."[47]

Truman was not in a position to reevaluate the moral, diplomatic, or political assumptions underlying the atomic weapon program. Time was short, he had little foreign policy experience, and Roosevelt had done little to prepare him to assume the presidency. As historian Martin Sherwin notes, "To compensate for his lack of experience, his inadequate knowledge, and his profound concern over his ability to do the job so suddenly thrust upon him, he relied upon those advisers who offered decisive advice." Historian Barton J. Bernstein believes that the question of using the bomb was never really open. "It was not a carefully weighed decision but the implementation of an assumption."[48]

By July 1945, the Tinian facilities had been completed, an adequate stock of parts had been built up, the bombing crews were on the island, and scientists were arriving. The first test unit was dropped on 23 July.[49] By late spring 1945, the 509th crews had become accomplished at dropping inert bombs and had made an easy transition to pumpkin bombs. At Tinian, the number of test runs was kept to a minimum; they were performed merely to see how components would function in the South Pacific climate.[50] Most of them were made over the ocean within sight of Tinian, or on Japanese-held islands, such as Rota.

Securely packaged and waterproofed implosion bomb materials arrived at Tinian on 23 July. The pit followed soon afterward, signed over on 25 July to de Silva, the official courier, and Raemer Schreiber, the G-pit team representative. De Silva and Schreiber traveled with the plutonium hemispheres by plane. Everything arrived on Tinian on 28 July after a slightly rough trip, in which one of the planes had to be replaced. The Fat Man bomb cases F-31 and F-32 arrived on 2 August. After the active materials had been stored in a case with a dessicant to combat the humidity, only the final assembly was left: the crew had to put the initiator in place, make sure that the two hemispheres fit together perfectly, place the resulting sphere in the center of the capsule, and then insert the capsule and trap-door HE blocks. For Little Boy, the first piece of fissionable material arrived on 26 July aboard the cruiser *Indianapolis*. Two days later, the first of the ^{235}U target inserts arrived by plane, followed a day later by the second and last of the inserts.[51] Unit L11, the active Little Boy model, was ready for use on 2 August.

The Fat Man test program on Tinian began on 1 August, when unit F13 was dropped without active material. It contained plaster blocks (instead of the HE), electronic fuzing, electric detonators, a Raytheon detonating unit, informers to test the simultaneity of the detonators, and a pyrotechnical smoke device to show that the fuzing had functioned.[52] The final model arrived rather late, since the decision on detonators was not made until late February 1945. Because Raytheon was far behind schedule in its manufacture of the X-unit (Chapter 16), these units were not available for test bombs until July. Full-scale HE blocks were not available until after the Trinity test in July. Assembly equipment and methods for Fat Man were not used until the Tinian operation was under way, and the first complete assembly could not be made until July 1945.[53]

The Bombing Missions

On 5 August, Parsons received the go-ahead signal to bomb Japan on the next day. As chief ordnance officer on site, he made the decision to arm Little Boy soon after take-off. He determined that this could be done safely, and that it was a risk worth taking to avoid a possible nuclear explosion on Tinian, for example, if the airplane were to crash on take-off. Groves learned of this change in plan after the mission was under way.[54] Piloted by Tibbets, The *Enola Gay* took off at 2:45 a.m. on 6 August 1945. Fifteen minutes later, Parsons and Morris Jepson began arming procedures by placing powder in the gun. Before climbing to

Fig. 19.1. Harold Agnew, Luis Alvarez, Lawrence Johnston, and Bernard Waldman, on Tinian Island, with diagnostic cannister dropped at both Hiroshim and Nagasaki. LA Photo, TR 612.

the bombing altitude, Parsons inserted the red plugs that completed the arming circuit and made the bomb active.

The flight to Hiroshima was uneventful. Tibbets even took time to visit each member of the crew personally. No easy feat since he had to crawl through a narrow tunnel that connected the front of the plane with the rear. Fifty miles from their assigned aiming point, the Aioi Bridge, the crew began final preparations for the drop. Nineteen miles from their target, a Japanese spotter reported the Tibbets's flight to the Japanese military command at Hiroshima. Although the message was broadcast over commercial radio, no one took the warning seriously.

Fig. 19.2. The *Enola Gay*, the U.S. bomber that carried Little Boy to Hiroshima. LA Photo, 841011.

The Japanese were accustomed to raids by large numbers of bombers. The Enola Gay roared on exactly on schedule.[55]

Moments before Hiroshima came into view, Parsons checked the electronic fuzes and found them working. Approximately one minute before the drop, Tibbets ordered his crew to put on safety goggles to protect them from the anticipated brightness of the burst. At 8:15:17, Little Boy was released from an altitude of 31,060 feet. Tibbets immediately put the plane in a shallow dive while turning 155 degrees to the right. This maneuver was designed by Luis Alvarez to place the *Enola Gay* as far away as possible from the burst during the bomb's drop to its detonation altitude of 1,750 feet above the city, opening the door to the atomic age.[56]

The *Enola Gay* was struck by two shock waves from the burst. The first came directly from the burst itself. The second was a shock wave reflected off the ground. Both jolts caught the crew by surprise and caused them a few momemnts of worry. Tibbets, among others, thought they had been shot at. The Enola Gay returned to Tinian without incident where Tibbets was immediately given a distinguished service cross by the Air Force Commander in the Pacific, Carl Spaatz.[57]

After the drop, Ramsey, then working on the final assembly of the Fat Man, received a coded message from Parsons on the *Enola Gay*. Parsons and Ramsey each had copies of the code, which consisted of statements formulated in advance covering all possible outcomes, each designated by a letter and number. After translation, the message was: "Clear cut results, in all respects successful. Exceeded TR test in visible effects. Normal conditions obtained in aircraft after delivery was accomplished. Visual attack on Hiroshima at 052315Z with only one tenth cloud cover. Flack and fighters absent."[58] Ramsey sent it via the usual army coded channel and went to sleep.

Ramsey was awakened several hours later by an urgent message from Groves requesting a report on the mission. The message was clearly several hours late.[59] Ramsey immediately re-sent the message by every channel he could find, to assure Groves that all had gone well. Ramsey believes that the reason the first message was delayed was that Gen. Charles MacArthur had ordered his command center to hold all messages, including Ramsey's, so that he could announce the results.[60]

After Truman heard the results, he announced that the blast had had the equivalent of 20,000 tons of TNT. This result came as a surprise to Ramsey, for at that very moment the blast measurement group was calculating the yield of the bomb. Truman's figure was derived from the message Parsons had sent. Parsons had visually assessed the blast from several miles and concluded that it was greater than Trinity. Because the Trinity yield had been estimated to be 18,000 tons of TNT, someone had decided on the value of 20,000 tons. Any further mention of the yield was suppressed, so that quibbling about the tonnage would not decrease the bomb's impact on the Japanese.[61] The only concrete information available at that time came from the blast gauge cannisters dropped by Alvarez, Johnston, and Agnew in the instrument plane. According to Alvarez, the data were not used to calculate a definite yield at that time. In 1953, Frederick Reines computed a yield of about 13,000 tons.[62] A 1985 calculation by Los Alamos physicist John Malik, based on those data plus radiological data from Hiroshima, set the yield at 15,000 tons (plus or minus 20 percent).[63]

Waldman flew on the observer plane, the *Great Artiste*, carrying a Fas-

Fig. 19.3. Leaflets dropped from B-29's over Japan in order to warn that Hiroshima and other cities might be bombed to total destruction. LA Photo, PUB-76156-1.

tax camera. He later reflected on the signal procedure for beginning the filming: "When the bomb was dropped, it released a little microswitch, and the microswitch started a 400-cycle note on the radio. So when we heard the 400-cycle note, we had 40 seconds to impact. This was sufficient apparently to do this. So we made the circle and we were aiming toward where the bomb was. All I could do was push a button, nothing else; so, well, I pushed the button." Unfortunately, when the film came back from being developed at an Army installation, half the emulsion was gone. They never learned whether there had been any pictures on the film.[64]

The first Fat Man bomb was originally scheduled to be dropped on 11 August, but the trap-door design and the around-the-clock schedule made it possible to advance the drop date to 9 August, when the weather was supposed to be better than on the next five days.[65] Maj. Charles Sweeney flew the strike plane, *Bock's Car*, carrying Fat Man, with Ashworth as weaponeer.[66]

Fig. 19.4. View of Hiroshima city after Little Boy was dropped. LA Photo, TR 720.

During the evening of 8 August, Luis Alvarez, Robert Serber, and Philip Morrison sat in the 509th's Officer's Club discussing the upcoming (Nagasaki) mission and speculating on what they might do to speed the end of the war. Alvarez worried that the Japanese government might choose to gamble that the United States had only two atomic bombs and continue the war. He proposed writing a letter to a former Berkeley colleague, Ryokichi Sagane, a professor of physics at the University of Tokyo, stating in the letter that "it was obvious that we could build as many more as we might need to end the war by force." Alvarez wrote the letter, and Serber and Morrison edited it. This message and two carbon copies were attached to the three parachute gauges that would be dropped the next day to monitor Fat Man's performance. The letters were found and turned over to military authorities, but never passed on to Sagane.[67]

While the Hiroshima mission had gone fairly smoothly, the Nagasaki mission was fraught with problems. One difficulty was that a faulty fuel pump on *Bock's Car* left the crew short 600 gallons of fuel. Another occurred on the way to the rendezvous point. The bomb's arming circuits indicated that the bomb was fully armed! Lt. Philip Barnes, Ashworth's assistant, calmly checked the circuits and found a wiring error and some incorrectly set switches. He reset the switches and corrected the wiring.[68] Yet another problem was bad weather, in spite of the forecast. The weather changed the flight patterns of the bomb, and the instrument and observer planes, with the result that the rendezvous point was missed. The main target, Kokura, was obscured by haze. After several passes, the crew flew to its secondary target, Nagasaki.

Nagasaki was also very cloudy, but there was not enough gasoline left to find another target city. So Ashworth asked Sweeney to approach by radar, even though he had been specifically ordered to approach visually. (Radar methods were still in their infancy and unreliable.) At the last minute, the clouds opened and the bombing run was made visually. During the final seconds of the run, Kermit Beahan, the bombadier, saw a hole in the clouds ahead. He took control of the aircraft and began aiming at his backup target, the Mitsubishi Arms Manufacturing Plant. As the bomb dropped, Beahan, in his excitement, shouted "Bombs Away!"[69]

Unlike the crew of the *Enola Gay*, the crew of *Bock's Car* were expecting to be buffeted by shock waves. Even with this knowledge, the sight of the waves approaching the plane made the tail gunner, Pappy Dehart, incoherent as he tried to warn the crew. Even more puzzling was the fact that the plane was hit not by two shock waves but five. No one had taken into account that the mountains surrounding Nagasaki would act as reflectors.

The target was missed by several miles because of cloudy conditions, and Fat Man's poor ballistics, which made it difficult to drop accurately. Ashworth's coded message from Okinawa to Tinian read as follows: "Over primary target, Kokura, weather was so bad that it had to be abandoned after several runs. Mission attacked Nagasaki with 90% radar approach with corrections made in last 10% time (30 seconds) and target hit. According to consensus of opinion, visible effect was equal to or greater than Hiroshima mission."[70] Upon communication with the observer planes, Ashworth ascertained that the bomb had missed the main part of the city, but had scored a direct hit on the Mitsubishi Steel and Arms Works on the city's north side, destroying the area's military effectiveness and lowering the number of casualties. The crew returned to Tinian after an emergency refueling stop on Okinawa.[71]

Recording data from the Nagasaki explosion was also problematic. The Fastax operator was supposed to be Serber, but while gathering his gear in the dark from the supply tent, he mistakenly picked up an extra life raft instead of a parachute. At the airfield, the military crew refused to let him board the aircraft without the parachute and the plane took off without him, even though his Fastax camera was the sole reason for the flight. As a result, no Fastax photos were taken of the Nagasaki blast. However, Waldman had given ordinary movie cameras to the crew and they obtained some movies of the blast. Blast gauge cannisters were dropped at Nagasaki, but the signals went off scale. The yield was determined by radiochemical and fireball yield data. Ashworth was proved correct: the blast was greater than the Hiroshima one. Evaluated at 21 kilotons ± 10 percent, the Nagasaki blast was remarkably similar to the Trinity blast, which was evaluated at 20 kilotons.[72]

Meanwhile, Los Alamos scientists continued to produce plutonium metal, which they formed into at least one more set of hemispheres. Oppenheimer cabled Groves's office on 9 August 1945 that a first-quality HE assembly would leave Kirtland on the morning of 11 August. The next active sphere was to leave on the evening of 12 August and yet another first-quality HE unit on 14 August.[73] However, all scheduled shipments to Tinian were reviewed by Oppenheimer and Groves to prevent unnecessary shipments. In a hasty phone call to Los Alamos, Groves caught Bacher just after he had signed the receipt for more active material and placed it in a car.[74] The next active sphere never reached Tinian. The plutonium assembly line that had taken three years to put in motion had come to a halt.

Groves did not allow the Project A team to leave Tinian until "the occupation of Japan, from a tactical standpoint, is complete."[75] A team of inspectors left Tinian via Guam to assess bomb damage, but the rest of the assembly crew had to stay there for about a month to close down the operations there. They occupied themselves with leisure activities, such as trading cowrie shells, while at Los Alamos partying replaced the exhausting technical activities of the past three years.[76]

Epilogue

The Japanese began surrender negotiations one day after the Nagasaki bombing. Communities everywhere experienced the war's end with heartfelt relief. Los Alamos scientists were particularly proud of the unique role they had played in bringing the war to a close. The relief – and pride – were short-lived, for most of those who had worked on the bomb suffered loss of focus, while confronting an array of difficult choices, for example, whether to feel guilty for adding atomic bombs to the world's arsenal, and whether to continue working at Los Alamos. For a short time, the technical work of the laboratory slowed down, almost to a halt.

Responses to the war's end at Los Alamos varied a great deal. Laura Fermi recalls children parading through the streets, banging on pots and pans and joyfully making mini-explosions, while their parents grappled with the sobering implications of their achievement.[1] Depression typically followed a short period of relief. Richard Feynman recalls that while he sat on a jeep and pounded on drums during one of the many end-of-the-war parties held at Los Alamos, he noticed that Robert Wilson was not jubilant. Feynman also became depressed soon afterward.[2] Only a few of the scientists saw hope in the fact that the bomb was so destructive – believing that nuclear weapons might actually end all wars because the second use of so terrible a weapon was unlikely.[3]

Oppenheimer was one of those who became severely pessimistic in

* This epilogue is by Paul Henriksen.

the aftermath of Hiroshima and Nagasaki. In a 17 August statement to Secretary of War Stimson on the future of nuclear weapons, he stressed four points: (1) such weapons would improve qualitatively and quantitatively over the next few years; (2) adequate defenses against nuclear weapons would not be developed; (3) the United States would not be able to retain hegemony in nuclear weapons, and even if it could, there would be no guarantee that it would not be destroyed; (4) wars could not be prevented, even if better nuclear weapons would be developed.[4]

He decided to resign as laboratory director after his successor had been identified, and to leave the laboratory. But he was at first uncertain whether to return to teaching. He considered his options while vacationing in the Pecos mountains, and his first idea was to teach either at Caltech or Columbia.[5] After declining an offer from Harvard, he accepted a post at Caltech on 16 October 1945. However, when Berkeley extended him a leave of absence, he returned there in August 1946, with an arrangement that allowed him to spend three days a month at Caltech. He remained at Berkeley until April 1947, when he became director of the Institute for Advanced Study in Princeton.[6] That year, with his appointment to the General Advisory Committee of the Atomic Energy Committee, Oppenheimer restored his close connection with Los Alamos. The chilling climax of Oppenheimer's career in atomic energy – the Oppenheimer security hearings in 1954 – has been treated in many works, although the definitive interpretation is still unfolding.

Many employees of Los Alamos took vacations after Japan capitulated. They considered their employment opportunities while adjusting to a peacetime schedule. Between 1 November 1945 and 8 February 1946, all the division leaders chose to leave the laboratory, but not before the new divisions and their leaders had been identified. Darol Froman took over Bacher's G-Division, renamed M-Division.[7] Eric Jette became head of Chemistry and Metallurgy. John Manley assumed responsibility for the Physics Division. George Placzek led the new Theoretical Division. Max Roy took over the Explosives Division, and Roger Warner the Ordnance (Z) Division. The facilities at Wendover Field were moved to Oxnard Field (later Sandia Base) to make them more accessible. Kirtland Field became the air base for the B-29 squadron assigned to help with drop testing. Ralph Carlisle Smith became head of the new Documentary Division, created in August 1946.[8]

In reflecting on who should replace Oppenheimer as director of the laboratory, Groves decided that the new director should have sufficient prestige to lead distinguished scientists, but should not be one of the more prominent wartime leaders, for he "wanted him to feel that this was a great opportunity." Norris Bradbury, Kistiakowsky's second in command in X-Division, had both the proper academic background and high standing within the project. As a Naval Reserve officer, Bradbury could speak to both the scientists and the military.[9] In early October, Bradbury accepted the directorship for a trial six-month term. But he soon realized that to ensure the laboratory's permanence he would have to become a permanent staff member. He was to serve in that post a quarter of a century.

Bradbury's immediate responsibility was to keep the laboratory alive at a time when many staff members wished to leave. His greatest initial task was to formulate a postwar mission for the laboratory. Groves felt that more and better bombs were needed; he hoped the laboratory could maintain the nucleus of its weapon staff.[10] That the contractor was still absentee and that the army was still administering the Los Alamos community were among the many confusing factors that confounded the problem of postwar organization and made work at Los Alamos uncertain in that transition period. It had not even been decided yet whether the laboratory would remain at its present site.

Bradbury attempted to define the postwar mission in a speech to the Coordinating Council on 1 October 1945. Expressing hope that atomic weapons would never be used and that weaponeering might end after a few years, he stated that to keep America's international bargaining position strong, it was crucial for Los Alamos to continue to develop and stockpile weapons. He planned more bomb tests, like the one at Trinity. As he explained to the council:

> Properly witnessed, properly publicized, further TR's may convince people more than any manifesto that nuclear energy is safe only in the hands of a wholly cooperating world. It may also be pointed out that I believe that further TR's may be a goal which will provide some intellectual stimulus for people working here. Answers can be found; work is not stopped short of completion; and lacking the weapon aspect directly, another TR might even be FUN.[11]

He proposed studies on the feasibility of the Super and proposed a program on peacetime applications of nuclear energy.[12]

Bradbury envisioned staffing the laboratory with civilians – roughly 600, with families, and 500 single people. That meant paring the community down from approximately 3,000 technical workers to about 1,000.[13] He realized that to remain attractive to highly qualified scientists the

Fig. E.1. Presentation of the Army-Navy "E" Award. LA Photo, 68-998.

laboratory would have to maintain an intellectually stimulating environment. To this end, he specified that fundamental research would be done in every division and that salaries would be competitive with those at other laboratories.

Bradbury became the second director of the Los Alamos laboratory on 17 October. On the morning of 16 October, Oppenheimer's last day, most of the Los Alamos residents gathered in front of Fuller Lodge to attend a short ceremony to commemorate the end of Project Y. Everyone who had been on the Project Y payroll was given the Army–Navy E Award, a small silver pin. At the ceremony's climax, Groves formally presented Oppenheimer a certificate of appreciation, appropriately from the secretary of war. In a short acceptance speech, Oppenheimer expressed his personal post-Hiroshima sentiments.

> If atomic bombs are to be added as new weapons to the arsenals of a warring world, or to the arsenals of nations preparing for war,

> then the time will come when mankind will curse the names of Los Alamos and Hiroshima. The peoples of this world must unite, or they will perish. This war, that has ravaged so much of the earth, has written these words. The atomic bomb has spelled them out for all men to understand .... By our works we are committed, committed to a world united, before this common peril, in law, and in humanity.[14]

Numerous scientists and historians have examined the ramifications of this unite-or-perish rule, especially in relation to political issues (such as the Cold War, the nuclear arms race, and holistic approaches to restoring balance to the planet). At present, a small group of scholars is exploring them. This volume attempts to contribute to that emerging effort.

20

The Legacy of Los Alamos

The United States would not have been able to complete the atomic bomb project without its vigorous economy and substantial industrial facilities. However, the scientific resources of the nation were just as important, given the existing gaps in scientific knowledge at the time Los Alamos opened its doors. President Roosevelt's decision to support atomic bomb research preceded the first demonstration of a divergent chain reaction, the development of an industrial-scale method for separating ^{235}U, and determination of plutonium's chemical and physical properties. In organizing the American atomic bomb project, Vannevar Bush drew on a sizable community of well-trained scientists having a wide repertoire of techniques and approaches. In bringing these tools to bear on the wartime problem of building the atomic bomb, the Los Alamos scientists developed a new approach to research.

What were the elements of this approach? First, the research was bound even more tightly than was conventional science to the behavior of artifacts and apparatus. The bombs had to explode, the detonators to fire, and the shape of the gadgets was constrained by that of the B-29 bomb bays. The technology had, in principle, to be totally reliable. Malfunctioning meant failure – it could no longer be construed as but another step in the process of understanding the physical world. In a context in which the lack of funding was not a constraint on research,

* This chapter is based on a draft by Lillian Hoddeson and Catherine Westfall, to which Roger Meade and Paul Henriksen contributed sections.

one result was that solutions were often approached in several ways at once. Another was that scientific equipment was typically overdesigned to avoid risk. Speed was also emphasized throughout – the project was driven by Groves's deadline of completing the bomb by summer 1945. In view of this time constraint, the researchers were unable to achieve full scientific understanding before they turned to designing and producing the necessary apparatuses. Trial and error, scale models, and iteration were all part of the Los Alamos strategy. And the organization of the research army was carefully developed in a series of multidisciplinary teams in which engineers and scientists worked hand in hand, as did theorists and experimentalists. Because research projects tended to be larger and more complex than their prewar counterparts, group leaders had to employ greater managerial skills in coordinating a widely divergent multidisciplinary staff. Scientific decisions were increasingly made by committees that reflected both the scientific and military concerns. During World War II, the military and industrial community were also brought into collaboration with the scientific community.[1]

At Los Alamos, these factors joined with the quintessentially American cultural idioms of the pioneer and tinkerer to produce a pragmatically oriented, conservative research team, in which the empirical methodology for solving problems used by the engineer and craftsman was absorbed into the more analytical, fundamental research tradition. Reliability replaced elegance; the emphasis shifted from ideas to artifacts. As a result, the traditions of scientist, engineer, and military person became fuzed. Attempting to characterize the more practical course that American research took from this point on, historian and physicist Sylvan Schweber reflects that the scientists had "to translate their understanding of the microscopic world into useful macroscopic devices."[2]

Historians of European laboratories Dominique Pestre and John Krige point to the emergence of a new kind of reseacher in the American context, one

> who can be described at once as physicist, i.e., in touch with the evolution of the discipline and its key theoretical and experimental issues, as conceiver of apparatus and engineer, i.e., knowledgeable and innovative in the most advanced techniques ... and entrepreneur, i.e., capable of raising large sums of money, of getting people with different expertise together, of mobilizing several kinds of human, financial, and technical resources.

In addition, the new kind of scientific researcher was characterized by "a pragmatic and utilitarian approach notable for its clear stress on 'getting numbers out,' an approach which preferred results and practical efficacy to means and aesthetic harmony." Strong postwar American science leaders, like Ernest Lawrence, Luis Alvarez, Edward Lofgren, Edwin McMillan, Robert Bacher, Wolfgang Panofsky, and Robert R. Wilson – many of whom had led groups or divisions at Los Alamos – fit the new model, which eventually was adopted worldwide.[3]

The Los Alamos Approach to Research

Prewar Research Models

The Los Alamos approach was built on prewar models of research. One prototype was Ernest Lawrence's laboratory at the Berkeley Radiation Laboratory in the 1930s, where Lawrence gathered together theoretical and experimental physicists, chemists, and engineers having a wide spectrum of skills and experience. He encouraged the use of empirical solutions to problems, such as the Edison approach, scale models, and the shotgun method. The collegial environment of Lawrence's laboratory stimulated creative problem solving, as did the dogged determination and self-confidence Lawrence was able to instill in his workers. Robert Wilson, a Berkeley graduate student in the late 1930s, explains that Lawrence taught them to "think how you want" to design a piece of equipment, and then work "as hard as you can on any point of weakness until you solve all your problems." The most important lesson was, "You don't say no, ever." This attitude, together with the resource of electronics equipment and other supplies obtained from industry, allowed Radiation Laboratory scientists to build, with strong encouragement and direction from Lawrence, a series of pioneering cyclotrons of increasing size, and to conduct research in nuclear chemistry, nuclear physics, and biomedicine.[4]

Elements of the Los Alamos methodology can also be found in prewar industrial laboratories, like those found in the American Telephone and Telegraph Company, General Electric, and du Pont. The practical research done within the pseudo-academic atmospheres created at such large science-based firms tended to be securely anchored in the concrete properties and behavior of real materials having potential applications. During the first half of the twentieth century, such industrial laboratories learned how to organize their diverse staffs of scientists and engineers in hierarchically arranged multidisciplinary teams representing a variety of skills and expertise.[5]

The MED applied these earlier models on a larger scale. Whereas in the 1930s Lawrence's laboratory had only a handful of buildings, fewer than sixty workers, and a yearly budget that never exceeded $125,000, the MED was about the size of the modern automobile industry, with a total budget of some $2.2 billion.[6] Faced with unprecedented resources and the need to minimize risk and maximize options, MED scientists designed a multiplicity of approaches into their program. The most notable and costly example of multiple approaches was the ^{239}Pu program, created as a backup for ^{235}U production. The decision to create the plutonium program was justified by the complementary uncertainty of producing the two fissionable isotopes – ^{235}U, although relatively well known, was difficult to separate physically from ^{238}U, and ^{239}Pu, although easy to separate chemically from ^{238}U, was almost completely unknown. To save time, the research and production of uranium and plutonium proceeded simultaneously. One of many examples of the use of the shotgun approach can be found in the chemical investigation of plutonium by Glenn Seaborg and others.

The Formation of Los Alamos

Los Alamos scientists were faced with solving many problems of physics, chemistry, metallurgy, and ordnance growing out of the fast-neutron fission chain reaction. Testing was limited by the scarcity of fissionable material and the danger of setting off an unintended explosive chain reaction. Security tended to be tighter and the military presence greater than at the other MED projects. The Los Alamos agenda required devising new organizational procedures responsive to the sometimes conflicting concerns of the military and the scientists.

In the first year of the project, the prime physics objective was to obtain critical mass and efficiency estimates to guide bomb design. With meager preliminary information and unreliable equipment, the experimentalists labored to improve nuclear measurements while theorists struggled to compensate for inconsistent or incomplete data and the limitations of existing theoretical models. In the meantime, the chemists and metallurgists attacked the problems of building plutonium weapon components, which required them to devise a plutonium purification process to meet a stringent standard and to make plutonium metal for the first time. The ordnance group worked to develop the uranium and plutonium gun assemblies. Their special challenge was the plutonium gun, which required speeds of unprecedented magnitude. The laboratory also investigated implosion as a backburner alternative for the gun assembly.

After Oppenheimer announced the bad news about spontaneous fission in the summer of 1944, the outstanding technical objective became developing the plutonium implosion weapon. (Although not yet complete, the uranium gun was never perceived as a major technical challenge.) Fortunately, the main gun assembly objectives of the first year had been met, so that the entire laboratory could now be mobilized to facilitate implosion research. Second, a substantial amount of preliminary research had already been done on implosion. The laboratory could now focus its hierarchical research organization on creating a workable implosion weapon – on solving the host of knotty problems that remained, such as how to focus detonation waves from high explosives, how to achieve symmetrical collapse of spherical metal shells, and how to use the jetting phenomenon to initiate the nuclear explosions. To solve these problems, the Los Alamos Laboratory would have to further refine its approach to research.

Experimental and Theoretical Physics

At the time Project Y began, the accuracy of critical mass, efficiency, and other crucial calculations was limited by imprecise nuclear constant measurements. For example, only a few plutonium fission cross sections had been measured, with questionable precision. To increase the accuracy of such measurements, experimenters consulted with theorists, who helped to design experiments and interpret results. They worked with chemists and electronics experts to devise clever experimental procedures and squeeze optimal performances from accelerators and detectors. They made liberal use of the empirical trial-and-error strategies, especially the shotgun and redundancy approaches. Measurements were taken over wide energy ranges using a variety of methods and instruments. Data were checked whenever possible by making similar measurements using different equipment. Because the resulting research program was larger and more complex than prewar projects, a number of physicists obtained experience in group management, and a cadre of young physicists became accustomed to well-funded, multidisciplinary group research.

Despite the ambitious program of experiments conducted in the first year of the project, Physics (P) Division leader Robert Bacher was worried in the spring of 1944 about the group's unexpected results, most notably the high rate of spontaneous fission in pile-produced plutonium. The division was also concerned about inconsistent ^{235}U ν measurements; the possibility that radiative capture to form ^{236}U might compete seriously with fission in ^{235}U; the difficulty in obtaining consistent ^{239}Pu fission spectrum measurements, which also threw ^{239}Pu ν measurements into doubt; and changes in ^{239}Pu fission cross section owing to changing half-life estimates.

Oppenheimer relied heavily on the Theoretical (T) Division's estimates of critical mass and efficiency, which were needed immediately to design the bombs and produce fissionable material. These calculations could not be carried out exactly in the available time, however. The members of T-Division therefore had to create approximate numerical solutions and develop a sense of how the results depended on parameters, to enable extrapolation into new physical regimes. They had to balance the need for speed against the need for accuracy. The scale-model approach, so widely used in other parts of the program, was of limited use in obtaining information on critical mass or efficiency: experimentalists could not simply check the theory by exploding a series of small-scale atomic bombs! Although the critical mass could be extrapolated quite accurately from neutron multiplication measurements of simulated versions of the weapon core, not enough fissionable material was available for such measurements until late 1944. As illustrated in neutron diffusion calculations, T-Division's primary strategy was to make the best possible calculations based on as many known factors as possible, employing extrapolation, approximation, and simplification, as needed.

Like other divisions in the laboratory, T-Division checked results by solving key problems by as many different methods as possible. Members of the division worked in close collaboration with experimentalists, constantly checking and reevaluating theoretical predictions against experimental findings. The most brilliant example of theoretical problem solving under these conditions was the development of the Bethe–Feynman formula for efficiency. The determination of efficiency depended on a highly complex calculation of the time evolution of numerous phenomena, including neutron diffusion. Nonetheless, Bethe and Richard Feynman identified the crucial physical parameters in the calculation. Guided by their extraordinary physical insight and understanding of physics, they were able to use their limited knowledge of the phenomena and available data to produce a workable approximate formula for efficiency. As this feat shows, finding the best approximations to theoretical problems hinged on a talent for integrating and interpreting information in relation to a deep understanding of the laws of physics, a skill possessed only by the most talented scientists.

Despite the difficulties they faced, P- and T-Divisions made considerable progress by summer 1944. Experimentalists had provided reliable measurements of ν and fission cross sections, and theory groups had produced workable estimates of critical mass and efficiency.

Chemistry and Metallurgy

Problem solving in CM-Division proceeded in a manner similar to that in the two physics divisions. To find a way to purify plutonium and make plutonium metal, CM-Division did not begin with detailed, time-consuming chemical studies. Instead, the chemistry group simply tried out various chemicals and procedures until an adequate purification scheme had been found. Similarly, with little time to spend on exploring the many fascinating and perplexing properties of plutonium metal, the metallurgy group empirically tried numerous approaches and materials until a suitable method had been developed for making the substance. To minimize mistakes and maximize options, in light of the lack of information about plutonium metal, the metallurgy group conducted a variety of overlapping metal-production schemes, a strategy that paid off in the spring of 1944 when Morris Kolodney unexpectedly found, using the relatively exotic electrolysis method, that plutonium has a melting point much lower than that of uranium.

The metallurgists had to base plutonium study on the best possible assumptions, for little was known about the element. Because uranium had been available for experimentation before plutonium had even been discovered, and was therefore much better understood, the group made the reasonable but incorrect assumption that plutonium would be metallurgically similar to uranium. This assumption temporarily hindered the production of plutonium metal. For the same reason, P-Division assumed that the nuclear properties of plutonium would be similar to those of uranium, an assumption that caused no major problems.

Like the physicists, the chemists and metallurgists were expected to gather new information at the limits of understanding, so that overall laboratory goals could be met. For example, the groundbreaking exploration of the complex physical characteristics of plutonium metal made it possible to fabricate the combat spheres.

However, CM-Division played an additional, sometimes frustrating role at the laboratory. Like Lawrence's laboratory (where Kennedy had worked before the war), but unlike the other MED projects, the Los Alamos Laboratory was organized as a physics laboratory and was managed, for the most part, by physicists. CM-Division was expected to offer a wide, varied range of support functions for the physics and other divisions – for example, chemical analyses and production of various compounds and metals – in addition to doing its own research. For this reason, chemists and metallurgists had to juggle their time, personnel, and other resources particularly carefully.

At the end of the first year, the division had devised absolutely reliable uranium metal production procedures and was ready to accommodate a range of last-minute needs, for example, a uranium alloy. By ensuring that chemistry and metallurgy problems were kept to a minimum, Kennedy and Smith's division had the resources in the spring of 1945 to implement full-scale plutonium production, from the purification of Hanford material through the fabrication of spheres. At the same time the division was able to divert a major percentage of resources to help complete the initiator for the implosion weapon.

The Gun Program

The gun program also embodied the Los Alamos research style. The Ordnance Division (E, later O) had to find reliable solutions to its technical problems quickly. Moreover, almost all the disciplines represented at Los Alamos played a role in developing the bomb gadget – mathematical physics, chemistry, engineering, and naval ordnance practice. (Only naval ordnance technicians were qualified to fire the guns.) The problems of the gun device were compounded by the necessity of fabricating the fissile materials into geometric shapes capable of withstanding the violence of being shot down a gun tube, being stopped abruptly in a target, and holding together in a supercritical arrangement long enough to detonate.

Although theory could provide some guidance, only the systematic trial and error of field testing could provide accurate information about target and projectile materials and geometries. Thus, in developing gun models, O-Division used a variety of empirical approaches developed by engineers, craftsmen, and inventors. The shotgun approach drove the experimental techniques. The Edison approach was used extensively. Small-scale work at 20 mm hastened the development of designs. Iteration was used simultaneously to prove assembly designs. Finally, numerical analysis, particularly by T-Division, allowed designers to make progress in determining critical masses of active material for the gun gadget. By early 1944, a gun that appeared capable of assembling ^{239}Pu had been developed.

To develop concepts for the nuclear gun, division members could draw on the considerable existing expertise in gun-building developed by the military. However, building an actual gun was nontrivial. The plutonium gun posed the greatest challenge because it required an assembly speed not found in existing ordnance.

The uranium gun, ever the stepchild of the plutonium model, did not receive as much attention as the plutonium gun during the laboratory's first year. Because this simpler gun required a much slower assembly speed and had no unusual quirks that posed research problems, any design for the plutonium gun, including target and projectile geometries, could (with minor changes) be scaled to meet the needs of the ^{235}U gun. Thus, like P-, T-, and CM-Divisions, O-Division completed its primary goals during the first year of the project.

Implosion

It was fortunate that the greatest problems of guns, nuclear physics, and chemistry could all be solved during the first year of Project Y, because the spontaneous fission crisis required all-out focus on implosion during the second year. It was also helpful that implosion study had already been undertaken with some degree of vigor, for there was no flexibility in the schedule between August 1944 and August 1945.

Implosion research began at Los Alamos as an application of the strategy of overlapping approaches. Neddermeyer's small initial implosion effort was conceived as an alternative to the gun program, should unexpected problems develop. Because the problem was poorly understood from the start, implosion researchers relied on the Edison approach, small-scale models, and approximate theories. Neddermeyer set off a series of crude model shots, differing from each other in basic parameters, such as size, shape, and material. Progress was slow in this initial phase owing to the low priority granted to implosion, which was reflected in limited personnel and resources. But the early work nevertheless exposed many of the problems that would have to be solved over the next

Fig. 20.1. View of Trinity site with tower in background. The garbage cans were used to protect equipment from the elements.

two years, for example, asymmetry. Although Neddermeyer's informal effort could not have succeeded in the time scale of World War II, it formed the basis for the subsequent large-scale program.

The program expanded most rapidly during two bursts of increased priority, the first in the fall of 1943 following von Neumann's and Teller's scientific suggestions. Implosion experimentation became better organized, while participation from the theoretical division increased. The serious experimental problems revealed in this period provided motivation for innovations, including the electric detonator and the explosive lens, both suggested in May 1944.

We may ask why in the fall of 1943, did Oppenheimer, along with his Governing Board and Groves, decide to invest more heavily in implosion research? For at this time, the laboratory was still banking on gun assembly for both plutonium and uranium weapons, and gun development was progressing smoothly, with no apparent obstacles. Although the documentary evidence does not answer this question, several explanations suggest themselves. First, after von Neumann lent his vigorous support to the implosion concept, and after he proposed a faster and more efficient approach to the implosion, this assembly concept was recognized as a technically efficient approach to the fission bomb. Second, the fact that the gun program was so rapidly solving its essential problems paradoxically allowed Oppenheimer to feel secure with respect to fulfilling the laboratory's main mission, and he could therefore spare a number of researchers from the gun program for implosion research. Third, from a purely intellectual point of view, the physical phenomenon of implosion was extremely challenging to leading theoreticians of the laboratory, who as research scientists were interested in pursuing a problem that went beyond any previous experience. Despite its focused wartime mission, Los Alamos remained a research laboratory that could afford to invest in the study of challenging but risky technology.

The second sizable expansion of the implosion program came in the summer of 1944, following the announcement of significant spontaneous fission in pile-produced plutonium. Because of the militarylike organization of the laboratory, Oppenheimer was able to turn implosion abruptly into the main mission of the laboratory and reallocate resources and personnel to this new first priority. Because the two new implosion divisions established at this time – G- and X-Divisions – were large-scale operations that had to function effectively in a goal-oriented manner, Oppenheimer chose not to make Neddermeyer head of the new implosion divisions. Instead he placed Bacher and Kistiakowsky in this position; both were of the new breed of scientist-organizer that emerged out of the wartime program.

The new implosion divisions formed at this time applied the full range of empirical approaches to the implosion problem – including Edison, shotgun, redundancy, scale-model, and iteration. Seven experimental diagnostics were used to explore the physics and engineering problems of implosion. Lens molds, explosive materials, detonator designs, and the hydrodynamics of implosion were among the many issues explored by iterative approximate procedures. T-Division used its ten International Business Machines to analyze implosion problems numerically. Christy's suggestion of the solid-core gadget offered a conservative brute-force design that could be relied on to work sufficiently well if supplemented by electric detonators, explosive lenses, and a modulated initiator.

The program's "moment of truth," the impressively powerful display at Trinity, demonstrated that a well-funded, large-scale, mission-oriented, multidisciplinary research laboratory employing the new blend

Fig. 20.2. Silhouette of Oppenheimer overseeing final preparation of the Trinity device. LA Photo, 83737.

of prewar and MED strategies could handle a problem that only one year earlier had looked impossible.

Impact of Project Y

Although the historical study of the ways in which scientists mapped the new legacy of large-scale research onto postwar science is still in its early stages, the patterns of influence are beginning to emerge.[7]

During the war, federal funding for research and development multiplied tenfold, from less than \$50 million to \$500 million a year. Afterward, the increase continued with a high of \$1 billion in 1950, and more than \$3 billion by 1956.[8] Three separate government agencies were formed to support physical research, the Office of Naval Research, the National Science Foundation, and the Atomic Energy Commission.[9] Although before World War II, scientists had in general been wary of political control accompanying federal funding, this attitude changed drastically after the war, when, warmed by wartime successes, many scientists grasped at the opportunities made possible by government support. Leaders of wartime projects were in an exceptionally favorable position to exploit their wartime contacts to obtain funding for large research studies.[10]

As projects increased in size as well as cost, the model of hierarchically organized multidisciplinary, government-supported research found wider use, not only in national laboratories, but in designing other large-scale, federally funded projects, for example, on space, microwaves, and lasers. Project Y served as a prototype. The Atomic Energy Commission, formed in 1947, inherited from the MED a system of national laboratories: Los Alamos, Oak Ridge, Hanford, and Argonne. All focused on applied projects, including atomic weapon and reactor development.[11] These laboratories all had in common ample federal funding, a research organization built on the coordination of multidisciplinary research groups working collaboratively, and a gradually fading distinction between the experimental scientist and the engineer. Also included in the national laboratory system were two laboratories focused on basic research: Lawrence's Radiation Laboratory in Berkeley and Brookhaven National Laboratory on Long Island. Throughout the 1950s and into the late 1960s, these laboratories would sponsor the nation's largest accelerator projects.[12]

The pragmatic orientation of research required during the war remained as a long-lasting impact. For example, as Schweber has noted, during the war theorists became accustomed to focusing on experimental results, and afterward they continued to tie their work closely to observable phenomena. More abstract topics not amenable to experimental study tended to be ignored. This tendency was then reinforced by the rapid development of expensive and powerful new experimental tools.[13] Paul Forman also notes that federal support, especially from the military, "effectively rotated the orientation of academic physics toward techniques and applications." Seidel adds that with federal support, the physics community contracted three new obligations: to provide manpower, to offer military advice as needed, and to be ready to turn expensive equipment, such as accelerators, "into instruments of war."[14]

The application of the Los Alamos approach to research at the nuclear weapons laboratories was direct and massive. It remains for historians having access to the documentation to analyze this crucial episode in American history. How did the transfer affect nonweapons research?

Technological contributions cover the full range of science and technology, from chemistry, physics, and the science of explosives to the revolutions in electronics and microelectronics. For example, the basic properties of plutonium metal were outlined, the correct formula for uranium hydride was identified, fundamental properties of many explosives were discovered, a theory for the equation of state for high-density matter was developed, numerous nuclear constants were measured, and many new observations were made of the fission process. New phenomena were uncovered, such as low-energy resonances in ^{235}U. New problems were identified, for example, the need to understand more about the fission process, especially at higher energies. Transfer of information from the MIT Radiation Laboratory enabled Los Alamos to refine the development of amplifiers, scaling circuits, and multidiscriminators. Although the Rad Lab deserves credit for many electronics advances, such as decreasing the response time in electronics from milliseconds to microseconds, Los Alamos helped turn the new electronics technology into a science.

To help transmit this science to a wider community, Los Alamos wartime researchers Matthew Sands and William C. Elmore wrote *Electronics: Experimental Techniques*, which became a landmark text, not only for experimental physicists but also for chemists, biologists, and medical professionals.[15] The new electronics extended the range of research. For example, in the area of particle physics, new techniques for detecting elementary particles and radiation (e.g., electron collection) increased the counting rate available for physics experiments by several orders of magnitude. The electron multipliers used as γ detectors at Trinity were adapted in particle detectors and expanded in their capability by the advent of phosphorescent material. Accelerator advances, such as improved time-of-flight equipment and monoenergetic neutron sources, improved the experimental capability of accelerators. Immediately after completion of his atomic bomb work, Edwin McMillan achieved a milestone in accelerator history with his invention of the principle of phase stability, a development without which more powerful postwar larger circular accelerators could not have been built.[16]

The potential of the computer for solving highly complex problems (e.g., those of hydrodynamics) was greatly expanded by the T-Division group responsible for the International Business Machines; and several Los Alamos theorists, most prominently Nicholas Metropolis, figured in the development of postwar computers.[17] Some of the materials made available at Los Alamos and other parts of the MED advanced postwar research in unexpected ways. Just after the war, for example, the work of Herbert Anderson on tritium, which could now be manufactured to produce ^{3}He, became a crucial material in the postwar development of low-temperature physics. With Aaron Novick, Anderson studied the nuclear magnetic moment of tritium and subsequently ^{3}He, into which tritium decays. Similarly, the development of pure isotopes at Los Alamos made possible the crucial discovery in the 1950s by Emmanuel Maxwell and Bernard Serin of the "isotope effect" in superconductors. This discovery set John Bardeen on the path to the development of the microscopic theory of superconductivity – the BCS theory of Bardeen, Leon Cooper, and J. Robert Scrieffer.[18]

Each of these important impacts on postwar research tells its own story about the degree to which the technical work at Los Alamos during World War II helped shape the course of modern science. The largest impact cannot be simply recorded, for it is intangible: the impact of the Los Alamos wartime experience on the American scientific spirit. Project Y created a community of scientists joined by a permanent bond of camaraderie deriving from their unique common experience of working on the first atomic bombs. This experience built confidence that scientists working together in groups could solve world problems using a scientific approach – the feeling that through common coordinated work they could conquer any of the most difficult technical problems. The desire to use the solutions in the interests of attaining world peace was widespread in this special community. Subgroups of the Los Alamos community coalesced in projects having broad idealistic goals (such as the Federation of Atomic Scientists and the international accelerator for world peace) or simply ingroups aimed at doing good physics (for example, the groups around Bethe and Feynman at Cornell, or around Bacher at Caltech).[19] These groups kept alive the Los Alamos esprit de corps. Freeman Dyson, a young British physicist who joined the Cornell group shortly after the war, tried to capture this spirit when he wrote, "It was youth, it was exuberance, it was a shared ambition to do great things together in science without any personal jealousies or squabbles over credit."[20] This spirit fueled the development of large-scale cooperative research from the 1950s to the present day[21] – a new course for physics research, with frightening new worries and brilliant new potential.

Notes

Preface

1 E.g., the excellent work by Richard Rhodes, *The Making of the Atomic Bomb* (New York: Simon and Schuster, 1986), or Gregg Herken, *The Winning Weapon: The Atomic Bomb in the Cold War, 1945–1950* (Princeton: Princeton University Press, 1981).

2 E.g., Luis Alvarez, *Adventures of a Physicist* (New York: Basic Books, 1987); and Rudolf Peierls, *Bird of Passage* (Princeton: Princeton University Press, 1985).

3 E.g., Henry DeWolf Smyth, *Atomic Energy for Military Purposes*, (Princeton: Princeton University Press, 1948); Richard Hewlett and Oscar Anderson, *The New World, 1939/1946* (University Park, Pennsylvania: Pennsylvania State University Press, 1962); and David Hawkins, "Manhattan District History, Project Y, The Los Alamos Project," LAMS-2532 (Los Alamos: Los Alamos Scientific Laboratory, University of California, 1961), later reprinted as *Project Y: The Los Alamos Story* (Los Angeles: Tomash Publishers, 1983).

4 G. Ortiz to R. L. Harris, 14 April 1982.

1. Introduction

1 The process was difficult because the mass difference of the ^{235}U and the predominant (greater than 99 percent) isotope ^{238}U is only 1 1/4 percent.

2 The neutrons in the plutonium production reactors came from fissioning ^{235}U. Reactor production of ^{239}Pu and subsequent recovery in almost trivial concentration from fission products and residual ^{238}U were viewed as conceivable only because the physical separation processes for uranium isotopes were also such massive undertakings.

3 A detonation wave in high explosive moves at several kilometers per second, whereas large projectiles move about one twentieth as fast.

4 "Thin Man" and "Fat Man" were terms believed to have been coined by Army Air Forces personnel who tried to make their telephone conversations concerning modifications to the B-29 bombers that would carry the weapons sound as though they were referring to a plane to accommodate Roosevelt and Churchill. Little Boy, which followed, was much smaller than Thin Man. Norman Ramsey, "History of Project A," A84.019, 29-6.

5 J. L. Heilbron and Robert W. Seidel, *Lawrence and His Laboratory: A History of the Lawrence Berkeley Laboratory*, vol. 1. (Berkeley: University of California Press, 1989); Leonard S. Reich, *The Making of American Industrial Research: Science and Business at GE and Bell, 1876–1926* (New York: Cambridge University Press, 1985); David Hounshell and John Kenly Smith, Jr., *Science and Corporate Strategy: Du Pont R&D, 1902–1980* (New York: University Press, 1988). At the MIT Radiation Laboratory, radar, a British development, was refined and its applications widely extended using a similar but far more limited approach. On the implications of the blending of scientific and military traditions, see Peter Galison, "The Physics Legacy of World War II," in Robert Seidel and Paul Henriksen eds., *The Transfer of Technology from Wartime Los Alamos to Peacetime* (Los Alamos, Los Alamos National Laboratory, 1992), pp. 17–21; Caroll W. Pursell, ed., *The Military Industrial Complex* (New York: Harper and Row, 1972); and A. Hunter Dupree, "The Great Instauration of 1940: The Organization of Scientific Research for War," in Gerald Holton, ed., *The Twentieth Century Sciences* (New York: W. W. Norton, 1972), pp. 443–67. See also articles in Volume 18 of *Historical Studies in the Physical and Biological Sciences*, guest edited by Robert Seidel (1987); Robert Cuff, "An Organization Perspective on the Military-Industrial Complex," *Business History Review 52*, no. 2 (Summer 1978), pp. 250–67; Paul A. C. Koistinen, "The 'Industrial-Military Complex' in Historical Perspective: The Inter-War Years," *Journal of American History 56*, (March 1970), pp. 819–39; Koistinen, "Mobilizing the World War II Economy: Labor and the Industrial-Military Alliance," *Pacific Historical Review 42* (November 1973), pp. 443–78; Benjamin F. Cooling, ed., *War Business and American Society: Historical Perspectives on the Military–Industrial Complex* (Port Washington, N.Y.: Kennikat Press 1977); Thomas A. Meeker, *The Military Industrial Complex: A Source Guide to the Issues of Defense Spending and Policy Control* (Los Angeles: Center for the Study of Armament and Disarmament, California State University, 1973).

6 The impact of World War II on postwar science was explored at length at the Los Alamos Meeting held in May 1987, organized by Robert Seidel, on The Transfer of Technology from Wartime Los Alamos in the Postwar Period, whose proceedings are to be published as a Los Alamos report. Beginning attempts to study the impact on big science are the articles in P. Galison and B. Hevly, eds., *Big Science: The Growth of Large-Scale Research* (Stanford: Stanford University Press, 1992); contributions to M. de Maria, M. Grilli, and F. Sebastiani, eds., *The Restructuring of Physical Sciences in Europe and the United States, 1945–1960* (Singapore: World Scientific, 1989); articles in *Historical Studies in the Physical and Biological Sciences, 18/1* (1987), guest edited by Seidel. Present efforts include: the history of Fermilab by Lillian Hoddeson, Catherine Westfall, Adrienne Kolb, Mark Bodnarczuk, and Kyoung Paik, a case-study in the emergence of big science; Henriksen's doctoral dissertation in progress on the transition of the Los Alamos laboratory from wartime to peacetime research; and Seidel's current study with J. L. Heilbron on the postwar work of the Lawrence-Berkeley laboratory. See also S. Schweber, "Big Science in Context: Cornell and MIT," in P. Galison and B. Hevly, eds., *Big Science*, pp. 149–83.

7 Nathan Reingold, ed., *Science in America Since 1820* (New York: Science History Publications, 1976); Robert V. Bruce, *The Launching of Modern American Science, 1846–1876* (New York: Alfred A. Knopf, 1987).

8 Some implications of this shift to pragmatic research are explored in S. Schweber, "Some Reflections on the History of Particle Physics in the 1950s," in L. Brown, M. Dresden, and L. Hoddeson, eds., *Pions to Quarks: Particle Physics in the 1950s* (Cambridge: Cambridge University Press, 1985), pp. 668–93. This shift from abstract to concrete had already begun in a number of areas during the 1930s, e.g., in the theory of solids. See L. Hoddeson, G. Baym, and M. Eckert, "The Development of the Quantum-Mechanical Theory of Metals, 1928–1933," *Reviews of Modern Physics 59*, no. 1 (1987), pp. 287–327; and S. S. Schweber, "The Young John Clark Slater and the Development of Quantum Chemistry," *Historical Studies in the Physical and Biological Sciences 20*, no. 2 (1990), 339–406.

9 The emergence of managerial research leaders in American science is examined in A. Needell, "Lloyd V. Berkner on Organizing American Science for Social Purposes," in M. de Maria, M. Grilli, and F. Sebastiani, eds., *The Restructuring of Physical Sciences in Europe and the United States, 1945–1960* (Singapore: World Scientific, 1989), pp. 85–95.

10 Daniel J. Kevles, *The Physicists: The History of a Scientific Community in Modern America* (New York: Alfred A. Knopf, 1978) pp. 26–7, 38–9; Katherine Sopka, *Quantum Physics in America 1920–1935* (Cambridge, Mass.: Harvard University Press, 1976), p. 1.2; Reich, *The Making of American Industrial Research*, pp. 40–1.

11 Kevles, *The Physicists*, pp. 5, 10, 12, 39, and 42; Sopka, *Quantum Physics in America*, pp. 1.9; 1.12–1.13; Stanley Coben, "The Scientific Establishment and the Transmission of Quantum Mechanics to the United States, 1919–32," *American Historical Review 76* (April 1971), pp. 442–3. For more information on the American physics community before World War II, see Nathan Reingold, ed., *The Sciences in the American Context: New Perspectives* (Washington, D.C.: Smithsonian Institution Press, 1979); R. Avis, H. T. David, and R. H. Stuewer, eds., *Springs of Scientific Creativity: Essays on Founders of Modern Science* (Minneapolis: University of Minnesota Press, 1983); A. Hunter Dupree, *Science in the Federal Government: A History of Policies and Activities to 1940* (Cambridge, Mass.: Harvard University Press, 1957); Nathan Reingold, *Science American Style* (New Brunswick, N.J.: Rutgers University Press, 1991); Roger Steuwer, ed., *Nuclear Physics in Retrospect: Proceedings of a Symposium on the 1930s* (Minneapolis: University of Minnesota Press, 1979); Albert E. Moyer, *American Physics in Transition* (Los Angeles: Tomash, 1983).

12 Sopka, *Quantum Physics in America*, pp. 1.15–1.16; Kevles, *The Physicists*, p. 77.

13 Kevles, *The Physicists*, p. 82; Sopka, *Quantum Physics in America*, pp. 1.12–1.14; Coben, "The Scientific Establishment," pp. 443–4.

14 Reich, *The Making of American Industrial Research*, pp. 4, 6, 40; George Wise, *Willis R. Whitney, General Electric and the Origins of U.S. Industrial Research* (New York: Columbia University Press, 1985), pp. 3–4, 315–16; Lillian Hoddeson, "The Emergence of Basic Research in the Bell Telephone System, 1875–1915," *Technology and Culture 22* (1981), pp. 512–44; Hounshell and Smith, *Science and Corporate Strategy*, pp. 11–110; Neil H. Wasserman, *From Invention to In-*

novation: Long-Distance Telephone Transmission at the Turn of the Century (Baltimore: Johns Hopkins University Press, 1985), pp. 120, 123–5.

15 Theodore B. Dolmatch, *Information Please Almanac* (New York: Information Please Publishing, 1978) p. 517; Kevles, *The Physicists*, pp. 111–38; Robert Seidel, "The Origins of Academic Research in California: A Study of Interdisciplinary Dynamics in Institutional Growth," *Journal of College Science Teaching 6* (September 1976), pp. 14–15; S. Weart, "The Physics Business in America, 1919–1940: A Statistical Reconnaissance," in Reingold, *The Sciences in the American Context*, p. 302.

16 Weart, "The Physics Business in America, 1919–1940," pp. 309, 315.

17 Kevles, *The Physicists*, p. 197; Coben, "The Scientific Establishment," pp. 459, 462–4; Sopka, *Quantum Physics in America*, pp. 2.25, 4.23–4.26; Robert Serber, "The Early Years," in *Oppenheimer* (New York: Charles Scribner's Sons, 1969), p. 38; Gerald Holton, "The Success Sanctified the Means," in Everett Mendelsohn, ed., *Transformation and Tradition in the Sciences* (London: Cambridge University Press, 1984), p. 168. For analyses of the American physics community during the 1930s, see Charles Weiner, "Physics in the Great Depression," *Physics Today 23* (October 1970), pp. 31–7; and Weart, "The Physics Business in America, 1919–1940," pp. 295–335. For a history of Lawrence Berkeley Laboratory, see Heilbron and Seidel, *Lawrence and His Laboratory*.

18 Current historical research is developing evidence to demonstrate how and why the approach became a model for postwar American big science. Studies by Heilbron and Seidel on Berkeley and by Hoddeson, Westfall, Bodnarczuk, and Kolb on Fermilab and the Superconducting SuperCollider are among the works in progress that promise to explain how the wartime research became a prototype for more recent big scince.

19 Kevles, *The Physicists*, pp. 252–8; Spencer Weart, *Scientists in Power* (Cambridge, Mass: Harvard University Press, 1979) p. 312. For a full history of the Science Advisory Board, see Lewis E. Auerbach, "Scientists in the New Deal," *Minerva 111* (1965), pp. 457–82.

20 Robin Rider, "Alarm and Opportunity: Emigration of Mathematicians and Physicists to Britain and the United States, 1933–1945," *Historical Studies in the Physical Sciences 15* (1984), pp. 107–70; C. Weiner, "A New Site for the Seminar," in Donald Fleming and Bernard Bailyn, eds., *The Intellectual Migration* (Cambridge, Mass.: Harvard University Press, 1969), p. 190. Also, Paul Hoch, "The Reception of Central European Refugee Physicists of the 1930s: USSR, UK, USA," *Annals of Science 40* (1983), pp. 206–46; Jarrell C. Jackman and Carla M. Borden, *The Muses Flee Hitler: Cultural Transfer and Adaptation, 1930–1945* (Washington, D.C.: Smithsonian Institution Press, 1983), which includes Gerald Holton, "The Migration of Physics to the United States" pp. 169–88.

21 Gerald Holton, "The Formation of the American Physics Community in the 1920s and the Coming of Albert Einstein," *Minerva 19* (1981), pp. 569–81.

22 At Fermilab he hired a number of scientists who had served in the wartime Los Alamos community. Wilson's first administrative assistant was Rose Bethe, who had helped Oppenheimer set up the Los Alamos housing office; his second administrative assistant was Priscilla Duffield, who had served in the same capacity under Oppenheimer in wartime Los Alamos. C. Westfall and L. Hoddeson, "Frugality and the Founding of Fermilab," in Nathan Reingold and David van Keuren, eds., *Science and the Federal Patron* (forthcoming). R. R. Wilson interview by Lillian Hoddeson, 12 January 1979, Fermilab History Collection.

2. Early Research on Fission: 1933–1943

1 Richard G. Hewlett and Oscar E. Anderson, Jr., *The New World 1939/1946: A History of the United States Energy Commission*, vol. 1. (University Park: Pennsylvania State University Press, 1962), p. 724.

2 For a more detailed analysis of the road to Los Alamos than we are able to provide in the summary that follows, see the rich literature on the story of how the discovery of fission led to the Manhattan Project. The long-time classics include: Hewlett and Anderson, *The New World*; Margaret Gowing, *Britain and Atomic Energy, 1939–1945* (London: Macmillan, 1964); Daniel J. Kevles, *The Physicists: The History of a Scientific Community in Modern America*, (New York: Alfred A. Knopf, 1977); Alwyn McKay, *The Making of the Atomic Age* (Oxford: Oxford University Press, 1984); Mark Walker, *German National Socialism and the Quest for Nuclear Power*, (Cambridge: Cambridge University Press, 1989); Spencer R. Weart, "The Discovery of Fission and a Nuclear Physics Paradigm," in William R. Shea, ed., *Otto Hahn and the Rise of Nuclear Physics* (Dordrecht: D. Reidel, 1983), and *Scientists in Power* (Cambridge, Mass.: Harvard University Press, 1979); Richard Rhodes, *The Making of the Atomic Bomb* (New York: Simon and Schuster, 1986). There are many more works on the topic, and this list is but a selected sample. A more complete review of recent works can be found in Robert Seidel, "Books on the Bomb," *ISIS 81* (September 1990), pp. 519–37. A number of important works are currently in progress, including Stanley Goldberg, *The Private Wars of Leslie R. Groves*, forthcoming, and the comprehensive collaborative Japanese study of the atomic bomb project now under way. See, for example, Masakatsu Yamasaki, "Origin of the Idea of the Atomic Bomb," *Historia Scientiarum* no. 38 (1989), pp. 1–28. Goldberg argues that the decision to go forward with a full attempt to build the atomic bomb was made by Bush in April in 1941 S. Goldberg, "Inventing a Climate of Opinion: Vannevar Bush and the Decision to Build the Bomb," *ISIS 83* (1992), pp. 429–52. Among the important recent works on large scientific or industrial laboratories that served as models for the wartime laboratories are J. R. Heilbron and Robert Seidel, *Lawrence and His Laboratory: A History of the Lawrence Berkeley Laboratory*, vol. 1 (Berkeley: University of California Press, 1990); and David A. Hounshell and John Kenly Smith, Jr., *Science and Corporate Strategy: Du Pont R&D, 1902–1980* (New York: Cambridge University Press, 1988).

3 The discovery of nuclear fission has been treated by many authors, including Weart, "The Discovery of Fission"; Heilbron and Seidel, *Lawrence and His Laboratory*, vol. 1, pp. 430–64; Hans G. Graetzer and David L. Anderson, eds., *The Discovery of Nuclear Fission* (New York: Van Nostrand Reinhold, 1971); Horst Wohlfarth, *40 Jahre Kernspaltung: Eine Einführung in die Originalliteratur* (Darmstadt: Wissenschaftliche Buchgesellschaft, 1979); Fritz Krafft, *Im Schatten der Sensation: Leben and Wirken von Fritz Strassmann* (Weinheim: Verlag Chemie, 1981); Roger Stuewer, "The Origins of the Liquid-Drop Model of the Nucleus," in Yehuda Elkana and Wolf Lepernies, eds., *Fifty Years of Nuclear Fission* (forthcoming).

4 Weart, "Discovery of Fission," pp. 93–5, 104; I. Noddack, "On the Element 93," *Angew. Chem.* *47* (1934), p. 654.

5 Weart, "Discovery of Fission," pp. 93–109; Gowing, *Britain and Atomic Energy*, p. 24; Heilbron and Seidel, *Lawrence and His Laboratory*, p. 447.

6 Weart, "Discovery of Fission," pp. 109–13; Otto Hahn and Fritz Strassmann,

"Uber den Nachweis und das Verhalten der bei der Bestrahlung des Uranmittels Neutronen Entstehenden Erdalkalzmetalle," *Naturwissenschaften 27* (1939), pp. 11–15; Heilbron and Seidel, *Lawrence and His Laboratory*, p. 436.

7 Weart, "Discovery of Fission," p. 113; Otto Frisch, *What Little I Remember* (Cambridge, England: Cambridge University Press, 1979), pp. 115–16; Rhodes, *The Making of the Atomic Bomb*, pp. 257–8; Roger Stuewer, "Bringing the News of Fission to America," *Physics Today 38* (1985), p. 50.

8 Frisch, "The Discovery of Fission," *Physics Today 20* (1967), p. 47

9 Frisch, *What Little I Remember*, p. 117; Weart, *Scientists in Power*, p. 64; Meitner and Frisch, "Physical Evidence for the Division of Heavy Nuclei under Neutron Bombardment," *Nature 143* (1939), pp. 239–40; Frisch, "Disintegration of Uranium by Neutrons: A New Type of Nuclear Reaction," *Nature 143* (1939), p. 276.

10 Weart, *Scientists in Power*, pp. 64–5, 81; Heilbron and Seidel, *Lawrence and His Laboratory*, pp. 440–1; Stuewer, "Bringing the News of Fission to America," *Physics Today 38* (1985), p. 52.

11 Weart, *Scientists in Power*, pp. 70–2, 108; Hewlett and Anderson, *The New World*, p. 13.

12 Szilard had long been a proponent of chain reaction studies, having in the early 1930s realized the possible military applications of nuclear chain reactions. In 1934, five years before uranium fission was discovered, he had obtained a secret patent in England for the chain reaction. He then struggled unsuccessfully to obtain funding and scientific support for a full range of experiments to produce a chain reaction.

13 Weart, *Scientists in Power*, pp. 70, 72, 76–7, 82–7; H. L. Anderson, E. T. Booth, J. R. Dunning, E. Fermi, G. N. Glasoe, and F. G. Slack, "The Fission of Uranium," *Phys. Rev. 55* (1939), pp. 511–12; Anderson, "Introduction to Paper 129," Fermi, *Collected Papers*, vol. 2 (Chicago: University of Chicago Press, 1965), pp. 1–2.

14 Weart, *Scientists in Power*, pp. 86–91, 304; H. Halban, F. Joliot, and L. Kowarski, "Number of Neutrons Liberated in the Nuclear Fission of Uranium." *Nature 143* (1939), p. 680.

15 Weart, *Scientists in Power*, pp. 91–2, 109; Gowing, *Britain and Atomic Energy, 1939–1945*, pp. 34–7; Walker, *German National Socialism and the Quest for Nuclear Power*, pp. 15–21.

16 Weart, *Scientists in Power*, pp. 108–13. Herbert Anderson, then Fermi's assistant, explained Fermi's discomfort with Szilard's way of working. Whereas Fermi worked hard and expected all his associates to do the same, Szilard liked to reserve his time for thinking, hiring assistants to take care of chores such as stuffing uranium into cans. H. L. Anderson, "The First Chain Reaction," in Robert Sachs, ed., *The Nuclear Chain Reaction – Forty Years Later: Proc. University of Chicago Commemorative Symposium* (Chicago: University of Chicago Press, 1984), pp. 27–9.

17 Weart, *Scientists in Power*, pp. 109–10, 116–24; Gowing, *Britain and Atomic Energy*, pp. 37–9.

18 Gowing, *Britain and Atomic Energy*, pp. 27–8; N. Bohr and J. A. Wheeler, "The Mechanism of Nuclear Fission," *Phys. Rev. 56* (1939), p. 426; Heilbron and Seidel, *Lawrence and His Laboratory*, p. 449.

19 Weart, *Scientists in Power*, pp. 141, 142; Heilbron and Seidel, *Lawrence and*

His Laboratory, p. 450; Walker, *German National Socialism and the Quest for Nuclear Power*, p. 21.

20 Weart, *Scientists in Power*, pp. 113, 125, 127.

21 Rudolf Peierls, *Bird of Passage: Recollections of a Physicist* (Princeton: Princeton University Press, 1985); Peierls interview by Westfall, 19 Sept. 1986, OH-141.

22 Weart, *Scientists in Power*, pp. 141, 147; Frisch, *What Little I Remember*, pp. 123–4; Peierls, *Bird of Passage*, pp. 153–4; Gowing, *Britain and Atomic Energy*, pp. 40–1.

23 Peierls interview by Hoddeson, 20–21 March 1986, OH-111; Frisch, *What Little I Remember*, p. 126; Gowing, *Britain and Atomic Energy*, pp. 41, Appendix 1.

24 Gowing, *Britain and Atomic Energy*, pp. 41–3, and Appendix 1; Peierls, *Bird of Passage: Recollections of a Physicist*, p. 154.

25 Gowing, *Britain and Atomic Energy*, pp. 45–8; Peierls, *Bird of Passage: Recollections of a Physicist*, pp. 146, 154; Weart, *Scientists in Power*, p. 148. Tizard quotation on p. 148.

26 Hewlett and Anderson, *The New World*, pp. 16–25; Weart, *Scientists in Power*, p. 143.

27 Weart, *Scientists in Power*, pp. 127–8, 143–4; H. L. Anderson and E. Fermi, "Production and Absorption of Slow Neutrons by Carbon," Report A-21, in Fermi, *Collected Papers*, vol. 2, pp. 32–40.

28 Weart, *Scientists in Power*, pp. 127, 128, 145; Anderson, "Introduction to Paper 136," in Fermi, *Collected Papers*, p. 31; Rhodes, *The Making of the Atomic Bomb*, p. 455; Walker, *German National Socialism and the Quest for Nuclear Power*, pp. 26–7.

29 Heilbron and Seidel, *Lawrence and His Laboratory*, pp. 456–64, 512–3; G. T. Seaborg, *Journal of Glenn T. Seaborg, 1 July 1939–17 April 1942*, (Berkeley: Lawrence Berkeley Laboratory, 1982), pp. 662–93. Weart, *Scientists in Power*, pp. 145, 169–70; Gowing, *Britain and Atomic Energy*, p. 59.

30 Heilbron and Seidel, *Lawrence and His Laboratory*, pp. 459–60; Hewlett and Anderson, *The New World*, pp. 25–6, 56; Weart, *Scientists in Power*, p. 146; Walker, *German National Socialism and the Quest for Nuclear Power*, p. 24.

31 Gowing, *Britain and Atomic Energy*, pp. 51, 56–8, 64; Hewlett and Anderson, *The New World*, pp. 256–7; David Brinkley, *Washington Goes to War* (New York: Alfred A. Knopf, 1988), pp. 46–9.

32 Gowing, *Britain and Atomic Energy*, pp. 65–6.

33 Weart, *Scientists in Power*, pp. 167–9; Gowing, *Britain and Atomic Energy*, pp. 58–9.

34 Peierls, *Bird of Passage: Recollections of a Physicist*, pp. 157–8; Gowing, *Britain and Atomic Energy*, pp. 56–8.

35 Hewlett and Anderson, *The New World*, pp. 24, 31.

36 Gowing, *Britain and Atomic Energy*, pp. 67–8.

37 Gowing, *Britain and Atomic Energy*, pp. 68, 70.

38 Rhodes, *The Making of the Atomic Bomb*, p. 356.

39 Heilbron and Seidel, *Lawrence and His Laboratory*, pp. 460–2; Hewlett and Anderson, *The New World*, pp. 33–5.

40 Seaborg, *Journal*, p. 693

41 Heilbron and Seidel, *Lawrence and His Laboratory*, pp. 462–3; Hewlett and Anderson, *The New World*, p. 33.

42 Gowing, *Britain and Atomic Energy*, pp. 76–7.

43 Hewlett and Anderson, *The New World*, pp. 41–9; Heilbron and Seidel, *Lawrence and His Laboratory*, p. 512.

3. The Early Materials Program: 1933–1943

1 Richard G. Hewlett and Oscar E. Anderson, Jr., *The New World 1939/1946: A History of the United States Energy Commission*, vol. 1 (University Park: Pennsylvania State University Press, 1962), pp. 41, 45, 51; Henry D. Smyth, *Atomic Energy for Military Purposes* (Princeton: Princeton University Press, 1948), p. 51.

2 A. H. Compton, *Atomic Quest: A Personal Narrative* (New York: Oxford University Press, 1956), pp. 57–9. Quotation from James Phinney Baxter III, *Scientists against Time* (Cambridge, Mass.: MIT Press, 1968), p. 427; Alice Kimball Smith and Charles Weiner, eds., *Robert Oppenheimer: Letters and Recollections* (Cambridge, Mass.: Harvard University Press, 1980), p. 223; Smyth, *Atomic Energy for Military Purposes*, p. 51.

3 Hewlett and Anderson, *The New World*, pp. 49–51; Richard Rhodes, *The Making of the Atomic Bomb* (New York: Simon and Schuster, 1986), pp. 388–9.

4 Hewlett and Anderson, *The New World*, pp. 54–6; A. Wattenberg, "Introduction to Papers 151 and 152," in Enrico Fermi, *Collected Papers*, vol. 2 (Chicago: University of Chicago Press, 1965), p. 137.

5 Glenn T. Seaborg, "History of the Met Lab Section C-1," vol. 1, hereafter GTSI, pp. 9–11, 33; Smyth, *Atomic Energy for Military Purposes*, p. 51.

6 For security reasons, Breit's title was "Coordinator of Rapid Rupture."

7 A. H. Compton to G. B. Kistiakowsky, 2 Jan. 1942, RG 77, Records of Argonne National Laboratory, Box 9; Hewlett and Anderson, *The New World*, p. 61.

8 A. H. Compton, "Progress Report on Contract OEMse-410," 1 Aug. 1942, Met Lab, NDN.45, DC V# 45054; Hewlett and Anderson, *The New World*, pp. 50–1, 62–3.

9 Hewlett and Anderson, *The New World*, pp. 57–9; J. R. Heilbron and Robert Seidel, *Lawrence and His Laboratory: A History of the Lawrence Berkeley Laboratory*, vol. 1 (Berkeley: University of California Press, 1990), p. 517.

10 Stephane Groueff, *Manhattan Project: The Untold Story of the Making of the Atomic Bomb* (Boston: Little, Brown, 1967), pp. 114–17.

11 Hewlett and Anderson, *The New World*, p. 63; Smyth, *Atomic Energy for Military Purposes*, p. 182.

12 Martin J. Sherwin, *A World Destroyed: The Atomic Bomb and The Grand Alliance* (New York: Vantage Books, 1975), pp. 71–2. Also, Peierls interview by Westfall, 19 Sept. 1986, OH-141.

13 Hewlett and Anderson, *The New World*, pp. 64–5, 102; Smyth, *Atomic Energy for Military Purposes*, pp. 169–70.

14 Vincent Jones, *Manhattan: The Army and the Atomic Bomb* (Washington, D.C.: Government Printing Office, 1985), pp. 31, 37; Kenneth D. Nichols, *The Road to Trinity* (New York: Morrow, 1987).

15 Stanley Goldberg, work in progress on biography of Groves.

16 Sherwin, *A World Destroyed*, p. 199.

17 V. Bush, *Pieces of the Action* (New York: William Morrow, 1970), pp. 61–2; Hewlett and Anderson, *The New World*, pp. 46, 73, 75, 81–3; Smyth, *Atomic Energy for Military Purposes*, pp. 81–5.

18 Hewlett and Anderson, *The New World*, p. 85; Anderson, "The First Chain Reaction," in Robert G. Sachs, ed., *The Nuclear Chain Reaction – Forty Years Later* (Chicago: University of Chicago, 1984), pp. 10–38.

19 F. G. Foote, "Early Uranium Metallurgy in Chicago," *Journal of Nuclear Materials 100* (1981), p. 27; Hewlett and Anderson, *The New World*, p. 28; J. C. Warner, "Early Methods for Producing Uranium Metal," *Uranium Technology*, J. E. Vance and J. C. Warner, NNES, Division 7, vol. 21, pp. 143–4.

20 Stephane Groueff, *Manhattan Project: The Untold Story of the Making of the Atomic Bomb* (Boston: Little, Brown, 1967), p. 73; Foote, "Early Uranium Metallurgy in Chicago," p. 27; Warner, "Early Methods for Producing Uranium Metal," pp. 143–4.

21 "Production and Absorption of Slow Neutrons by Carbon," Report A-21, pp. 31–40; "Fission Cross Section of Unseparated Uranium for Fast Rn + Be," Report C-83, pp. 107–8; "Preliminary Report on the Exponential Experiment at Columbia," Report CP-26, March–April 1942, pp. 137–43, all in Fermi, *Collected Papers*, vol. 2.

22 "Status of Research Problems in Experimental Nuclear Physics," excerpt from report C-133 for week ending 20 June 1942, pp. 200–1; "Feasibility of a Chain Reaction," Report CP-383, 26 Nov. 1942, p. 265, all in Fermi, *Collected Papers*, vol. 2; and Anderson, "The First Chain Reaction," pp. 31–2.

23 Anderson, "The First Chain Reaction," pp. 31–3; Compton, *Atomic Quest*, pp. 136–7; Anderson, "Introduction to Papers 180 and 181," in Fermi, *Collected Papers*, vol. 2, p. 268.

24 The article Fermi published anonymously after the war giving the details of the procedure, "Experimental Production of a Divergent Chain Reaction," *Am. J. Phys. 20* (1952), pp. 536–58, contains a graph showing the almost straight line extrapolation to layer 56 as the critical layer. For a detailed account of how the first chain reaction was achieved, see A. Wattenberg, "The Birth of the Nuclear Age," *Physics Today 46*, no. 1, 1993, pp. 44–51. Anderson, "Introduction to Nos. 180 and 181," pp. 268–9, and "Work Carried Out by the Physics Division," excerpt from CP-387 for month ending 15 December 1942, pp. 270–1, in Fermi, *Collected Papers*; Hewlett and Anderson, *The New World*, p. 112.

25 Warner, "Early Methods for Producing Uranium Metal," pp. 143–4; Hewlett and Anderson, *The New World*, pp. 65–6.

26 J. C. Warner, "Early Methods for Producing Uranium Metal," pp. 146–7, 155–7. A temperature of 650° C was sufficient to initiate the thermite-type reaction and melt both the slag and the metal. By this time, the researchers understood that the melting point of uranium metal was about 1,130° C, not 1,850° C, as reported earlier in the handbooks. Foote, "Early Uranium Metallurgy in Chicago," p. 27. Hewlett and Anderson, *The New World*, p. 293. Although the Metal Hydrides and Westinghouse processes were also expanded into industrial production, the process developed at Ames was the most economical and provided the bulk of material used in the full-scale plutonium production plants.

27 Hewlett and Anderson, *The New World*, pp. 291–2.

28 Hewlett and Anderson, *The New World*, pp. 84–5, 208, 218; W. P. Eatherly, "Nuclear Graphite – The First Years," *J. Nuclear Materials 100* (1981), p. 56; David Hounshell and John Kenly Smith, Jr., *Science and Corporate Strategy: Du Pont R&D, 1902–1980* (New York: University Press), pp. 338–46.

29 Hewlett and Anderson, *The New World*, pp. 90, 183; Glenn Seaborg, "Forty Years of Plutonium Chemistry," in William T. Carnall and Gregory R. Choppin, eds., *Plutonium Chemistry*, ACS Symposium Series 216 (Washington, D.C.: American Chemical Society, 1983), pp. 2–4.

30 Seaborg, "Forty Years of Plutonium Chemistry," p. 6; Hewlett and Anderson, *The New World*, p. 90; Seaborg, GTSI, pp. 39–40, 129; Groueff, *The Manhattan Project*, p. 150.

31 Seaborg, GTSI, pp. 40, 61, 71–2, 80, 169; Hewlett and Anderson, *The New World*, pp. 91, 186; Wahl interview by Westfall, 11 Nov. 1985, OH-81.

32 GTSI, pp. 13, 184, 189–93, 199, 202–4, 217–18, 228–9, 232–4.

33 GTSI, pp. 325–6, p. 620; also Hewlett and Anderson, *The New World*, p. 109.

34 Hewlett and Anderson, *The New World*, pp. 108–9.

35 GTSI, p. 368.

36 GTSI, p. 367.

37 Lewis suggested gaseous diffusion.

38 GTSI, pp. 377–8.

39 "Conclusions of Reviewing Committee, 4 Dec. 1942," RG 227, S-1 Files, Breit-Briggs, A. H. Compton folder.

40 GTSI, p. 394; Hewlett and Anderson, *The New World*, pp. 113, 236.

41 "Conclusions of Reviewing Committee, 4 Dec. 1942," RG 227, S-1 Files, Breit-Briggs, A. H. Compton folder. Also Hewlett and Anderson, *The New World*, pp. 111–13.

42 Hewlett and Anderson, *The New World*, pp. 120–2; Smyth, *Atomic Energy for Military Purposes*, pp. 180–1.

43 Hewlett and Anderson, *The New World*, pp. 113–14.

44 Hewlett and Anderson, *The New World*, pp. 114–15.

45 GTSI, p. 557; Seaborg, "History of the Met Lab Section C-1, May 1943 to April 1944," LBL, 1978, pp. 49–50; Hewlett and Anderson, *The New World*, pp. 186, 188, 190, 207, 215, 305.

46 Hewlett and Anderson, *The New World*, pp. 130, 150, 163, 297.

4. Setting Up Project Y: June 1942 to March 1943

1 A. H. Compton to Bush and Briggs, 20 Dec. 1941, RG 77, Records of Argonne National Laboratory, Box 9; also, Compton to Kistiakowsky, 20 Jan. 1942, RG 77, Records of Argonne National Laboratory, Box 9; Compton, Progress Report on OEMsr-410, Met Lab-NDN.450, DC V 45054, National Archives.

2 Compton to Kistiakowsky, 2 Jan. 1942, RG 77, Records of Argonne National Laboratory, Box 9; Richard G. Hewlett and Oscar E. Anderson, Jr., *The New World 1939/1946: A History of the United States Energy Commission*, vol. 1 (University Park: Pennsylvania State University Press, 1962), p. 61.

3 Hewlett and Anderson, *The New World*, p. 103; Rudolf Peierls, *Bird of Passage: Recollections of a Physicist* (Princeton: Princeton University Press, 1985), p. 171.

4 Oppenheimer to Compton, 8 Dec. 1941, Ernest O. Lawrence Papers (EOL), University of California, Berkeley, 19:27.

5 *Robert Oppenheimer: Letters and Recollections*, Alice Kimball Smith and Charles Weiner, eds. (Cambridge, Mass.: Harvard University Press, 1980.), p. 223. Oppenheimer and two of his students had contributed to Lawrence's development of the electromagnetic process earlier in 1941. United States Atomic Energy Commission (AEC), *In the Matter of J. Robert Oppenheimer: Transcript of Hearing before Personnel Security Board* (Washington, D.C.: U.S. Government Printing Office, 1954), p. 4.

6 *In the Matter of J. Robert Oppenheimer*, pp. 9–10.; Compton, *Atomic Quest: A Personal Narrative* (New York: Oxford University Press, 1956), pp. 125–7.

7 Smith and Weiner, *Robert Oppenheimer*, p. 223; AEC, *In the Matter of J. Robert Oppenheimer*, p. 4. I. I. Rabi, Robert Serber, Victor F. Weisskopf, Abraham Pais, Glenn T. Seaborg, *Oppenheimer* (New York: Charles Scribner's Sons, 1969).

8 Hewlett and Anderson, *The New World*, p. 103; Manley interview by Norberg, 9 June 1976, OH-146, 36-37. Los Alamos Questionnaire M.D.H-1, A-84-019, 20-4.

9 J. H. Manley, "Assembling the Wartime Labs," *Bulletin of the Atomic Scientists 30*, no. 5 (May 1974), pp. 43–5.

10 Margaret Gowing, *Britain and Atomic Energy, 1939–1945* (London: Macmillan, 1964), pp. 127–8; Peierls, *Bird of Passage*, pp. 171–2; Peierls interview by Hoddeson, 20–21 March 1986, OH-111.

11 David Hawkins, *Project Y: The Los Alamos Story* (Los Angeles: Tomash, 1983), p. 14.

12 Papers of J. R. Oppenheimer, Government File, Box 182, folder – Los Alamos, Correspondence 1942–1946.

13 E.g., Confidential Report on Chicago Conference of Fast Neutron Group of Section S-1, 9 June 1942, RG 227, S1-materials, Folder Manley, J.H.–Correspondence NARA.

14 Hewlett and Anderson, *The New World*, p. 103.

15 Oppenheimer to Ernest Lawrence, 19 May 1942, Papers of J. R. Oppenheimer, Government File, Box 182, folder – Los Alamos, Correspondence 1942-46; personnel data sheet in Manley files, Technical Personnel, A-84-019, 63-10; Hewlett and Anderson, *The New World*, p. 103. For details on the conference, see 9 June 1942, confidential Report on Chicago Conference of Fast Neutron Group of Section S-1, RG 227, S1-materials. Folder Manley, J.H.–Correspondence NARA. N. P. Heydenburg, Confidential Report on Chicago Conference of Fast Neutron Group of S-1, 9 June 1942, RG 227, S-1 Materials, Carnegie Institution, folder: Manley J. H. Correspondence, NARA.

16 The length of the meeting is somewhat uncertain. The beginning is dated from Oppenheimer to Manley, 1 July 1942, JRO papers, Box 50 Case file Folder–Manley, J. H. (LC). We thank Ruth Harris for kindly locating this and later letters concerning this important meeting in July. See also Edward Teller and Allen Brown, *The Legacy of Hiroshima* (Garden City, New York: Doubleday, 1962), p. 38. Serber recalls that the meeting lasted roughly a week; Serber interview by Hoddeson, 25–26 Feb. 1986, OH-110. According to Teller, the meeting continued for a few weeks. Bethe recalls it lasted at least six weeks. Private communication. Bethe,

15 Feb. 1989. The attendees included Oppenheimer, Van Vleck, Serber, Teller, Konopinski, Frankel, Bethe, Nelson, and Bloch.

17 Oppenheimer to Manley, 14 July 1943, Oppenheimer papers.

18 Gowing, *Britain and Atomic Energy*, pp. 83–4; Peter Goodchild, *J. Robert Oppenheimer: Shatterer of Worlds* (Boston: Houghton Mifflin Company, 1981), p. 51.

19 Private communication, Bethe, 16 Feb. 1989.

20 Alfred O. Nier, et al., "Nuclear Fission of Separated Uranium Isotopes," *Phys. Rev. 57* (15 March 1940), p. 546; and A. O. Nier, "Further Experiments on Fission of Separated Uranium Isotopes," *Phys. Rev. 57* (15 April 1940), p. 748.

21 Hewlett and Anderson, *The New World*, p. 22; Samuel Glasstone, *Sourcebook on Atomic Energy*, (Princeton: D. Van Nostrand, Company, Inc., 1967), (11.25, 13.51, 13.74, 13.75).

22 Serber interview by Hoddeson, 25–26 Feb. 1986, OH-110.

23 Goodchild, *Oppenheimer*, p. 51.

24 Serber interview by Hoddeson, 25-26 Feb. 1986, OH-110. Goodchild, *Oppenheimer*, p. 51.

25 Some of Bethe's research on the fusion processes in the sun provided a model of such a reaction. Goodchild, *Oppenheimer*, pp. 51–52.

26 Teller and Brown, *The Legacy of Hiroshima*, pp. 36–7. There is a rapidly expanding literature on the hydrogen bomb. Three first-hand accounts merit close reading: Hans Bethe,"Comments on the History of the H-Bomb," *Los Alamos Science* (Fall 1982), pp. 43–53; Edward Teller, "The Work of Many People," *Science 121* (25 Feb. 1955), pp. 267–75; and J. Carson Mark, "A Short Account of Los Alamos TheoreticalWork on Thermonuclear Weapons, 1946–1950," *Los Alamos Scientific Report* LA-5647-MS (July 1974). Recent historical work on the Super includes Peter Galison and Barton Bernstein, "In Any Light: Scientists and the Decision to Build the Superbomb, 1952–1954," *Historical Studies in the Physical and Biological Sciences 19*, vol. 2 (1989), 267–347.

27 Teller and Brown, *The Legacy of Hiroshima*, pp. 36–7.

28 Teller and Brown, *Legacy of Hiroshima*, p. 38.

29 Goodchild, *Oppenheimer*, p. 53.

30 Serber interview by Hoddeson, 25–26 Feb. 1986, OH-110. Teller and Brown, *Legacy of Hiroshima*, p. 39.

31 Papers of J. Robert Oppenheimer, Box 20, Bethe, Hans A., Folder: excerpts from an obituary by Hans Bethe for J. R. Oppenheimer for *Science*, NARA; private communication Bethe, 15 Feb. 1989; Hawkins, *The Los Alamos Story*, p. 86; Teller and Brown, *Legacy of Hiroshima*, p. 39. Konopinski spent most of his time taking notes on what was discussed, but it is not known whether these notes survived. Konopinski interview by Henriksen, 21 March 1986, OH-112.

32 Robert Serber, *The Los Alamos Primer: The First Lectures on How to Build an Atomic Bomb* (Berkeley: University of California Press, 1992), p. xxxi.

33 Smith and Weiner, *Robert Oppenheimer*, p. 227.

34 H. Bethe, "Ultimate Catastrophe?," *Bulletin of Atomic Scientists 32* (June 1976), pp. 36–7.

35 Serber interview by Hoddeson, 25–26 Feb. 1986, OH-110. Konopinski interview by Henriksen, 21 March 1986, OH-112. Oppenheimer to Teller, 11 Sept. 1942, printed

in Smith and Wiener, *Robert Oppenheimer*, p. 228. Private communication Bethe, 15 Feb. 1989.

36 Teller and Brown, *Legacy of Hiroshima*, p. 39. Also Smith and Weiner, *Robert Oppenheimer*, pp. 226–7.

37 Bethe, "Comments on the History of the H-Bomb," pp. 42–54.

38 Private communication Bethe, 15 Feb. 1989.

39 Oppenheimer to Manley, 25 July 1942, A-84-019, 63-1.

40 Hewlett and Anderson, *The New World*, p. 104.

41 Manley to Oppenheimer, 1 July 1942, A-84-019, 63-1; Oppenheimer to Manley, 25 July 1942, A-84-019, 63-1; also Oppenheimer to Manley, 1 July 1942, A-84-019, 63-1.

42 Manley and Oppenheimer, "Outline of Fast Neutron Projects," 11 June 1942, A-84-019, 64-1.

43 Van de Graaffs reached several million volts from static electric charge stored on hollow conducting spheres. Although lower in energy than cyclotrons, particle beams from a Van de Graaff were at constant voltage for long periods of time.

44 In electronics, modulation generally means to change one wave by interaction with another, e.g., to add a signal to a carrier wave. A modulated neutron source would be capable either of varying the energy of the neutron output, or of turning the beam on and off.

45 Manley and Oppenheimer, "Outline of Fast Neutron Projects"; Glasstone, *Source Book on Atomic Energy*, p. 417; M. S. Livingston and J. P. Blewett, *Particle Accelerators* (New York: McGraw-Hill, 1962), p. 160; see also, R. Bacher interview by Hoddeson and Kerr, 30 July 1984, OH-45.

46 Manley and Oppenheimer, "Outline of Fast Neutron Projects". Taschek interview by Westfall, 24 Oct. 1986, OH-146.

47 Manley and Oppenheimer, "Outline of Fast Neutron Projects," in Robert R. Wilson, John H. Manley, and H. H. Barschall, "Los Alamos Technical Series Volume 3: Section A, (Nuclear Physics), " LA-1009, 19 March 1947, p. 96; N. P. Heydenburg and R. C. Meyer, "Fission Cross-Section of ^{235}U and ^{238}U for Fast Neutrons," 19 Dec. 1942, A-84-019, 63-9, p. 1.

48 A. H. Compton, memorandum of discussion, 2 Sept. 1942, A-84-019, 63-2; Notes on CF Conference Week of 21 Sept. 1942, A-84-019, 64-1; J. R. Oppenheimer, Minutes of the Last Meeting of the Fast Neutron Group, 23 Sept. A-84-019, 64-1; Oppenheimer to Manley, 12 Oct. 1942, A-84-019, 63-1.

49 W. Bennett to Manley, 21 Sept. 1942, A-84-019, 63-6; Oppenheimer, Notes on CF Conference Week of 21 Sept. 1942; and Manley to Oppenheimer, 6 Nov. 1942, A-84-019, 63-1.

50 Oppenheimer to Manley, 25 July 1942, A-84-019, 63-1; and F. Bloch, Stanford Cyclotron Project: Official Report 5, 13 Oct. 1942, A-84-019, 63-7; Hawkins, *The Los Alamos Story*, p. 96.

51 The term "barn," 10^{-24} cm^2, was proposed by Baker and Holloway in 1942. As they later explained, it served as a code word, which was desirable at the time, and seemed appropriate because "a cross section of 10^{-24} cm^2 for nuclear purposes was really as big as a barn." M. Holloway and R. Baker, "Note on the Origin of the Term 'Barn'," LAMS-523, (13 March 1944).

52 O. Chamberlain, J. W. Kennedy, E. Segrè, "Fission Cross Section of ^{235}U for Intermediate Velocity Neutrons," 24 Oct. 1942, A-84-019, 63-1; and Oppenheimer

to McKibben, 4 Sept. 1942, A-84-019, 63-8. Also J. H. Manley, CF Group, Progress Report No. 1, A-84-019, 63-10; Oppenheimer to Manley, 25 July 1942, A-84-019, 63-1; and Manley to Oppenheimer, 6 Nov. 1942, A-84-019, 63-1.

53 N. P. Heydenburg and R. C. Meyer, Fission Cross-Section of ^{235}U and ^{238}U for Fast Neutrons, 9 Dec. 1942, A-84-019, 63-9, p. 8.

54 Oppenheimer to Manley, 1 July 1942, A-84-019, 63-1; R. R. Wilson, J. H. Manley, and H. H. Barschall, "Los Alamos Technical Series Vol. 3: Section A (Nuclear Physics)," LA-1009, 19 March 1947, p. 96.

55 Wilson, et al., LA-1009, pp. 80–1, 84, and 96–7.

56 A. O. Hanson, "Measurement of Fission Cross-Sections by the Use of a Coincidence Proportional Counter," p. 1, A-84-019, 63-8. Also J. L. McKibben to Compton, 9 April 1943, A-84-019, 63-8; D. L. Benedict and A. O. Hansen, "Measurement of the 25 [^{235}U] Cross Section for 0.53 MeV Neutrons by the Manganese Solution Method," sec. 4 of final report on the Wisconsin fast-fission project, ca. 9 April 1943, A-84-019, 63-8; Wilson, et al., LA-1009, p. 97; and Robert McFarlane, private communication, 10 Dec. 1987.

57 C. Wiegand interview by Westfall, 29 May 1987, OH-157; Heydenburg to Manley, 9 Nov. 1942, A-84-019, 63-9.

58 Bacher interview by Hoddeson and Kerr, 30 July 1984, OH-45, p. 13. Oppenheimer to Manley, 25 July 1942, A-84-019, 63-1; and Manley to Bacher, 27 July 1942, A-84-019, 63-3.

59 Bacher to Compton, 9 Nov. 1942, A-84-019, 63-3. Also Bacher, "Progress Report on the Work on Fast Neutrons at Cornell University," A-84-019, 63-3; N. Feather, "The Time Involved in the Process of Nuclear Fission," *Nature 143* (8 April 1939), pp. 597–8; and T. Snyder and R. W. Williams, *Time for Emission of Fission Neutrons*, LA-50, 21 Jan. 1944, p. 2.

60 Bethe to Manley, 8 Dec. 1942, A-84-019, 63-3; and Manley to Bethe, 11 Dec. 1942, A-84-019, 63-3. Also Bethe to Oppenheimer, 6 Oct. 1942, Papers of Robert Oppenheimer, Box 20. Wilson interview by Westfall, 25 May 1987, OH-154, pp. 41–2.

61 A. H. Compton to Conant, 18 Sept. 1942, RG 227, NARA.

62 Materials, especially heavy elements, were commonly referred to in the project by code numbers formed of the last digits, in order, of the atomic number and atomic weight; or they were simply referred to by their atomic weights. Thus $^{233}_{92}$U became "23," or 233; $^{235}_{92}$U became "25," or 235; $^{238}_{92}$U became "28," or 238 and $^{239}_{94}$Pu became "49," or 239.

63 Serber interview by Hoddeson, 26 March 1979, OH-9, and telephone conversation with Hoddeson, 16 Oct. 1985. The Tolman–Serber paper on implosion is possibly in the Met Lab papers in the National Archives, but has not been located. Hawkins confirms that "the idea of something like an implosion, as an alternative to gun assembly, had entered several heads before the beginning of Los Alamos." Hawkins, *The Los Alamos Story*, pp. 80–1, 4.

64 Telephone conversations, Bethe with Hoddeson, 11 July 1985, Konopinski with Hoddeson, 15 Oct. 1985. McMillan interview by Seidel, 17 April 1985, OH-63; As Bethe recently commented, Tolman may have made the suggestion, but Oppenheimer might not have considered it important enough to discuss in the group.

65 Oppenheimer to H. D. Smyth, 14 April 1945, Smyth Papers, American Philosophical Society.

66 Seidel, unpublished memorandum, 1985.

67 L. Groves, *Now It Can Be Told: The Story of the Manhatten Project* (New York: Harper and Brothers, 1962), pp. 62–3. Groves was not dissuaded by the negative reactions to Oppenheimer by Lawrence and members of the Military Policy Committee. Lawrence had impetuously promised the directorship of Los Alamos to McMillan. McMillan interview by Seidel, 17 April 1985, OH-63. For Lawrence's opposition to Oppenheimer, see Lee DuBridge interview by James Culp, Oct. 1981.

68 Testimony of J. R. Oppenheimer and Colonel John Lansdale, *In the Matter of J. Robert Oppenheimer; Transcript of Hearing before Personnel Security Board and Texts of Principal Documents and Letters* (Cambridge: MIT Press, 1971), p. 12.

69 Manley to Compton, 19 Sept. 1942, A-84-019, 63-10.

70 Los Alamos was Project Y, Hanford Engineer Works was Project W and the Clinton Engineer Works at Oak Bridge was Project X. Compton to McMillan, 14 Sept. 1942 and 17 Sept. 1942, McMillan Files, Lawrence Berkeley Laboratory. We wish to thank McMillan for permitting R. Seidel to examine his files pertaining to this period. A report of the meeting dated 23 Sept. 1942 is in the AEC Files. McMillan interview by Seidel, 17 April 1985, OH-63; Oppenheimer to Compton, 29 Sept. 1942, McMillan Files.

71 Oppenheimer testimony, *In the Matter of J. Robert Oppenheimer*, pp. 12, 261.

72 Oppenheimer to Manley, 12 Oct. 1942, A-84-019, 63-1; William Lauren, *The General and the Bomb* (New York: Dodd, Mead, 1988).

73 Manley to Oppenheimer, 12 Oct. 1942, A-84-019, 63-1.

74 Undated document in Manley, Technical Personnel file, "General Remarks" (the document contains general plans for the Los Alamos laboratory, probably written in Dec. 1942 or Jan. 1943), A-84-019, 63-10.

75 The personnel estimates for July 1943 were theoretical 20, experimental 50, executive 5, secretarial 5, miscellaneous technical 10, shop 15, and helpers 25. Oppenheimer to Manley, 12 Oct. 1942, A-84-019, 63-1.

76 Smith and Weiner, *Robert Oppenheimer*, p. 229; Vincent Jones, *Manhattan: The Army and the Atomic Bomb* (Washington, D.C.: Government Printing Office, 1985), p. 77.

77 John H. Dudley, "Ranch School to Secret City," in Lawrence Badash, Joseph Hirschfelder, and Herbert P. Broida, eds., *Reminiscences of Los Alamos, 1943–1945* (Dordrecht, Holland: D. Reidel Publishing Company, 1980) pp. 4–5. This account is the only one that agrees with Oppenheimer's testimony before the Gray Board for his security clearance hearings. At that time, he said that he recommended the site to Groves, who acquired it quickly. *In the Matter of J. Robert Oppenheimer*, MIT Press, p. 28. Los Alamos correspondence sheds light on the date when the Boys Ranch was visited. Oppenheimer wrote to Manley on 23 Oct. that he expected a visit from Dudley shortly and was prepared to visit the sites he recommended. Another letter from Oppenheimer to Manley tells of a meeting between Dudley, Groves, and Oppenheimer just before 6 Nov. 1942 at which the question of the site needed only a visit from Oppenheimer and McMillan to be settled. This account contradicts Dudley's mention of a last-minute change of mind by Oppenheimer. Oppenheimer to Manley, 6 Nov. 1942, A-84-019, 63-1.

78 The overall contributions of scientists to World War II are explained in James Phinney Baxter III, *Scientists against Time* (Cambridge, Mass.: MIT Press, 1968);

and Irwin Stewart, *Organizing Scientific Research for War* (Boston: Little, Brown, 1948).

79 Oppenheimer to Manley, 25 July 1942, A-84-019, 63-1.

80 Oppenheimer to Conant, 21 Oct. 1942, A-84-019, 4-10. Conant to Oppenheimer, 28 Oct. 1942, A-84-019, 4-10. A native Californian, McMillan was working for the U.S. Navy Radio and Sound Laboratory in San Diego at the time he became involved in the Manhattan Project in 1942. At the Chicago meeting that year, with Oppenheimer, Manley, Fermi, and Lawrence (a meeting he attended at Compton's request), McMillan had to choose whether to remain at the Navy laboratory, return to Berkeley and work with Lawrence, work with Compton in Chicago, or help set up a new laboratory. McMillan chose the last option.

81 Parsons to Ashbridge, 23 June 1944, A-84-019, 18-7. This transfer almost fell through because no suitable apartments were available.

82 N. Ramsey interview by Henriksen, 13 Sept. 1985, OH-66.

83 E.g., one in Chicago on 21 Jan. 1943. McKibben to Manley, 14 Jan. 1943, A-84-019, 63-8. Other letters concerning the meeting are: Manley to Williams, 12 Jan. 1943, A-84-019, 63-4; Manley to McKibben, 12 Jan. 1943, A-84-019, 63-8; Williams to Manley, 28 Jan. 1943; and reply from Manley to Williams on 2 Feb. 1943, A-84-019, 63-4.

84 McMillan interview by Seidel, 17 April 1985, OH-63; E. O. Lawrence to H. D. Smyth, 16 Feb. 1943, H.D. Smyth Papers, American Philosophical Society, Philadelphia.

85 R. R. Wilson, "My Fight Against Team Research," in G. E. Holton, ed., *The Twentieth Century Sciences: Studies in the Biography of Ideas* (New York: Norton, 1972), p. 475.

86 Manley, "Assembling the Wartime Labs," p. 45; G. M. Almy, "A Century of Physics at the University of Illinois, 1868–1968," a talk given before the History of Science Society in Dec. 1967, revised, 1975, Center for History of Physics, American Institute of Physics, p. 20.

87 R. Bacher interview by Hoddeson, 30 July 1984, OH-45, p. 21.

88 Oppenheimer to Manley, 25 Feb. 1943, A-84-019, 64-1.

89 Bacher interview by Hoddeson, 30 July 1984, OH-45, p. 21.

90 Williams to Oppenheimer, 2 March 1943, A-84-019, 63-4.

91 Hawkins, *The Los Alamos Story*, p. 9.

92 Duffield interview by Balibrera, 1982, OH-36, pp. 1–2.

93 Marjorie Bell Chambers, "Technically Sweet Los Alamos" (Ph.D. diss., University of New Mexico, 1974), pp. 60–2; Hawkins, *The Los Alamos Story*, p. 59.

94 Chambers, "Technically Sweet Los Alamos," p. 62.

95 Chambers, "Technically Sweet Los Alamos," pp. 87–97; Hawkins, *The Los Alamos Story*, pp. 9, 59.

96 Chambers, "Technically Sweet Los Alamos," pp. 85–6.

97 Manley, "Assembling the Wartime Labs," p. 43. For the smaller Van de Graaff machine, see D. B. Parkinson, R. G. Herb, E. J. Bernet, and J. L. McKibben, "Electrostatic Generator Operating under High Air Pressure – Operational Experience and Accessory Apparatus," *Phys. Rev. 53* (1938), p. 642.

98 G. M. Almy, "A Century of Physics at the University of Illinois, 1868–1968."

99 McMillan interview by Seidel, 17 April 1985, OH-63.

100 R. Wilson, "A Recruit for Los Alamos," *Bull. Atom. Sci.* (March 1975), pp. 41–7.

101 Hawkins, *The Los Alamos Story*, pp. 47–8.

102 Bacher remembers these meetings. Bacher interview by Hoddeson, 30 July 1984, OH-45; Hawkins, *The Los Alamos Story*, pp. 47–8.

103 Conant to Oppenheimer, 20 Nov. 1942, A-84-019, 4-10. AAA ratings were the highest, but were only for use in emergencies and could not be granted to an entire project. Vincent Jones, *Manhattan: The Army and the Atomic Bomb* (Washington, D.C.: Government Printing Office, 1985), p. 57.

104 David Brinkley, *Washington Goes to War* (New York: Alfred A. Knopf, 1988), pp. 67–72, gives a vivid picture of Nelson.

105 Jones, *Manhattan*, p. 82.

106 Oppenheimer to Mitchell, 17 Feb. 1943, A-84-019, 1-9.

107 Groves approved the contract on 16 June 1944. "Statement of Approval for Contract W-7405 ENG-36," A-83-0033, 1-1. A letter contract providing $1 million to begin work was dated 1 Jan. 1943. A copy is in A-84-019, 1-7.

108 Contract W-7405 ENG-36, A-83-0033, 1-1, p. 2.

5. Research in the First Months of Project Y: April to September 1943

1 Meeting Minutes, 6 March 1943, A-83-0013, 1-1.

2 Duffield interview by Hoddeson, 17 May 1978, OH-136.

3 Present were Oppenheimer, Condon, Mitchell, William Dennes (administrative aide to Oppenheimer), Mack, McMillan, Manley, Serber, Teller, Wahl, Williams, and Wilson, with Bethe and Neddermeyer arriving late. Planning Board Minutes, 8 April 1943, VFA-213.

4 David Hawkins, *Project Y: the Los Alamos Story* (Los Angeles: Tomash Publishers, 1983), pp. 23–4; Richard Hewlett and Oscar Anderson, Jr., *The New World, 1939/1946* (University Park, Penn.: Pennsylvania State University Press, 1962), p. 235–6.

5 Serber interview by Hoddeson, 26 March 1979, OH-9, p. 24.

6 Robert Serber, *The Los Alamos Primer: The First Lectures on How to Build an Atomic Bomb* (Berkeley: University of California Press, 1992.)

7 Serber, *Los Alamos Primer*, p. 3.

8 Serber, *Los Alamos Primer*, pp. 9, 10.

9 Serber, *Los Alamos Primer*, pp. 11, 12.

10 Serber, *Los Alamos Primer*, pp. 13, 19–20.

11 Serber, *Los Alamos Primer*, pp. 21, 22.

12 Unfortunately, no documentation of the Berkeley work is available, and the Serber lectures serve as the best record of work there.

13 Serber, *Los Alamos Primer*, pp. 25–27.

14 Serber, *Los Alamos Primer*, pp. 29–32.

15 Serber, *Los Alamos Primer*, p. 31.

16 Serber, *Los Alamos Primer*, pp. 33–36.

17 Serber, *Los Alamos Primer*, p. 38.

18 Serber, *Los Alamos Primer*, p. 43–44.

19 Serber, *Los Alamos Primer*, p. 46.

20 Serber, *Los Alamos Primer*, pp. 49–50.

21 ^{239}Pu would produce a high-neutron background because of its much higher alpha emission rate; owing to its very short half-life (~24,000 years), plutonium produced many more neutrons than did ^{235}U from the same level of light-element impurities.

22 Serber, *Los Alamos Primer*, pp. 52–54.

23 Serber, *Los Alamos Primer*, p. 56.

24 As historians Richard G. Hewlett and Oscar E. Anderson have pointed out, "Experienced craftsmen in the nation's arsenals needed only orders and blueprints to turn to their lathes." *The New World, 1939/1946. A History of the United States Energy Commission*, vol. 1 (University Park, Penn.: Pennsylvania State University Press, 1962), p. 245.

25 Serber, *Los Alamos Primer*, pp. 56–63.

26 Oppenheimer in 11 March letter of invitation to Bacher, Holloway, Rabi, Segrè, Allison, Bethe, Bloch, Fermi, and Condon (on 12 March), A-84-019, 11-12. Documents reveal further meetings from Tuesday, 27 April through Saturday, 1 May. Memo from Teller to Participants of Conference, 26 April 1943, Program of Conference, with attached "Convention Schedule" A-84-019, 11-12. A "Partial List of Scientists Attending the First Los Alamos Conference in April 1943" is Robert Bacher, Kenneth Bainbridge, Hans Bethe, Felix Bloch, Owen Chamberlain, Robert Christy, Edward Condon, Enrico Fermi, Richard Feynman, Stanley Frankel, Al Graves, Joseph Kennedy, John Manley, Joseph McKibben, Edwin McMillan, Seth Neddermeyer, Eldred Nelson, I. I. Rabi, Emilio Segrè, Robert Serber, Cyril Smith, Hans Staub, Edward Teller, Richard Tolman, Arthur Wahl, Victor Weisskopf, John Williams, and Robert Wilson. J. R. Oppenheimer papers, Gov't File, Box 182, folder – Los Alamos – Correspondence 1942-1946, Ms. Div. Library of Congress.

27 Hawkins, *The Los Alamos Story*, pp. 18–22.

28 Feynman, "Theoretical Department," A-83-002, 19-15; Bethe, "Theoretical Group Organization," LA-6, 6 May 1943; Bethe, "Theoretical Group Program," LA-7, 6 May 1943.

29 Hawkins, *The Los Alamos Story*, p. 19.

30 Ibid.

31 Also reports by J. H. Williams for Group P-2 and R. Wilson for Group P-1. in P-Division Progress Report 15 July 1943, LAMS-4.

32 Governing Board Minutes, 15 July 1943, A-83-0013, 1-17; Hawkins, *The Los Alamos Story*, p. 64.

33 Williams, P-Division Progress Report, Group P-2, 15 Aug. 1944, LAMS-7.

34 Wilson, P-Division Progress Report, Group P-1, 15 July 1943, LAMS-4.

35 Hawkins, *The Los Alamos Story*, p. 92.

36 P-Division Progress Report, Group P-1, 1 Sept. 1943, LAMS-9.

37 Hans Bethe, "Comments on the History of the H-Bomb," *Los Alamos Science 3*, no. 3 (Fall 1982) pp. 42–54.

38 Robert Serber, *The Los Alamos Primer: The First Lectures on How to Build an Atomic Bomb* (Berkeley: University of California Press, 1992), pp. 56–7.

39 Tolman to Oppenheimer, 29 March 1943, A-84-019, 18-2.

40 Critchfield to Parsons, 1 June 1943, A-84-019, 18-2; Rose was even invited to Los Alamos during the summer of 1943. Parsons to Rose, 23 June 1943, A-84-019, 18-2.

41 Oppenheimer to Rose, 17 May 1943, A-84-019, 18-2.

42 Born in Chicago, Parsons grew up in the Fort Sumner–Santa Rosa, New Mexico area. After high school, he attended the U.S. Naval Academy and later the academy's postgraduate school, where he studied ordnance engineering. During the 1930s, Parsons was stationed at the Naval Ordnance Proving Ground in Dahlgren, Virginia, where he became the protégé of Dr. L. T. E. Thompson. Thompson, who helped Goddard in the early development of rockets, taught Parsons physics and introduced him to such luminaries as Eugene Wigner and Frederick Seitz. As a result of this exposure, Parsons began reading physics treatises before bed each night, a habit he continued throughout his involvement with Project Y. Parsons saw combat early in World War II when he went to the Pacific to develop proximity fuzes for the navy. In addition to his combat experience, Parsons served as special assistant to the director, OSRD, Section T (Proximity Fuzes for Shells) before coming to Los Alamos. J. Hirschfelder, "The Scientific and Technological Miracle at Los Alamos," in Lawrence Badash, Joseph Hirschfelder, and Herbert Broida, eds., *Reminiscences of Los Alamos* (Dordrecht: Reidel, 1980), pp. 82–3.

43 McMillan, "Early Days at Los Alamos," in Badash et al., *Reminiscences*, pp. 13–18; "The College Pump," *Harvard Magazine 88* (November–December 1985), p. 120.

44 Born in Shreve, Ohio, Critchfield grew up in Washington, D.C., and attended George Washington University, where he became a protégé of Gamow and Teller. In 1943, while working for the Geophysical Laboratory on a project to perfect sabots, Critchfield was approached by both Oppenheimer and Teller and persuaded to join the project.

45 Critchfield interview by Hoddeson, 5 Aug. 1980, OH-16, pp. 1–26 and by Hoddeson and Kerr, 2 Aug. 1984, OH-46, pp. 1–6.

46 Hirschfelder, "The Scientific and Technological Miracle," p. 72.

47 Private communication Serber, 7 June 1988.

48 Teller memo, 26 April 1943, announcing schedule of discussion at Conferences for Tuesday 27 April through Saturday 1 May, A-84-019, 11-12; also Oppenheimer to Smyth, 14 April 1945, B-9 collection, file 201, Smyth, H. D.

49 Hawkins, *The Los Alamos Story*, p. 14.

50 Oppenheimer to Smyth, 14 April 1945, B-9 collection, 201 Smyth, H. D.

51 Hawkins, *The Los Alamos Story*, p. 125.

52 L. T. E. Thompson to Neddermeyer on implosion experiments, 23 June 1943, A-84-019, 4-1.

53 Thompson to Oppenheimer, 25 June 1943, A-84-019, 4-1.

54 Kistiakowsky, "Reminiscences of Wartime Los Alamos," and McMillan, "Early Days at Los Alamos," both in Badash et al., *Reminiscences*, pp. 49 and 17.

55 McMillan interview by R. Seidel, 17 April 1985, OH-63.

56 Hawkins, "Manhattan District History, Project Y, the Los Alamos Project," p. 125.

57 Critchfield interviews by Hoddeson, 2 Aug. 1984, OH-46, pp. 6–7. McMillan, "Early Days at Los Alamos," p. 17. High explosives are detonated by an "explosive train," such as a blasting cap. Primacord, a small-diameter (about 0.22-inch) detonating fuse containing PETN inside a fabric wrapping, was used extensively early in the project to set off the high explosive. An electric blasting cap would detonate the Primacord. The cap contained a small amount of primary HE that was set off by bridgewire, heated by an electric current. The detonation was then picked up by a larger charge of secondary HE, typically PETN, and the detonation carried over to the Primacord, or through a "booster" tetryl pellet, to the main HE charge. U.S. Army, TM 9-1300-214, pp. 3-1 to 3-2.

58 Neddermeyer, "Collapse of Hollow Steel Cylinders," LA-18. Streib also worked on the theory of implosions in one dimension, using various simplifying assumptions. J. F. Streib, "One-Dimensional Motion of a Detonated Explosive," LA-23, 13 Sept. 1943.

59 Oppenheimer to Smyth, 14 April 1945, B-9 collection, 201 Smyth, H. D.

6. Creating a Wartime Community: September 1943 to August 1944

1 David Hawkins, *Project Y: The Los Alamos Story* (Los Angeles: Tomash Publishers, 1983) pp. 23–4.

2 K. Mark, "Roof over Our Heads," p. 40 of pp. 31–45, in "The Atom and Eve,"unpublished manuscript, VFA-201. Edited article is reprinted (but without the quotation) in Jane S. Wilson and Charlotte Serber, *Standing By and Making Do: Women of Wartime Los Alamos* (Los Alamos: Los Alamos Historical Society, 1988), pp. 29–41.

3 Conant to Oppenheimer, 28 Oct. 1942, A-84-019, 4-10.

4 Oppenheimer to Conant, 23 March 1943, A-84-019, 4-10.

5 Hawkins, *The Los Alamos Story* (Los Angeles: Tomash, 1983), pp. 34–5. Hughes began his new job on 30 June 1943. Oppenheimer to Laboratory, n.d., A-83-003, 1-1.

6 The Personnel Office also handled personnel security, draft deferments, and the placement of military personnel working within the laboratory. Hawkins, *The Los Alamos Story*, p. 35.

7 Governing Board Minutes, 11 Nov. 1943, 18 Nov. 1943, 3 May 1943 and 27 May 1943, A-83-0013, 1-33, 1-34, 1-2, 1-9. Hawkins, *The Los Alamos Story*, pp. 487–8.

8 Hawkins, *The Los Alamos Story*, graphs on pp. 484–8.

9 Hawkins, *The Los Alamos Story*, pp. 484–8.

10 First meeting of the Planning Board, VFA 213. L. Groves, *Now It Can Be Told: The Story of the Manhattan Project* (New York: Harper and Brothers, 1962); William Lawren, *The General and the Bomb* (New York: Dodd, Mead, and Company, 1988), p. 130.

11 Governing Board Minutes, 15 July 1943, A-83-0013, 1-17.

12 The security vouchers were sent to General Groves in August. Oppenheimer to Groves, 7 Aug. 1943, A-84-019, 12-7.

13 Oppenheimer to division and group leaders and members of the Coordinating Council, 23 June 1943, A-83-0003, 1-1.

14 Oppenheimer to group and division leaders and members of the Coordinating Council, 10 July 1943, A-83-0003, 1-1.

15 Governing Board Minutes, 6 May 1943, A-83-0013, 1-3.

16 Governing Board Minutes, 5 Aug. 1943, A-83-0013, 1-20.

17 Groves, *Now It Can Be Told*, p. 141; and Groves's testimony at the Oppenheimer hearings, U.S. Atomic Energy Commission (AEC), *In the Matter of J. Robert Oppenheimer; Transcript of Hearing before Personnel Security Board and Texts of Principal Documents and Letters* (Cambridge, Mass.: MIT Press, 1971), pp. 160–80; memo from Groves to A. H. Compton and Oppenheimer, A-84-019, 34-11.

18 N. Ramsey interview by Henriksen, 13 Sept. 1985, OH-66.

19 Letter from Capt. J. H. McKinley, Corps of Engineers to A. H. Compton at the Met Lab, 30 Jan. 1945.

20 J. Wilson, "Not Quite Eden," in Wilson, ed., "The Atom and Eve," unpublished manuscript, pp. 47–62; and in Wilson and Serber, *Standing By and Making Do*, p. 46.

21 In the early days of the laboratory, people from the towns surrounding Los Alamos manned the guard posts and admitted residents who had typewritten passes signed by Oppenheimer. The inner fence was patrolled by two mounted MPs, while the outer fence was patrolled by four mounted MPs. Capt. A. L. Cernaghan commander C.M.P., special orders, 24 May 1943, A-84-019, 12-7.

22 White for staff members, blue for technicians and secretaries, red for the clerks and warehouse employees, and yellow for people who did not work for the University of California but were required to be in the Tech Area. Pass System in the Tech Area, Spring 1943, A-84-019, 12-7; Bacher's papers, 4 April 1945, A-83-0003, 1-1.

23 Oppenheimer to Groves, 7 June 1943, A-84-019, 12-7; Harman to Oppenheimer, 19 May 1943, A-84-019, 12-7.

24 Capt. Peer de Silva replaced the overworked Lieutenant Clark in October 1943 and was brought to Los Alamos to work more closely with the Security Committee in the Tech Area, but a year later the committee was still complaining about the lack of communication between the civilian and military security personnel. Governing Board Minutes, 28 Oct. 1943, A-83-0013, 1-31. Governing Board Minutes, 21 Oct. 1943, A-83-0013, 1-30, p. 3; Governing Board Minutes, 14 Oct. 1943, A-83-0013, 1-29.

25 Oppenheimer to all civilian employees of the Tech Area, n.d., A-84-019, 39-4; Oppenheimer to the Coordinating Council, 11 April 1944, A-84-019, 12-7.

26 Typical infractions were technicians receiving information and indiscriminately passing it on to other technicians, scientists taking books and documents out of the Tech Area to their living quarters, classified waste not being burned, and classified documents not locked up at the end of the day. Each evening a group of WACs searched the Tech Area and confiscated any secret documents found left out. Offending parties were often punished by being asked to check for infractions by others for a week. The inspectors found an average of three to six serious breaches of security per day. Hawkins to all Group Leaders, n.d. (probably summer 1943), A-84-019, 12-7.

27 Governing Board Minutes, 19 Aug. 1943, A-83-0013, 1-22. AEC, *In the Matter of J. R. Oppenheimer*, p. 259; Governing Board Minutes, 23 Sept. 1943, A-83-0013,

1-26, pp. 1, 2. Governing Board Minutes, 30 Sept. 1943, A-83-0013, 1-27, p. 5. As Groves explained to the Governing Board, the mail would be opened at an unspecified point outside Los Alamos by a person unknown to anyone on the project, who in turn did not know anyone on the project. Classified mail would not be censored. Letters, to be mailed unsealed, could only be written in English or in one of several specified European languages, and could not be written in code or symbols. They were not to include specific information, such as the location or size of the project, the names and professions of people, or any details about the work. It was also forbidden to mention that the mail was being censored. The Governing Board objected to only the last of these regulations, so Groves compromised by allowing people to send a card to their correspondents telling them that letters were being censored. Governing Board Minutes, 4 Nov. 1943, A-83-0013, 1-32. Major Ralph Carlisle Smith, lawyer and head of the laboratory's patent office, remembers no outcry from anyone during the town meeting at which the military announced these regulations. Governing Board Minutes, 28 Oct. 1943, A-83-0013, 1-31. The minutes contain a list of the censorship regulations.

28 Hawkins, *The Los Alamos Story*, pp. 40–1, 486. The average monthly increase of SEDs in the period from Fall 1943 to August 1944 was thirty-seven per month. V. Fitch, "The View from the Bottom," *Bulletin of the Atomic Scientists* (February, 1975), pp. 43–6, esp. 44. M. H. Trytten would find the SEDs, and then Groves would transfer them to Los Alamos.

29 Hawkins, *The Los Alamos Story*, p. 41. Army regulations limited the number of promotions to a specified fraction of each rank.

30 Fitch, "The View from the Bottom," pp. 43–6; the peculiar status of the SED was established at a meeting of the Governing Board on 20 January 1944. Governing Board Minutes, 20 Jan. 1944, A-83-0013, 1-39.

31 Fitch, "The View from the Bottom," pp. 43–6. R. Davisson interview by Hoddeson, 10 April 1983, OH-51. Many in the SED, such as Fitch, Benjamin Bederson, and Richard Davisson, went on to scientific careers.

32 Fitch, "View from the Bottom," p. 44.

33 Margaret Gowing, *Britain and Atomic Energy 1939–1945* (London: Macmillan, 1964), pp. 260–1, 265–6.

34 Ruth Moore, *Niels Bohr: The Man, His Science, and the World They Changed* (Cambridge, Mass.: MIT Press, paperback ed., 1985), pp. 324–31; Edith Truslow and Ralph Carlisle Smith, *Project Y: The Los Alamos Story: Part 2 beyond Trinity* (Los Angeles: Tomash, 1983), p. 306, and Appendix 3 (which contains biographical information on the British Mission).

35 Governing Board Minutes, 9 Sept. 1943, A-83-0013, 1-24.

36 These totals do not include the SED, WACs, and employees of the post administration. If these groups are included, the ratio of women to men drops to about 24 percent.

37 C. Serber, "Labor Pains," in Jane Wilson, ed., "The Atom and Eve," unpublished manuscript, n.d., pp. 63–82, esp. 64; in *Standing By and Making Do*, p. 57.

38 L. Hornig interview by Henriksen, 25 June 1986, OH-128.

39 L. and B. Langer interview by Henriksen, 23 March 1986, OH-113, p. 31.

40 Building A became the new administration building, B was used for laboratories and offices for the ordnance program, C was the ordnance shop, D the dust-free chemistry and metallurgy laboratory, and Q the headquarters for the health

group. Smaller buildings, G and O, were built in the same period by the Force Account crews employed directly by the government, rather than by contractors. Hawkins, *The Los Alamos Story*; and Edith Truslow, "Manhattan District History: Nonscientific Aspects of Project Y, 1942-1946," LA-5200, March 1973, pp. 11–19. Two more building spurts occurred between January and March 1944 (25,000 square feet) and between April and September 1944 (75,000 square feet)

41 Hawkins, *The Los Alamos Story*, pp. 483–94. The information in Table 5-4 was taken from a Memo from the LASL Director to the U.S.A.E.C manager, office of Santa Fe Directed Operations, 11 September 1947, General Background Data concerning the Los Alamos Scientific Laboratory. LANL, E Bay, Row 7, Box 40, file "classified file, 1947." Notes on the use of the sites are from Bacher's papers, A-83-0003, 1-1.

42 Private communication R. Schreiber, 29 Nov. 1988.

43 Hawkins, *The Los Alamos Story*, pp. 56, 483–94; Governing Board Minutes, 3 Feb. 1944, A-83-0013, 1-41, p. 6.

44 Hawkins, *The Los Alamos Story*, p. 57.

45 Hawkins, *The Los Alamos Story*, pp. 57–8; Oppenheimer to Long, 14 Aug. 1944, A-84-019, 6-1. Parsons, who had controlled C Shop, believed that the shops serving the Ordnance Division should be under the control of an engineer rather than Long. Oppenheimer, however, felt that the primary responsibility of the shops was to produce equipment for the research and development program, not for the final gadget, and was opposed to putting engineers in charge. Oppenheimer to Parsons, 8 Aug. 1944, A-84-019, 6-1.

46 The vouching system did not work in the shops, because the men were not known by anyone at Los Alamos. E. Long to D. Dow, 28 Feb. 1944, A-84-019, 6-1; Dow to Oppenheimer, 29 Feb. 1944, A-84-019, 6-1.

47 Before the reorganization C Shop and V Shop, output increased gradually to about 14,000 and 10,000 man-hours per month, respectively. After the reorganization, the number of man-hours of work in both shops nearly tripled. V Shop output peaked in March 1945 at 28,000 man-hours. C Shop peaked at 42,500 man-hours in June 1945 in preparation for Trinity. Most of V Shop's expanded work load came from CM- and G-Divisions, while C Shop served the needs of X-Division. Hawkins, *The Los Alamos Story* charts 9, 10, and 11, pp. 491–3.

48 Maj. S. Stewart to Underhill, 25 March 1943, A-84-019, 18-7; Oppenheimer to Stewart, 25 March 1943, A-84-019, 18-7. Report on temporary housing, J. A. D. Muncy to Major Stewart, 11 May 1943, A-84-019, 18-7.

49 Sundt apartments came in several styles, but the most common were rectangular duplexes and four-family, two-story quadruplexes with red sloping roofs and green walls with interior walls so thin that a person stationed in one of the apartments could listen for babies crying in any of the other three.

50 Mark, "A Roof over Our Heads," pp. 31–46.

51 "Housing Manual for Laboratory Employees and Supervisors," Personnel Department, Los Alamos Scientific Laboratory, 1956, VFA-261, p. 19.

52 Governing Board Minutes, 10 June 1943, A-83-0013, 1-7, 1-12.

53 "Housing Manual," p. 22. This rent system was in effect until early in 1947. Since nearly everyone worked more than forty hours, hourly employees paid a lower percentage of their salary in rent than salaried Tech Area scientific workers, who did not receive overtime pay. Administrative Board Minutes, 26 Oct 1944, A-83-0013, 1-57, p. 5.

54 Governing Board Minutes, 5, 19, and 26 Aug., and 9 Sept. 1943, A-83-0013, 1-20, 1-22, 1-23, and 1-24. See also Hawkins, *The Los Alamos Story*, pp. 36–7.

55 Governing Board Minutes, September 1943, A-83-0013, 1-24 through 27.

56 Governing Board Minutes, 21 Oct. 1943 and 28 Oct. 1943, A-83-0013, 1-30 and 1-31.

57 Governing Board Minutes, 18 Nov. 1943, A-83-0013, 1-34, p. 3. Oppenheimer to Groves, 24 Nov. 1943, A-84-019, 18-7. Marjorie Bell Chambers, "Technically Sweet Los Alamos: The Development of a Federally Sponsored Scientific Community" (Ph.D. diss., University of New Mexico, 1974), p. 110.

58 Governing Board Minutes, 26 Nov. 1943, A-83-0013, 1-35, p. 1; Chambers, "Technically Sweet Los Alamos," pp. 109–10.

59 "Housing Manual," p. 20; E. Jette, *Inside Box 1663: Life in Los Alamos during the Manhattan Project* (Los Alamos: Los Alamos Historical Society, 1977), p. 56. The Morgan duplexes were small and plain, with a thin adjoining wall, allowing little privacy. They had pitched roofs and clapboard siding and were set in neat parallel rows in military order. R. Marshak, "Secret City," in J. Wilson, ed., "The Atom and Eve," pp. 1–20, quote on p. 9. Machinists recruited from Texas refused to live in the units, touching off a battle in the Town Council, when the laboratory tried to move fourteen families from Sundt apartments to the Morgan area.

60 Governing Board Minutes, 6 July 1944, A-83-0013, 1-51, p. 2.

61 Stock room and receiving clerks were apparently hard to find without housing as well. Governing and Administrative Board Minutes, 26 Oct., 6 July, 20 July 1944, and 6 May 1943, A-83-0013, 1-57, 1-51, 1-52, p. 1, and 1-3.

62 Governing Board Minutes (last meeting), 29 June 1944, A-83-0013, 1-50.

63 Hawkins, *The Los Alamos Story*, pp. 36–7; Chambers, "Technically Sweet Los Alamos," p. 113. The 100 small, prefabricated, single-family McKee houses became available in three stages between July and December 1944. They were considered more desirable than the Morgan duplexes, because they were single units offering some degree of privacy. Mark, "A Roof over Our Heads," pp. 31–46. McKee houses had smooth outer walls and flat roofs and about 400 square feet of floor space.

64 Governing Board Minutes, 6 March 1943, A-83-0013, 1-1.

65 Hempelmann was there learning how to operate the 45-inch cyclotron and set up a clinic to treat cancer patients with radioactive phosphorus. His background in radiology and physics was on a par with most of the other medical experts in the country; he had spent a few months at the Berkeley Radiation Laboratory working with John Lawrence (Ernest's brother) and a month in New York's Memorial Hospital learning about radiation therapy. At the suggestion of John Lawrence, Oppenheimer asked Hempelmann to come to Los Alamos to handle the radiological health problems. Hempelman interview by Henriksen, 31 Jan. 1986, OH-106.

66 Louis Hempelmann, "History of the Health Group," March 1943 to Nov. 1945, A-84-019, 8-5, p. 4.

67 Hawkins, *The Los Alamos Story*, p. 59.

68 Hempelmann, "History of the Health Group," p. 3.

69 Hempelman interview by Henriksen, 31 Jan. 1986, OH-106, p. 4 .

70 MED doctors knew that plutonium has a lower α activity than radium. They did not know that it is more difficult to absorb from the digestive tract and harder to eliminate from the body once absorbed. Fortunately, these two differences offset

each other, so that if plutonium enters the body, it is usually excreted by both the lungs and digestive system before it is absorbed. Hawkins, *The Los Alamos Story*, p. 54; Hempelmann interview by Henriksen, 31 Jan. 1986, OH-106.

71 Hempelmann, "History of the Health Group, pp. 7–8.

72 Hempelman interview by Henriksen, 31 Jan. 1986, OH-106. Governing Board Minutes, 17 Aug. 1944, A-83-0013, 1-54, p. 1.

73 At the time of the Trinity test, eight military members of the group were sent overseas and not replaced.

74 Letter from Oppenheimer to all group and division leaders, 26 Sept. 1944, A-84-019, 6-1.

75 Priscilla Duffield credits R. Wilson with the idea of forming a town council. Duffield interview by LANL, 1982, OH-36, p. 9. The scientists and their families chose the first council from among themselves, and then let the Governing Board make the official appointments. A. K. Smith, "Law and Order," in J. Wilson, ed., The Atom and Eve," unpublished manuscript, pp. 83–104, p. 87, in *Standing By and Making Do*, p. 75 William R. Dennes was named the first chairman.

76 Condon to J. M. Harman, 3 April 1943, A-84-019, 18-7.

77 Town Council Minutes, 7 May 1943. Candidates for the first election were Robert Christy, William Dennes, Benjamin Diven, Leo Lavatelli, Julian Mack, Mary Mack, John Manley, Kitty Oppenheimer, W. Schafer, Helen Stokes, John Williams, and Robert Wilson.

78 Town Council Minutes, 21 May 1943. One of the council's duties was to advise David Hawkins, the liaison to the Post Administration, who sat in on most of the council meetings as an unofficial member. Governing Board Minutes, A-83-0013, 1-3.

79 Governing Board Minutes, 22 July 1943, A-83-0013, 1-18. Letter in the Town Council Minutes for 22 July 1943.

80 Smith, "Law and Order," pp. 84, 96; Weisskopf interview by Hoddeson, 10 March 1978, OH-6, pp. 11–13. Also *Standing By and Making Do*, pp. 76–7, 83.

81 Town Council Minutes, 13 Dec. 1943, 27 March 1944, 3 April 1944.

82 Town Council Minutes, 8, 22, and 25 May, and 26 June 1944.

83 Town Council Minutes, 3 and 10 April, 8 May, and 5 June 1944.

84 Hawkins to Oppenheimer on the Outing Committee, in Town Council Minutes, 2 May 1944; Town Council Minutes, 22 May and 3 July 1944; Jette, *Inside Box 1663*, pp. 60–1.

85 R. C. Smith interview by Kerr, 5 April 1983, OH-40, pp. 6–8.

86 Town Council Minutes, 24 and 28 May 1945, and Executive Committee of the Town Council (Grubman, Greisen, and Allison) to Tyler and Oppenheimer, 24 May, 1945; Town Council Minutes, 19 Feb. and 6 June 1945. Weisskopf was chairman of the Town Council in early 1945. Weisskopf to Ross, 14 Feb. 1945, in Town Council Minutes; Smith, "Law and Order," pp. 83–104.

87 Town Council Minutes, 1 and 15 May, and 5 June 1944. One of the few instances of agreement between the council and post was the judgment that small refrigerators should not be replaced with larger electric models. Town Council Minutes, 8 and 29 May, 5, 19, and 26 June, 3, 24, and 31 July, 7 and 14 Aug. 1944, and 4 Sept. 1944.

88 R. Marshak, "Secret City," pp. 1–20 in J. Wilson, ed., "The Atom and Eve," unpublished manuscript, p. 17

89 Marshak, "Secret City," p. 16; S. Barnett, "Operation Los Alamos," pp. 105–22, in J. Wilson, ed., "The Atom and Eve," unpublished manuscript.

90 B. Brode, "Tales of Los Alamos," in Lawrence Badash, Joseph Hirschfelder, and Herbert Broida, eds., *Reminiscences of Los Alamos, 1943–1945* (Dordrecht: Reidel, 1980), pp. 133–60. The chapter is a condensation of Brode's articles for the *LASL Community News*, 2 June to 22 Sept. 1960. The original articles give more personal details.

91 The median age of Los Alamos employees was twenty-seven, and the average age of the scientific employees was twenty-nine. The school opened with 140 students; the first graduating class consisted of two students. Of the first 40 students in the upper school in the fall of 1943, only 8 were the children of staff members.

92 Jette, *Inside Box 1663.*

93 Marshak, "Secret City," pp. 13–14.

94 Town Council Minutes, 26 Feb. 1945.

95 Governing Board Minutes, 24 June and 19 Aug. 1943, A-83-0013, 1-14 and 1-22.

96 Governing Board Minutes, 8 July and 5 Aug. 1943, A-83-0013, 1-16 and 1-20.

97 Governing Board Minutes, 21 Oct. 1943, A-83-0013, 1-30.

98 Langer interview by Henriksen, 21 March 1986, OH-113.

99 Most of the trees on the ski hill were cleared with the help of plastic explosives administered by Kistiakowsky.

100 Town Council Minutes, 22 Jan. 1944.

101 Santa Fe was the most convenient place to buy alcohol, which could not be sold legally in Los Alamos, a military post. Residents with access to the chemistry or physics laboratories would often brew up "Tech Area punch," a potent concoction consisting of fruit juice and pure alcohol.

102 Marshak, "Secret City," p. 12.

103 Langer interview by Henriksen, 21 March 1986, OH-113.

7. The Gun Weapon: September 1943 to August 1944

1 Parsons to Thompson, 25 June 1943, A-84-019, 18-2.

2 Minutes of Conference on Gun, 16 July 1943, A-84-019, 18-2.

3 David Hawkins, *Project Y: The Los Alamos Story* (Los Angeles: Tomash, 1983), pp. 114–15; untitled, B-9, 310.1.

4 Robert Serber, *The Los Alamos Primer: The First Lectures on How to Build an Atomic Bomb* (Berkeley: University of California Press, 1992), p. 57

5 J. Serduke, LANB-C-66.

6 Hirschfelder to Parsons, n.d., A-84-019, 18-2.

7 Hirschfelder to Parsons, n.d., A-84-010, 18-2.

8 P. Serao and J. Pierce, "Sensitivity of Performance to Combustion," in Herman Krier and Martin Summerfield, eds., *Interior Ballistics of Guns* (New York: American Institute of Aeronautics and Astronautics, 1979), pp. 261–2.

9 Progress Reports of Group E-8, 1 March 1944, LAMS-73, and 15 June 1944, LAMS-106; Hirschfelder to Parsons, 7 June 1944, A-84-019, 77-8.

10 E-Division Progress Report, 1 March 1944, LAMS-73.

11 Hirschfelder to McMillan, 13 July 1944, A-84-019, 18-2.

12 E. M. McMillan, "Early Days at Los Alamos," in Lawrence Badash, Joseph Hirschfelder, and Herbert Broida, eds., *Reminiscences of Los Alamos, 1943–1945* (Dordrecht: Reidel, 1980), pp. 17–18.

13 McMillan, "Early Days," in *Reminiscences*, pp. 17–18.

14 R. R. Wilson, "Addendum to RRW letter to AM," 12 Nov. 1991.

15 Private communication Critchfield, 16 Dec. 1988.

16 In October, the committee merged with the initiator committee. Parsons to Oppenheimer, 11 Feb. 1944; Oppenheimer to Committee Members, 12 Feb. 1944; and McMillan to Committee, 26 Feb. 1944, all in B-9 files, folder 334 Steering Committee.

17 E-Division Progress Report, 1 March 1944, LAMS-73, p. 32.

18 Hawkins, *The Los Alamos Story*, p. 116.

19 "Development of Subcaliber Projectiles for the Hispano-Suiza Gun, NDRC-233, VFA-558. *NDRC-233*, pp. 3–4.

20 E-Division Progress Report, 15 April 1944, LAMS-80, pp. 3, 32–33.

21 The time between the emission of a neutron and the resulting fission was estimated to be on the order of 10^{-8} sec; Serber, *The Los Alamos Primer*, pp. 51–2.

22 Serber, *The Los Alamos Primer*, pp. 52.

23 Prestwood to Oppenheimer, Kennedy and Segrè, 3 April 1943, A-84-019, 15-8.

24 Oppenheimer to Latimer, 8 July 1943, A-84-019, 15-8.

25 Allison to Whitaker, 7 July 1943, A-84-019, 15-8.

26 Hamilton to Oppenheimer, 17 July 1943, A-84-019, 15-8.

27 Oppenheimer to Groves, 27 July 1943, A-84-019, 15-8; Groves to Oppenheimer, 16 Aug. 1943, A-84-019, 15-8.

28 Oppenheimer to Groves, 27 July 1943, A-84-019, 15-8.

29 Hamilton to Oppenheimer, 17 July 1943, A-84-019, 15-8.

30 Memorandum of Meeting, 21 July 1943, A-84-019, 15-8.

31 Thomas to Oppenheimer, 31 July 1943, A-84-019, 15-8.

32 Governing Board Minutes, 19 Aug. 1943, A-83-0013, 1-22.

33 Segrè to Oppenheimer, 28 Oct. 1943, A-84-019, 15-8.

34 Ruhoff to the District Engineer, 24 Sept. 1943, A-84-019, 15-8.

35 Oppenheimer to Thomas, 5 Oct. 1943, A-84-019, 15-8.

36 Oppenheimer to Groves, 27 July 1943, A-84-019, 15-8; R. C. Hewlett and O. E. Anderson, Jr., *The New World, 1939/1946: A History of the United States Energy Commission*, vol. 1 (University Park: Pennsylvania State University Press, 1962), p. 282. For an explanation of Anglo-American atomic energy relations at the time, see also pp. 255–84.

37 Oppenheimer to Thomas, 5 Oct. 1943, A-84-019, 15-8; Thomas to A. H. Compton, 1 Oct. 1943, A-84-019, 15-8.

38 Segrè to Oppenheimer, 28 Oct. 1943, A-84-019, 15-8.

39 Thomas to Groves and Conant, 6 Nov. 1943, A-84-019, 9-4.

40 Thomas to Conant and Groves, 6 Nov. 1943, A-84-019, 9-4.

41 Thomas to Conant and Groves, 6 Nov. 1943, A-84-019, 9-4.

42 Penneman interview by Westfall, 23 Oct. 1985, OH-76.

43 Polonium Meeting Minutes, 16 Dec. 1943, A-84-019, 15-8; Thomas to Groves and Conant, 4 Jan. 1944, A-84-019, 9-4.

44 Thomas to Conant and Groves, 4 Jan. 1944, A-84-019, 9-4.

45 Thomas to Groves and Conant, 10 Feb. 1944, A-84-019, 9-4; Thomas to Groves, 6 April 1944, A-84-019, 15-8.

46 Private communication Penneman, 19 Dec. 1985.

47 Thomas to Groves and Conant, 10 Feb. 1944, A-84-019, 9-4.

48 Thomas to Groves and Conant, 8 April 1944, A-84-019, 9-4.

49 Oppenheimer to Groves, 26 Aug. 1944, A-84-019, 15-8.

50 Parsons, 9 Feb. 1944, A-84-019, 17-4.

51 Oppenheimer to Parsons, 8 Feb. 1944, A-84-019, 17-4.

52 Oppenheimer to Division Leaders and Members of the Steering Committee, 8 Feb. 1944, A-84-019, 17-4; and Oppenheimer to Parsons, 8 Feb. 1944, A-84-019, 17-4.

53 E-Division Progress Report, 15 May 1944, LAMS-73, p. 4.

54 E-Division Progress Report, 15 April 1944, LAMS-80, pp. 4, 33.

55 Critchfield moved over to G-Division, where he then worked on the implosion initiator.

56 Birch to Parsons, 30 Nov. 1944, and 9 Feb. 1945, A-84-019, 78-10.

57 Serber, *The Los Alamos Primer*, pp. 56–9; Hawkins, *The Los Alamos Story*, pp. 64, 80.

58 Hawkins, *The Los Alamos Story*, pp. 64, 80.

59 E-Division Progress Report, 1 March 1944, LAMS-73, pp. 3–4, 32.

60 E-Division Progress Report, 15 April 1944, LAMS-80.

61 Hawkins, *The Los Alamos Story*, p. 80.

8. The Implosion Program Accelerates: September 1943 to July 1944

1 Oppenheimer to von Neumann, 27 July 1943, A-84-019, 35-8; von Neumann testimony in U.S. Atomic Energy Commission (AEC), *In the Matter of J. Robert Oppenheimer: Transcript of Hearing before Personnel Security Board* (Washington, D.C.: U.S. Government Printing Office, 1954), p. 644; and Tolman to Oppenheimer, 23 July 1943, R.G. 227, S-1 Files, Tolman Papers, Folder - Oppenheimer J. R., NARA.

2 Shaped charges are pieces of high explosive that have a conical depression and are lined by a heavy metal. When the high explosive is detonated, the liner projects forward in a rapidly moving jet of molten metal, followed by a slug containing most of the liner, also molten, useful for piercing armor and other fortifications. In 1888, Charles Munroe, a botanist at the Naval Academy, invented the shaped charge, according to Joseph Hirschfelder, as a method of exploring the fine details of leaves and plants. Munroe had placed leaves on a small steel plate, and above these a piece of igloo-shaped plastic explosive. Detonation of the explosive set up shock waves that were focused by the variations in the thickness of the structure of the

leaves, causing a raised etched image of the leaves to form on the plate. Between the two World Wars, scientists in Germany added to Munroe's experiment a steel liner placed inside the igloo-shaped explosive. The resulting jets of steel, traveling at high velocity, were capable of penetrating many feet of concrete, an effect then utilized by the Germans in penetrating the Maginot Line in the early part of World War II. Subsequently, von Neumann would help to work out the theory of the focusing of shaped charges. Also "Military Explosives," TM 9-1300-214, Department of the Army Technical Manual, Sept. 1984, pp. 2–20.

3 J. von Neumann, "Theory of Detonation Waves," Progress Report, OSRD No. 549. More on von Neumann's background can be found in the von Neumann papers at the Library of Congress, the Herman Goldstein papers at the American Philosophical Society, the Goldstone papers at Hampshire College, and records of the Princeton University Mathematics Department. Conversation with William Asprey, 26 May 1987 at Los Alamos.

4 Oppenheimer teletype to Groves, 6 Aug. 1943, RG 227, S-1 Files, Tolman Papers, folder – Oppenheimer, J. R., NARA.

5 Parsons to von Neumann, 14 Aug. 1943, A-84-019, 35-8; Parsons to Purnell, 23 Aug. 1943, A-84-019, 35-8. Von Neumann in AEC *In the Matter of J. Robert Oppenheimer*, p. 644.

6 J. von Neumann, Remarks Concerning Detonation Waves and Shock Waves, 16 Sept. 1942, A-85-025, 1-3; David Hawkins, *The Los Alamos Story* (Los Angeles: Tomash Publishers, 1983), p. 125; Bethe interview by Hoddeson, 3 Oct. 1986, OH-140.

7 Hawkins, *The Los Alamos Story*, p. 126. According to Donald Mueller, a member of Neddermeyer's group, Neddermeyer had independently arrived at the same considerations and had calculated the compression about a month before von Neumann.

8 Critchfield interview by Hoddeson and Kerr, 2 Aug. 1984, OH-46, pp. 8–9.

9 In fact, the hollow implosion device, as conceived then, could not have been built with the wartime technology of high-explosive and fissile material. But there was no way to predict the future development of this novel field. Ironically, the implosion device used at Trinity and Nagasaki had a short assembly time and depended totally on an initiator.

10 Governing Board Minutes, 23 Sept. 1943, A-83-0013, 1-26.

11 Minutes of Conference, 1 Oct. 1943, File 334 of meetings, A-84-019, 38-8.

12 Neddermeyer to Parsons, 4 Oct. 1943, "HE Program," A-84-019, 4-1.

13 Fine to Tolman, 12 Oct. 1943, on "Discussion with Dr. von Neumann," RG 227, S-1 files, Tolman Papers, von Neumann folder, NARA.

14 Von Neumann to Parsons, 9 Oct. 1943. RG 227, S-1 Files, Tolman papers, von Neumann folder, NARA. Teller to von Neumann, 14 Oct. 1943. RG 227, S-1 Files, Tolman Papers, von Neumann folder, NARA.

15 These plans eventually fell through. "Conference of HE Group" 19 Oct. 1943, A-84-019, 4-1; Minutes of the Liaison Committee, 30 Nov. 1943, Papers of J. R. Oppenheimer, Box 291, Feb.–Dec. 1943, Manuscript Division Library of Congress.

16 Memorandum, "HE Program Discussed at Meeting," 25 Oct. 1943, A-84-019, 4-1; see also 28 Oct. 1943 memo from Neddermeyer to HE group, "Division of Responsibilities for Starting HE Program," A-84-019, 4-1.

17 Governing Board Minutes, 28 Oct. 1943, A-83-0013, 1-31.

18 Governing Board Minutes, 28 Oct. 1943, A-83-0013, 1-31.

19 A certain minimum enrichment is necessary for the successful use of 25 in an implosion. The limited energy of the implosion system to compress the active material is wasted on any nonfissile diluent, whose presence not only increases the total amount of ^{235}U required for criticality but also wastes neutrons through capture and scattering during the energy-producing phase.

20 Governing Board Minutes, 28 Oct. 1943, A-83-0013, 1-3.

21 S. Neddermeyer to members of HE Group, 28 Oct. 1943, "Division of Responsibilities for Starting HE Program," A-84-019, 4-1.

22 Parsons to Oppenheimer, "Ordnance Engineering Program on HE Development," 29 Oct. 1943, A-84-019, 4-1; Bethe to Oppenheimer, "Preliminary Specifications for Implosion Method," 30 Oct. 1943, B-9, 201 Bradbury, X-1 General.

23 Hawkins, *The Los Alamos Story*, p. 69. For further information on the electromagnetic method, see Richard G. Hewlett and Oscar E. Anderson, Jr., *The New World, 1939/1946: A History of the United States Energy Commission*, vol. 1 (University Park: Pennsylvania State University Press, 1962), pp. 151, 301. Electromagnetically enriched ^{235}U was actually used in the gun, despite an Oppenheimer proposal, turned down by Groves, to use the ^{235}U far more efficiently in composite implosion capsules. Bethe interview by Hoddeson, Baym, and Redman, 18 Aug. 1988. Bethe pointed out that there would have been some delay in making such capsules, and that Groves was in fact correct, in that the two bombs used ended the war. A second Fat Man was ready about mid-August but was not needed.

24 Governing Board Minutes, 4 Nov. 1943, A-83-0013, 1-32, pp. 2–3.

25 Oppenheimer to H. D. Smyth, 14 April 1945, 201 Smyth, H. D., B-9 files.

26 Hawkins, *The Los Alamos Story*, p. 126.

27 Governing Board Minutes, 28 Oct. 1943, A-83-0013, 1-31.

28 See Richard Rhodes, *The Making of the Atomic Bomb* (New York: Simon and Schuster, 1986), p. 376. According to Rhodes, Kistiakowsky was "tall, big-boned, boisterous, with a flat Slavic face and abiding self-confidence." Born in Kiev in 1900, he had joined the White Army at eighteen and fought in the Russian Revolution. In 1925, he fled to Berlin, where he took a doctorate in chemistry at the University of Berlin. He then sought a career in the United States and came to Princeton on a fellowship. He soon achieved faculty status and in 1930, at Conant's urging, moved to Harvard, where he was made professor in 1937.

29 Governing Board Minutes, 28 Oct. 1943, A-83-0013, 1-31.

30 Kistiakowsky, "Reminiscences of Wartime Los Alamos," in Lawrence Badash, Joseph Hirschfelder, and Herbert Broida, eds., *Reminiscences of Los Alamos, 1943/1945* (Dordrecht: Reidel, 1980), p. 45.

31 Oppenheimer to Conant in Washington, 1 Nov. 1943, 471.6 Implosion Type, RMC.

32 Kistiakowsky to Parsons, 24 Nov. 1943, "Program and Requirements of the Research and Development Phase of the HE Project," A-84-019, 4-1; also Governing Board Minutes, 26 Nov. 1943, A-83-0013, 1-35.

33 Kistiakowsky to Parsons, 24 Nov. 1943, "Program and Requirements of the Research and Development Phase of the HE Project," A-84-019, 4-1; also Governing Board Minutes, 26 Nov. 1943, A-83-0013, 1-35.

34 Kistiakowsky to Parsons, 24 Nov. 1943, "Program and Requirements of the Research and Development Phase of the HE Project," A-84-019, 4-1; also Governing Board Minutes, 26 Nov. 1943, A-83-0013, 1-35.

35 McKibbin cards and documents, A-83-010; Kistiakowsky, "Reminiscences," p. 49; Hawkins, *The Los Alamos Story*, p. 126.

36 Bainbridge's group was eventually divided into sections under R. W. Henderson, W. Schaffer, and Lewis Fussell, and his place as head of the instrumentation group, E-2, was taken by Lyman Parratt. Hawkins, *The Los Alamos Story*, p. 112. The implosion group, E-5, was divided into the sections Kistiakowsky had specified in late November. In addition, David Busbee was made responsible for S-Site, a new larger casting facility that was to be built, and J. FitzPatrick was put in charge of maintenance and service.

37 Despite this decision, Busbee, with Parsons's support, went ahead with the design, construction, and equipping of the new high-explosive casting plant at S-Site using large-scale single-pour castings of explosive; see explosives section below. Minutes of the Liaison Committee, 30 Nov. 1943, Papers of J. R. Oppenheimer, Box 291, Feb.–Dec. 1943, Ms. Division, Library of Congress.

38 The tests suggested possible causes of the asymmetry and lower velocity, e.g., detonation synchronization problems, chemical inhomogeneities, the geometric arrangement, and intrinsic characteristics of the high explosives. Only much later would Los Alamos scientists realize that, because of edge and other effects, cylindrically symmetric implosions could not adequately simulate spherically symmetric ones. Although cylindrical lenses were tried, they were abandoned in winter 1944–45, because they failed to prevent jetting. Electric detonators were not available; Primacord was not consistent enough. L. Redman, private communication.

39 The description of the method appears in three documents, two British, by Tuck, scientific adviser, Ministry of Defense, and the radiological section of the Armaments Research Department, and the third American, by John Clark and Leslie Seely at Aberdeen Proving Ground. J. L. Tuck and the Armaments Research Department, "Studies of Shaped Charges by Flash Radiography. 1. Preliminary," 15 March 1943, A. C. 3654; Tuck and the Armaments Research Department, "2. The Munroe Effect," June 1943, A. C. 4130; L. B. Seely and J. C. Clark, "High Speed Radiographic Studies of Controlled Fragmentation. I. The Collapse of Steel Cavity Charge Liners," 16 June 1943, B. R. L. 368. Von Neumann came across two reports by Tuck and handed them over to Fine in Washington on 20 Nov. suggesting that they be sent "to Y as quickly as possible." Fine noted on this letter, "AC-3654 Sent to Parsons on 24 Nov., AC-4130 Sent to Parsons on 22 Nov. PJF." J. von Neumann to Fine, 20 Nov. 1943. RG 227, S-1 Files, R. C. Tolman Papers, von Neumann folder, NARA.

40 Donald Mueller, a physicist who had done some graduate work at Princeton and Cornell, had earlier been with the Hartford Empire Glass company in Connecticut and was also a teacher of meteorology in Albuquerque. His focus had been nuclear physics; at Princeton in the 1930s he helped to develop the first pressurized Van de Graaff machine. Oppenheimer learned about Mueller from contacts at the University of New Mexico, where Mueller was teaching, and invited him to join Project Y, along with his fiancée Frances Clough, who was asked to teach in the nursery school. Mueller arrived on 20 Oct. 1943; Frances joined him eight months later. His early jobs included designing a cubicle, with a 2-inch-thick concrete wall, to protect the X-ray film, and a building to house the equipment, through which a hole was drilled to allow the X rays to pass.

41 E-Division Progress Report, 1 March 1944, LAMS-73, p. 35.

42 E-Division Progress Report, 15 April 1944, LAMS-80, pp. 37–8.

43 E-Division Progress Report, 1 March 1944, LAMS-73, pp. 5, 35, and 39–40. E-Division Progress Report, 15 April 1944, LAMS-80, pp. 5, 25–28.

44 For instance, by analyzing the mass motion during collapse and the subsequent expansion in three phases (lens action, Munroe jets, and expansion jets), Greisen tried to remove the "apparent disagreement" between X-ray results and those obtained by Linschitz in the terminal observations program.

45 Greisen, "Report on Factors Which Have Limited the Output of the Anchor Ranch X-ray Group," 11 Aug. 1944, A-84-019, 62-3.

46 A. Wayne Campbell, an ERL physicist, invented this technique early in 1943 for "photographing the detonation of shells, and the jets from various cavity charges." In the original design, the argon gap surrounded a spherical Comp B charge. Campbell did not realize that in the fall of 1943 Kistiakowsky had brought news about this flash technique to Los Alamos. Thus on his arrival at Los Alamos in late April 1945, Campbell was surprised to learn of the extensive use to which his invention had been put in the implosion program. "The Flash Photography of Detonating Explosives, to 1 May 1943," 3 June 1943, NDRC B-1488, Div. 8; "The Application of Flash Photography to the Study of Explosion Phenomena," 28 Jan. 1946, NDRC-B-5616; A. C. Graves, ed., *Miscellaneous Physical and Chemical Techniques: Experimental Techniques*, National Nuclear Energy Series, Div. V., Los Alamos Project, V. III (New York: McGraw-Hill Publishers, 1952), p. 38.

47 Koski, "Flash Photography of Collapsing Cylinders," LAMS-77, 19 April 1944; G. H. Messerly, "A Rotating Drum Camera for the Optical Study of Detonations," NDRC-B-285; also "The Use of the Rotating Drum Camera for the Measurement of the Velocities of Shell or Bomb Fragments"; 15 July 1944, NDRC-B-3900, OSRD Nos. 1964, 1965, 2071, 3022, and 3079; "Construction and Operation of the Rotating Mirror Camera," 2 Jan. 1946, NDRC-B-5614, Div. 8; D. P. MacDougall, G. H. Messerly, "A Rotating Drum Camera for the Optical Study of Detonations," 8 July 1942, Serial No. 285 OSRD Report No. 682; Payman, Shepherd and Woodhead, "High Speed Camera for Measuring the Rate of Detonation in Solid Explosives," Safety in Mines Research Board Paper No. 99, H. M. Stationery Office, London (1937).

48 B. Brixner interview by Henriksen, 27 Jan. 1986, OH-105.

49 Brixner interview by Henriksen, 27 Jan. 1986, OH-105; Brixner, a member of Julian Mack's photography group, had, before joining Los Alamos, been a general photographer in the Soil Conservation Service of the U.S. Department of Agriculture. In Mack's photography group, he helped build and maintain cameras, and even cleaned cameras dug out of gravel after test implosions.

50 Brixner interview by Henriksen, 27 Jan. 1986, OH-105; Streib, "Photographic Observation of Collapsing Cylinders, Progress Report of Feb. 7, 1944," LAMS-60, 14 March 1944.

51 With a Ph.D. from Johns Hopkins in nuclear chemistry in 1942 and experience there with uranium isotope separation, Koski had worked at the Hercules Powder Co. in Wilmington, Md., on the sensitivity of high explosives and on the stability of smokeless powders. He had studied such explosives as RDX, PETN, TNT, and nitroglycerine under the auspices of the NDRC. Through this work he met Kistiakowsky, who then called on Koski to help with Project Y. W. Koski interview by Hoddeson, 23 May 1986, OH-121; McKibbin cards; Koski, Los Alamos Notebook 29, entries on 18 and 19 Jan. 1944, pp. 3–4.

52 Brixner interview by Henriksen, 27 Jan. 1986, OH-105; J. Streib, based on work

done by W. Koski, J. Streib, and assistants, "Photographic Observation of Collapsing Cylinders," 7 Feb. 1944, A-84-019, 76-18. The experimental setup is described on pp. 2–3. Also, monthly progress reports of Koski's work on flash photography from Feb. 1944 through early April 1945, gathered in W. Koski, "Flash Photography, Section X-1C, Results of Studies Made by the Flash Photography Method in 1944–1945," 3 April 1945.

53 M. Patapoff, "The Observation of Collapsing Cylinders by the Rotating Mirror Camera" 5 June 1944, A-84-019, 76-19. The technical problems of the first rotating prism camera are given in J. C. Hoffman to G. B. Kistiakowsky, "Trouble shooting the first rotating prism camera," 6 July 1944, A-84-019, 76-19. Also J. G. Hoffman, F. Martin, and A. W. Campbell, "The Rotating Mirror Camera," in Thoma Snyder, David Lipkin, Julian Mack, *et al.*, "Los Alamos Technical Series Volume 1: Part 3, General Equipment and Techniques, Section C (Chapters 21 through 23)," LA-1032, 15 Dec. 1947, chapter 22, sections 1.3 and 4, pp. 18–19.

54 E-Division Progress Report, 1 March 1944, LAMS-73, pp. 17, 34, 36–38.

55 Parsons's summary in E-Division Progress Report, 15 April 1944, LAMS-80, pp. 4–5; Koski reported on 14 April that collapsing cylinders initiated at four points did not collapse symmetrically and that the collapse times ranged from 20 to 29 μs, depending on the size and mass ratios. Koski, "On Flash Photography of Collapsing Cylinders," 14 April 1944, A-84-019, 76-18; also Kistiakowsky in E-Division Progress Report, 15 April 1944, LAMS-80, pp. 11, 12, 19, 20.

56 Koski, "Flash Photography of Collapsing Cylinders," 19 April 1944, LAMS-77, pp. 2–6.

57 R. R. Wilson, "Addendum to RRW letter to AM," 12 Nov. 1991.

58 E-Division Progress Report, LAMS-80, 15 April 1944, pp. 4, 12, 37.

59 H. Linschitz, "Terminal Observations Progress Report," for the period 13 May to 15 July 1944, 28 July 1944, A-84-019, 77-5, covers recovery techniques, wave interactions, and symmetry (i.e., Kauzmann's work), the mechanism of action of partitions, and a number of special experiments including flow of plugs in spheres, spall of liners, mixing of liners and tamper, Primacord tests, and tracer spheres.

60 Rossi has recently published his vivid recollections of the RaLa experiment, as well as other features of his experience at Los Alamos during World War II, in his autobiography: Bruno Rossi, *Moments in the Life of a Scientist* (Cambridge: Cambridge University Press, 1990).

61 One curie of material is the quantity that has the same number of radioactive disintegrations per second as one gram of radium, 3.7×10^{10} per sec. Governing Board Minutes, 4 Nov. 1943, A-83-0013, 1-32.

62 James S. Allen and Bruno Rossi, "Time of Collection of Electrons in Ionization Chambers," LA-115, 29 July 1944.

63 Rossi's handwritten notes on his copy of the 1 Dec. 1943 progress report indicate his early thoughts on detecting equipment for RaLa as well as a half dozen other projects. Bruno Rossi, private collection.

64 Since the source had to be small enough to be placed at the center of the implosion assembly without significantly altering the geometry, the basic problem facing the radiochemists was how to produce the most intense source in the smallest possible volume. This size restriction would be even more crucial after December 1944, when the hollow implosive assembly gave way to the Christy gadget in the RaLa program (Chapter 15). Normal chemical extraction methods involved using an inert carrier; however, to avoid diluting the source material, which would increase

the size of the source, the Oak Ridge chemists had to devise separation schemes to produce a "carrier-free" product. Since any inert chemical carrier had to be removable, the addition of inactive barium as a carrier was ruled out. The chemists suggested lead and sodium sulfate to precipitate both insoluble lead sulfate and barium sulfate, with lead being removed later. Coryell to Doan, 28 March 1944, "Plans for making RaLa," A-84-019, 32-18; C. D. Coryell and H. A. Levy to R. L. Doan, 28 March 1944, "Conditions Required for Carrying out the Lanthanum Preparations," A-84-019, 16-1; also A. H. Compton to Oppenheimer, 3 May 1944, A-84-019, 32-18.

65 Kistiakowsky to Parsons, "Meeting Held in Oppenheimer's Office, 8 April 1944," 12 April 1944, A-84-019, 16-1. Present were Oppenheimer, Rossi, Bethe, Teller, Parratt, Neddermeyer, Alvarez, and Kistiakowsky.

66 Oppenheimer to A. H. Compton, 7 April 1944, A-84-019, 16-1.

67 A. Compton to J. R. Oppenheimer, 8 April 1944, A-84-019, 16-1; A. H. Compton to Whitaker, 8 April 1944, "In re: Radioactive Sources for Y," A-84-019, 16-1.

68 Doan to Oppenheimer, "Ra La," 15 April 1944, A-84-019, 32-18.

69 Barton C. Hacker notes that chemists doing the separation were in fact overexposed, but not at the firing site. B. Hacker, *The Dragon's Tail: Radiation Safety in the Manhattan Project, 1942–1946* (Berkeley: University of California Press, 1987).

70 Oppenheimer to Groves, 13 April 1944; Oppenheimer to Col. K. D. Nichols, 21 April 1944, A-84-019, 16-1.

71 Attended by Alvarez, Bethe, Kistiakowsky, Neddermeyer, Oppenheimer, Parratt, Parsons, Rossi, Teller, and Weisskopf.

72 Kistiakowsky, "Memorandum on Conference of 15 April 1944," and "Subject: Discussion of the Ra-Ba-La Experiments," A-84-019, 16-1.

73 Kistiakowsky to Oppenheimer, "Ra-Ba-La Experiments," 24 April 1944. A-84-019, 32-18.

74 J. R. Oppenheimer to H. Staub, on "Ra-La Committee, 26 April 1944," A-84-019, 16-1.

75 Kistiakowsky to Parsons, 2 May 1944, "Tank for mobile laboratory for the Ra-Ba-La tests," A-84-019, 40-2; Alvarez interview by Hoddeson, 28 April 1986, OH-120.

76 A. H. Compton to J. R. Oppenheimer, transmitting a letter from H. S. Brown to Alvarez, 3 May 1944, A-84-019, 16-1.

77 L. W. Alvarez to J. R. Oppenheimer, "Site for Radio Lanthanum Experiments," 22 May 1944, A-84-019, 16-1.

78 Teletype, Alvarez to Oppenheimer, A. H. Compton, Whitaker, and Brown, 11 May 1944, A-84-019, 32-18; Oppenheimer to S. K. Allison, 11 May 1944, A-84-019, 16-1; H. S. Brown and H. A. Levy to L. W. Alvarez and H. Staub, "Data Concerning Shipment of Active Barium–Lanthanum," 31 May 1944, A-84-019, 32-18. The latter document includes a diagram of the packing container for the barium-lanthanum mixture.

79 Alvarez interview by Hoddeson, 28 April 1986, OH-120.

80 Teletype, Murphy to the Commanding Officer, U.S. Engineer Office, 1 June 1944 and 2 June 1944, A-84-019, 16-1.

81 Later in the program, phosphate precipitation was abandoned in favor of an oxalate or hydroxide precipitation. The oxalate precipitation method had to be performed as rapidly as possible, because the intense radiation present destroyed

the oxalate ion and caused the lanthanum to go back into solution. The nature of this radiative destruction process was the subject of considerable investigation by R. A. Penneman, in Milton Burton's group at the Chicago Metallurgical Laboratory. Eventually, hydroxide was used to separate the lanthanum, with the barium remaining soluble. Then oxalate with a trace of fluoride was used to give a filterable precipitate containing the ^{140}La for collection in the base of the cone-shaped tip.

82 Alvarez interview by Hoddeson, 28 April 1986, OH-120. Rossi and his group had serious reservations about the way in which this dress rehearsal was conducted, questioning, in particular, whether Alvarez had handled the radioactive source safely. In late August, a new RaLa group under Rossi would take over the Rala method, while Alvarez assumed responsibility for detonator research.

83 Oppenheimer to Alvarez, "Shipment of RaLa," 25 July 1944, A-84-019, 16-1, 32-18; Oppenheimer to Parsons, Kistiakowsky, Alvarez and Dodson, "Ra-Ba," 11 Aug. 1944, A-84-019, 32-18; teletype from Oppenheimer to A. H. Compton, 16 Aug. 1944, A-84-019, 16-1.

84 Among those eventually discarded were: the "D–D" method, suggested by I. I. Rabi, based on collisions between added deuterium, which – as it is compressed – produces higher-energy neutrons; the microwave penetration method, suggested by Alvarez; the high-hydrided active material method; and the "method of studying jets."

85 R. R. Wilson remembers conceiving this idea at a Governing Board meeting. The idea was patented in his name. R. R. Wilson, "Addendum to RRW letter to AM," 12 Nov. 1991.

86 Telephone interview with Kerst by Hoddeson, 8 Aug. 1985.

87 Greisen, "Minutes of Meeting Held in Oppenheimer's Office, Wednesday, June 28, 1944, for the Purpose of Discussing the Feasibility of Use of the Betatron for a Study of the Implosion Process," LAMS-107, 3 July 1944.

88 McMillan interview by Westfall, 31 Oct. 1985, OH-77.

89 An electrically conducting nonmagnetic metal moving in a static magnetic field induces a time-dependent magnetic field.

90 E-Division Progress Reports, 15 April 1944, LAMS-80, p. 24.

91 N. F. Ramsey, "Summary of Meeting of Project Y Technical Board on July 13, 1944," 15 July 1944, LAMS-113.

92 A different "pin" method had been used by Clark and Seely at Aberdeen Proving Ground for measuring detonation velocities in explosives. When two pins connected to an electrical detector are placed in a stick of explosive, the velocity could be obtained by measuring the time it took for the signal to traverse the distance between the two pins. The idea for transplanting this concept to implosion tests likely occurred to a number of people, including Frisch and Titterton.

93 Jacob Wechsler, technician for Frisch, recalls that the pin group was formed in June.

94 The previous organization, had been divided into eight sections: (1) Critical Masses and ν' Calculations (Ehrlich, Olum, Serber, Frankel, Nelson, Inglis), (2) Efficiency (Serber, Frankel, Nelson, Feynman, Ashkin, Weisskopf, Teller, Roberg, Konopinski, Metropolis), (3) Subcritical Experiments (Nelson), (4) Ordnance (Inglis, Teller Group, Konopinski), (5) Effects (von Neumann, Christy, Bethe, Teller), (6) Experiments (Feynman, Weisskopf, Olum, Konopinski), (7) Instruments (Richman, Olum, Konopinski), and (8) Making Tea (M. Frankel). None considered implosion.

95 Untitled document, 11 Jan. 1944, A-84-019, 6-10.

96 R. P. Feynman, N. Metropolis, and E. Teller, "Equation of State of Elements Based on the Generalized Fermi–Thomas Theory," *Phys. Rev.* 75 (May 1949), pp. 1561–73. H. A. Bethe, Feynman Memorial Lecture, Los Alamos, 16 Aug. 1988. Hawkins, *The Los Alamos Story*, p. 81.

97 Hawkins, *The Los Alamos Story*, p. 81.

98 Bethe recalls that these machines had been ordered largely to satisfy Parsons's wish to improve the calculations of the possibility of predetonation in the gun assembly. Bethe interview by Hoddeson, 3 Oct. 1986, OH-140; Bethe, Feynman Memorial Lecture, Los Alamos, 16 Aug. 1988; T-Division Progress Report, 29 Feb. 1944, LAMS-61.

99 Rudolf Peierls, *Bird of Passage: Recollections of a Physicist* (Princeton: Princeton University Press, 1985), p. 187. Peierls recently recalled that in his method, "The shock waves represent a boundary condition. You would run the machines integrating the equation up to the point where the shock is, and then put in by hand or by a machine operation ... the conditions for the shock wave." Peierls interview by Hoddeson, 20–21 March 1986, OH-111.

100 Oppenheimer to Groves, 14 Feb. 1944, Library of Congress, J. R. Oppenheimer, Box 36, Groves.

101 Feynman also tells how he and others on the staff would often repair the machines themselves to save time. Richard Feynman, "Los Alamos from Below," in Badash, et al., eds., *Reminiscences*, pp. 105–32, pp. 124–25.

102 The thermal agitation method is now known as the von Neumann-Richtmyer method of artificial viscosity.

103 Hawkins, *The Los Alamos Story*, p. 82. The first postwar step to improve the design of the implosion system was to adopt a conservative levitated design, which would in fact have worked perfectly well at Trinity and given a significantly higher yield. In a levitated system more instabilities would be expected. L. Redman, private communication.

104 McKibbin cards; R. C. Smith, "Report of Foreign Personnel at Project Y." 5 July, 2 Aug. 1944 A-85-001, 1-1.

9. New Hopes for the Implosion Weapon: September 1943 to July 1944

1 TNT, which melts at about 80° C, can be readily liquefied in kettles heated by steam. Casting is the convenient way of loading HE into bombs and shells. In World War I, inexpensive ammonium nitrate was used as an extender for filling artillery shells. "Military Explosives," TM 9-1300-214, Department of the Army Technical Manual, Sept. 1984, pp. 2–20. These pages contain a useful chronology of explosives history. Also Luis Alvarez, *Alvarez: Adventures of a Physicist* (New York: Basic Books, 1987), p. 131; and Les Redman, private communication.

2 There are two classes of high explosive: a primary explosive is readily detonated by heat, impact, friction, or the like; a secondary explosive is usually detonated by a shock wave and, if ignited when unconfined, it usually burns without detonation.

3 PETN was used at Los Alamos at the small Anchor Ranch casting room from October 1943 as a 50-50 mixture with TNT, which is called Pentolite, and was

cast around metal cylinders for the study of cylindrical implosions. Pentolite was also used in some early two-dimensional lenses.

4 Three British forms of RDX were: Composition A, essentially RDX ground up with wax; Composition B, made by stirring RDX into TNT and beeswax; and Composition C, a malleable plastic explosive. Torpex was a high explosive developed in England, composed of RDX, molten TNT and aluminum powder. M. Roy interview by Hoddeson, 2 April 1986, OH-115; D. MacDougall interview by Hoddeson, 9 April 1986, OH-116; and J. Russell interview by Hoddeson, 16 July 1986, OH-133. Composition B [British Specification C. S./1078.B, 30 Jan. 1939], a castable, and thus more usable, RDX-based high explosive would be a major component of the Los Alamos implosion system. "Progress Report on Composition B to 15 Jan. 1943," NDRC-B-1167. Russell interview by Hoddeson, 16 July 1986, OH-133. Roy interview by Hoddeson, 2 April 1986, OH-115; MacDougall interview by Hoddeson, 9 April 1986, OH-116.

5 Russell interview by Hoddeson, 16 July 1986, OH-133. In early 1944 Russell, then working at Tennessee Eastman, decided to enter military service. He obtained a commission in the U.S. Naval Reserve and was sent to Oak Ridge, arriving there in March 1944. Initially he worked there on chemical aspects of the Calutron process. Early in 1945, he was interviewed by Norris Bradbury, who offered him a job at Los Alamos. Russell joined Project Y and immediately was sent to "where the Comp B was," at S-Site.

6 ERL also assisted in the design and building of the explosives plant at Kingsport. Roy interview by Hoddeson, 2 April 1986, OH-115; MacDougall interview by Hoddeson, 9 April 1986, OH-116. Initially Conant headed Division B, while Kistiakowsky led a section of Division B. Because the NDRC could not let contracts, the Office of Scientific Research and Development (OSRD) was created under Vannevar Bush, and Division B was converted to Division 8 of the NDRC, with Kistiakowsky as its head. MacDougall interview by Hoddeson, 9 April 1986, OH-116.

7 E.g., "Controlled Fragmentation and Shaped Charges," covering, e.g., "Optical Studies of Coned Charges," and a "Fragment Velocity Camera," 15 Aug. to 15 Sept. 1942, AM-1155, Div. B of NDRC; "Controlled Fragmentation and Shaped Charges," 15 July 1943–15 Aug. 1943, AM-1166; "Shaped Charges,"15 Nov. 1943–15 Dec. 1943, AM-1130; "Detonation, Fragmentation and Air Blast," 15 Sept. 1943–15 Oct. 1943, AM-1169; "Shaped Charges," 15 Aug. 1943–15 Sept. 1943, AM-1127; "Experimental Studies of Cone Collapse and Jet Formation: Part I, Recovery of Cones from Low-powered Charges" – a Report on "the Nature of the Cavity Effect," 29 Nov. 1943, NDRC-B-2070, Div. 8; "The Effective Diameter of the Jet from a Cavity Charge," 5 Aug. 1943, NDRC-B-1679, Div. 8; "Shaped Charges," period 15 April 1944 to 15 May 1944, AM-1135; "Studies Made of Penetration," 15 March–15 April, AM-135; "Studies Made of Penetration," 15 Jan.–15 Feb. 1944, AM-1138, on "jet velocities," 15 Feb.–15 March 1944, AM-113.

8 E.g., "Composition B (Cyclotol), up to 15 Jan. 1943," NDRC-B-1167.

9 E.g., "The Effect of Charge Wrapping on the Optical Records of Detonating Explosives," 14 May 1943, NDRC-B-1411.

10 E.g., "Preparation and Testing of Explosives," covering subjects such as "RDX Compositions" (work on A, B, and C), "Preparation and Properties of Explosives, Preparation of HMX, Stability of PETN and Pentolite," 15 Nov.–15 Dec. 1942, PT-4; "Preparation and Testing of Explosives," 15 March–15 April 1943, PT-8;

"Preparation and Testing of Explosives," 15 May–15 June 1943 PT-10; "Rates of Burning of High Explosives," 15 July–15 Aug. 1943, PT-12; MacDougall and Eyster, "Report on the Explosives Compositions," 15 Feb. 1943–15 March 1943, PT-7; MacDougall, Eyster and Weltman, "PEP-2 (PETN Plastic Explosive)," 11 Feb. 1944, NDRC-B-3240, Div. 8; "Preparation and Testing of Explosives," 15 April–15 May 1944, PT-21.

11 E.g., "Interim Report for 15 July 1943 to 15 Aug. 1943, on optical studies of detonation rates," NDRC-B-Fs-12; "The Effect of Particle Size on the Detonation Velocity of Ammonium Picrate" 20 Aug. 1943, NDRC-B-175; MacDougall, Messerly, Campbell et al., "The Estimation of Detonation Pressure from the Shock Wave Velocity in Lead," 14 Dec. 1945, NDRC-B-5612; Messerly, Boggs et al., "Initiation Studies in Solid Explosives," 14 Dec. 1945, NDRC-B-5617.

12 W. Kauzmann and E. J. Huber "Tracer Research at the ERL," 14 Feb. 1944, OSRD-3283.

13 "Detonation and Fragmentation," 15 April–15 May 1945, AM-1392; "Detonation and Fragmentation," 15 June 1945–15 July 1945, covers "Optical Studies of Detonation" (velocities of shocks produced, comparison of observed behavior of shocks in air, optical measurements of detonation, and the radiation emitted by detonating explosives), "Pit Fragmentation Studies," and "Studies of Shell Fragmentation," AM-1609; R. C. Connor and W. A. Noyes, *Chemistry, A History of the Chemistry Components of the NDRC, 1940–1946* (Boston: Little, Brown, 1948).

14 The casting house was run by Esther Strip. Campbell interview by Hoddeson, 29 July 1986.

15 McKibbin cards.

16 MacDougall interview by Hoddeson, 9 April 1986, OH-116, MacDougall to Lt. Col. J. Lansdale, Jr., 29 Aug. 1944, A-84-019 47-3. The latter reference includes a list of those at Bruceton working on Project Q.

17 David Hawkins, *The Los Alamos Story* (Los Angeles: Tomash Publishers, 1983), p. 127.

18 E-Division Progress Report, 15 April 1944, LAMS-80, p. 35.

19 E-Division Progress Report, 1 March 1944, LAMS-73.

20 D. Hawkins, *The Los Alamos Story*, p. 212.

21 Kistiakowsky, "Reminiscences of Wartime Los Alamos," in Lawrence Badash, Joseph Hirshfelder, and Herbert Broida, eds., *Reminiscences of Los Alamos 1943–1945* (Dordrecht: Reidel Publishing Co., 1980), p. 50.

22 Kistiakowsky, "Reminiscences of Wartime Los Alamos," p. 50.

23 Hawkins, *The Los Alamos Story*, p. 212.

24 Anchor Ranch was still equipped with only four two-gallon candy kettles.

25 Hawkins, *The Los Alamos Story*, p. 213.

26 The design was based on RDX and Baratol. M. J. Poole, Chairman, Sub-Committee 4 of the Explosives Research Committee (Physics and Physical Chemistry), of Ministry of Supply Advisory Council on Scientific Research and Technical Development, "Suggestions for Improvement of ... Bombs," 2 Sept. 1942, A. C. 2644 (equivalently B.M. 1061). Roy interview by Hoddeson, 2 April 1986, OH-115.

27 Roy, who had been working under Kistiakowsky at NDRC since 1941, in Division B and later Division 8, and who became Kistiakowsky's Washington contact after Kistiakowsky came to Los Alamos, sent the Poole document to Los Alamos early in the project. For reasons unknown to Roy, the document was not received at

Los Alamos until much later in the program. Kistiakowsky, in a letter of 21 May 1945, notes that the document "has just been received." Roy suggests that the delay might have occurred because the report was treated as super secret.

28 See MacDougall interview by Hoddeson, 9 April 1986, OH-116, in which he states that Boggs's discovery was independent. ERL fired a simplified experimental shot soon afterward, and observed some convergence. The progress report refers on p. 5 to "as in Dr. Bethe's diagram," indicating instructions from Los Alamos.

29 Greisen and Bradbury interview by Hoddeson, 11 Dec. 1985, OH-89; McKibbin cards.

30 Primacord was developed in 1936 by the Ensign-Bickford Company in Connecticut, based on a French patent. "Military Explosives," TM-9-1300-214, 2-13.

31 Alvarez arrived 25 April 1944, McKibben cards; Alvarez, *Adventures*, p. 123.

32 Alvarez, *Adventures*, pp. 132–4.

33 Alvarez interview by Hoddeson, 28 April 1986, OH-120.

34 Alvarez, *Adventures*, p. 134; Lawrence Johnston Memoir, LANL Archives VFA-117, p. 2; also Alvarez interview by Hoddeson, 28 April 1986, OH-120.

35 Alvarez interview by Hoddeson, 28 April 1986, OH-120.

36 Alvarez, *Adventures*, p. 134; also Johnston memoir, VFA-117.

37 Johnston memoir, VFA-117, p. 84.

38 Alvarez, *Adventures*, p. 134.

39 Whereas primary explosives like lead azide can be set off by minute amounts of energy, PETN has a threshold below which the detonators will not ignite, even when the bridgewire is subjected to a current surge sufficient to melt the wire. Military regulations required an effective mechanical interruption of the detonation train between the primary explosive and the main charge, so that firing of the primary explosive cannot set off the main charge until the "safety gate" is opened, for example, by centrifugal force in a rotating artillery shell during flight.

40 Johnston memoir, VFA-117, pp. 1–2; Hawkins, *The Los Alamos Story*, pp. 205–6.

41 Kistiakowsky to Alvarez, Bainbridge, Bradbury, Fitzpatrick, Fussell, Parratt and Bacher, "Electric Detonators," 16 Aug. 1944, A-84-019, 57-20.

42 Governing Board Minutes, 27 Jan. 1944, A-83-0013, 1-40.

43 Governing Board Minutes, 3 Feb. 1944, A-83-0013, 1-41.

44 They also began to debate the merits of conducting the final test in a confinement chamber. Governing Board Minutes, 20 Jan. 1944, A-83-0013, 1-39. For a discussion of the confinement chamber, see the section on Jumbo below.

45 Ramsey to Oppenheimer, 12 Feb. 1944, A-84-019, 22-7.

46 Oppenheimer, Memorandum on Test of Implosion Gadget, 16 Feb. 1944, A-84-019, 22-7.

47 Oppenheimer to Groves, 10 March 1944, A-84-019, 22-7.

48 Hawkins, *The Los Alamos Story*, p. 232.

49 Kistiakowsky, "Reminiscences of Wartime Los Alamos," p. 56; F. M. Szasz, *The Day the Sun Rose Twice* (Albuquerque: University of New Mexico Press, 1984), p. 26.

50 Groves to Oppenheimer, 1 Nov. 1944, A-84-019, 22-1.

51 Bethe interview by Hoddeson, 3 Oct. 1986, OH-140.

52 Critchfield interview by Hoddeson and Kerr, 2 Aug. 1984, OH-46.

53 Kistiakowsky to Bradbury, "Group E-5," 17 Aug. 1944, A-84-019, 6-1.

54 Kistiakowsky to Oppenheimer and Parsons, 15 June 1944, A-84-019, 4-1 and 6-1. Kistiakowsky to Parsons, "Organization of the H.E. Project," 21 June 1944, A-84-019, 6-1. In a second memorandum to Parsons that day, Kistiakowsky listed eleven field facilities for the HE organization. Kistiakowsky to Parsons, "Field facilities distribution," 21 June 1944, A-84-019, 6-1. Further details on the reorganization appear in Kistiakowsky's memorandum to the staff of E-5 Group of 26 June 1944. The section leaders of the new E-5 group were: Greisen of the Anchor Ranch X-ray program; Linschitz of the Terminal Observations program; Koski of the Flash Cylinder program; and J. G. Hoffman of the Rotating Prism Camera program. Kistiakowsky to staff of E-5 Group, "Reorganization of E-5," 26 June 1944, A-84-019, 6-1. Greisen Los Alamos Notebook 160, p.1.

10. The Nuclear Properties of a Fission Weapon: September 1943 to July 1944

1 David Hawkins, *Project Y: The Los Alamos Story*, (Los Angeles: Tomash Publishers, 1983), pp. 77–9.

2 Inglis first studied the critical mass of an infinite cylinder, both tamped and untamped. These calculations were extended by Frankel and Nelson to untamped finite-length cylinders and then by Inglis to long finite-tamped cylinders.

3 Hawkins, *The Los Alamos Story*, p. 78.

4 Hawkins, *The Los Alamos Story*, p. 80.

5 Bethe, Feynman Memorial Lecture, Los Alamos, 16 Aug. 1988.

6 Hawkins, *The Los Alamos Story*, p. 79.

7 Governing Board Minutes, 30 Sept. 1943, A-83-0013, 1-27, pp. 3–4.

8 Also P-Division Progress Report, #22, 1 June 1944, LAMS-111, p. 20

9 In the original report, the λ_f in this equation is written λ_t, apparently a typographical error.

10 Robert Serber, *The Los Alamos Primer: The First Lectures on How to Build an Atomic Bomb* (Berkeley: University of California Press, 1992), pp. 42–3.

11 Chapter 12 describes early spontaneous fission research; see section C for development of the Water Boiler. P-Division Progress Reports, 15 July 1943, LAMS-4; 15 Aug. 1943, LAMS-7; 15 Sept. 1943, LAMS-12; and Hawkins, *The Los Alamos Story*, p. 90.

12 Governing Board Minutes, 6 March 1943, A-83-0013, 1-1. Bacher interview by Westfall, 14 Dec. 1987, OH-169.

13 Taschek interview by Westfall, 24 Oct. 1986, OH-142. Wilson, Manley, and Barschall, "Nuclear Physics," LA-1009, 19 March 1947, p. 9.

14 Chapter 12 describes the spontaneous fission research. Bacher interview by Westfall, 14 Dec. 1987, OH-169.

15 H. A. Bethe to J. R. Oppenheimer, 6 Oct. 1942, Papers of Robert Oppenheimer, Box 20; Bacher to A. H. Compton, 9 Nov. 1942, A-84-019, 63–3; Bacher interview by Hoddeson, 30 July 1984, OH-45, p. 13.

16 J. H. Williams, "Cross Sections for Fission of 25, 49, 28, 11, 37, 00, 02, Boron and Lithium, LA-150, 5 Oct. 1944.

17 Higinbotham interview by Westfall, 26 May 1987, OH-156, pp. 6 and 7.

18 Ibid., pp. 12, 14, 17, 19–20.

19 R. Wilson, "Nuclear Physics," LA-1009, p. 10. P-Division Progress Report, 1 Jan. 1944, LAMS-48, p. 5. Higinbotham interview by Westfall, 26 May 1987, OH-156, p. 23; and Taschek interview by Westfall, 24 Oct. 1986, OH-142.

20 P-Division Progress Reports, 15 Oct. 1943, LAMS-21, p. 2; 1 Oct. 1943, LAMS-20, p. 4. Also P-Division Progress Report, 1 Nov. 1943, LAMS-25, p. 1; 1 Jan. 1944, LAMS-48, p. 2; 1 March 1944, LAMS-58, p. 2; and 15 Nov. 1943, LAMS-28, p. 3; private communication Bacher, 19 Aug. 1988. Taschek interview by Westfall, 24 Oct. 1986, OH-142.

21 Higinbotham interview by Westfall, 26 May 1987, OH-156; and Taschek interview by Westfall, 24 Oct. 1986, OH-142.

22 P-Division Progress Report, 1 Jan. 1944, LAMS-48, p. 3; J. Williams, "Fission and Capture Cross Sections for Fast Neutrons," in "Nuclear Physics," LA-1009, pp. 92–5.

23 Williams, "Fission and Capture Cross Sections for Fast Neutrons," pp. 10, 78.

24 Governing Board Minutes, 7 Oct. 1943, A-83-0013, 1-28.

25 Wilson, "Nuclear Physics," LA-1009, p. 8

26 Bacher interview by Westfall, 14 Dec. 1988, OH-169; and Segrè interview by Westfall, 31 Oct. 1985, OH-78.

27 Wilson interview by Westfall, 25 May 1987, OH-154, pp. 29, 50–2. P-Division Progress Reports, 1 Oct. 1943, LAMS-20, p. 5; 1 Jan. 1944, LAMS-48, p. 5; and Luis Alvarez, *Alvarez: Adventures of a Physicist* (New York: Basic Books, 1987), pp. 126–7.

28 P-Division Progress Report, 1 Nov. 1943, LAMS-25, p. 5.

29 J. Manley and H. Barschall, "Fast Neutron Scattering," in "Nuclear Physics," LA-1009, p. 68; and P-Division Progress Report, 15 Oct. 1943, LAMS-21, p. 4. Also P-Division Progress Report, 1 Sept. 1943, LAMS-9, p. 5. Taschek interview by Westfall, 24 Oct. 1986, OH-142, p. 18.

30 P-Division Progress Reports, 15 Nov. 1943, LAMS-28, p. 5; and 1 Nov 1943, LAMS-25, p. 3.

31 P-Division Progress Report, 1 June 1944, LAMS-111, p. 3 and J. Manley and H. Barschall, "Fast Neutron Scattering," in "Nuclear Physics," LA-1009, p. 71.

32 Manley and Barschall, "Fast Neutron Scattering," in "Nuclear Physics," LA-1009, pp. 19, 23; also pp. 18, 21, 63.

33 Governing Board Minutes, 18 Nov. 1943, A-83-0013, 1-34, p. 4; and "Nuclear Physics," LA-1009. Wilson, "Neutrons Per Fission from 29 Compared with 25 – Slightly Delayed Neutrons from 49 to 25," LA-104, 4 July 1944; P-Division Progress Report, 15 Jan. 1944, LAMS-50, p. 2; and Wilson interview by Westfall, 25 May 1987, OH-154, p. 48.

34 Bacher interview by Hoddeson, 30 July 1984, OH-45, p. 36. P-Division Progress Report, 1 Dec. 1943, LAMS-33, p. 1. Wilson interview by Westfall, 25 May 1987, OH-154, p. 46.

35 P-Division Progress Report, 15 Nov., LAMS-28, p. 3.

36 P-Division Progress Report, 15 Jan. 1944, LAMS-50, p. 5. P-Division Progress Report, 1 Jan. 1944, LAMS-48, p. 5.

37 P-Division Progress Report, 15 Jan. 1944, LAMS-50, p. 2.

38 P-Division Progress Report, 1 May 1944, LAMS-95, p. 4; and Bacher interview by Westfall, 14 Dec. 1987, OH-169, p. 6.

39 Quotes from P-division Progress Report, 15 Jan. 1944, p. 4, and D. Inglis, "Experiments Related to the Fission Process, " in "Nuclear Physics," LA-1009, pp. 112 and 113; T. Snyder and J. Williams, "Number of Neutrons Per Fission for 25 and 49," LA-102, 30 June 1944, p. 2; R. Walker, "Absolute Calibration of a Ra-Be Neutron Source, " LA-400, 28 Sept. 1945.

40 Williams, "Cross Sections for Fission of 25, 49, 28, 11, 00, 02, B and Li," LA-150, 5 Oct. 1944, pp. 6, 9.

41 Williams, "Cross Sections for Fission of 25, 49, 28, 11, 00, 02, B and Li." pp. 10, 11.

42 D. Inglis, "Experiments Related to the Fission Process," and R. Williams, "Competition between Capture and Fission," p. 138 in "Nuclear Physics," LA-1009, pp. 123, 131.

43 Bacher interview by Westfall, 14 Dec. 1987, OH-169, p. 7.

44 Kerst interviews by Hoddeson, 8 Aug. 1985, OH-54, and 28 April 1986, OH-118.

45 "An Enriched Homogeneous Nuclear Reactor," *Review of Scientific Instruments*, 22 (July 1951), p. 489.

46 Hawkins, *The Los Alamos Story*, p. 104.

47 P-Division Progress Report, 15 Aug. 1943, LAMS-7.

48 Governing Board Minutes, 5 Aug. 1943, A-83-0013, 1-20.

49 Governing Board Minutes, 9 and 30 Sept. 1943, A-83-0013, 1-24 and 1-27. The slow pace of the ^{235}U production would probably not have allowed the high power operation, but that was not evident until early in 1944, well after this decision was made.

50 R. Christy, "Critical Mass of ^{235}U in Water Solution with Water Reflector," LAMS-18, 4 Oct. 1943.

51 P. King, "Design and Description of Water Boiler Reactors," paper presented at the International Conference on the Peaceful Uses of Atomic Energy, 30 June 1955.

52 Dodson to Kennedy, CM-Division Progress Report, 1 Oct. 1943.

53 P-Division Progress Report, 1 Nov. 1943, LAMS-25; CM-Division Progress Report, 15 Nov. 1943.

54 P-Division Progress Report, 15 Jan. 1944, LAMS-50.

55 Governing Board Minutes, 11 Nov. 1943, A-83-0013, 1-33.

56 Since changes in temperature as small as 0.001° C would produce noticeable changes in the reactivity of the solution, accurate temperature control for the Water Boiler room was important. The Water Boiler was enclosed in a special room maintained at a constant temperature with an electronic thermostat control designed by Matthew Sands of the electronics group. Temperature changes in the boiler were held to within 0.01° C. King, "Los Alamos Technical Series V. 5: Critical Assemblies, Part 2," LA-1034, 19 Dec. 1947, p. IV-7; P-Division Progress Report, 1 May 1944, LAMS-95, p. 2; CM-Division Progress Report, 1 May 1944, LAMS-86.

57 P-Division Progress Report, 1 Feb. 1944, LAMS-53.

58 P-Division Progress Report, 15 May 1944, LAMS-96; and Baker, H. K. Daghlian, Friedlander, "Water Boiler," LA-134, 8 Sept. 1944, pp. 22–5.

59 P-Division Progress Reports for 1 March, 15 March, 1 April, and 1 May 1944, LAMS-58, 75, 78, and 95.

60 Hawkins, *The Los Alamos Story*, p. 106; King, "Critical Assemblies, Part 2," LA-1034, V. 5, chapter 4.

61 P-Division Progress Report, 15 May 1944, LAMS-96.

62 P-Division Progress Report, 1 June 1944, LAMS-111.

63 King, "Critical Assemblies, Part 2," LA-1034, V. 5, chapter 4, p. IV-17.

64 P-Division Progress Report, 1 June 1944, LAMS-111.

65 F. L. Bentzen, R. E. Carter, J. Hinton, King, J. C. Nevenzel, R. Schreiber, J. W. Starner, and P. H. Watkins, "High-Power Water Boiler," AECD-3065, 19 Sept. 1945, p. 3; P-Division Progress Report, 1 July 1944, LAMS-117.

66 King, Schreiber, and others at Purdue had measured the tritium-deuterium reaction cross section and found it to be very large.

67 Governing Board Minutes, 24 Feb. 1944, A-83-0013, 1-43. Bethe gives an account of the situation in "Comments on the History of the H-Bomb," *Los Alamos Science 3*, no. 3 (Fall 1982), pp. 42–54. Also, R. C. Hewlett and O. E. Anderson, Jr., *The New World, 1939/1946: A History of the United States Energy Commission*, vol. 1 (University Park: Pennsylvania State University Press, 1962), p. 240.

68 Governing Board Minutes, 24 Feb. 1944, A-83-0013, 1-43.

69 Governing Board Minutes, 24 Feb. 1944, A-83-0013, 1-43.

11. Uranium and Plutonium: Early 1943 to August 1944

1 Uranium and plutonium production at Chicago and Oak Ridge provide a more complex example of this strategy. The ease of working with uranium balanced the difficulty of working with newly discovered plutonium, and the ease of chemically separating plutonium from uranium balanced the difficulty of separating ^{235}U from natural uranium.

2 Seaborg, Met Lab Diary, V. I (hereafter GTSI), p. 434.

3 Kennedy and Smith continued in this capacity and were officially named director and associate director, respectively, of the chemistry and metallurgy section in early 1944; Thomas to Groves and Conant, 4 Jan. 1944, A-84-019, 9-4; Governing Board Minutes, 3 May 1943, A-83-0013, 1-2; Warner to Thomas, "April Meeting with C. A. Thomas on Final Purification and Metallurgy of 49," 17 April 1944; L. Brewer, manuscript from talk at R. D. Baker Memorial Symposium, 21 April 1986, VFA-472.

4 GTSI, pp. 482–4.

5 David Hawkins, *The Los Alamos Story*, (Los Angeles: Tomash Publishers, 1983), p. 135; R. C. Hewlett, and O. E. Anderson, Jr., *The New World, 1939/1946. A History of the United States Energy Commission*, vol. 1 (University Park: Pennsylvania State University Press, 1962), p. 236.

6 1 April 1944, LAMS-72; 1 Nov. 1944, LAMS-155.

7 Oppenheimer to Groves, 27 May 1943, A-84-019, 9-4; Hawkins, *The Los Alamos Story*, pp. 20–1; Governing Board Minutes, 5 May 1943, A-83-0013, 1-2.

8 Kennedy to Oppenheimer, 21 June 1943, A-83-019, 4-10; Governing Board Minutes, 22 July 1943, A-83-0013, 1-18.

9 Groves to Compton and Oppenheimer, 17 June 1943, A-84-019, 70-9; Hewlett and Anderson, *The New World*, p. 239.

10 Allison to Oppenheimer, 3 March 1943, A-84-019, 4-10.

11 Thomas to Conant and Groves, 23 Aug. 1943, A-84-019, 9-4; Seaborg Met Lab Diary, V. II (hereafter GTSII), pp. 110, 132.

12 Smith interview by Westfall, 16 Dec. 1985, OH-92 and 93, pp. 10–11

13 Governing Board Minutes, 8 July 1943, A-83-0013, 1-16, p. 3.

14 J. E. Burke, "Recollections of Processing Uranium Hydride and Plutonium at Wartime Los Alamos," *J. Nuclear Materials 100* (1981), pp. 11–16, esp. p. 11; Governing Board Minutes, 8 July 1943, A-83-0013, 1-16.

15 Burke, "Recollections," p. 11; Hawkins, *The Los Alamos Story*, pp. 138–9.

16 Progress Report, Metallurgy Group No. 1, 19 July 1943, A-84-019, 76-3; Smith interview by Westfall, OH-92 and 93, p. 5–6. Governing Board Minutes, 9 Sept. 1943, A-83-0013, 1-24.

17 M. L. Perlman and S. I. Weissman, "Preparation of Compacts of High-Density Uranium Hydride," Los Alamos Scientific Laboratory report LA-33, 16 Oct. 1943.

18 Governing Board Minutes, 14 Oct. 1943, A-83-0013, 1-29.

19 Because of the prohibitive cost of the necessary quantities of high-quality calcium, Ames researchers switched to magnesium as a reductant in March 1943. For detailed information on the development of foundry scale uranium metal production, see J. C. Warner, "Early Methods for Producing Uranium Metal," and "Methods for Production of Uranium Metal," in J. E. Vance and J. C. Warner, eds., *Uranium Technology: General Survey*, National Nuclear Energy Series, Division VII, vol. 2A (Oak Ridge, Tenn.: U.S. Atomic Energy Commission Technical Information Service, 1951), pp. 142–61. This source also has detailed information on the extraction of uranium from ores, the preparation of U_3O_8 and other uranium compounds, and the large-scale uranium fabrication developed to produce uranium slugs for the production piles. Smith interview by Westfall, 16 Dec. 1985, OH-92 and 93, pp. 17–18.

20 About two liters of a partly separated solution of uranium nitrate was sent from Berkeley to Los Alamos in fall 1943. Wahl interview by Westfall, 11 Nov. 1985, OH-81.

21 GTSII, p. 301.

22 Thomas to Groves and Conant, 5 Oct. 1943, A-84-019, 9-4. Wahl interview by Westfall, 11 Nov. 1985, OH-81.

23 GTSI, pp. 491, 587, 593.

24 GTSI, pp. 593, 646, 661, 670; GTSII, pp. 23, 28, 54, 57, 70, 97, 149, 153.

25 GTSII, p. 124.

26 Hawkins, *The Los Alamos Story*, p. 64.

27 GTSII, 44.

28 GTSII, p. 111; Wilson interview by Westfall, 25 May 1987, OH-154, p. 7.

29 GTSII, pp. 3, 86, and 156.

30 GTSII, pp. 177, 211.

31 GTSII, p. 177.

32 GTSII, pp. 157, 176, 196.

33 R. A. Penneman, "The Role of W. H. Zachariasen in Actinide Research," in N. M. Edelstein, ed., *Actinides in Perspective* (Oxford: Pergamon Press, 1982).

34 Thomas to Franck, 1 Oct. 1943, A-84-019, 15-6; Franck to Oppenheimer, 29 Oct. 1943, A-84-019, 15-6; Oppenheimer to Bacher, Dodson, Kennedy, Samaras, Segrè, Smith, Wahl, Williams, Wilson, 14 Oct. 1943, A-84-019, 15-6; Governing Board Minutes, 23 Sept. 1943, A-83-0013, 1-26.

35 GTSII, p. 123; Hewlett and Anderson, *The New World*, pp. 209–10.

36 Hewlett and Anderson, *The New World*, pp. 211–12.

37 Hewlett and Anderson, *The New World*, pp. 211–12.

38 To study the kinetics of hydride formation, they placed a small length of uranium wire in a small pyrex tube. After the evacuated tube, heated in an electric furnace, reached the selected reaction temperature, hydrogen was introduced. With the exception of a few preliminary runs, they used a glass train. They then measured the absorption rate using a gas burette. Burke, "Recollections," p. 12.

39 In the 10-g experiments, a thermocouple was placed along the wall of the bomb liner to record temperatures in conditions resembling an actual reduction.

40 C. S. Smith, W. F. Arnold, E. D. Selmanoff, S. Marshall, "Casting of Uranium," in "Uranium Metallurgy," p. 67.

41 Hewlett, and Anderson, *The New World*, pp. 128–9, 135–6, 162–3.

42 Hewlett and Anderson, *The New World*, pp. 164, 243; Governing Board Minutes, 11 Nov. 1943.

43 Governing Board Minutes, 20 Jan. and 17 Feb. 1944 A-83-0013.

44 P-Division Progress Report, 15 May 1944, LAMS-96, p. 1; Hewlett and Anderson. CM-Division Progress Report, 1 Sept. 1944, LAMS-127.

45 For a description of the final cube fabrication process, which was changed at the last minute owing to a change in specifications, see Chapter 16. Progress Reports, Group C-8, 1 May 1944, LAMS-86; Burke, "Recollections," pp. 12–13.

46 1 April 1944, LAMS-72; 1 May 1944, LAMS-86; 1 June 1944, LAMS-97.

47 Progress Reports, Group C-5, 1 May 1944, LAMS-86; 1 June 1944, LAMS-97; Private communication R. Penneman, 14 July 1987.

48 Thomas to Groves and Conant, 4 Jan. 1944, A-84-019, 9-4; Thomas to Groves and Conant, 8 April 1944, A-84-019, 9-4; GTSII, pp. 301, 357–60, 402–3, 497.

49 GTSII, pp. 57, 279.

50 All density measurements hereafter are given in grams per cubic centimeter.

51 Thomas to Groves and Conant, 4 Jan. 1944, A-84-019, 9-4; GTSII, pp. 242, 328.

52 GTSII, pp. 137, 284.

53 "Proposed Schedule of Metallurgical Operations and Tests," 9 Feb. 1944, A-84-019, 15-6.

54 C. S. Smith, "Some Recollections of Metallurgy at Los Alamos, 1943–1945," *J. Nuclear Materials 100* (1981), pp. 3–10, esp. pp. 4–5.

55 GTSII, pp. 491, 495; Zachariasen to Allison, 28 Jan. 1944, Zachariasen Papers.

56 J. W. Kennedy, "Notes on Visit to Chicago," Feb. 1944, A-84-019, 61-10. Penneman interview by Westfall, 18 Oct. 1985, OH-75.

57 GTSII, p. 496; Thomas to Groves and Conant, 8 April 1944, A-84-019, 9-4; Werner to Thomas, "April Meeting with Dr. C. A. Thomas on Final Purification and Metallurgy of 49," 17 April 1944, A-84-019, 61-10.

58 Wet chemistry involves liquid, usually aqueous solutions. Dry chemistry involves solid and gas phases only.

59 GTSII, p. 542; Warner to Thomas, "April Meeting with Dr. C. A. Thomas on Final Purification and Metallurgy of 49," 17 April 1944, A-84-019, 61-10; Wahl interview by Westfall, 11 Nov. 1985, OH-81; Penneman interview by Westfall, 23 Nov. 1985, OH-76; Wahl, "Purification of Plutonium," Los Alamos Scientific Laboratory Report LA-75, 1 May 1944, p. 3; Hawkins, *The Los Alamos Story*, p. 135.

60 Wahl, "Chemistry of Uranium," and Garner, "Tetrafluoride and Other Dry Compounds of Plutonium," in "Chemistry of Uranium and Plutonium," LA-1100, 20 April 1947, pp. 70, 126–9.

61 Thomas to Groves and Conant, 4 April 1944, A-84-019, 9-4.

62 Warner to Thomas, "April Meeting with Dr. C. A. Thomas on Final Purification and Metallurgy of 49," 17 April 1944, 21 April 1944, A-84-019, 61-10. Penneman interview by Westfall, 23 Oct. 1985, OH-76.

63 Purification and Reduction Meeting Minutes, 27 Apr. 1944, A-84-019, 76-7.

64 C. Smith, "Some Recollections," pp. 3–10, esp. p. 7.

65 Thomas to Groves and Conant, 13 June 1944, A-84-019, 15-6. Penneman interview by Westfall, 18 Oct. 1985, OH-75.

66 Cyril Smith, "Plutonium Metallurgy at Los Alamos During 1943–45," in A. S. Coffinberry and W. N. Miner, eds., *The Metal Plutonium* (Chicago: University of Chicago Press, 1961), p. 29.

67 Smith, "Some Recollections," pp. 3–10, esp. p. 5; Thomas to Groves and Conant, 13 June 1944, A-84-019, 15-6. Penneman interview by Westfall, 23 Oct. 1985, OH-76.

68 Smith, "Some Recollections," pp. 3–10, esp. p. 5; Thomas to Groves and Conant, 13 June 1944, A-84-019, 15-6.

69 C. S. Garner, "Tetrafluoride and Other Dry Compounds of Plutonium," in LA-1100, p. 111.

70 Although Wahl's scheme was simpler for large-scale production than that proposed by Seaborg, it had these disadvantages. A. C. Wahl, "Chemical Purification of Plutonium," in LA-1100, pp. 70–1. Wahl interview by Westfall 12 Nov. 1985, OH-82.

71 As Wahl later explained, the new "A" process "consisted of reducing Pu-IV or Pu-VI nitrate with iodide, precipitating Pu-III oxalate, oxidizing with bromate and nitric acid, precipitating sodium plutonyl acetate, dissolving in nitric acid, ether-extracting the plutonyl nitrate from an ammonium nitrate solution, reducing with iodide and precipitating Pu-III oxalate." A. C. Wahl, "Chemical Purification of Plutonium," in LA-1100, pp. 71–2.

72 Purification and Reduction Meeting Minutes, 6 July 1944, A-84-019, 76-7. Penneman interview by Westfall 18 Oct. 1985, OH-75; Wahl interview by Westfall, 12 Nov. 1985, OH-82.

73 Seaborg, "History of the Met Lab Section C-1," V. III (hereafter, GTSIII), p. 145.

74 Hewlett and Anderson, *The New World*, pp. 182–5, 210–12.

12. The Discovery of Spontaneous Fission in Plutonium and the Reorganization of Los Alamos

1 W. F. Libby, "Stability of Uranium and Thorium for Natural Fission," *Phys. Rev. 55* (1939), p. 1269.

2 N. Bohr and J. A. Wheeler, "The Mechanism of Nuclear Fission," *Phys. Rev. 56* (1940), pp. 426–50.

3 Segrè interview by Hoddeson, 23 July 1985, OH-55.

4 K. Petrzhak and G. N. Flerov, *Akademiia Nauk SSSR 28* (1940), p. 500; G. N. Flerov and K. A. Petrzhak, "Spontaneous Fission of Uranium," *Journal of Physics 3*, no. 4–5 (1940), pp. 275–80; G. N. Flerov, "Soviet Research into Nuclear Fission before 1942," in *50 Years with Nuclear Fission* (La Grange Park, IL: American Nuclear Society, 1989), pp. 53–9.

5 J. Heilbron and R. Seidel, *Lawrence and His Laboratory: A History of the Lawrence Berkeley Laboratory* (Berkeley: University of California Press, 1990). Segrè, an Italian-born nuclear physicist, had taken his doctoral degree in Rome in 1928, conducted postdoctoral studies in Hamburg and Amsterdam, and taught in Rome and Palermo before emigrating in 1938 to the United States, where he took a position at the Radiation Laboratory in Berkeley. There in the spring of 1941 – together with Seaborg, Kennedy, and Wahl – he discovered ^{239}Pu and established its principal properties, including fission under slow-neutron bombardment. Segrè interview by Hoddeson, 23 July 1985, OH-55; Glenn T. Seaborg, "The Discovery of Plutonium in the Cyclotron," in A. S. Coffinberry and W. N. Miner, eds., *The Metal Plutonium* (Chicago: University of Chicago Press, 1961), pp. 9–12.

6 Monthly reports of the Segrè-Kennedy group in 1942, Segrè's papers; Segrè Notebook (NB) covering work from May 1940 to December 1944, p. 28, Segrè's private collection. Segrè interview by Hoddeson, 23 July 1985, OH-55, p. 7; and Farwell interviews by Hoddeson, 23 July 1985, OH-58 and 30 July 1985, OH-53. For a summary of the Berkeley work, see Segrè NB, e.g., 6 June 1943.

7 David Hawkins, *The Los Alamos Story* (Los Angeles: Tomash Publishers, 1983), p. 94.

8 Initially the group designed and built slow air-filled ionization chambers coupled to linear amplifiers sensitive to pulses in the millisecond range. Later they would develop faster nitrogen or argon-filled chambers for collecting ions slowly or electrons rapidly, and linear amplifiers that would respond on the order of microseconds to the pulses from the collected electrons or ions.

9 The early experiments were carried out in a private residence on Union Street used mainly by the music department of the University. Segrè interview by Hoddeson, 23 July 1985, OH-55; and Farwell interviews by Hoddeson 23 July 1985, OH-58 and 30 July 1985, OH-53; Chamberlain interview by Hoddeson, 31 July 1985, OH-57.

10 Segrè interview by Hoddeson, 23 July 1985, OH-55; Chamberlain interview by Hoddeson, 31 July 1985, OH-57.

11 D. Hawkins, *The Los Alamos Story*, p. 94.

12 Segrè NB, p. 51. But later at Los Alamos, as described below, the measurements converged to 40 f/g hr, which is still the accepted value. At present, measured spontaneous fission rates in f/g hr are $^{238}\text{U} = 20\ (1.5 \times 10^{16})$, $^{235}\text{U} = 1\ (3 \times 10^{17})$, $^{239}\text{Pu} = 40\ (7 \times 10^{15})$, $^{240}\text{Pu} = 1.6 \times 10^{6}\ (2 \times 10^{11})$. The values in parentheses are the neutron emission half-lives in years. Despite having a half-life greater than

the age of the universe, neutron emission from ^{240}Pu becomes significant because of the quantity of material present.

13 The captured neutrons were lost to the fission chain reaction; consequently, if the capture cross section proved to be a large fraction of the total neutron cross section of ^{235}U, much more material would be required to make a critical mass.

14 Indeed, having a somewhat higher fission cross section, ^{239}Pu is a somewhat better fuel for a bomb than ^{235}U.

15 The argument for the case of the excited uranium nuclei of mass numbers 236 and 239 is laid out by Niels Bohr in his February 1939 letter to the *Physical Review*, "Resonance in Uranium and Thorium Disintegration and the Phenomenon of Nuclear Fission," *Physical Review 55* (1939), pp. 418–19. For a more general overview of the understanding of nuclear fission theory in that period, see L. A. Turner, "Nuclear Fission," *Reviews of Modern Physics*, vol. 12 (1940), pp. 1–29.

16 G. Seaborg, "History of Met. Lab., Sec. C-1, April 1942 to April 1943," Lawrence Berkeley Laboratory, PUB 112, Feb. 1977.

17 Segrè interview by Hoddeson, 3–4 Dec. 1985, OH-104.

18 D. McKibbin card file of Los Alamos scientists in the Los Alamos Archives. Farwell interviews by Hoddeson, 23 July 1985, OH-58 and 30 July 1985, OH-53; Wiegand interview by Hoddeson, 29 May 1987, OH-157.

19 Site TA-18, where "Dwight Young's" log cabin (building PL-29), used then by security guards, still stands. Hawkins, *The Los Alamos Story*, p. 94. Farwell interviews by Hoddeson, 23 July 1985, OH-58 and 30 July 1985, OH-53; Wiegand interview by Hoddeson, 29 May 1987, OH-157.

20 Segrè interview by Hoddeson, 3–4 Dec. 1985, OH-104.

21 Hawkins, *The Los Alamos Story*, p. 95. In fact, spontaneous neutron emission from polonium does not occur and the polonium–beryllium initiator is not impossible.

22 The meeting was attended by Bethe, Hawkins, Kennedy, McMillan, Mitchell, Parsons, Oppenheimer, and Williams.

23 Hawkins, *The Los Alamos Story*, pp. 95, 73.

24 Hawkins, *The Los Alamos Story*, p. 95.

25 Subsequently, the radiochemistry group of the Chemistry-Metallurgy Division also explored the neutron emission accompanying α emission by studying the α emitter radon – a substance relatively easy to purify – but found no spontaneous neutrons. These studies were dropped when polonium purification was achieved. Hawkins, *The Los Alamos Story*, p. 95.

26 P-Division Progress Report, 15 July 1943, LAMS-4, pp. 4–5 and 1 Aug. 1943, LAMS-5; Farwell's Pajarito Data Book, no. 6, hereafter cited as PDB6. Wiegand interview by Hoddeson, 29 May 1987, OH-157.

27 PDB6; E. Segrè and C. Wiegand, "Boron Trifluoride Neutron Detector for Low Neutron Intensities," *Rev. Sci. Inst. 18* (1947), pp. 86–9. Farwell interviews by Hoddeson, 23 July 1985, OH-58 and 30 July 1985, OH-53.

28 Wiegand interview by Hoddeson, 29 May 1987, OH-157.

29 P-Division Progress Report, 1 Aug. 1943, LAMS-5; PDB6.

30 The designators were formed by combining the last digits of the atomic number and atomic weight of the isotope.

31 P-Division Progress Report, 1 Oct. 1943, LAMS-20.

32 W. Maurer and H. Pose, "Neutronemission des Urankerns als Folge seiner spontanen Spaltung," *Zeit. f. Physik 121* (1943), pp. 285–92. One should note the remarkable use of contemporary German wartime publications! P-Division Progress Report, #9, 15 Nov. 1943, LAMS-28.

33 E. G. Segrè, C. E. Wiegand, "Average Number of Neutrons Emitted Per Spontaneous Fission by Plutonium 240," LA-491, 11 November 1946.

34 Los Alamos is approximately 7300 feet above sea level (with only approximately 73 percent of sea level atmosphere above) and hence is subject to a higher cosmic-ray intensity than Berkeley, which is at sea level. Although a substantial fraction of the cosmic-ray neutrons are too slow to induce fission in ^{238}U, they can do so in ^{235}U. P-Division Progress Report, 1 Dec. 1943, LAMS-33. Segrè pointed out that the group's progress reports in this period were usually one month behind because they were too busy to keep them timely. Segrè interview by Hoddeson, 23 July 1985, OH-55; Chamberlain interview by Hoddeson, 31 July 1985, OH-57. Segrè NB, p. 81.

35 P-Division Progress Report, #7, 15 Oct. 1943, LAMS-21. Mary Miller, an WAC working with the group, became a specialist at making thin layers of active material for these experiments. Segrè to Fermi, 26 Nov. 1943, in Segrè's personal papers. Segrè interview by Hoddeson, 23 July 1985, OH-55.

36 Summary on 31 Jan. 1944, PDB6.

37 The two other hours went into checking the biases and whether the chambers were fully responsive to all possible pulses.

38 B_2O_3 is a material that absorbs cosmic-ray neutrons. P-Division Progress Report, 1 Dec. 1943, LAMS-33.

39 Farwell interviews by Hoddeson, 23 July 1985, OH-58 and 30 July 1985, OH-53; Wiegand interview by Hoddeson, 29 May 1987, OH-157; PDB6.

40 The irradiation is the mean neutron flux multiplied by the total exposure time.

41 S. K. Allison to A. H. Compton, 25 Jan. 1944, A-84-019, 15-6.

42 Oppenheimer to Kennedy and Parsons, 6 March 1944, A-84-019, 15-6.

43 Farwell interviews by Hoddeson, 23 July 1985, OH-58 and 30 July 1985, OH-53; PDB6.

44 PDB6, pp. 137, 141.

45 P-Division Progress Report, 15 April 1944, LAMS-93, p. 15.

46 Owen Chamberlain, George W. Farwell, and Emilio Segrè, "Plutonium 240 and Its Spontaneous Fission," LAMS-131, 8 Sept. 1944, containing data on the irradiation of the pile-produced plutonium samples.

47 Wiegand interview by Hoddeson, 27 May 1987, OH-157. Farwell interviews by Hoddeson, 23 July 1985, OH-58 and 30 July 1985, OH-53.

48 Wiegand interview by Hoddeson, 27 May 1987, OH-157.

49 The 261 number was actually high. PDB6; P-Division Progress Report, 15 June 1944, LAMS-114; P-Division Progress Report, 1 July 1944, LAMS-117.

50 PDB6; Farwell interviews by Hoddeson, 23 July 1985, OH-58 and 30 July 1985, OH-53.

51 For a more detailed account of this incident, see Stanley Goldberg, "Groves and the Scientists: Compartmentalization and the Struggle to Build the Bomb," presented at the 1990 Meeting of the History of Science Society, Seattle, Washington, in press. Bacher interview by Hoddeson and Kerr, 30 July 1984, OH-45.

52 Bacher interview by Hoddeson and Kerr, 30 July 1984, OH-45.

53 Segrè interview by Hoddeson, 3–4 Dec. 1985, OH-104.

54 Segrè remembered telling Fermi that the group thought the difference between the Berkeley and Clinton plutonium was due to ^{240}Pu: "We told him this [probably between 1 and 11 May], and the next week Fermi came back and told me, 'You know I think I know what it is. It is ^{240}Pu.' And then I told Fermi, 'Well, you know I told you last week.' And then he said, 'Ah, you know you are pulling a fast one on me' .... I tried to make him remember. And then finally he remembered and he said, 'By golly you are right.' You see, he had heard it, forgotten it, rediscovered it and then this happened." Segrè interview by Hoddeson, 3–4 Dec. 1985, OH-104. Fermi's experimental suggestion to re-irradiate plutonium had in fact been made half a year earlier. He evidently then forgot it, heard about it again from Segrè, forgot it again, rediscovered it, and then had the exchange that Segrè recalled.

55 R. C. Smith to Lt. Col. Whitney Ashbridge, "Report of Foreign Personnel at Project Y," 2 Aug. 1944, A-85-001, 1-1.

56 Trips file; "Findings of Trip to L.A." 4 July 1944, handwritten, presumably by Conant, RG 227, Bush-Conant Files, NARA. I would like to thank Ruth Harris for finding this document and bringing it to my attention.

57 Conant, "Findings of Trip to L.A." The report was not entirely pessimistic. Successful realization of the uranium gun bomb was virtually certain, although insufficient quantities of available uranium meant that few bombs could be made by the summer of 1945. Conant noted: "Assuming one is confident of 10 kg '25' by Jan. 1 and 50 kg July 1 and 30 kg a month thereafter. We are confident that one bomb can be dropped on the evening on Aug 1 [1945] with every prospect of a success .... A similar bomb could be dropped every six weeks thereafter."

58 P-Division Progress Report, 1 July 1944, LAMS-117. The first irradiated sample arrived on 1 July 1944. Oppenheimer to R. Bacher, R. Dodson, E. Segrè, R. Wilson on "Shipment of Material," 28 June 1944, A-84-019, 46-3.

59 PDB6, pp. 128–31.

60 It was already difficult to separate ^{235}U from ^{238}U, differing by three units of mass. In the case of the plutonium, only one unit of mass was in question. Oppenheimer to the Washington (D.C.) Liaison Office, 11 July 1944.

61 Farwell interviews by Hoddeson, 23 July 1985, OH-58 and 30 July 1985, OH-53.

62 Teletype, Oppenheimer to Groves, 14 July 1944. A-84-019, 15-7.

63 R.C. Hewlett, and O.E. Anderson, Jr., *The New World, 1939/1946: A History of the United States Energy Commission*, vol. 1 (University Park: Pennsylvania State University Press, 1962), p. 251. Ralph C. Smith recalls that Charles Thomas, a chemical adviser to General Groves, was a frequent visitor to Los Alamos. Smith, "Some unedited notes on section of history related to discovery of spontaneous fission in pile-produced plutonium," 17 March 1986, unpublished.

64 J. H. Manley, "A New Laboratory Is Born," in Lawrence Badash, Joseph Hirschfelder, and Herbert Broida, eds., *Reminiscences of Los Alamos, 1943–1945* (Dordrecht: Reidel, 1980), p. 33.

65 Manley, "A New Laboratory Is Born," p. 33.

66 Hewlett and Anderson, and Anderson, *The New World*, pp. 251–2.

67 Oppenheimer to Groves, 18 July 1944, A-84-019, 15-6.

68 Oppenheimer to Groves, 18 July 1944, A-84-019, 15-6.

69 Administrative Board Minutes, 20 July 1944, A-83-0013, 1-52.

70 Thomas to Groves, 21 July 1944, A-84-019, 15-6.

71 Hewlett and Anderson, *The New World*, pp. 251, 684, note 45. N. F. Ramsey, "Summary of Meeting of Project Y Technical Board on July 13, 1944," LAMS-113, 15 July 1944.

72 PDB6.

73 Farwell interviews by Hoddeson, 23 July 1985, OH-58 and 30 July 1985, OH-53.

74 R. C. Smith to Ashbridge, "Report of Foreign Personnel at Project Y," 2 Aug. 1944, A-85-001, 1-1.

75 Smith to Ashbridge, 2 Aug. 1944, A-85-001, 1-1.

76 PDB6 and also p. 114 of Segrè's NB, with a summary of all the work on p. 141.

77 Farwell, "Further Investigation of Spontaneous Fission in Plutonium," LA-490, 25 April 1946.

78 Farwell interviews by Hoddeson, 23 July 1985, OH-58 and 30 July 1985, OH-53; PDB6.

79 The results of the spontaneous fission experiments are summarized in Chamberlain, Farwell, and Segrè, "Plutonium 240 and Its Spontaneous Fission," LAMS-131. Further work done by the group on spontaneous fission is summarized in Farwell, "Further Investigation of Spontaneous Fission in Plutonium," LA-490.

80 Bacher interview by Hoddeson, 30 July 1984, OH-45.

81 Administrative Board Minutes, 20 July 1944, A-83-0013, 1-52. Bacher remembers the discussion taking place in June before he went on vacation in July. Oppenheimer to Bacher and Kistiakowsky, 14 Aug. 1944, A-84-019, 36-10; Administrative Board Minutes, A-83-0013, 1-52. See also Administrative Board Minutes, 20 July 1944, A-83-0013, 1-52 and Hawkins, *The Los Alamos Story*, p. 155–7.

82 Parsons to Oppenheimer 1 July 1944, A-84-019, 6-1.

83 Parsons was offered the position of Associate Director on 14 Aug. 1944. Oppenheimer to Parsons, 14 Aug. 1944, A-84-019, 6-1.

84 Hawkins, *The Los Alamos Story*, p. 197.

85 This reorganization was done in two stages. On 17 August Kistiakowsky wrote to Bradbury that Oppenheimer and Parsons "have approved a change in our status by which I cease to be the acting Group Leader for Group E-5, and you become Group Leader of E-5." Kistiakowsky to Bradbury, Group E-5, 17 Aug. 1944, A-84-019, 6-1. Bradbury's E-5 group then became X-1.

86 Kistiakowsky, "Reminiscences of Wartime Los Alamos," in L. Badash, et al., *Reminiscences of Los Alamos, 1943–1945* (Dordrecht: Reidel, 1980), pp. 49–65.

87 Hawkins, *The Los Alamos Story*, p. 172.

88 Administrative Board Minutes, 3 Aug. 1944, A-83-0013, 1-53. Kennedy to Oppenheimer, 30 Aug. 1944, "Personnel and space changes effected as a result of our change in program," A-84-019, 61-9.

89 Hawkins, *The Los Alamos Story*, p. 156.

90 Bacher interview by Hoddeson, 3 March 1986, OH-134.

91 Administrative Board Minutes, 6 and 20 July 1944, A-83-0013, 1-51 and 1-52.

92 Parsons chaired the committee composed of the following permanent members: Ashworth, Bacher, Bainbridge, Brode, Galloway, Henderson, Kistiakowsky, Lockridge, and Ramsey.

93 Although the main C Shop problems had been worked out by the time the conference began, there were always more shop projects than man-hours. Hawkins, *The Los Alamos Story*, pp. 157–8.

94 Members included Birch, Brode, Bradbury, Fussell, Fowler, and Morrison. Hawkins, *The Los Alamos Story*, pp. 157–8.

95 Historical studies of industrial laboratories reveal similar patterns, but much detailed study is needed to solidify our understanding. See, e.g., L. Hoddeson, "Innovation and Basic Research in the Industrial Laboratory: the Repeater, Transistor, and Bell Telephone System," in A Sarlemijn and P. Kroes, eds., *Between Science and Technology* (North Holland: Elsevier Science, 1990), pp. 181–214.

96 Hirschfelder to McMillan, 13 July 1944, A-84-019, 18-2.

13. Building the Uranium Bomb: August 1944 to July 1945

1 Birch, born 22 August 1903 in Washington, D.C., received his Ph.D. in physics from Harvard in 1932 after taking a master's degree from the University of Strasbourg, France. At the request of Parsons and Bainbridge, he came to Los Alamos with his wife and three children. He held the rank of lieutenant commander in the Naval Reserve.

2 W. S. Parsons, Personnel Report, 22 June 1945, A-84-019, 29-1; Parsons to Purnell, 23 July 1943, A-84-019, 35-8; and A-83-010, McKibbin cards.

3 Parsons, Personnel Report, 22 June 1945, A-84-019, 29-1; and A-83-010, McKibbin cards.

4 David Hawkins, *The Los Alamos Story* (Los Angeles: Tomash Publishers, 1983), pp. 192–5.

5 Langer interview by Henriksen, 21 March 1986, OH-113.

6 The electromagnetic plant had a variety of racetrack-shaped separation units. The Alpha plant, built in early 1943, contained five such tracks, each of which contained ninety-six tanks for separating ^{235}U from ^{238}U. The much smaller Beta plant built in 1944 consisted of seventy-two tanks. It received feed from the alpha tanks and further enriched the product. To increase the production of the Alpha plant, Groves decided in September 1943 to make an addition to the Alpha plant, Alpha II, consisting of four tracks, each with ninety-six tanks. Richard C. Hewlett and Oscar E. Anderson, Jr., *The New World, 1939/1946. A History of the United States Energy Commission*, vol. 1 (University Park: Pennsylvania State University Press, 1962), pp. 140–1, 149–52, 167, 294, and 296.

7 Hewlett and Anderson, *The New World*, pp. 298–300.

8 CM-Division Progress Reports, Group CM-8 section, 1 Nov. 1944, LAMS-155.

9 CM-Division Progress Reports, Group CM-11 section, 1 Oct. 1944, LAMS-146; 1 Nov. 1944, LAMS-155.

10 When early polonium samples were delivered to Los Alamos, the chemists were puzzled to find that assays showed much less polonium than Monsanto reported sending. Further checking revealed that the polonium had migrated to shipping cartridges; in some cases, as much as 40 percent was lost. CM-Division Progress Report, Group C-4 section, 1 June 1944, LAMS-97.

11 Oppenheimer to Thomas, 20 Sept. 1944, A-84-019, 15-8.

12 Technical and Scheduling Conference Agenda, 1 Jan. 1945, A-83-0013, 2-50 and 2-52.

13 Birch to Parsons, 1 Jan. 1945, A-84-019, 68-2.

14 Oppenheimer to Parsons and Birch, 7 Dec. 1944, A-84-019, 18-2.

15 Parsons to Moore, 23 April 1945, A-84-019, 18-2.

16 Parsons to Birch, 28 July 1945, A-84-019, 18-2; and Larkin to Bradbury, 29 Dec. 1945, A-84-019, 18-2. Bradbury's reply to Larkin appears as a handwritten note on the bottom of Larkin's memo and is dated 3 Jan. 1946.

17 Birch to Bacher, 10 Feb. 1945, A-84-019, 57-24.

18 Ibid.

19 Holloway to Birch and Ramsey, 17 Feb. 1945, A-84-019, 57-24.

20 Ibid.

21 Born in Walla Walla, Washington, Brode received his undergraduate degree from Whitman College in 1921 and his physics doctorate from the California Institute of Technology in 1924. He spent the next year at Oxford University as a Rhodes Scholar (the only one on the project). When the war started, Brode went to work for Section T of the OSRD and then came to Los Alamos in the summer of 1943, where Parsons put him in charge of the Fuze Group, E-3 (later O-3).

22 Only two take-off crashes were suffered in 3,549 sorties, or 0.056 percent. Report entitiled "Statistics of Crashes of B-29's at Takeoff of XXI Bomber Group in July 1945," VFA 350 and Gordon Thomas and Max Morgan Witts, *Enola Gay* (New York: Pocket Books, 1978), pp. 185–6.

23 Hewlett and Anderson, *The New World*, p. 374; Oppenheimer to C. S. Smith and Bacher, 23 April 1945, A-84-019, 16-9.

24 Hewlett and Anderson judge that the thermal diffusion plant "accelerated by about a week the flow of weapons-grade ^{235}U" from the electromagnetic plant. Although gaseous diffusion proved to be the most efficient method of isotope separation, the electromagnetic method alone could provide the necessary enrichment during wartime. Hewlett and Anderson, *The New World*, p. 624; also pp. 298, 374.

25 Hewlett and Anderson, *The New World*, p. 15.

26 TWX, Oppenheimer to Groves, 19 July 1945; Groves to Oppenheimer, 19 July 1945. A-84-019, 17-5.

14. Exploring the Plutonium Implosion Weapon: August 1944 to February 1945

1 C. S. Smith, "Some Recollections of Metallurgy at Los Alamos, 1943–45," *Journal of Nuclear Materials 100* (1981), pp. 3–10, esp. p. 8. A. C. Wahl interview by Westfall, 12 Nov. 1985, OH-82.

2 For more detail about the RaLa experiments see Bruno Rossi, *Moments in the Life of a Scientist* (Cambridge: Cambridge University Press, 1990).

3 Oppenheimer to Rossi, Shipment of Ra-La, 31 Aug. 1944, A-84-019, 40-2.

4 Rossi notebook, 202, p. 30.

5 Bacher interview by Hoddeson, 30 July 1984, OH-45.

6 Rossi 1 memoir; Rossi notebook No. 202, pp. 38–40.

7 "Notes on the RaLa Method for Observing Implosions," A-84-019, 40-2.

8 The meeting was attended by Ackerman, Alvarez, Bacher, Bethe, Bradbury, Dodson, Kistiakowsky, Koontz, Long, Oppenheimer, Parsons, Peierls, Rossi, Smith, Staub, Teller, and Weisskopf.

9 The issue of expanding the RaLa program – from three to five shots per month – emerged in mid-February 1945. This expansion required a second RaLa site to be built in Bayo Canyon. Rossi and Staub initially worried that the quality of the work might fall off with such an expansion, particularly if the group was not enlarged. Allison to Oppenheimer, RaLa Program, 20 Feb. 1945, A-84-019, 32-18; Rossi to Bacher, 5 March 1945, A-84-019, 16-2. However the new site was built, and a new subgroup of G-6, under James Allen, set up to run it. The frequency of RaLa shots now increased, sometimes to over one a week.

10 A more detailed description of the pin method appears in D. Froman, "Contact Electrical Method of Studying Implosion," 7 Dec. 1944, LA-182.

11 McMillan to Bacher, Program and personnel for magnetic method, 4 Oct. 1944, A-84-019, 44-18. A number of SEDs worked with the magnetic method group, among them John Fuller and Richard Davisson. Davisson made particularly important contributions to the electronics.

12 Creutz, who joined the laboratory in Oct. 1944, had worked on uranium problems at Princeton with Eugene Wigner, and then at the University of Chicago, as an OSRD project group leader on various metallurgical issues, including fabrication of uranium, corrosion protection in the casting of uranium, aluminum fabrication, and preparation of beryllium and thorium. Creutz interview by Hoddeson, 30 April 1986, OH-119.

13 David Hawkins, *The Los Alamos Story* (Los Angeles: Tomash Publishers, 1983), p. 202. Kerst interview by Hoddeson, 8 Aug. 1985, OH-45 and 28 April 1986, OH-118.

14 Oppenheimer to Washington Liaison Office, 24 Aug. 1944, A-84-019, 14-3.

15 Conant to Oppenheimer, 25 Aug. 1944, A-84-019, 14-3.

16 Oppenheimer to Washington Liaison Office, 26 Aug. 1944, A-84-019, 14-3.

17 Teletype, Moreland to Oppenheimer, 26 Aug. 1944, A-84-019, 14-3.

18 Kerst, "Minutes of Meeting of Betatron Experiment of September 7, 1944," LAMS-132, 11 Sept. 1944.

19 In fact, the compression was better tested by the betatron than by the RaLa method, because the RaLa method, giving a line integral of the density, emphasized points near the center, while the betatron method gave the average density, a value of greater usefulness to the theorists.

20 Kerst, "Minutes of Meeting of Betatron Experiment of September 7, 1944," LAMS-132, 11 Sept. 1944.

21 Neddermeyer to Bacher, Estimated Personnel and Facilities Required for Betatron Experiment, 14 Sept. 1944, A-84-019, 48-7.

22 Plans for Installation of Betatron at Y, A-84-019, 14-3. He reported here that installation would require about 200 machine shop man-hours, 500 electronics shop man-hours and a "considerable concrete structure at the shooting site. This would be done on outside contract and completed 50 days from date of authorization at an unreasonable [*sic*] cost which might be $250,000."

23 Oppenheimer to Groves, 15 Sept. 1944, A-84-019, 14-3.

24 S. B. Ritchie, Ordnance Department, to G. E. Beggs, Jr., 3 Oct. 1944, A-84-019, 14-3.

25 This informer differs from the one proposed in the gun program. Critchfield to Bacher, 26 Aug. 1944, A-84-012, 1-7.

26 Greisen, "X-ray Results on Implosion of Gas-Tamper Spheres," 1 Dec. 1944, LA-178.

27 The basic reason for the smaller jets was that Torpex, like other aluminized high explosives, releases the energy from burning of its aluminum content relatively slowly, in particular, after the implosion shock wave has moved inward; this energy release does not contribute to the implosion. Torpex was eventually discarded by the implosion program.

28 Hornig had worked earlier with Mary Nachtrieb on the fundamental chemistry of plutonium. As a member of section X-1, she worked on X-ray and terminal observation study of lenses. Machinist David Greenglass was one of those involved in making the lenses for these studies, where he had access to drawings of lenses passed on to the Soviet Union; e.g., Norman Moss, *Klaus Fuchs: The Man Who Stole the Atom Bomb* (New York: St. Martins Press, 1987). For a discussion of both Greenglass and the Rosenbergs, see Louis Nizer, *The Implosion Conspiracy* (New York: Doubleday, 1973).

29 L. Hornig interview by Henriksen and Hoddeson, 25 June 1986, OH-128.

30 Wahl, "Chemical Purification of Plutonium," in "Chemistry of Uranium and Plutonium," LA-1100, 20 April 1947; C. A. Thomas and J. C. Warner, "The Chemistry, Purification, and Metallurgy of Plutonium," Pub. 112, p. 19.

31 The standard size was set at 160 g because this was less than the Water Boiler's critical mass and was also a multiple of the 80-g batch size planned for Hanford deliveries. Eventually, Hanford increased the batch size to 160 g. CM-Division Progress Report, 1 Nov. 1944, LAMS-155; Wahl, "Chemical Purification of Plutonium," in "Chemistry of Uranium and Plutonium," LA-1100, pp. 73–4, 90–1.

32 A greatly truncated purification method, which eliminated the dangerous process of ether extraction but produced plutonium with the necessary purity, was developed after the war. Wahl interview by Westfall, 11 Nov. 1985, OH-81. Wahl, "Chemical Purification of Plutonium," in "Chemistry of Uranium and Plutonium," LA-1100, p. 73.

33 I. B. Johns and G. H. Moulton Jr., "Large-Scale Preparation of the Anhydrous Fluorides of Plutonium," LA-193, 20 Dec. 1944; Garner, "Tetrafluoride and other Dry Compounds of Plutonium," in "Chemistry of Uranium and Plutonium," LA-1100, 20 April 1947, p. 111.

34 CM-Division Progress Report, Group CM-5, 1 Oct. 1944, LA-146; and 1 Nov. 1944, LA-155.

35 Garner, "Tetrafluoride and Other Dry Compounds of Plutonium," in "Chemistry of Uranium and Plutonium," LA-1100, pp. 149–56.

36 Private communication Penneman, 3 Oct. 1985.

37 The centrifuge method was developed by the energetic, slapdash Theodore T. Magel; the gram scale reduction was developed by Richard D. Baker, a slower moving, more careful metallurgist who had already gained considerable expertise with bomb reductions using uranium.

38 Under these conditions, a centrifuge works best because the droplets are immediately collected using centrifugal force.

39 Electrorefining methods are now used as a final step to produce extremely pure plutonium metal. Thomas and Warner, "The Chemistry, Purification, and Metallurgy of Plutonium," Pub. 112, p. 294; Smith, "Some Recollections," pp. 3–10.

40 Thomas and Warner, "The Chemistry, Purification, and Metallurgy of Plutonium," Pub. 112, p. 295; private communication Wahl, 7 Aug. 1986.

41 Thomas and Warner, "The Chemistry, Purification, and Metallurgy of Plutonium," Pub. 112, p. 216; Smith, "Plutonium Metallurgy at Los Alamos During 1943–45," in A. S. Coffinberry and W. N. Miner, eds., *The Metal Plutonium* (Chicago: University of Chicago Press, 1961), p. 31; 1 Oct. 1944, LAMS-146; 1 Nov. 1944, LAMS-155.

42 Thomas and Warner, "The Chemistry, Purification, and Metallurgy of Plutonium," Pub. 112, p. 319.

43 Although the melting point of plutonium is about 630° C, bomb reductions at this scale had to be heated to higher initial temperatures to offset heat loss to the container so that the melting point could be reached and maintained in the product metal, allowing coalescence into a unitary button; Thomas and Warner, "The Chemistry, Purification, and Metallurgy of Plutonium," Pub. 112, pp. 319, 321; private communication Penneman, 12 Oct. 1985; private communication Redman, 20 Oct. 1986.

44 Thomas and Warner, "The Chemistry, Purification, and Metallurgy of Plutonium" Pub. 112, pp. 18–23, 25, 469.

45 Thomas and Warner, "The Chemistry, Purification, and Metallurgy of Plutonium," Pub. 112, p. 340.

46 Thomas to Groves and Conant, 13 June 1944, A-84-019, 15-6; Smith, "Some Recollections," pp. 3–10, esp. p. 8; Smith interviews by Westfall, 16 Dec. 1985, OH-92 and 93. Leo Brewer, manuscript from talk given at R. D. Baker Memorial Symposium, 12 April 1986, VFA-472.

47 Group CM-8 section in LAMS-155, 1 Nov. 1944.

48 CM-Division Progress Reports, Group CM-8, 1 Oct. 1944, LAMS-146; 1 Nov. 1944, LAMS-155; Smith, "Plutonium Metallurgy at Los Alamos, 1943–45," p. 34.

49 The phase specifications for early metal are uncertain. Later research has shown that mixed-phase systems containing α can be malleable, which suggests that in World War II they actually dealt with α phase metal with increasing amounts of δ phase, which they labeled β and γ.

50 Smith, "Some Recollections," pp. 3–10, esp. p. 7.

51 Thomas and Warner, "The Chemistry, Purification, and Metallurgy of Plutonium," Pub. 112, p. 361.

52 Thomas and Warner, "The Chemistry, Purification, and Metallurgy of Plutonium," Pub. 112, pp. 216ff.

53 Thomas to Groves and Conant, 8 April 1944, A-84-019, 9-4.

54 L. Brewer, manuscript from talk at R. D. Baker Memorial Symposium, 12 April 1986, VFA-472; Smith, "Plutonium Metallurgy at Los Alamos During 1943–45," p. 31; Smith, "Refractories," in vol. 10, chap. 5, Metallurgy, LA-1259, 1 Dec. 1946, p. 4. Smith interviews by Westfall, 16 Dec. 1985, OH-92 and 93.

55 Thomas and Warner, "The Chemistry, Purification, and Metallurgy of Plutonium," Pub. 112, p. 217; Smith, "Refractories," vol. 10, chap. 5, LA-1259.

56 Thomas and Warner, "The Chemistry, Purification, and Metallurgy of Plutonium," Pub. 112, p. 225.

57 Thomas and Warner, "The Chemistry, Purification, and Metallurgy of Plutonium," Pub. 112, pp. 218, 220, 353–4; Smith, "Refractories," vol. 10, chap. 5, LA-1259, p. 4.

58 Smith and Warner, "The Chemistry, Purification, and Metallurgy of Plutonium,"

Pub. 112, pp. 368–71; CM-Division Progress Reports, Groups CM-6 and CM-11, 1 Oct. 1944, LAMS-146; 1 Nov. 1944, LAMS-155.

59 F. K. Pittman, "Chemistry of the Recovery of Plutonium," in "Chemistry of Uranium and Plutonium," LA-1100, p. 170. Wahl interview by Westfall, 11 Nov. 1985, OH-81.

60 Pittman, "Chemistry of the Recovery of Plutonium," in "Chemistry of Uranium and Plutonium," LA-1100, pp. 200–10, 192.

61 Smith, "Some Recollections," pp. 3–10, esp. p. 6.

62 Wahl interview by Westfall, 11 Nov. 1985, OH-81.

63 Wahl interview by Westfall, 12 Nov. 1985, OH-82.

64 The name "DP" is said to be short for "D-prime," derived from the first plutonium separation facility, D building. In GI argot, the building was known as "Depot Plutonium." Private communication, Penneman, 16 Dec. 1985.

65 Neutron emission rates were measured by two devices developed at Los Alamos, the Szilard-Chalmers Counter and the Boron Trifluoride Multiple Chamber Counter. Thomas and Warner, "The Chemistry, Purification, and Metallurgy of Plutonium," Pub. 112, p. 469.

66 Richard C. Hewlett, and Oscar E. Anderson, Jr., *The New World, 1939/1946: A History of the United States Energy Commission*, vol. 1 (University Park: Pennsylvania State University Press, 1962), p. 317.

67 Hewlett and Anderson, *The New World*, pp. 305–8.

68 MED History, Book IV, Part 4, V. Part II: Clinton Laboratories, p. 4.10.

69 Hewlett and Anderson, *The New World*, p. 310.

70 The radioassay group, which provided data for control measures as well as for plutonium bookkeeping, commonly employed wives of Los Alamos researchers who wanted to work and had some chemistry training.

71 Wahl, "Chemical Purification of Plutonium," in "Chemistry of Uranium and Plutonium," LA-1100, p. 74; Wahl interview by Westfall, 12 Nov. 1985, OH-82. Thomas and Warner, "The Chemistry, Purification, and Metallurgy of Plutonium," Pub. 112, p. 446.

15. Finding the Implosion Design: August 1944 to February 1945

1 Conant to Groves, "Report on Visit to Los Alamos," 18 Oct. 1944, dated 20 Oct. 1944. He reported the Los Alamos estimates, which by either method suggested the yield would be about 8,000–10,000 tons TNT equivalent "for delivery to enemy on 1 July 1945." Conant made a point here of urging that plutonium be used in the Trinity test, a question apparently not yet settled, so that all the available production uranium would be on hand for the gun bomb. He also stressed that using plutonium "simplifies the recovery problem in case of an unsuccessful trial shot of the implosion bomb at Trinity." The reason plutonium was preferable for recovery of the fissile material was that the implosion weapon would have a natural uranium tamper, and plutonium mixed with uranium in recovered material could be separated chemically. However, if ^{235}U were tested in an implosion that failed, the uranium isotopes might mix in regions, making necessary tedious isotope separation to recover the ^{235}U.

2 X-2, under Kenneth Bainbridge, was an engineering group aimed primarily at Trinity preparations. Bradbury had been Assistant Experimental Officer under Parsons at Dahlgren Proving Ground. Parsons secured his services for Los Alamos in June 1944. Bradbury interview by Redman, 2 Dec. 1987, OH-166.

3 A standard method of measurement was to embed metal pins in the rate stick and observe when the electrically conducting detonation wave passed each pin or pair of pins.

4 Kistiakowsky, "Reminiscences of Wartime Los Alamos," in Lawrence Badash, Joseph Hirschfelder, and Herbert P. Broida, eds., *Reminiscences of Los Alamos, 1943–1945* (Dordrecht: D. Reidel, 1980), pp. 53–4.

5 Russell interview by Hoddeson, 16 July 1986, OH-133.

6 Gurinsky came to Los Alamos from the Met Lab in mid-November 1944, when he was no longer needed there as a plutonium chemist. The three other "surplus" men from the Met Lab were Melvin Brooks, Larry Foster, and Bernard Turovlin. The Special Problems unit later became Section X-3D, and then X-3B.

7 E-Building Project Q report, 1-15 Dec. 1944, and following reports. The stick detonation velocity of Baratol-1 was about 4,800 m/sec; that of Baratol-2 about 4,900m/sec. After TNT became commercially available about 1904, a wide variety of TNT-based explosives was developed. Mixtures with inorganic nitrates were popular. The barium nitrate mixture, Baratol, had a lower detonation velocity and pressure, which led to larger fragment size, and, as opposed to cheaper sodium nitrate, it was nonhygroscopic, which gave it a longer shelf life; these two qualities made Baratol attractive as a bursting charge for grenades. Baratol was used both in the United States and Great Britain as a military explosive. The 1913 U.S. Patent, No. 1 056 389, was on a Baratol composition. Private communication, John B. Ramsay, 1987, who has collected extensive information on the history of Baratol.

8 Parsons to Oppenheimer, 16 Oct. 1944, Design of Non-lens Assembly, A-84-019, 39-8.

9 The participants were Allison, Alvarez, Bacher, Bainbridge, Baker, Bradbury, Critchfield, Fermi, Galloway, Henderson, Kistiakowsky, Peierls, Parsons, Smith. Oppenheimer to group, 28 Oct. 1944, A-84-019, 13-3.

10 Oppenheimer to Bacher and Kistiakowsky, 3 Nov. 1944; Kistiakowsky, Notice of Meeting, 13 Nov. 1944; A-84-019, 13-3.

11 Parsons to Oppenheimer, Home Stretch Measures, 19 Feb. 1945, A-84-019, 13-3.

12 Bethe interview by Hoddeson, 3 Oct. 1986, OH-140 and Peierls interview by Hoddeson, 20–21 March 1986, OH-111. However, others, like John Russell, who was integrally involved in the wartime lens development, question whether this "elegant solution to the problem" was indeed necessary or desirable. Russell interview by Hoddeson, 16 July 1986, OH-133.

13 Alvarez, *Adventures of a Physicist* (New York: Basic Books, 1987), pp. 134–5.

14 Donald Hornig arrived at Los Alamos on 30 May 1944, with his wife Lilli; McKibbin cards. D. Hornig was then a twenty-four-year-old physical chemist who had taken his degree at Harvard with E. Bright Wilson, director of the underwater explosives research laboratory in Woods Hole. Kistiakowsky had been one of his professors. Hornig was diverted by the war into piezoelectric measurements of shock and detonation waves. At the Woods Hole Oceanographic Institution, he had worked on interaction of blast waves in air with the ground.

15 Alvarez interview by Hoddeson, 28 April 1986, OH-120.

16 Alvarez, *Adventures*, p. 135.

17 Including mechanical safety gates in the bomb system (Chapter 9) was to be avoided at all cost. The energy needed to set off the spark gap detonator was much less than that required to set off the bridgewire detonator.

18 McKibbin cards. Lofgren arrived at Los Alamos on 26 November 1944.

19 The high-speed rotating mirror camera had been invented for Navy use by Ira Bowen, Robert Millikan's long-time colleague in cosmic ray studies. Michelson had used a similar camera in his pioneering first measurements of the velocity of light. Alvarez, *Adventures*, p. 135.

20 Problems 4 and 5, completed in August, studied the acceleration of a spherical shell by a convergent detonation wave. Problem 6, finished in September, studied the acceleration by a convergent shock wave. Problem 7, started in September, to study a converging shock wave, would not be completed until January.

21 Serber interview by Hoddeson, 25–26 Feb. 1986, OH-110; and Christy interview by Hoddeson, 14 April 1986, OH-117. Christy received his Ph.D. from the University of California, Berkeley, in 1941; he then became an instructor in physics at the Illinois Institute of Technology in 1941–42, and later a research associate at the University of Chicago in 1942–43, where he worked on reactor design. Christy joined Los Alamos in April 1943. Although he had no group under him, because of his professional stature he was treated as a group leader.

22 E.g., Feynman and Serber and their groups were assigned to the critical assemblies group, G-1; Keller to work on general problems and Weisskopf on instrumentation in the X-ray method group, G-2; Feynman as well to the magnetic method group, G-3, the RaLa method group, G-6, and the electric detonator group, G-7; Keller to work on general problems and Serber on instrumentation in the betatron group, G-5; Christy to the initiator group, G-10; and Penny to the optics group, G-11.

23 Other examples can be found in the literature on industrial research, e.g., see David A. Hounshell and John Kenly Smith, *Science and Corporate Strategy: Du Pont R&D, 1902–1980* (Cambridge, England: Cambridge University Press, 1988).

24 Critchfield interview by Hoddeson and Kerr, 2 Aug. 1984, OH-46; Bacher interview by Hoddeson, 30 July 1984, OH-45.

25 Bacher interview by Hoddeson, 30 July 1984, OH-45.

26 Critchfield interview by Hoddeson and Kerr, 2 Aug. 1984, OH-46; Bacher interview by Hoddeson, 30 July 1984, OH-45.

27 Memo from Oppenheimer to Bethe, Fermi, and Christy, with carbon copies to Bacher, Baker, Critchfield, and Dodson, 9 Feb. 1945, A-84-019, 17-4.

28 Critchfield interview by Hoddeson and Kerr, 2 Aug. 1984, OH-46, pp. 37–38; David Hawkins, *Project Y: The Los Alamos Story* (Los Angeles: Tomash Publishers, 1983). Niels Bohr met with this committee when he visited Los Alamos. Others who at times attended meetings of the initiator committee included Cyril Smith, Richard Dodson, Richard Baker, Bacher, Critchfield, and Parsons.

29 Thomas to Oppenheimer, 28 Dec. 1944, A-84-019, 15-8.

30 Dodson to Kennedy, 6 Feb. 1945, A-84-019, 61-13.

31 Oppenheimer to Thomas, 6 June 1945, A-84-019, 15-8.

32 Bainbridge, "Prelude to Trinity," *Bulletin of the Atomic Scientists* (April 1975), p. 46. According to Henderson, Groves rejected the California site, since it was in use by General Patton, whom Groves refused to even approach about use of the site. Henderson interview with Henriksen and B. Perkins, 7 March 1989.

33 The committee included Oppenheimer, Stevens, Silva, Captain Wilbur F. Schaffer, Robert Henderson, Roy Carlson, Lewis Fussell, and Ensign George T. Reynolds. Bainbridge, "Trinity," LA-6300-H, May 1976, p. 3. Ferenc Morton Szasz, *The Day the Sun Rose Twice: The Story of the Trinity Site Nuclear Explosion, July 16, 1954* (Albuquerque: University of New Mexico Press, 1984), p. 28. Hawkins, *The Los Alamos Story*, p. 233.

34 Bainbridge, "Prelude to Trinity," *Bulletin of the Atomic Scientists* (April 1975), pp. 42–46, esp. p. 46.

35 Administrative Board Minutes, 26 Oct. 1944, A-83-0013, 1-57, p. 1.

36 Bainbridge, private communication, 1988.

37 Bainbridge, "A Foul and Awesome Display," *Bulletin of the Atomic Scientists* (May 1975), p. 40.

38 Bainbridge, private communication, 1988.

39 Hawkins, *The Los Alamos Story*, pp. 240–1. Also see map on p. 505.

40 Bainbridge, "A Foul and Awesome Display," p. 40.

41 Notes by Conant, "Summary of Trip to Y," Dec. 1944, RG 227, Bush Conant Files, NARA.

42 Hawkins, *The Los Alamos Story*, p. 160.

43 The meeting was attended by group leaders and others at Bacher's invitation.

44 Hawkins, *The Los Alamos Story*, pp. 168–9.

45 Hewlett and Anderson, *The New World*, pp. 313–14.

46 Hawkins, *The Los Alamos Story*, p. 234. Chapter 18 discusses Trinity.

16. Building the Implosion Gadget: March 1945 to July 1945

1 David Hawkins, *The Los Alamos Story* (Los Angeles: Tomash Publishers, 1983), pp. 158, 169.

2 Hawkins, *The Los Alamos Story*, p. 205.

3 Thomas, however, remained an important ally as research director at Monsanto.

4 Critchfield interview by Hoddeson and Kerr, 2 Aug. 1984, OH-46.

5 Bacher interview by Hoddeson, 30 July 1984, OH-45.

6 Bethe interview by Hoddeson, 3 Oct. 1986, OH-140.

7 Penneman private communication, 18 Dec. 1985.

8 Russell interview by Hoddeson, 16 July 1986, OH-133.

9 Roy joined Los Alamos and X-Division in March 1945. He had earlier been the Los Alamos contact in the Washington OSRD Division 8.

10 Although the molds were scheduled for early April, the first one arrived at Los Alamos on 29 April, and delivery in useful quantities came in May. Molds were also supplied to Yorktown Naval Mine Depot, near Norfolk, Virginia, and to McAlester Naval Ammunition Depot, Oklahoma. Yorktown worked on Baratol and both navy depots made Comp B castings. Some inert castings of plaster and cement, made both at the navy depots and in Los Angeles, were used for practice assembly of the device. However, high-explosives castings made by the navy installations were not of adequate quality for an implosion system. Technical and Scheduling Conference

Agenda, 5 Dec. 1944 to 10 Aug. 1945, 18 June 1947; Kistiakowsky to Bradbury, 14 April 1945, A-84-019, 13-3.

11 Safe procedures were followed, based on ERL experience, and in fact there were no wartime explosions in the plant at Los Alamos.

12 Russell interviews by Hoddeson, 18 July 1986, OH-133 and by Redman, 6 Jan. 1988, OH-184. Immediately after the war, such minor repair work became routine for stockpile charges. In the postwar era, lenses and inner charges were machined under far better conditions than were available for such work during the war.

13 Kistiakowsky, Booster Detonator Units for FM–Henderson Assembly Drawing Y-1773C2, 4 April 1945, A-84-019, 82-16; R. S. Warner to W. F. Schaffer, Detonator-Booster Assembly Operation, 6 April 1945, A-84-019, 82-16.

14 This delay, together with difficulties encountered in designing a radio informer system to inform ground observers of how the X-units were performing in free fall, was to limit the use of the Raytheon product to but a few experimental bomb drops. By 1 July, only 15 of the Raytheon circuits had arrived at Los Alamos.

15 Russell interview by Hoddeson, 16 July 1986, OH-133.

16 Hawkins, *The Los Alamos Story*, p. 200.

17 McMillan interview by Westfall, 31 Oct. 1985, OH-77; Creutz interview by Hoddeson, 30 April 1986; and Davisson interview by Hoddeson, 17 Nov. 1986, OH-151.

18 C. S. Garner, "Tetrafluoride and Other Dry Compounds of Plutonium," in "Chemistry of Uranium and Plutonium," LA-1100, 20 April 1947, p. 110. Wahl interview by Westfall, 12 Nov. 1985, OH-82.

19 Wahl, "Chemical Purification of Plutonium," in "Chemistry of Uranium and Plutonium," LA-1100, 20 April 1947, p. 73.

20 In June 1943, the CM-Division had about 20 people, close to the number thought to be needed before the decision was made to conduct plutonium purification and metallurgy at Los Alamos. At its peak in 1945, the chemistry-metallurgy group had over 400 members, including researchers and technicians. Hawkins, *The Los Alamos Story*, p. 137; Richard C. Hewlett, and Oscar E. Anderson, Jr., *The New World, 1939/1946. A History of the United States Energy Commission*, vol. 1 (University Park: Pennsylvania State University Press, 1962), pp. 313–14, 320. Wahl interviews by Westfall, 11 Nov. 1985, OH-81 and 12 Nov. 1985, OH-82.

21 Wahl, "Chemical Purification of Plutonium," in LA-1100, p. 76.

22 C. Smith, Los Alamos Technical Series, vol. 10, Metallurgy, chap. 5, LA-1259, 1 Dec. 1946, p. 4.

23 Hawkins, *The Los Alamos Story*, p. 223.

24 C. Smith, "Some Recollections of Metallurgy at Los Alamos, 1943–45," *Journal of Nuclear Materials 100* (1981), pp. 3–10, esp. p. 8.

25 Quotes from Smith, "Some Recollections," p. 9.

26 Smith interview by Westfall, 16 Dec. 1985, OH-92 and OH-93.

27 Smith, "Some Recollections," pp. 3–10, esp. p. 9.

28 Smith interview by Westfall, 16 Dec. 1985, OH-92 and OH-93.

29 Smith, "Some Recollections," pp. 3–10, esp. p. 9.

30 In addition to carrying problem 12 – a solid gadget of standard design at finite temperature, to collapse – they repeated problem 10 for a shell of active material, but with finite temperature, carried out three studies of solid gadgets with various tampers (14, 15, 17) and one of a preassembled core (16). Richard Feynman, "Los

Alamos from Below," in Lawrence Badash, Joseph Hirschfelder, and H. P. Broida, eds., *Reminiscences of Los Alamos, 1943–1945* (Dordrecht: D. Reidel, 1980), pp. 125–6; R. P. Feynman, *Surely You're Joking, Mr. Feynman* (New York: W. W. Norton, 1985), pp. 108–11.

31 Otto Frisch was considered but not chosen, because members of the British Mission were not allowed to hold official positions with the pit teams. Bacher later added Baker to the G-Engineers. Bacher also worked directly with this group.

32 R. E. Schreiber, private communication, 16 Aug. 1988.

33 Private communication, A. Van Vessem and R. Schreiber, 26 July 1988.

17. Critical Assemblies and Nuclear Physics: August 1944 to July 1945

1 David Hawkins, *The Los Alamos Story* (Los Angeles: Tomash Publishers, 1983), pp. 180–1.

2 Ibid.; R-Division Progress Report, 1 Sept. 1944, LAMS-135.

3 R-Division Progress Report, 1 Oct. 1944, LAMS-151, p. 1.

4 R. Serber, "Preliminary Report on Multiplication by Spheres of Beta-Stage Material," LAMS-154, 1 Nov. 1944; A. O. Hanson, R. Serber, and J. H. Williams, "Multiplication by Small Spheres of Active Material," LAMS-227, 11 April 1945; H. T. Richards, "Compact Mock Fission Neutron Sources," LA-201, 13 Jan. 1945.

5 T was calculated by measuring the activity of fission fragments in thin cellophane sheets placed between plates of ^{235}U at the center of the sphere of ^{235}U. When the mock-fission source was placed in the center of the sphere, the cellophane collected the fission fragments. Measuring the cellophane's activity, the ratio of the cellophane catcher activities was then sure to be the same as T.

6 R-Division Progress Reports, 1 Oct. 1944, LAMS-151, p. 2, and 1 Nov. 1944, LAMS-163, p. 2.

7 R-Division Progress Reports, 1 Oct. 1944, LAMS-151, p. 2 and 1 Nov. 1944, LAMS-163, p. 2.

8 R. E. Carter, J. Hinton, and L. D. P. King, "Critical Mass Measurements for a 25 Sphere in Tu and WC Tampers," 30 Oct. 1945, LA-442. Anderson, "Neutron Multiplication in Spheres of 25," LA-402, 6 Oct. 1945. Wilson's summary in R-Division Progress Report, 1 Nov. 1944, LAMS-163.

9 R-Division Progress Report, 1 Dec. 1944, LAMS-175. The Technical and Scheduling Conference decided, however, that a 2.5-inch sphere was not large enough to reveal errors in the theoretical treatment of inelastic scattering and transport cross section.

10 R-Division Progress Report, 1 March 1945, LAMS-222.

11 Quotations from J. E. Burke, "Recollections of Processing Uranium Hydride and Plutonium at Wartime Los Alamos," in *Journal of Nuclear Materials 100* (1981), p. 13.

12 R-Division Progress Report, 1 March 1945, LAMS-222.

13 Hawkins, *The Los Alamos Story*, p. 199.

14 Hawkins, *The Los Alamos Story*, pp. 270–6.

15 Hawkins, *The Los Alamos Story*, pp. 177–82; R-Division Progress Report, 1 Sept. 1944, LAMS-135.

16 R-Division Progress Report, 1 Oct. 1944, LAMS-151, pp. 3 and 5.

17 R-Division Progress Reports, 1 Sept. 1944, LAMS-135, pp. 8 and 10.

18 To help overcome this problem, they used a screen to collimate the fission fragments, so that the particles were prevented from emerging at small angles to the emitting surface. The collection gas was a mixture of carbon dioxide and purified argon. A fast amplifier with a clipping time of about 0.2 μs amplified the detected pulses.

19 R-Division Progress Report, 1 Sept. 1944, LAMS-135, pp. 4–5; B. MacFarlane, private communication, 10 Dec. 1987.

20 R-Division Progress Report, 1 Nov. 1944, LAMS-163, p. 2.

21 R-Division Progress Report, Group R-2, 1 March 1945, LAMS-222; Hawkins, *The Los Alamos Story*, pp. 181–2.

22 Hawkins, *The Los Alamos Story*, pp. 181–2.

23 See S. Glasstone, *The Effects of Nuclear Weapons*, 3d ed. (Washington: U.S. Dept. of Defense, 1977), pp. 387ff.

24 Serber interview by Hoddeson, 25–26 Feb. 1983, p. 126.

25 T-Division Progress Report, May 1945, LAMS-260, p. 5.

26 E. Konopinski, C. Marvin, and E. Teller, "Ignition of the Atmosphere with Nuclear Bombs," LA-602, 14 Aug. 1946; Bethe, private communication, 15 Feb. 1989; Konopinski interview by Henriksen, 21 March 1986, OH-112.

27 Frisch to Oppenheimer, "Proposal for a Mechanically Controlled Fast Chain Reaction," 17 Oct. 1944, and addition on 24 Oct. 1944, B-9, file 471.6, Core-Gadget-Spec. Efficiency.

28 O. Frisch, *What Little I Remember* (Cambridge: Cambridge University Press, 1979), p. 159.

29 Charles P. Baker, Otto R. Frisch, and Bernard T. Feld, "Los Alamos Technical Series Volume 5: Critical Assemblies Part 4," LA-1036, 30 Dec. 1947, p. 8.

30 F. L. Bentzen, R. E. Carter, J. Hinton, L. D. P. King, J. C. Nevenzel, R. E. Schreiber, J. W. Starner, and P. H. Watkins, "High-Power Water Boiler, AECD-3065," 19 Sept. 1945, p. 1.

18. The Test at Trinity: January 1944 to July 1945

1 K. Bainbridge, "A Foul and Awesome Display," *Bulletin of the Atomic Scientists* (May 1975), p. 41.

2 It was Bainbridge's least favorite experiment. Bainbridge, "A Foul and Awesome Display," p. 41.

3 Bainbridge, "A Foul and Awesome Display," p. 40.

4 Bainbridge, "A Foul and Awesome Display," p. 41. Some of these deliberations took place in R-Division at least as late as 31 March 1945.

5 Manley to Bainbridge, 19 March 1945, A-84-019, 53-12.

6 Damage from the air concussion would be indistinguishable from the ground vibrations, however, and could be significant. L. D. Leet, "100-Ton Test – Ground Vibrations," LA-439, 15 Nov. 1945.

7 Memo by Bainbridge, 8 June 1945, A-84-019, 54-4.

8 Mack, a physicist from the University of Wisconsin at Madison, was originally hired to organize the Tech Area shops and services, but he soon switched to optical research and development. Brixner was an Albuquerque native then completing a job of preparing planimetric maps of the New Mexico region with the U.S.D.A. Soil Conservation Service.

9 Mack, "July 16th Nuclear Explosion–Space–Time Relationships," LA-531, 2 April 1946, appendix 1, list of cameras used at Trinity.

10 B. Brixner, "Optical Engineering at Los Alamos – a History," copy provided by Brixner to LANL Archives, VFA-197.

11 Bainbridge memo on measurements proposed for gadget test, 1 Sept. 1944, A-84-019, 54-2. The spectrograph was a drum camera coupled to an oscilloscope that measured the energy released by the explosion in the region of the radiation spectrum absorbed by a black body.

12 David Hawkins, *Project Y: The Los Alamos Story* (Los Angeles: Tomash Publishers, 1983), pp. 197, 209, 240.

13 Hawkins, *The Los Alamos Story*, p. 243.

14 A nuclear explosion tries to stop itself as soon as it begins because the heat energy generated by the reaction expands the active material and makes its configuration less critical. The faster the neutrons are generated the more fissions will have taken place before the bomb blows itself apart and the greater the yield.

15 Bridge, DeWire, and Snyder, "Measurement of Neutron Multiplication Rate," LA-540.

16 B. Rossi, *Moments in the Life of a Scientist* (New York: Cambridge University Press, 1990).

17 Trinity site interview with Robert Krohn, Berlyn Brixner, John Manley, and Joseph McKibben, by Bearss, June 1986, OH-43.

18 Segrè to Bainbridge, Fermi, Moon, and Wilson, 21 March 1945, A-84-019, 53-16.

19 Segrè to Bainbridge, Fermi, Moon, and Wilson, 21 March 1945, A-84-019, 75-29.

20 Bainbridge, "Trinity," LA-6300-H, p. 53.

21 P. B. Moon, Nuclear physics measurements in the vicinity of the test gadget, ca. July 1944? A-84-019, 54-2.

22 P. B. Moon, Nuclear physics measurements in the vicinity of the test gadget, ca. July 1944 A-84-019, 54-2.

23 Moon to Bainbridge, 8 Aug. 1944, A-84-019, 54-2. Bainbridge memo on measurements proposed for gadget test, 1 Sept. 1944, A-84-019, 54-2. The experiment was so easy to do that it was guaranteed to be tried at Trinity. By February 1945 this experiment had decreased in importance drastically, but remained on the agenda.

24 Hawkins, *The Los Alamos Story*, pp. 243–4.

25 Three of the cellophane catcher cameras were constructed by John Williams's group. One was suspended from a barrage balloon 300 feet up and 300 feet away from Ground Zero; the other two were placed on the ground 300 and 600 feet away. Bainbridge, "Trinity," LA-6300-H, p. 54.

26 Hawkins, *The Los Alamos Story*, p. 244.

27 He also thought that tanks should be able to enter the contaminated area soon after the explosion to retrieve the necessary samples. Bainbridge memo, on measurements proposed for gadget test, 1 Sept. 1944, A-84-019, 54-2.

28 H. Anderson interview by Henriksen, 22 April 1986, OH-122. Los Alamos chemists

had not had time to work on fission product radiochemistry, which was centered at the Met Lab and the Clinton Lab. Private communication Spence, 10 Jan. 1989.

29 Reynolds to Bainbridge on Trinity blast measurements, 14 April 1945, A-84-019, 53-12.

30 Waldman interview by Henriksen, 4 Nov. 1985, OH-79.

31 Private communication Bainbridge.

32 J. C. Hoogterp, "100-Ton Test – Box Gauges for Blast Pressures," LA-288, 7 July 1945.

33 Overview of the present state of the Trinity measurements, Manley to Bainbridge, 19 March 1945, A-84-019, 53-12; Manley to Bainbridge, 19 March 1945, "Trinity Measurements – Present Status," A-84-019, 53-11.

34 Jorgensen, "100-Ton Test - Mechanical Impulse Meter," LA-284, 31 May 1945; T. Jorgensen, Jr., "July 16th Nuclear Explosion: Impulse Gauge," LA-355, 11 Sept. 1945.

35 Bainbridge, "A Foul and Awesome Display," p. 41.

36 Herbert L. Anderson, "100-Ton Trial – Radioactivity Measurements," LA-282, 25 May 1945.

37 R. L. Walker, "100-Ton Test–Piezo Gage Measurements," LA-286, 22 June 1945.

38 Waldman to Parsons, 24 May 1945, in D. Williams, and P. Yuster, "Los Alamos Technical Series Volume 24: Trinity, Appendices 55 through 71," LA-1027 DEL, 17 Dec. 1963, Appendix 63.

39 Hoogterp, "100 Ton Test: Box Gauges for Blast Pressures," LA-288.

40 For example, the signal interfered with Barschall and Martin's excess velocity method, clamping the amplifiers on some of the microphones at the wrong time and allowing signals that were not from the main blast to come through

41 Bainbridge, "A Foul and Awesome Display," p. 41.

42 Bainbridge, "A Foul and Awesome Display," p. 42.

43 Hubbard, Abstract, "100-Ton Test–Meteorological Report," LA-285, 7 June 1945.

44 Hubbard, "Los Alamos Journal," 4 July 1945, p. 81, 12 and 13 July 1945, pp. 88 and 90, A-82-007, 1-1.

45 Groves, Thomas Farrell, Oppenheimer, Bainbridge, Col. Benjamin Holzman. Tolman, and Hubbard were present.

46 Hubbard, "Los Alamos Journal," p. 104, A-82-007, 1-1; and Leslie Groves, *Now It Can Be Told: The Story of the Manhattan Project* (New York: Harper and Brothers, 1962), p. 293. Groves had also made arrangements to meet with Oppenheimer at 1:00 a.m. They drove to the control dugout 5 miles from the bomb to discuss the weather in private. They apparently made no decisions at this time except to wait as long as possible until postponing again. Groves, *Now It Can Be Told*, p. 294.

47 Hubbard, "Los Alamos Journal," p. 110; Bainbridge, "A Foul and Awesome Display," pp. 45–6.

48 Hubbard, "Los Alamos Journal," p. 119.

49 The Trinity charges were assembled at V site by a group headed by Lt. Schaffer. Those for Creutz's full-scale test were assembled in Pajarito Canyon by a group headed by H. S. North. Capt. W. F. Schaffer, "Work Preceding and Including Assembly at Trinity," in Bainbridge, "Trinity," LA-6300-H, pp. 39–41.

50 A. D. Van Vessem interview with Henriksen and B. Perkins, 22 March 1989.

51 Hawkins, *The Los Alamos Story*, p. 215.

52 Kistiakowsky to Oppenheimer, 23 Feb. 1944, A-84-019, 22-2l; E-Division Progress Report, 1 March 1944, LAMS 73, p. 42.

53 Stewart to Oppenheimer and Mitchell, 1 March 1944, A-84-019, 22-2.

54 Summary of HE implosion project, in E-Division Progress Report, 15 April 1944, LAMS-80, p. 13.

55 Kistiakowsky memo on conclusions reached at the meeting on Jumbo, 23 May 1944, dated 24 May 1944, A-84-019, 22-2.

56 Oppenheimer to Groves, telegram on 6 July 1944, A-84-019, 22-2; Bainbridge to De Silva, 11 July 1944, A-84-019, 22-1.

57 Oppenheimer to Groves, telegram of 14 July 1944, A-84-019, 22-2; Parsons to Groves, telegram of 17 July 1944, A-84-019, 34-11.

58 Bainbridge to De Silva, 20 July 1944, A-84-019, 22-1; Oppenheimer to Lansdale, telegram on 14 Aug. 1944, A-84-019, 22-1; Parsons to Oppenheimer, 24 Aug. 1944, A-84-019, 22-1.

59 One of the heads contained a large manway forging welded into the hemispherical shell formed by five "orange peel" segments. The other was similar, but without the manway. Four semicylindrical plates joined by a circumferential weld on the center line of the vessel and by longitudinal welds along their edges, formed the central cylindrical body. The outside surface was then machined before the laminated tension banding plates were wrapped around the center.

60 Ramsey interview by Henriksen, 13 Sept. 1985, OH-66. For the dating on the decision see Bainbridge to Mack, 3 March 1945, A-84-019, 56-5; Froman to Wilson, 14 March 1945, A-84-019, 56-5; Bainbridge to Captain Davalos on construction at Trinity, 15 March 1945, A-84-019, 22-1.

61 Bainbridge to D. Inglis, re Henderson's report on the present status of the Jumbo Pressure Vessel, 14 Sept. 1945, A-84-019, 56-5. R. W. Henderson checked Jumbo's condition on 27 Aug. 1945.

62 Bainbridge, "A Foul and Awesome Display," p. 43. Jumbo met its demise in April 1946. An order had been given to "test" Jumbo by detonating somewhat more than one ton of TNT inside it. However Richard A. Blackburn, 1st Lt. Ordnance, Assistant Production Officer in charge of demolition at Sandia Base, got permission to set off instead in Jumbo some 500-lb demolition bombs that had been declared unserviceable. But the bombs were loaded in the bottom instead of being suspended in the exact middle. When they detonated they blew the top and bottom out of Jumbo, spewing fragments nearly one mile away.

63 Bainbridge, "Trinity," LA-6300-H, pp. 16–17.

64 Bainbridge, "Trinity," LA-6300-H, p. 16.

65 Hawkins, *The Los Alamos Story*, pp. 240–1.

66 The mattresses would not have protected the gadget, but they helped the men to feel better.

67 McKibben to L. Ramin, 17 Jan. 1989. Copy in author's possession.

68 McKibben to Ramin, 17 Jan. 1989.

69 Bainbridge, "A Foul and Awesome Display," pp. 45–46.

70 Description of the Trinity explosion by Elizabeth and Alvin C. Graves, written about 2 hours after the explosion in Carrizozo, A-84-019, 53-10; Otto Frisch, Eyewitness Report of Nuclear Explosion, 16 July 1945, A-83-0002, 6-4; McMillan,

Impressions of the Trinity Test, 19 July 1945, A-84-019, 65-5; Weisskopf to Taylor, 24 July 1945, A-84-019, 10-7; Larkin to Taylor, 27 July 1945, A-84-019, 22-6; M. M. Shapiro to T. O. Jones, Description of Trinity test, as observed from Coordinating Council area, 23 July 1945, A-84-019, 69-11.

71 V. Weisskopf to Lt. Taylor, "Eye Witness Account," 24 July 1945, VFA 470.

72 Maurice M. Shapiro to Captain T.O. Jones, "Observations of Trinity Test," 23 July 1945, VFA 470.

73 Comments on Trinity test shot trip by Ralph Carlisle Smith, 5 Sept. 1945, A-84-019, 10-7.

74 Fermi. Observations during the explosion at Trinity on July 16, 1945, A-83-0002, 6-4. Fermi was at the Base Camp at Trinity about 10 miles from the explosion. Herbert Anderson also recounts the story in, "Fermi, Szilard, and Trinity," *Bulletin of the Atomic Scientists* (Oct. 74), p. 47.

75 J. Malik, "The Yields of the Hiroshima and Nagasaki Weapons," LA-8819, 8 April 1981.

76 The photograph record is in "July 16th Nuclear Explosion: Space-Time Relationships," LA-531, a collection of photographs from movie frames of the explosion taken by the photographic crew at Trinity. Brixner, "Optical Engineering at Los Alamos – a History," copy provided by Brixner to LANL Archives, VFA-197.

77 Mack, "July 16th Nuclear Explosion: Space-Time Relationships," LA-531.

78 Szasz, *The Day the Sun Rose Twice* (Albuquerque: University of New Mexico Press, 1984), p. 136.

79 P. Aebersold and P. B. Moon, "July 16th Nuclear Explosion: Radiation Survey of Trinity Site Four Weeks after the Explosion," LA-359, 19 Sept. 1945.

80 Barton C. Hacker: *The Dragon's Tale: Radiation Safety in the Manhattan Project, 1942–46* (Berkeley: University of California Press, 1987), pp. 102–9.

81 J. M. Blair, D. H. Frisch, and S. Katcoff, "Detection of Nuclear-Explosion Dust in the Atmosphere," LA-418, 2 Oct. 1945.

82 Bainbridge to Leet, 26 Sept. 1945, A-84-019, 54-7.

83 Hawkins, *The Los Alamos Story*, p. 244.

84 Segrè to Bainbridge, Fermi, Moon, and Wilson, 21 March 1945, A-84-019, 53-16.

85 Bainbridge, "Trinity," LA-6300-H, p. 53.

86 John L. Magee's section in Bainbridge, "Trinity," LA-6300-H, p. 47.

87 Bainbridge, "Trinity," LA-6300-H, p. 54.

88 Hawkins, *The Los Alamos Story*, pp. 244, 277.

89 See Bainbridge, "Trinity," LA-6300-H, figure 3, p. 50, for the results of the gold foil experiments.

90 According to Alvarez, they obtained one good record of the blast. In 1953, Alvarez gave the data to Reines for a new calculation of the blast pressure. Reines calculated an energy release from the blast of 12.5 kilotons ± 1 kiloton for the Hiroshima explosion. Alvarez, *Adventures of a Physicist* (New York: Basic Books, 1987), pp. 141–3.

91 Hawkins, *The Los Alamos Story*, p. 245.

92 Hawkins, *The Los Alamos Story*, p. 243.

93 Hawkins, *The Los Alamos Story*, p. 426.

94 Hawkins, *The Los Alamos Story*, pp. 277, 427.

95 Another member of the group worked with some other figures that gave a higher yield that turned out to be close to the actual measured yield, but his results were not accepted.

96 Henry H. Barschall and Gordon M. Martin, "100-Ton Test–Measurement of the Velocity of Sound," LA-291, 15 June 1945; Hawkins, *The Los Alamos Story*, p. 244.

97 E. R. Graves and J. C. Hoogterp, "July 16th Nuclear Explosion – Measurement of Blast Pressure," LA-354, 28 August 1945; Hawkins, *The Los Alamos Story*, p. 244; Bainbridge, "Trinity," LA-6300-H, pp. 64, 68.

98 T. Jorgensen Jr., "July 16th Nuclear Explosion Impulse Gauge," LA-355, 11 Sept. 1945, pp. 5, 8–9.

99 Bradbury, Kistiakowsky, and Max Roy suggested to Parsons that the exploding bomb might blind enemy troops. Memo, 17 July 1945, A-84-019, 29-9.

100 Teletype, Oppenheimer to Groves, 19 July 1945; teletype, Groves to Oppenheimer in reply (paraphrase, not dated), A-84-019. 17-5.

19. Delivery: June 1943 to August 1945

1 It proved difficult to transfer Ramsey permanently from Edward Bowles's office connected to the MIT Radiation Laboratory to the Manhattan Project. Neither Groves nor Bowles had previously been turned down in a request for personnel, and neither wanted this to be the first time. Ramsey interview by Henriksen, 13 Sept. 1985, OH-66.

2 Ramsey, "History of Project A," A-84-019, 29-6. Ramsey had recently become familiar with American airplanes in a project to evaluate the types of radar equipment needed to equip airplanes with up-to-date systems. Ramsey interview by Henriksen, 13 Sept. 1985, OH-66.

3 At this point, the implosion bomb was not yet a serious issue. Its size was to be limited by the size of the airplane bomb bay; Ramsey, "History of Project A."

4 The B-29 also had four 2200 horsepower engines, a wingspread of 141 feet, 3 inches, and a tail height of 34 feet, 7 inches. It could fly to a maximum height of 38,000 feet and attain a maximum speed about 375 miles per hour at 33,000 feet. Lee Bowen, "Project Silverplate, 1943–1946," vol. 1 of the five-volume *A History of the Air Corps Atomic Energy Program, 1943–1953* (U.S. Air Force Historical Division), pp. 91–2.

5 Leslie R. Groves, *Now It Can Be Told* (New York: Harper and Brothers, 1962), p. 254. Ramsey interview by Henriksen, 13 Sept. 1985, OH-66.

6 Ramsey interview by Henriksen, 13 Sept. 1985, OH-66.

7 Ramsey, "History of Project A." Bowen, "Project Silverplate," p. 92. Navy personnel at Dahlgren knew nothing about the tests other than that a certain shape of bomb was being tested. They were to ask no questions and did not do so.

8 Ramsey, "History of Project A."

9 Ibid., Ramsey interview by Henriksen, 13 Sept. 1985, OH-66; and Waldman interview by Henriksen and Hoddeson, 4 Nov. 1985, OH-79.

10 The uranium gun had the same diameter as the plutonium gun, but was shorter. Ramsey, "History of Project A." Documents suggest that most tests at this time

were made with the longer bomb, because the shorter one was viewed as unproblematic, if the longer one could be dropped properly.

11 Ramsey, "History of Project A."

12 Waldman interview by Henriksen and Hoddeson, 4 Nov. 1985, OH-79.

13 Ramsey, "History of Project A."

14 Arnold put Colonel D. L. Putt in charge of the actual modification. Captain R. L. Roark, Project Officer in charge of the modification, supervised the changes, based on information supplied by Colonel R. C. Wilson and Ramsey, and up-to-date bomb models sent from Los Alamos. Wright Field engineer Charles Speer did much of the design work. Ramsey, "History of Project A," p. 4.

15 Major Clyde S. Shields and Captain David Semple were the pilot and bombardier, respectively, for the modified aircraft.

16 Ramsey, "History of Project A."

17 Ramsey interview by Henriksen, 13 Sept. 1985, OH-66.

18 Brixner interview by Henriksen, 27 Jan. 1986, OH-105.

19 Ibid.

20 Ibid.

21 The most serious mishap occurred while the crews were experimenting with the release mechanism. One of the cables went prematurely slack, dropping the bomb on the bomb bay doors while the B-29 was still climbing to altitude. The doors were opened to release the bomb, which did further damage to them on the way through. The test program's only modified B-29 was put out of commission for several months, stalling the test program.

22 Ramsey interview by Henriksen, 13 Sept. 1985, OH-66.

23 Ramsey, "History of Project A."

24 Ibid.

25 Ibid.

26 Parsons to Groves via Oppenheimer, 25 Sept. 1944, B-9, 635 Planning file.

27 Ramsey, "History of Project A."

28 The name Kingman possibly arose from the attempt of pilots who were flying Project Y personnel to Wendover to keep the location of the base secret from observers on the ground. Instead of flying directly toward Wendover, they would fly toward Kingman, Arizona, and then change course. Ramsey interview by Henriksen, 13 Sept. 1985, OH-66.

29 Colonel Clifford Heflin was the commanding officer responsible for the day-to-day operation of the camp. Major Shields, although not a member of Tibbets's group, flew many of the early flight tests and was in charge of the Flight Test Section. Captain Henry Roerkohl led the Ordnance Test Unit. Ramsey, "History of Project A."

30 Silsbee to Brode, Report on Kingman Project, Oct. 1944 to May 1945, 12 July 1945, A-84-019, 67-5, p. 8.

31 Ramsey, "History of Project A."

32 J. D. Gerrard-Gough and A. B. Christman, *The Grand Experiment at Inyokern* (Washington, D.C.: Naval History Division, 1978), p. 219.

33 Ramsey to Oppenheimer, Parsons, and Lauritsen, 28 May 1945, A-84-019, 67-3.

34 Silsbee to Brode, Report on Kingman Project, Oct. 1944 to May 1945, 12 July 1945, A-84-019, 67-5.

35 Ibid.

36 Ramsey to Oppenheimer, Parsons, Kistiakowsky, Brode, and Bainbridge, 9 Sept. 1944, A-84-019, 29-1; Kistiakowsky to Ramsey, 16 Sept. 1944, A-84-019, 29-1; Ramsey to Kistiakowsky, 18 Sept. 1944, A-84-019, 29-1.

37 Those attending the meeting were Colonel Tibbets, Captain Parsons, Lt. Colonel Gerald E. Bean, Lt. Colonel Charles E. Trowbridge, Lt. Colonel Carl Luetcke, Commander F. L. Ashworth, Major Charles Sweeney, Captain Charles Begg, Captain A. G. Coombs, Lt. H. I. McClenahan, and Lt. John Wright. Memo on conference, 17 Jan. 1945, A-84-019, 38-6. Ashworth interview by Henriksen, 11 Nov. 1985, OH-80. For vivid personal recollections of Tinian, see Harlow W. Russ, *Project Alberta: The Preparation of Atomic Bombs for Use in World War II* (Los Alamos: Exceptional Books, 1990).

38 Parsons to Oppenheimer, re. special military ordnance group for Advanced Base Assembly, 20 Jan. 1945, A-84-019, 38-6. A report on this conference is dated 17 January 1945: Conference on organization, work, and problems of Special Ordnance Group, in A-84-019, 38-6.

39 Vincent Jones, *Manhattan: The Army and the Atomic Bomb* (Washington D.C.: Government Printing Office, 1985), pp. 523–4. Ashworth interview by Henriksen, 11 Nov. 1985, OH-80.

40 Jones, *Manhattan*, p. 524.

41 Ashworth interview by Henriksen, 11 Nov. 1985, OH-80.

42 Ramsey interview by Henriksen, 13 Sept. 1985, OH-66.

43 Gerrard-Gough and Christman, *The Grand Experiment*, p. 220. Ramsey interview by Henriksen, 13 Sept. 1985, OH-66.

44 Kirkpatrick to Derry, 1 June 1945, A-84-019, 9-7; Kirkpatrick to Groves, 28 June 1945, A-84-019, 9-7.

45 "Preliminary Report on Operational Procedures," A-84-019, 29-14.

46 De Silva to Parsons, 11 July 1945, A-84-019, 9-7; Kirkpatrick to Groves, 28 June 1945, A-84-019, 9-7.

47 Quotation in Martin J. Sherwin, *A World Destroyed: The Atomic Bomb and the Grand Alliance* (New York: Vintage Books, 1977), p. 214. For the meeting notes of the Scientific Panel of the S-1 Committee, see Sherwin, pp. 295–305. Also A. Steiner, "Scientists, Statesmen, and Politicians: The Competing Influences on American Atomic Energy Policy 1945–46," *Minerva 12* (October 1974), pp. 469–509.

48 Sherwin, *A World Destroyed*, p. 148; B. J. Bernstein, "Roosevelt, Truman and the Atomic Bomb, 1941–1945: A Reinterpretation," *Political Science Quarterly 90*, (1975), pp. 23–69, esp. p. 62. The classic references on the use of the atomic bomb and surrender of Japan are in Barton J. Bernstein, ed., *The Atomic Bomb: the Critical Issues* (Boston: Little, Brown, 1976). The historiography is in Bernstein, "The Atomic Bomb and American Foreign Policy, 1941–1945: an Historiographical Controversy," *Peace and Change 2*, no. 1 (Spring 1974), 1-1-16; and J. Samuel Walker, "The Decision to Use the Bomb: A Historiographical Update," *Diplomatic History* 14 (Winter 1990), 97–114.

49 The second dummy bomb, unit L2, was dropped on 24 July with a slight problem in the electrical release. Lawrence Langer, LASL Notebook F-45.

50 Ramsey interview by Henriksen, 13 Sept. 1985, OH-66.

51 Ramsey, "History of Project A."

52 Ibid.

53 Ibid.

54 Groves, *Now It Can Be Told*, p. 317.

55 Gordon Thomas and Max Morgan Witts, *Enola Gay* (New York: Pocket Books, 1977), pp. 294–309.

56 Ibid.

57 Ibid.

58 Telegram from Washington Liaison Office to Commanding Officer USEO, Clear Creek NM (Los Alamos) 7 Aug. 1945, A-84-019, 29-15.

59 Groves, *Now It Can Be Told*, pp. 320–322.

60 Ramsey interview by Henriksen, 13 Sept. 1985, OH-66.

61 Ramsey interview by Henriksen, 13 Sept. 1985, OH-66.

62 Alvarez, *Adventures of a Physicist* (New York: Basic Books, 1987), p. 143.

63 John Malik, "The Yields of the Hiroshima and Nagasaki Nuclear Explosions," LA-8819, Sept. 1985.

64 Waldman interview by Henriksen and Hoddeson, 4 Nov. 1985, OH-79.

65 Ramsey, "History of Project A."

66 Bowen, "Project Silverplate," p. 138; and Ramsey, "History of Project A."

67 L. Alvarez, *Alvarez: Adventures of a Physicist* (New York: Basic Books, 1987), pp. 144–5.

68 F. W. Ashworth, private communication.

69 Frank W. Chinnock, *Nagasaki: The Forgotten Bomb* (New York: The World Publishing Company, 1969.)

70 Memorandum from Maj. Derry to J. R. Oppenheiemr, 10 Aug. 1945, paraphrasing telegram from Washington Liaison office to Commanding Officer USEO Clear Creek NM, A-84-019, 29-15.

71 Ramsey, "History of Project A,"; Ashworth private comm.

72 Malik, "Yields," LA-8819, p. 25.

73 Telegram from Oppenheimer to Derry in Washington D.C., 9 Aug. 1945, A-84-019, 29-13.

74 William Lawren, *The General and the Bomb* (New York: Dodd, Mead, and Company, 1988), p. 255.

75 Parsons to Oppenheimer, 23 Aug. 1945. Paraphrased teletype of 21 August, A-84-019, 12-2.

76 Alvarez, *Adventures*, p. 146.

Epilogue

1 E. Fermi, *Atoms in the Family* (Chicago: University of Chicago Press, 1954), pp. 240–2.

2 R. Feynman, *Surely You're Joking, Mr. Feynman* (New York: Bantam Books, 1985), p. 118.

3 A group of scientists interested in exploring the concept of international control of atomic weapons subsequently formed the Association of Los Alamos Scientists and

later the Federation of American Scientists. Fermi, *Atoms in the Family*, p. 245; Alice Kimball Smith, *A Peril and a Hope: The Scientists' Movement in America, 1945–1947* (Chicago: The University of Chicago Press, 1965); Alice Kimball Smith and Charles Weiner, eds., *Robert Oppenheimer: Letters and Recollections* (Cambridge, Mass.: Harvard University Press, 1980), p. 305; L. Alvarez, *Adventures of a Physicist* (New York: Basic Books, 1987), p. 147.

4 Oppenheimer to Sec. of War H. Stimson, 17 Aug. 1945, A. Smith and C. Weiner, pp. 293–4.

5 Smith and Weiner, *Letters and Recollections*, pp. 293–302.

6 Smith and Weiner, *Letters and Recollections*, pp. 305, esp. 308–9. Oppenheimer to Conant, 29 Sept. 1945, p. 308.

7 Bacher, Oppenheimer, and Lauritsen had in March 1945 discussed the postwar responsibility of G-Division, agreeing that the future work would be to improve the implosion, and possibly use ^{235}U, which now was available in ample supply.

8 Truslow and Smith, *Beyond Trinity*, pp. 272, 279.

9 Leslie Groves, *Now It Can Be Told: The Story of the Manhattan Project* (New York: Harper and Brothers, 1962), pp. 377–8.

10 Truslow and Smith, *Beyond Trinity*, part II of *Project Y: The Los Alamos Story* (Los Angeles: Tomash, 1983), p. 267.

11 Notes on talk given by Bradbury at Coordinating Council, 1 Oct. 1945, A-83-0002, 6-1, reprinted in Truslow and Smith, *Beyond Trinity*, p. 363.

12 David Hawkins, *Project Y: The Los Alamos Story* (Los Angeles: Tomash, 1983), Appendix A, pp. 360–8.

13 Truslow and Smith, *Beyond Trinity*, p. 363.

14 Smith and Weiner, *Letters and Recollections*, p. 310; Hawkins, *The Los Alamos Story* pp. 260–1; Eleanor Jette, *Inside Box 1663: Life in Los Alamos during the Manhattan Project* (Los Alamos: Los Alamos Historical Society, 1977), p. 111.

20. The Legacy of Los Alamos

1 Pre–World War II attempts to mobilize research for defense are treated in L. Auerbach, "Scientists in the New Deal: A Pre-War Episode in the Relations between Science and Government in the United States," *Minerva 111*, (1965), pp. 457–82; Daniel Kevles, *The Physicists: The History of a Scientific Community in Modern America* (New York: Alfred A. Knopf, 1978), pp. 102–54. For further treatment of the joining of engineers and physicists and the postwar legacy of Los Alamos and other wartime projects, see Peter Galison, "Bubble Chambers and the Experimental Workplace," in P. Achinstein and O. Hannaway, eds., *Observation, Experiment and Hypothesis in Modern Physical Science* (Cambridge, Mass.: MIT-Bradford, 1985), pp. 309–73; and P. Galison, "The Physics Legacy of World War II," in Robert Seidel and Paul Henriksen, eds., *The Transfer of Technology from Wartime Los Alamos to Peacetime* (Los Alamos: Los Alamos National Laboratory, 1992), pp. 17–21.

2 S. S. Schweber, "Cornell and MIT," in Peter Galison and Bruce Hevly, eds., *Big Science: The Growth of Large-Scale Research* (Stanford: Stanford University Press, 1992), pp. 149–83, quote on p. 174; also S. S. Schweber, "The young

John Clark Slater and the development of quantum chemistry," *Historical Studies in the Physical and Biological Sciences*, vol. 20, no. 2 (1990), pp. 339–406.

3 Armin Hermann, John Krige, Ulrike Mersits and Dominique Pestre, with Laura Weiss, *History of CERN. Volume II, Building and Running the Laboratory* (Amsterdam: North Holland, 1990), p. 799.

4 Wilson interview by Westfall, 25 May 1987, OH-154. For more information on the early history of the Lawrence Berkeley Laboratory, see J. L. Heilbron, R. W. Seidel, and B. R. Wheaton, *Lawrence and His Laboratory: Nuclear Science at Berkeley, 1931–1961* (Berkeley: Office for History of Science and Technology, 1981), and *Lawrence and His Laboratory: A History of Lawrence Berkeley Laboratory, vol. 1* (Berkeley: University of California Press, 1989), pp. 212, 223.

5 See, e.g., L. Hoddeson, "The Entry of the Quantum Theory of Solids into Bell Telephone Laboratories, 1925–40: A Case-Study of the Industrial Application of Fundamental Science," *Minerva 18* no. 3 (1980), pp. 422–47.

6 *Lawrence and His Laboratory,* pp. 212, 223; R. Rhodes, *The Making of the Atomic Bomb* (New York: Simon and Schuster, 1986), p. 605; Richard Hewlett and Oscar Anderson, *The New World, 1939/1946* (University Park: Pensylvania State University Press, 1962), p. 724.

7 The impact of World War II on the transformation from wartime to peacetime research in America is currently under study by numerous scholars, including Andrew Pickering, Daniel Kevles, Sylvan Schweber, Dominique Pestre, John Krige, Bruce Hevly, Peter Galison, Paul Forman, John Heilbron, and Robert Seidel. Among the contributions to this area are P. Forman, "Behind Quantum Electronics: National Security as Basis for Physical Research in the United States," *Historical Studies in the Physical and Biological Sciences 18*, no. 1 (1987), p. 216; R. Seidel, "The Postwar Political Economy of High Energy Physics," in L. Brown, M. Dresden, and L. Hoddeson, eds., *Pions to Quarks: History of Particle Physics in the 1950s* (New York: Cambridge University Press, 1989); Peter Galison, "The Physics Legacy of World War II," in Robert Seidel and Paul Henriksen, eds., *The Transfer of Technology from Wartime Los Alamos to Peacetime* (Los Alamos: Los Alamos National Laboratory, 1992), pp. 17–21; C. Westfall and L. Hoddeson, "Frugality and the Founding of Fermilab," in Nathan Reingold and David van Keuren, eds., *Science and the Federal Patron,* (Stanford University Press, in press); Armin Hermann, John Krige, Ulrike Mersits and Dominique Pestre, with Laura Weiss, *History of CERN: Volume II, Building and Running the Laboratory* (Amsterdam: North Holland, 1990), p. 799; A. Needell, "Lloyd V. Berkner on Organizing American Science for Social Purposes," in M. de Maria, M. Grilli, and F. Sebastiani, eds., *The Restructuring of Physical Sciences in Europe and the United States, 1945–1960* (Singapore, World Scientific, 1989), pp. 85–95; L. Brown, M. Dresden, and L. Hoddeson, eds., *Pions to Quarks: History of Particle Physics in the 1950s* (New York, Cambridge University Press, 1989); D. Devorkin, "Organizing for Space Research: The V-2 Rocket Panel," *Historical Studies in the Physical and Biological Sciences 18*, no. 1 (1987), pp. 11–24; A. Hunter Dupree, *Science in the Federal Government: A History of Policies and Activities to 1940* (Cambridge, Mass.: Harvard University Press, 1957); P. Galison, "Bubble Chambers and the Experimental Workplace," in P. Achinstein and O. Hannaway, eds., *Observation, Experiment, and Hypothesis in Modern Physical Science* (Cambridge, Mass.: MIT Press, 1985); J. Heilbron, R. Seidel, and B. Wheaton, *Lawrence and His Laboratory: Nuclear Science at Berkeley, 1931–1961* (Berkeley: Office for History of Science and Technology, 1981); R. G. Hewlett and O. E. Anderson, Jr.,

The New World, 1939/1946: A History of the United States Atomic Energy Commission, vol. 1 (University Park: The Pennsylvania State University Press, 1962); R. G. Hewlett and F. Duncan, *Atomic Shield, 1947–1952*. vol. 2: *A History of the United States Atomic Energy Commission* (University Park: Pennsylvania State University Press, 1969); L. Hoddeson, "Establishing Fermilab in the U.S. and KEK in Japan: Nationalism and Internationalism in High Energy Accelerators," *Social Studies of Science 13* (1983), pp. 1–48. Kevles, D. J., *The Physicists*; S. W. Leslie, "Playing the Game to Win: The Military and Interdisciplinary Research at Stanford," *Historical Studies in the Physical and Biological Sciences 18*, part 1, pp. 55–88; Needell, "Preparing for the Space Age: University-Based Research, 1946–1957," *Historical Studies in the Physical and Biological Sciences 18*, no. 1 (1987), pp. 89–109; N. Reingold, "Vannevar Bush's New Deal for Research: Or the Triumph of the Old Order," *Historical Studies in the Physical and Biological Sciences 17* (1987), pp. 300–44; H. M. Sapolsky, "The Origins of Office of Naval Research," a presentation to the Sixth Naval History Symposium, Annapolis, Maryland, Sept. 1983, LAA; S. Schweber, "Shelter Island, Pocono and Oldstone: The Emergence of American Quantum Electrodynamics after World War II," *Osiris 2* (1986); S. Schweber, "The Empiricist Temper Regnant: Theoretical Physics in the United States, 1920–1950," *Historical Studies in the Physical and Biological Sciences 17*, no. 1 (1986), pp. 55–98; Seidel, "Accelerating Science: The Postwar Transformation of the Lawrence Radiation Laboratory," *Historical Studies in the Physical Sciences 13*, no. 2 (1983), pp. 375–400; R. Seidel, "A Home for Big Science: The Atomic Energy Commission's Laboratory," in *Historical Studies in the Physical and Biological Sciences 16*, no. 1 (1986), pp. 135–75; R. Seidel, "From Glow to Flow: A History of Military Laser Research and Development," *Historical Studies in the Physical and Biological Sciences 18*, no. 1 (1987), pp. 111–47; C. Westfall, "The First 'Truly National Laboratory': The Birth of Fermilab," (Ph.D. diss., Michigan State University, 1988). Also, a number of conferences have treated this subject, e.g., the Fermilab International Symposium on Particle Physics in the 1950s held in May 1985; "Big Science: The Growth of Large-Scale Research," held at Stanford, August 1988; and "The Restructuring of Physical Sciences in Europe and the United States, 1945–1960," held in Rome, September 1988. Seidel, "Accelerating Science," p. 382.

8 D. J. Kevles, *The Physicists*, pp. 341, 369, 386.

9 For information on the founding of ONR, see Sapolsky, "The Origins of Office of Naval Research," a presentation to the Sixth Naval History Symposium, Annapolis, Maryland, September, 1983. On the National Science Foundation, see J. M. England, *A Patron for Pure Science: The National Science Foundation's Formative Years 1945–1955* (Washington D. C.: National Science Foundation, 1982). On founding of the Atomic Energy Commission, see R. G. Hewlett and O. E. Anderson, Jr., *The New World, 1939/1946: A History of the United States Atomic Energy Commission*, vol. 1 (University Park: Pennsylvania State University Press, 1962).

10 See, e.g., R. Seidel, "The Postwar Political Economy of High Energy Physics"; Luis Alvarez, *Alvarez: Adventures of a Physicist* (New York: Basic Books, Inc., 1987), p. 155.

11 For more information on the founding of Brookhaven National Laboratory, see A. Needell, "Nuclear Reactors and the Founding of Brookhaven National Laboratory," *Historical Studies in the Physical Sciences 14*, pt. 1 (1984), pp. 93–122; N. Ramsey, "Early History of Associated Universities and Brookhaven National

Laboratory"; Brookhaven National Laboratory, "The Founding of the Brookhaven National Laboratory by Associated Universities, Inc." (Brookhaven: Brookhaven National Laboratory, 1948).

12 For more information on the AEC National Laboratories, see Seidel, "A Home for Big Science," pp. 135–75; S. Schweber, "Big Science in Context: Cornell and MIT," in P. Galison and B. Hevly, *Big Science: The Growth of Large-Scale Research* (Stanford: Stanford University Press, 1992), pp. 149–83.

13 S. Schweber, "The Empiricist Temper Regnant," *Historical Studies in the Physical and Biological Sciences 17* no. 1 (1986), pp. 55–98.

14 Quotations from Forman, "Behind Quantum Electronics: National Security as Basis for Physical Research in the United States," *Historical Studies in the Physical and Biological Sciences 18*, no. 1 (1987), p. 216; and Seidel, "The Postwar Political Economy of High Energy Physics, in Brown, Dresden, and Hoddeson, eds., *Pions to Quarks: History of Particle Physics in the 1950s* (New York: Cambridge University Press, 1989).

15 William C. Elmore and Matthew Sands, *Electronics: Experimental Techniques, National Nuclear Energy Series*, vol. 1 (New York: McGraw-Hill Book Company, 1949). Comments by Sands in R. Seidel and P. Henriksen, eds., *The Transfer of Technology from Wartime to Peacetime Research* (Los Alamos: Los Alamos Report, 1992), Technology Transfer.

16 See Peter Galison, "Bubbles, Sparks, and the Postwar Laboratory," in *Pions to Quarks*, pp. 213–51; and Laurie M. Brown, Max Dresden, and Lillian Hoddeson, "Pions to Quarks: Particle Physics in the 1950s," in *Pions to Quarks*, pp. 3–39.

17 For more information on these developments, see Seidel and Henriksen, *The Transfer of Technology.*

18 See Chapter 8 in L. Hoddeson, E. Braun, J. Teichmann, and S. Weart, *Out of the Crystal Maze: Chapters in the History of Solid State Physics* (New York: Oxford, 1992).

19 A. Kolb and L. Hoddeson, "The Mirage of the World Accelerator for World Peace: Origins of the Superconducting Super Collider," unpublished ms.

20 Quotations, respectively, from R. R. Wilson, remarks at Los Alamos Technology Transfer Conference; and Freeman Dyson, *Disturbing the Universe* (New York: Harper and Row, 1979), p. 49.

21 R. Wilson interview by Lillian Hoddeson, 12 Jan. 1979, Fermilab History Collection, p. 133. For more information on the origins of Fermilab, see C. Westfall, "The First 'Truly National Laboratory," and Hoddeson, "Establishing Fermilab in the U.S. and KEK in Japan," pp. 1–48; Hoddeson, Westfall, Bodnarczuk and Kolb, *Fermilab, a Case Study in the Emergence of Big Science* (in progress).

Name Index

Subject Index

CPSIA information can be obtained
at www.ICGtesting.com
Printed in the USA
LVOW08*0004291116
514867LV00014B/407/P

9 780521 441322